AF800994

Lecture Notes in Applied and Computational Mechanics

Volume 91

This series aims to report new developments in applied and computational mechanics - quickly, informally and at a high level. This includes the fields of fluid, solid and structural mechanics, dynamics and control, and related disciplines. The applied methods can be of analytical, numerical and computational nature. The series scope includes monographs, professional books, selected contributions from specialized conferences or workshops, edited volumes, as well as outstanding advanced textbooks.

Indexed by EI-Compendex, SCOPUS, Zentralblatt Math, Ulrich's, Current Mathematical Publications, Mathematical Reviews and MetaPress.

More information about this series at http://www.springer.com/series/4623

Theodoros Triantafyllidis
Editor

Recent Developments of Soil Mechanics and Geotechnics in Theory and Practice

Springer

Editor
Theodoros Triantafyllidis
Institute of Soil Mechanics
and Rock Mechanics (IBF)
Karlsruhe Institute of Technology
Karlsruhe, Germany

ISSN 1613-7736 ISSN 1860-0816 (electronic)
Lecture Notes in Applied and Computational Mechanics
ISBN 978-3-030-28515-9 ISBN 978-3-030-28516-6 (eBook)
https://doi.org/10.1007/978-3-030-28516-6

This Springer imprint is published by the registered company Springer Nature Switzerland AG
The registered company address is: Gewerbestrasse 11, 6330 Cham, Switzerland

Preface

The transition from conventional to alternative or renewable energy production is a challenge for geotechnical engineering in many aspects. The geotechnical expertise is needed for the renaturation of the open cast mining operation areas and the release of the field as a recreation area around the residual lakes for the public. This step requires methods for soil improvement and a design of the slopes to resist wave and seismic actions.

The generation of energy from renewable sources—like wind—is booming worldwide. The development of larger turbines for repowering purposes as well as the larger blades causes large loadings on tower and foundation systems onshore as well as offshore, and special attention for limit state design as well as for serviceability states has to be paid in order to ensure a safe operation for the lifetime due to the high-cycle fatigue loading applied to the supporting structures.

For the realization of the foundation structures, smart design procedures and methods of installation have to be developed.

The simulation of installation processes for geotechnical engineering is an extremely challenging task and requires sophisticated descriptions of the soil behavior and numerical tools to cope with large deformations taking into account the interaction with multiphase materials under dynamic excitation.

In this book, the recent development in fundamental research on constitutive modeling of granular and clayey materials under consideration of historiotropy and dilatancy effects is addressed as well as computational analyses for high-cycle loading on geotechnical structures are presented.

The recent development of design tools and construction processes is briefly discussed, and methods for monitoring the ground deformations are outlined.

The editor likes to express his sincere thanks to all contributors for their efforts to prepare the manuscripts, and my special thanks are dedicated to Mrs. Meininger for the help to organize and prepare the submission to Springer.

The book contains some contributions, which are considered to be breakthrough papers for the scientific community in the field of soil mechanics and foundation engineering, and the editor likes to wish to all the readers of the book enjoyment and an increase of knowledge.

September 2019 Theodoros Triantafyllidis

Contents

Constitutive and Numerical Modelling

Constitutive Modelling of Multiporous Lumpy Soils 3
Ivo Herle, X. S. Shi, and Christian Karcher

A Constitutive Model with a Historiotropic Yield Surface for Sands . . . 13
Carlos Eduardo Grandas Tavera, Theodoros Triantafyllidis, and Lukas Knittel

Geomechanical Influences of Interface Dilatancy 44
A. P. S. Selvadurai

State-Dependent Dilatancy of Soils: Experimental Evidence and Constitutive Modeling 54
Merita Tafili and Theodoros Triantafyllidis

On FE Modeling of Two-Phase Media Using the *up*-Discretization 85
Andrzej Niemunis

Possibilities and Limitations of ALE Large Deformations Analyses in Geotechnical Engineering 97
Frank Rackwitz

Approaches for the Design of Foundations for Offshore Wind Turbines: A Review Based on Comparisons with HCA-Based Models . . . 113
Hauke Zachert and Torsten Wichtmann

Behaviour of Granular Soils Under Uni- and Multidimensional Drained High-Cyclic Loading 136
Torsten Wichtmann and Lukas Knittel

Soil Dynamics and Structural Dynamics

Shear Wave Based Screening Method for Liquefaction Evaluation 169
J. Yang

A Study on the Wave Dispersion in Concrete Due to Damage; Numerical Observations and Theoretical Predictions 183
Georgios Livitsanos, Abdul Bashit Ahmadi, Dimitrios G. Aggelis, Theodore V. Gortsas, and Demosthenes Polyzos

Longitudinal Shock Waves in Soil 199
Vladimir A. Osinov

Renaturation of Open Cast Mining Areas

Geotechnical Aspects in Designing the Permanent Slopes of Future Lakes in Opencast Mines in the Rhenish Lignite Mining Area 221
Dieter Dahmen

Coupled Numerical Simulations for Seismic Hazard Assessment of Large Opencast Slopes 237
Jan Machaček, Carlos Eduardo Grandas Tavera, and Theodoros Triantafyllidis

Infrastructure Projects and Monitoring Concepts

Prediction and Control of Ground Deformations for Mechanized Tunneling in Clays with Mixed Face Conditions 267
Andrew J. Whittle and Vasiliki Founta

Two Neighbouring Tunnels in Saturated Soil Under Blast Loading 281
Vladimir A. Osinov and Stylianos Chrisopoulos

In Situ Stress Assessment Based on Width and Depth of Brittle Borehole Breakouts 297
Eleni Gerolymatou and Alexandros Petalas

Foundation Engineering: Challenges and Achievements

Geotechnical Challenges for the Numerical Prediction of the Settlement Behaviour of Foundations in Rosenheim's Seeton 323
Roberto Cudmani, Daniel Rebstock, and Joshua Schorr

Diaphragm Wall Technique – Planning, Execution and Development over the Last 65 Years 347
Gebhard Dausch

Stability and Large Deformations of Slender Structures Supported by Soil Materials 355
Matthias Baeßler, Pablo Cuellar, Falk Lüddecke, and Albrecht Victor

Excavation Pits: Calculation Methods 370
Achim Hettler and Theodoros Triantafyllidis

Traces of Prof. Triantafyllidis in Bochum 383
Diethard König, Achim v. Blumenthal, and Meisam Goudarzy

Author Index 409

Constitutive and Numerical Modelling

Constitutive Modelling of Multiporous Lumpy Soils

Ivo Herle[1(✉)], X. S. Shi[2], and Christian Karcher[3]

[1] Institut für Geotechnik, Technische Universität Dresden, 01062 Dresden, Germany
ivo.herle@tu-dresden.de
[2] Department of Civil and Environmental Engineering, Hong Kong University of Science and Technology, Hong Kong, China
xiusongshi@ust.hk
[3] RWE Power AG, Köln/Essen, Germany
christian.karcher@rwe.com

Abstract. The deposition of excavated natural soils, e.g. in open pit mines or during land reclamation, produces in general lumpy soils with multimodal pore sizes. The mechanical behaviour of such soils is determined by interaction between the firm to stiff lumps and the soft to liquid soil material in the macrovoids between the lumps. The initial skeleton composed of lumps resembles a coarse grained soil. However, contrary to mineral grains, the macrograins (lumps) are not incompressible and time-independent. Due to rising overburden during the subsequent deposition of the excavated soil, a time-dependent deformation of the lumps takes place, being accompanied by a softening of the lumps' surface. The space between the lumps, originally occupied by air, becomes gradually filled with that soft material. A constitutive description of the lumpy soils should take into account all the mentioned effects. In this paper, main ideas and fundamentals for the constitutive modelling of the relevant phases of lumpy soils are outlined. With help of a homogenization method, a material model suitable for practical applications can be obtained.

1 Introduction

An excavation and a subsequent deposition of large volumes of natural soils can be encountered e.g. in open pit mining or in land reclamation. In case of open pit mining, bucket wheels disintegrate the original, often stiff soils into big lumps (Fig. 1 left). After their transport to the deposition location, the lumps are being spread in a material flow from a non-negligible height (Fig. 2 right) and thus further broken. An initial structure of the lumpy skeleton resembles a coarse grain soil (Fig. 3 left). The macroscopic voids between lumps are continuous and filled with air. One can observe steep slopes created during spreading which suggests a high friction angle of this artificial coarse grained soil.

Contrary to mineral grains, the lumpy macrograins are composed of numerous primary grains with their own (microscopic) porosity. If clay particles are

T. Triantafyllidis (Ed.): *Recent Developments of Soil Mechanics and Geotechnics in Theory and Practice*, LNACM 91, pp. 3–12, 2020.
https://doi.org/10.1007/978-3-030-28516-6_1

Fig. 1. Excavation of natural soil layers.

Fig. 2. Deposition of the excavated soil.

involved, additionally pores inside the clay aggregates can be distinguished. The primary grains are held together in a lump by suction, interlocking or natural cementation. The lumps are far from being permanent due to their particulate nature. They can change their shape and volume during loading or in time. Even more pronounced is their degradation due to environmental impacts, mainly oscillations in humidity or rainfall events [1,3].

The lump degradation by weathering starts from the surface of a lump. Getting into contact with water or excessive humidity, the suction at the surface is suppressed and the surface layer becomes saturated. Effective stresses vanish and the surface layer softens. A reconstituted soil is created and gradually squeezed into the pore space whereas the volume of the stiff lump diminishes

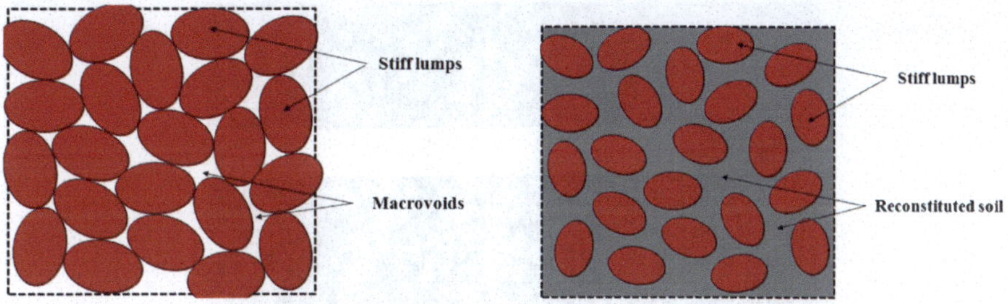

Fig. 3. Fresh (left) and weathered (right) lumpy soil [5].

(Fig. 3 right). Simultaneously, a total pressure on the lump is increasing due to overburden from a further deposition of excavated soils. A continuous transition from a granular soil to a reconstituted soil takes place in time and with depth.

A constitutive description of a lumpy landfill including the process of lumps deformation can be split into several particular models:

- granular soil composed of lumps and voids filled with air [1,5,11,16]
- fully reconstituted soil [4,6,9,13]
- mixture of lumps and reconstituted soil without consolidation [8,10,12]
- mixture of lumps and reconstituted soil considering overconsolidation of the lumps [7,14]
- mixture of lumps and reconstituted soil including coupled consolidation of both components [15]

A selection or combination of these models can be used for predictions of stability and deformation behaviour of landfills composed of double porosity soils.

2 Lumpy Soil as a Granular Soil

Fresh lumps are created by the disintegration of the excavated soil during its transport and deposition. Considering the lumps as soil grains, the resulting soil could be characterised as gravel with cobbles and boulders. Nevertheless, the lumpy grains are soft, having the uniaxial strength usually not higher than 1 MPa [1]. In order to reduce the grain size of the natural lumps for laboratory investigations, the original soil was cut into small pieces representing the downscaled lumps (Fig. 4).

The soft nature of the grains produces some untypical phenomena like densification of the soil during shearing, in spite of a high initial relative density [16]. Although the volume decreases during shearing, the shear strength is higher than in the reconstituted (normally consolidated) state. After normalization with Hvorslev's equivalent pressure, the state boundary surface of a lumpy soil can be defined (Fig. 5).

With increasing stress level, the preconsolidation pressure of the original soil material (before being excavated) is being approached. Even without any

Fig. 4. Downscaled natural lumpy soil [16].

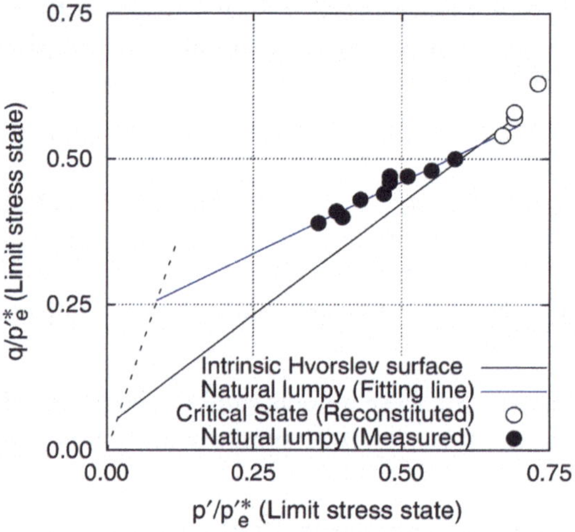

Fig. 5. Limit stress state of a downscaled natural lumpy soil [16].

degradation (weathering) effects, the lumps are excessively distorted and fill the macrovoids between each other. Although the compression behaviour may resemble a natural soil with a metastable structure, significant differences can be found when interpreting the data with respect to the intrinsic compression line [11], see Fig. 6.

At higher pressures, the mechanical response of a lumpy soil tends to the normally consolidated behaviour of the constituents within the lumps [1]. This process results in a pronounced reduction of the overall permeability and thus an increase in the consolidation time [5].

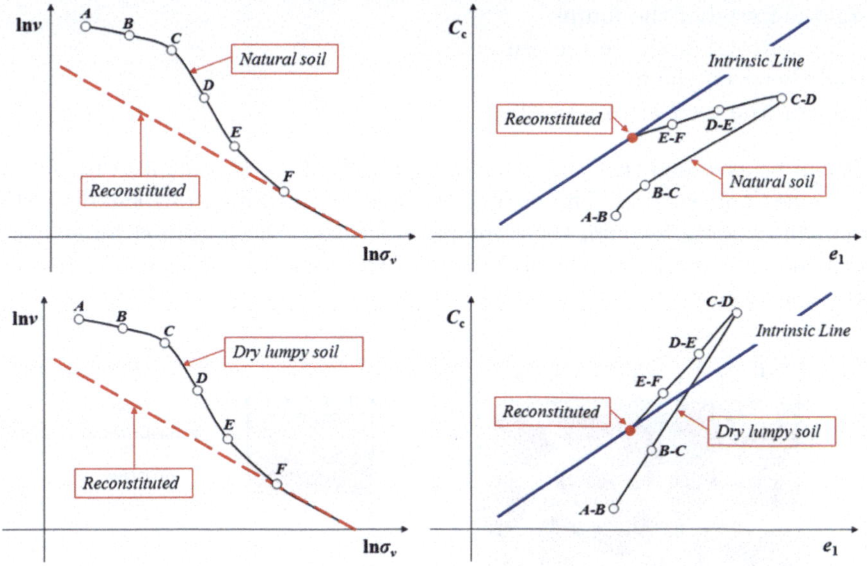

Fig. 6. Difference in the structure transition between the natural and dry lumpy soil [11].

3 Mixtures of Reconstituted Soils

Oscillations in atmospheric conditions are responsible for weathering and subsequent disintegration of the lumps. Asymptotically, after sufficient time, a reconstituted soil is created. Nevertheless, in cases of long transportation of the excavated material on conveyor belts the lumps may become fully disintegrated and reconstituted prior to their deposition. Under such circumstances, various soils in the reconstituted state may be mixed together during the transport.

The properties of mixtures of fully saturated reconstituted soils cannot be obtained by a simple interpolation between the constituents. The remolded shear strength can be estimated from the consistency limits and the actual water content of the constituents [9]. An evolution of the shear strength in time within artificial deposits (landfills) requires the knowledge of a statistically probable composition of the soil mixtures and their states [2]. Cone penetration testing may be useful for the determination of the soil composition and state of heterogeneous mixtures in the field [17].

4 Lumpy Soil Without Consolidation

When the voids between the stiff lumps are filled with a reconstituted material as a consequence of the lumps weathering, two constitutive phases—lumps and reconstituted soil—must be distinguished. The following four partial volumes should be taken into account:

1. solid material of the lumps
2. solid material of the reconstituted soil
3. voids between the lumps
4. voids in the reconstituted soil

It can be assumed that the behaviour of each phase is controlled by its average stresses and strains. The overall stresses and strains are related to those of the constituents through their volume fractions. The overall stiffness can be expressed as a function of the stiffnesses of the constituents using a stress concentration ratio.

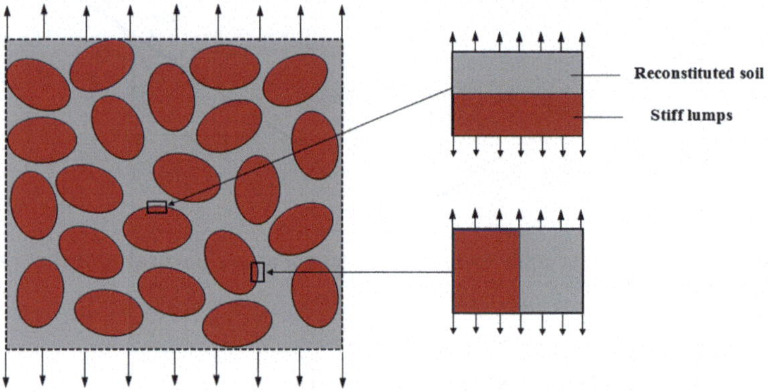

Fig. 7. Limit cases of the phase interaction for a two-phase composite soil [8].

Two limit cases can be defined for the contact between the lumps and the reconstituted soil at the mesoscale, see Fig. 7. These can be described as a parallel and a series configuration, respectively, of the two phases. Denoting K_r the stiffness of the reconstituted soil and K_l the stiffness of the lumps, the overall stiffness can be calculated as

$$K_p = n_e K_r + (1 - n_e) K_l \tag{1}$$

in the parallel case and as

$$\frac{1}{K_s} = \frac{n_e}{K_r} + \frac{1 - n_e}{K_l} \tag{2}$$

in the series case, where

$$n_e = \frac{n_t - n_l}{n_r - n_l} \tag{3}$$

is a function of the porosities (n_t: soil mixture, n_r: reconstituted soil, n_l: lumps) [10].

Assuming the soil structure as a random distribution of the parallel and series cases at different orientations, a generalised equation for the homogenised mixture may be formulated as:

$$\log K = n_e \log K_r + (1 - n_e) \log K_l \tag{4}$$

The homogenised shear stiffness can be obtained by a similar derivation [10,12].

The outlined approach was validated by laboratory experiments and numerical simulations. The numerical model comprised a soil element filled with a certain volume fraction of the lumps embedded in the reconstituted soil (Fig. 8) under isotropic and triaxial shear loading, respectively.

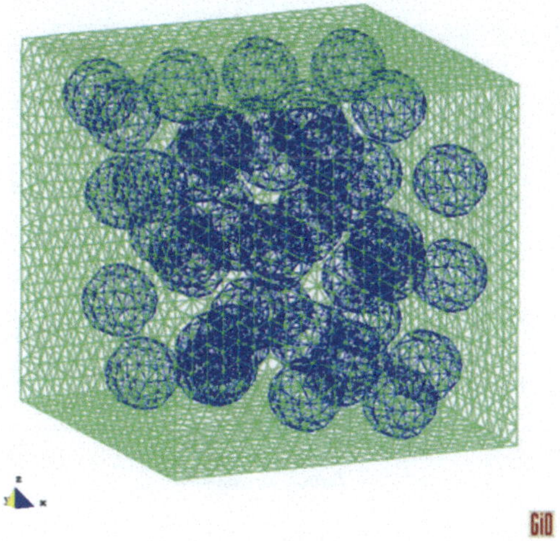

Fig. 8. Numerical model for the validation of the homogenisation method [8].

The lumps are often in an overconsolidated state with respect to low overburden pressures. Standard elastoplastic and hypoplastic models usually overestimate the shear strength at the low effective stress range. Thus, a realistic limit stress condition is needed in the constitutive description of the lumps. A nonlinear Hvorslev surface (see Fig. 5 in Sect. 2) helps to improve numerical predictions of the mechanical behaviour for the lumpy soils [7,14].

5 Lumpy Soil During Consolidation

The consolidation process of a lumpy soil includes a simultaneous consolidation of both (fully saturated) phases, the lumps and the reconstituted soil. A continuous exchange of water between the lumps and the reconstituted soil in the macrovoids takes place. The proposed model is based on two simplifications.

Firstly, there are no direct contacts between the lumps, i.e. the lumps are floating in the reconstituted soil. Secondly, a further disintegration of the lumps is negligible. Consequently, there is no exchange of the solid phase of the lumps into the reconstituted soil. The model enables to follow the non-uniform stress and strain distribution of both phases accompanied by a corresponding evolution of the porosities in time. The excess pore water pressure dissipation in the reconstituted soil is more rapid than in the lumps due to its higher permeability. As a result, there is a transfer of fluid from the lumps to the reconstituted soil, which accelerates the whole consolidation process. A good agreement between laboratory experiments and the model predictions of the compression behaviour (Fig. 9) and of the consolidation times (Fig. 10) could be shown [15].

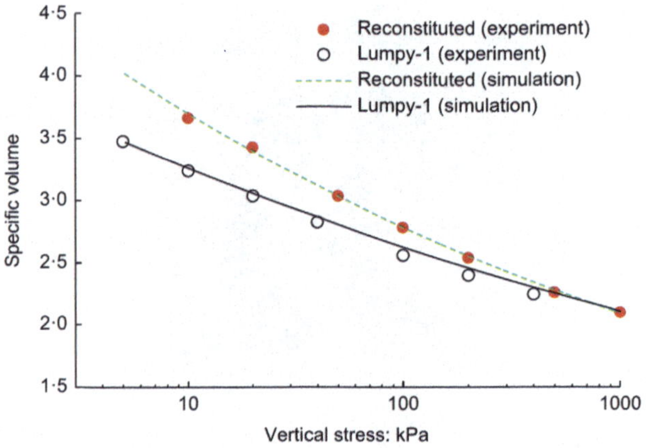

Fig. 9. Experiments and simulations of the compression behaviour of a lumpy soil during consolidation [15].

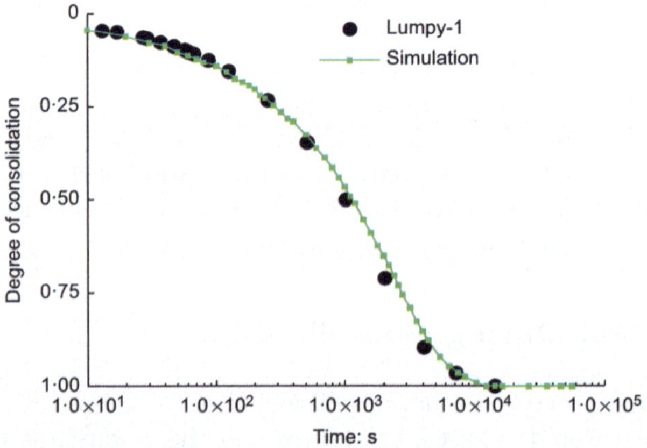

Fig. 10. Measured and calculated consolidation curve for the stress increment between $\sigma'_i = 10\,\text{kPa}$ and $\sigma'_{i+1} = 20\,\text{kPa}$ [15].

6 Concluding Remarks

The material description of multiporous lumpy soils is a challenge for the constitutive modelling. It has been shown that a hierarchical approach can be used for this purpose. A homogenisation model can combine the contributions of the lumps (composed of an overconsolidated soil) and of the reconstituted soil to the overall behaviour. If lumps are floating in the reconstituted soil, a consolidation model can well capture the time evolution of the compressibility.

In spite of a high complexity of the consolidation model [15], many relevant features of dumped lumpy soils are still missing. The following effects should be included if a general model is required for realistic field predictions:

- A part of the lumps can create a quasi-skeleton with voids filled by the reconstituted soil.
- A solid phase of the lumps can be gradually transferred to the reconstituted soil (weathering).
- Air phase is added with focus on the partial saturation of the interlump voids and on the initial suction within the lumps.

References

1. Herbstová, V., Herle, I.: Structure transitions of clay fills in North-Western Bohemia. Eng. Geol. **104**(3–4), 157–166 (2009). https://doi.org/10.1016/j.enggeo.2008.10.001
2. Herle, I., Mašín, D., Kostkanová, V., Karcher, C., Dahmen, D.: Experimental investigation and theoretical modelling of soft soils from mining deposits. In: International Symposium on Deformation Characteristics of Geomaterials, Seoul, South Korea, pp. 858–864 (2011)
3. Kostkanová, V., Herle, I., Bohác, J.: Transitions in structure of clay fills due to suction oscillations. Procedia Earth Planet. Sci. **9**, 153–162 (2014). https://doi.org/10.1016/j.proeps.2014.06.011
4. Schwiteilo, E., Herle, I., Dahmen, D., Karcher, C.: Change in consistency limits as a result of mixing processes in soils (Änderung der Konsistenzgrenzen bei Mischvorgängen an Böden). World Min. - Surf. Undergr. **68**(6), 376–384 (2016)
5. Shi, X., Herle, I.: Laboratory investigation of artificial lumpy materials. Eng. Geol. **183**, 303–314 (2014). https://doi.org/10.1016/j.enggeo.2014.10.020
6. Shi, X., Herle, I.: Modeling the compression behavior of remolded clay mixtures. Comput. Geotech. **80**, 215–225 (2016). https://doi.org/10.1016/j.compgeo.2016.07.007
7. Shi, X., Herle, I.: A model for natural lumpy composite soils and its verification. Int. J. Solids Struct. **121**, 240–256 (2017). https://doi.org/10.1016/j.ijsolstr.2017.05.033
8. Shi, X., Herle, I.: Laboratory investigation of two basic configurations for inhomogeneous soils. Eur. J. Environ. Civil Eng. **21**(2), 206–237 (2017). https://doi.org/10.1080/19648189.2015.1110056
9. Shi, X.S., Herle, I.: Compression and undrained shear strength of remoulded clay mixtures. Géotech. Lett. **5**(2), 62–67 (2015). https://doi.org/10.1680/geolett.14.00089

10. Shi, X.S., Herle, I.: Analysis of the compression behavior of artificial lumpy composite materials. Int. J. Numer. Anal. Meth. Geomech. **40**(10), 1438–1453 (2016). https://doi.org/10.1002/nag.2493
11. Shi, X.S., Herle, I.: Structure transition of lumpy landfills with increasing stress level. Q. J. Eng. Geol.Hydrogeol. **49**(4), 336–343 (2016). https://doi.org/10.1144/qjegh2015-093
12. Shi, X.S., Herle, I.: Numerical simulation of lumpy soils using a hypoplastic model. Acta Geotech. **12**(2), 349–363 (2017). https://doi.org/10.1007/s11440-016-0447-7
13. Shi, X.S., Herle, I.: Undrained shear strength and water content distribution of remoulded clay mixtures. Geotechnik **40**(1), 60–63 (2017). https://doi.org/10.1002/gete.201500031
14. Shi, X.S., Herle, I., Bergholz, K.: A nonlinear Hvorslev surface for highly overconsolidated soils: elastoplastic and hypoplastic implementations. Acta Geotech. **12**(4), 809–823 (2017). https://doi.org/10.1007/s11440-016-0485-1
15. Shi, X.S., Herle, I., Muir Wood, D.: A consolidation model for lumpy composite soils in open-pit mining. Géotechnique **68**(3), 189–204 (2018). https://doi.org/10.1680/jgeot.16.P.054
16. Shi, X.S., Herle, I., Yin, J.: Laboratory study of the shear strength and state boundary surface of a natural lumpy soil. J. Geotech. Geoenviron. Eng. **144**(12), 04018093 (2018). https://doi.org/10.1061/(ASCE)GT.1943-5606.0001987
17. Uhlig, M., Herle, I., Karcher, C.: Determination of time-dependent undrained shear strength of mining landfill. In: Wuttke, F., Bauer, S., Sánchez, M. (eds.) Energy Geotechnics, pp. 397–403. Taylor & Francis Group, Milton Park (2016)

A Constitutive Model with a Historiotropic Yield Surface for Sands

Carlos Eduardo Grandas Tavera[2(✉)], Theodoros Triantafyllidis[1], and Lukas Knittel[1]

[1] Institute of Soil Mechanics and Rock Mechanics (IBF), Karlsruhe Institute of Technology (KIT), Cottbus, Germany
[2] BTU, Cottbus, Germany
carlos.grandastavera@b-tu.de

Abstract. A simple, yet versatile yield surface in the stress space is combined with a hypoplastic equation to simulate the influence of recent deformation history on the mechanical behaviour of sand. This yield surface is used to describe the intensity of anelastic flow. In the model, the state is fully described by the current stress, the void ratio and a novel back stress-like tensor. This new state variable determines the shape and size of the yield surface and accounts for recent deformation/stress history. The direction of the anelastic flow upon shearing is obtained from a generalization of the Taylor's dilatancy rule. As a distinctive feature of the model, this dilatancy is able to reproduce the strong contractancy upon reversals observed in experiments without the need of additional state variables. The model corrects some known shortcomings of previous hypoplastic models like overshooting and the excessive accumulation of stress/strain upon strain/stress cycles of small amplitude (ratcheting). Laboratory tests are simulated to show the capabilities of the model to reproduce the soil behaviour under monotonic and cyclic loading conditions after different deformation histories.

1 Introduction

Besides pycnotropic (density dependency) and barotropic (pressure dependency), the mechanical behaviour of soils is also historiotropic, i.e. it is strongly influenced by the loading history (stress/strain path dependency). Despite having the same density and the same pressure, samples subjected to different loading histories exhibit different mechanical responses. This well known phenomenon can be observed in the experiments of Doanh et al. [1] on loose Hostun sand. The Fig. 1 shows different stress paths obtained during undrained triaxial compression and extension tests. At the beginning of the undrained tests, the samples have the same isotropic stress p = 100 kPa and nearly the same void ratio $e \approx 0.94$, but different recent histories. These recent histories were generated by first subjecting the sample to drained triaxial compression up to a certain stress ratio (marked as preloading with the numbers 1 to 10 in Fig. 1)

T. Triantafyllidis (Ed.): *Recent Developments of Soil Mechanics and Geotechnics in Theory and Practice*, LNACM 91, pp. 13–43, 2020.
https://doi.org/10.1007/978-3-030-28516-6_2

and then reducing the stress ratio until the isotropic stress of $p = 100$ (marked with the number 1) was reached. It can be observed that loose sand may behave like dense sand (dilative behaviour) and can reach large deviatoric stresses if the undrained shearing coincides with the direction of the previous deviatoric strain/stress history. The higher the stress ratio reached during drained triaxial compression the higher the stress ratio upon subsequent undrained triaxial compression. On the contrary, a strong reduction of the maximum deviatoric stress reached during undrained triaxial extension is observed when the sample was previously subjected to drained triaxial compression.

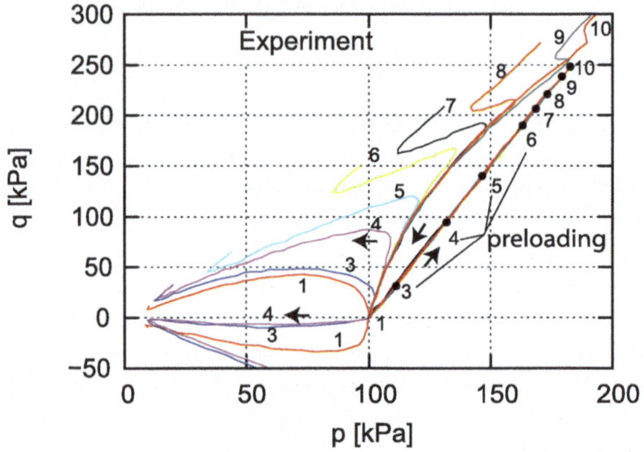

Fig. 1. Undrained triaxial compression and extension tests on Hostun RF loose sand preceded by different drained triaxial preloading. Experiment data adapted from [1].

The importance of recent history is not restricted to the monotonic behaviour of loose samples. Triaxial tests reported by Ishihara und Okada [2] show the effect of preshearing (application of a shear stress before the actual test) on the cyclic behaviour of medium dense Fuji River sand. In these tests, the samples were first subjected to undrained cyclic stresses until the pore water pressure increased up to a certain value. Then, the drainage was opened, the pore water dissipated, and the initial effective pressure was reached again (reconsolidation). Finally, subsequent undrained cyclic stresses were applied to observe influence of preshearing on the development of effective stresses and strains. Ishihara und Okada found that after *small* preshearing, the reduction of effective pressure and the accumulation of strains in the subsequent undrained stress cycles were less than those obtained during the preshearing phase. However, "samples subjected to *large* preshear on one side of triaxial loading, compression or extension, became stiffer on that side, but softer on the opposite side", [2]. Ishihara und Okada defined *small* and *large* preshearing in terms of stress (not of strain) and used the *Phase Transformation Line* (PTL) [3] as the boundary between small and large. Preshearing that reaches stress ratios below and above the PTL were called small and large, respectively. These experiments show also that the effects of stress histories reaching large stress ratios can be observed even after the unloading from the large shear stress to the isotropic axis (with a shear strain of approx. 0.5%) and after the reconsolidation (with a volumetric strain of approx. 0.7%).

A similar behaviour has been observed in the experiments of Wichtmann and Triantafyllidis [4] on dense Karlsruhe fine sand samples. After a preloading with stresses beyond the PTL, the accumulation of pore water pressure in the subsequent undrained cyclic test with small stress amplitudes is so strong, that the effective pressure in the sample almost vanished after few cycles. This surprising effect, by which a dense sand behaves almost like a loose sand upon undrained cyclic loading, may be attributed to the persistent influence of the stress history.

In order to describe the effects of recent stress/strain history on the mechanical behaviour of sand, we propose a constitutive model that combines the well known hypoplastic equation, [5] and [6], with a simple yield surface in stress space. The yield surface is used to compute the intensity of anelastic flow. The anelastic flow is more intense for stresses outside than for those inside the yield surface. The shape of the yield surface is determined by a novel state variable $\boldsymbol{\sigma}_B$, which can be regarded as *back-stress* and provides information about the recent stress loading history. The evolution of $\boldsymbol{\sigma}_B$ is provided separately for its isotropic and deviatoric portions. The direction of the anelastic flow upon shearing is obtained from a generalization of the Taylor's [7] dilatancy rule. As a distinctive feature of the model, this dilatancy is able to reproduce the strong contractancy upon reversals observed in experiments without the need of additional state variables. Furthermore, the hypoelastic stiffness is replaced by a hyperelastic one (similar as in the neohypoplastic model [8]). The model corrects some known shortcomings of previous hypoplastic models like overshooting and the excessive accumulation of stress/strain upon strain/stress cycles of small amplitude (ratcheting).

After defining the notation used throughout the paper in Sect. 2, the main hypoplastic equation and the hyperelastic stiffness are described in Sects. 3 and 4, respectively. The constitutive model is designed to reproduce two observed attractors: the Limiting Compression Curve (LCC) [9] upon monotonic compressive volumetric straining and the Critical State (CS) [10] and [11] after long monotonic shearing. In the CS, plastic deviatoric flow ($\operatorname{tr} \dot{\boldsymbol{\epsilon}} = \mathbf{0}$ with $\dot{\boldsymbol{\epsilon}} \neq \mathbf{0}$) is possible without changes in stress $\dot{\boldsymbol{\sigma}} = \mathbf{0}$ when a critical void ratio $e = e_c$ is reached. To describe these attractors, two different degrees of nonlinearities and flow rules are adopted: Y_I and $\mathbf{m}_I$ for radial compression and Y_D and $\mathbf{m}_D$ for shearing. The Sect. 5 presents a one dimensional model to simulate isotropic compression. In this simple model, loading/unloading and reloading phases are distinguished by the overconsolidation ratio, which is based on the concept of overstress. The overconsolidation ratio is a measure of how close the current stress and the yield stress are. In this simple model, the yield stress provides a mechanism to "memorize" previous stress paths. These concepts, i.e. the distinction between loading and unloading and the yield stress as a memory mechanism, are generalized in Sect. 6 to the stresses and strains in 3 dimensions (and for shear loading). For this purpose a yield surface in the stress space defined by a new state variable $\boldsymbol{\sigma}_B$ is introduced. A novel flow rule that depends on void ratio, stress and the direction of the strain rate is constructed from the well known Taylor's [7] dilatancy. In Sect. 7 the results of different element tests reported in the literature are compared with the simulations computed with the model to evaluate the performance of the proposed constitutive equation. In addition, a special set of tests conducted to evidence the influence of monotonic loading on subsequent cyclic behaviour is presented and contrasted with simulation results.

2 Notation

A fixed orthogonal Cartesian coordinate system with unit vectors $\{\mathbf{e}_1, \mathbf{e}_2, \mathbf{e}_3\}$ is used throughout the text. A repeated (dummy) index in a product indicates summation over this index taking values of 1, 2 and 3. A tensorial equation with one or two free indices can be seen as a system of three or nine scalar equations, respectively. We use the Notation Kronecker's symbol δ_{ij} and the permutation symbol e_{ijk}. Vectors and second-order tensors are distinguished by bold typeface, for example $\mathbf{N}, \boldsymbol{\sigma}, \mathbf{v}$. Fourth order tensors are written in sans serif font (e.g. L). The symbol $\cdot$ denotes multiplication with one dummy index (single contraction). For instance, the scalar product of two vectors can be written as $\mathbf{a} \cdot \mathbf{b} = a_k b_k$. Multiplication with two dummy indices (double contraction) is denoted with a colon, for example $\mathbf{A} : \mathbf{B} = \mathrm{tr}\,(\mathbf{A} \cdot \mathbf{B}^T) = A_{ij} B_{ij}$, wherein $\mathrm{tr}\,\mathbf{X} = X_{kk}$ reads trace of a tensor. The expression $()_{ij}$ is an operator extracting the component (i, j) from the tensorial expression in brackets, for example $(\mathbf{T} \cdot \mathbf{T})_{ij} = T_{ik} T_{kj}$. The tensor I is singular (yields zero for every skew symmetric tensor), but for symmetric argument $\mathbf{X}$, I represents the identity operator, such that $\mathbf{X} = \mathsf{I} : \mathbf{X}$. A tensor raised to a power, like $\mathbf{T}^n$, is understood as a sequence of $n-1$ multiplications $\mathbf{T} \cdot \mathbf{T} \cdot \ldots \mathbf{T}$. The brackets $\| \quad \|$ denote the Euclidean norm, i.e. $\|\mathbf{v}\| = \sqrt{v_i v_i}$ or $\|\mathbf{T}\| = \sqrt{\mathbf{T} : \mathbf{T}}$. The definition of Mc Cauley brackets reads $<x> = (x + |x|)/2$. The deviatoric part of a tensor is denoted by an asterisk, e.g. $\mathbf{T}^* = \mathbf{T} - \frac{1}{3}\,\mathbf{1}\mathrm{tr}\,\mathbf{T}$, wherein $(\mathbf{1})_{ij} = \delta_{ij}$ holds. The components of diagonal matrices (with zero off-diagonal components) are written as diag[, ,], for example $\mathbf{1} =$ diag[1, 1, 1]. The Roscoe's invariants for the axisymmetric case $\sigma_2 = \sigma_3$ and $\epsilon_2 = \epsilon_3$ are then defined as $p = -\mathbf{1} : \boldsymbol{\sigma}/3$, $q = -(\sigma_1 - \sigma_3)$, $\epsilon_v = -\mathbf{1} : \boldsymbol{\epsilon}$ and $\epsilon_q = -\frac{2}{3}(\epsilon_1 - \epsilon_3)$. The general definitions $q = \sqrt{\frac{3}{2}}\|\boldsymbol{\sigma}^*\|$ and $\epsilon_q = \sqrt{\frac{2}{3}}\|\boldsymbol{\epsilon}^*\|$ are equivalent to the ones from the axisymmetric case but may differ in sign. The isomorphic invariants for the axisymmetric case are defined as $P = -\boldsymbol{\sigma} : \vec{\mathbf{1}}$, $Q = -\boldsymbol{\sigma} : \vec{\mathbf{1}}^*$, $\epsilon_P = -\boldsymbol{\epsilon} : \vec{\mathbf{1}}$, and $\epsilon_Q = -\boldsymbol{\epsilon} : \vec{\mathbf{1}}^*$, with $\vec{\mathbf{1}} = \frac{1}{\sqrt{3}}$ diag[1, 1, 1] and $\vec{\mathbf{1}}^* = \sqrt{\frac{2}{3}}$ diag[1, −1/2, −1/2]. Dyadic multiplication is written *without* ⊗, e.g. $(\mathbf{ab})_{ij} = a_i b_j$ or $(\mathbf{T}\,\mathbf{1})_{ijkl} = T_{ij}\delta_{kl}$. Proportionality of tensors is denoted by tilde, e.g. $\mathbf{T} \sim \mathbf{D}$. The operator $(\sqcup)^{\rightarrow} = \sqcup/\|\sqcup\|$ normalizes the expression $\sqcup$, for example $\vec{\mathbf{D}} = \mathbf{D}/\|\mathbf{D}\|$. The sign convention of general mechanics with *tension positive* is obeyed. Objective Zaremba-Jaumann rates are denoted with a superimposed dot, for example the rate of the Cauchy stress is $\dot{\boldsymbol{\sigma}}$.

3 The Hypoplastic Equation

In the new hypoplastic model, the stress rate $\dot{\boldsymbol{\sigma}}$ is written as nonlinear function of the strain rate $\dot{\boldsymbol{\epsilon}}$ as follows

$$\dot{\boldsymbol{\sigma}} = \mathsf{E} : (\dot{\boldsymbol{\epsilon}} - Y\mathbf{m}\,\|\dot{\boldsymbol{\epsilon}}\|) \tag{1}$$

where the fourth rank tensor $\mathsf{E}(\boldsymbol{\sigma}, e)$ is the hyperelastic stiffness, $Y(\boldsymbol{\sigma}, e, \boldsymbol{\sigma}_B)$ is the so called degree of nonlinearity and $\mathbf{m}(\boldsymbol{\sigma}, e, \boldsymbol{\sigma}_B, \vec{\dot{\boldsymbol{\epsilon}}}^*)$ is the flow direction. In order to reproduce the influence of previous deformation paths on the current material response, we introduce a new stress-like state variable, $\boldsymbol{\sigma}_B$. Thus, the material state is fully determined by the current stress $\boldsymbol{\sigma}$, the current void ratio e, and the current *back-stress* $\boldsymbol{\sigma}_B$. In addition to the evolution of $\boldsymbol{\sigma}$ (1) and the evolution of the void ratio e,

$$\dot{e} = (1 + e)\,\mathrm{tr}\,\dot{\boldsymbol{\epsilon}}, \tag{2}$$

the constitutive model requires the evolution of the back-stress $\boldsymbol{\sigma}_B$, which is described in Sect. 6.7.

4 Hyperelasticity

Previous hypoplastic models include an empirical dependency of the stiffness tensor $\mathsf{E}^{\mathrm{hp}}(p)$ on the stress [5,12]. To reduce excessive ratcheting of hypoplasticity, Niemunis and Herle [13] introduced a new strain-like variable (the so called intergranular strain $\mathbf{h}$) which is fully determined by the recent history of deformation. During changes in the strain direction (detected by the angle between $\mathbf{h}$ and $\dot{\boldsymbol{\epsilon}}$), the nonlinear part of the model ($Y\mathbf{m}\,\|\dot{\boldsymbol{\epsilon}}\|$) vanishes and the overall stiffness increases. Since one cannot guarantee that there exists a one-to-one function $\boldsymbol{\sigma}(\boldsymbol{\epsilon})$, this model is called hypoelastic. Hypoelastic models have in general some disadvantages. For example, one can find a closed strain loop for which neither the stress nor the energy is recovered [6,14]. Since the model is path dependent due to the arbitrarily included pressure dependency, energy can be either created or dissipated depending on the sense of circulation in which the closed loop is applied[1]. This may become a serious shortcoming for modeling cyclic loading. To overcome this deficiency, we replaced the hypoelastic stiffness $\mathsf{E}^{\mathrm{hp}}(p)$ of previous hypoplastic models [5] by a hyperelastic one $\mathsf{E}(\boldsymbol{\sigma})$.

We use the hyperelastic stiffness proposed for paraelasticity [15] and neo-hypoplasticity [16]. This stiffness is a homogeneous function of stress of order $n \approx 0.6$, $\mathsf{E}(\lambda\boldsymbol{\sigma}) = \lambda^n \mathsf{E}(\boldsymbol{\sigma})$ (with $\lambda > 0$). The complementary energy is given by

$$\bar{\psi}(\boldsymbol{\sigma}) = P_0 c \bar{P}^{\alpha} \bar{R}^{2-n-\alpha}, \tag{3}$$

[1] Clockwise (CW) or counter clockwise (CCW).

where the stress invariants $P = -\boldsymbol{\sigma} : \vec{\mathbf{1}} = -\frac{1}{\sqrt{3}}\mathrm{tr}\,\boldsymbol{\sigma}$ and $R = \|\boldsymbol{\sigma}\|$ are normalized $\bar{P} = P/P_0$, $\bar{R} = R/P_0$ by the reference pressure P_0, say $P_0 = 1$ kPa, to attain unit consistency. The constants $c \approx 1 \cdot 10^{-4}$, $n \approx 0.6$, and $\alpha \approx 0.1$ are material parameters. The compliance C is obtained as the second derivative of the potential with respect to the stress, i.e. $\mathsf{C} = (\partial^2 \bar{\psi} / \partial \boldsymbol{\sigma} \partial \boldsymbol{\sigma})$. That is

$$P_0 \mathsf{C} = A_\alpha \vec{\mathbf{1}}\vec{\mathbf{1}} + B_\alpha(\vec{\mathbf{1}}\vec{\sigma} + \vec{\sigma}\vec{\mathbf{1}}) + C_\alpha \vec{\sigma}\vec{\sigma} + D_\alpha \mathsf{I} \tag{4}$$

wherein I represents the fourth order identity tensor $I_{ijkl} = \left(\delta_{ik}\delta_{jl} + \delta_{il}\delta_{jk}\right)/2$ and

$$A_\alpha = c(\alpha - 1)\alpha \bar{P}^{\alpha-2} \bar{R}^{2-n-\alpha} \tag{5}$$

$$B_\alpha = -c\alpha(2 - n - \alpha)\bar{P}^{\alpha-1} \bar{R}^{1-n-\alpha} \tag{6}$$

$$C_\alpha = c(2 - n - \alpha)(-n - \alpha)\bar{P}^{\alpha} \bar{R}^{-n-\alpha} \tag{7}$$

$$D_\alpha = c(2 - n - \alpha)\bar{P}^{\alpha} \bar{R}^{-n-\alpha} \tag{8}$$

For a given degree of homogeneity n of the stiffness with respet to the pressure, the exponent α can be related to the Poisson ratio ν_{iso} at the isotropic stress axis ($Q = 0$) using [15]:

$$\alpha = \frac{n^2 \nu_{\text{iso}} + n^2 - 5n\nu_{\text{iso}} - 2n + 6\nu_{\text{iso}}}{2\nu_{\text{iso}} - 1} \tag{9}$$

The hyperelastic stiffness is obtained after the inversion of the compliance. For convenience (see Sect. 5), and in order to take into account the influence of the void ratio on the stiffness, we introduce the factor $F(e) = (1 + e)/e$. Finally, the stiffness is written as

$$\mathsf{E} = \frac{1 + e}{e} \mathsf{C}^{-1} \tag{10}$$

5 Model for Isotropic Compression

Let consider the special case of isotropic compression, i.e. isotropic stress ($Q = 0$, $P \neq 0$) and isotropic strain rates ($\dot{\epsilon}_Q = 0, \dot{\epsilon}_P \neq 0$). We assume that upon isotropic compression, regardless of the initial void ratio and pressure, the Limiting Compression Curve (LCC) proposed by Pestana and Whittle [9] is asymptotically reached. Furthermore, we merge the concept of LCC with the pycnotropy function $e_i(P)$ of the hypoplastic model of Gudehus [17] and Bauer [18], which describes the loosest possible state at a given pressure P. States beyond e_i, that is $e > e_i$, are not allowed. The LCC (or $e_i(P)$) is described by the Bauer's [18] formula

$$e_i = e_{i0} \exp\left[-\left(\sqrt{3}P/h_{si}\right)^{n_{Bi}}\right], \tag{11}$$

where the maximum allowed void ratio at zero pressure e_{i0}, the so called hardness of solid phase h_{si}, and the exponent n_{Bi} are material constants.

The bulk modulus $K_i = \frac{\partial p}{\partial \epsilon_{\mathrm{vol}}} = \frac{1}{3}\frac{\partial P}{\partial \epsilon_P}$ along the LCC is obtained after solving (11) for P, replacing e_i by e, and differentiating after ϵ_P

$$K_i = \frac{1}{3}\frac{\partial P}{\partial e}\frac{\partial e}{\partial \epsilon_P} = \frac{h_{si}}{3n_{Bi}}\left(\sqrt{3}P/h_{si}\right)^{1-n_{Bi}}\frac{1+e}{e} \tag{12}$$

with $\frac{\partial e}{\partial \epsilon_P} = -(1+e)\sqrt{3}$ from (2).

If we assume that unloading and reloading are nearly elastic processes, we can describe them using the hyperelastic stiffness given in (10). The bulk stiffness K for hyperelasticity[2] yields

$$K = \frac{1}{3}\vec{\mathbf{1}} : \mathsf{E} : \vec{\mathbf{1}} = \frac{P_0}{3c(n-2)(n-1)}\left(\frac{P}{P_0}\right)^n \frac{1+e}{e} \tag{13}$$

The loading/unloading/reloading processes can be described by

$$\dot{P} = \begin{cases} 3K_i\dot{\varepsilon}_P & \text{for normal compression along LCC with } e = e_i \\ 3K\dot{\varepsilon}_P & \text{for unloading/reloading} \end{cases} \tag{14}$$

In order to simulate these processes with the hypoplastic Eq. (1), we must find a suitable definition of Y. We start by extracting the isotropic portion of the model by multiplying both sides of (1) by $-\vec{\mathbf{1}}$:

$$\dot{P} = -\vec{\mathbf{1}} : \dot{\sigma} = -\vec{\mathbf{1}} : \mathsf{E} : (\dot{\boldsymbol{\epsilon}} - Y\mathbf{m}\,\|\dot{\boldsymbol{\epsilon}}\|). \tag{15}$$

Since the deformation is purely volumetric $\dot{\boldsymbol{\epsilon}} = -\vec{\mathbf{1}} : \dot{\epsilon}_P$ and the stress remains isotropic $Q = 0$, the flow rule must be isotropic $\mathbf{m} = -\vec{\mathbf{1}}$. Therefore, (15) becomes

$$\dot{P} = 3K\left(\dot{\epsilon}_P - Y\,|\dot{\epsilon}_P|\right). \tag{16}$$

We propose the following expression for Y

$$Y = Y_I \mathrm{OCR}^{-n_O} \tag{17}$$

with

$$Y_I = \max\left[0, 1 - \left(\frac{e_i}{e}\right)^{n_O}\frac{K_i}{K}\right] \tag{18}$$

and the overconsolidation Ratio OCR defined as

$$\mathrm{OCR} = \frac{P_B}{P} \tag{19}$$

---

[2] Notice that the K can be obtained directly from the hyperelastic potential as $K = \frac{1+e}{e}\left[3\frac{\partial^2 \bar{\psi}}{\partial P^2}\right]^{-1}$. In this case, the potential simplifies to $\bar{\psi}(P, R) = P_0 c \bar{P}^{2-n}$ because $P = R$ holds for $Q = 0$.

where P_B is a state variable that resembles the preloading pressure, i.e. it accounts for previous stress loading.

The evolution of P_B upon volumetric compression must satisfy some empirical observations. In contrast to clay[3], there is no unique virgin compression line for sand. Sand samples prepared at the same initial pressure but different initial densities show different compression lines. However, for a sample prepared at a given initial density e_0 and pressure P_0 there is a single compression line, at which we may regard the sample state as "normal consolidated". Along this line the pressure P and the preconsolidation pressure P_B are identical, i.e. OCR $= P_B/P = 1$. Therefore, we propose the evolution of the preconsolidation pressure P_B due to volumetric compression to be

$$\dot{P}_B = 3K\,(1 - Y_I)\,\mathrm{OCR}^{-n_O}\,\dot{\epsilon}_P \tag{20}$$

where n_O is a material constant. During volumetric compression starting from $P_B = P$ (i.e. OCR = 1), Eqs. (20) and (16) become identical (hence, $\dot{P}_B = \dot{P}$ and OCR = 1). Upon unloading (volumetric extension), OCR increases and both $\dot{P}_B$ and Y tend to zero. In this case the model response (16) is nearly hyperelastic. This means, both processes in (14) can be described with a single equation Eq. (15). Instead of an explicit criterion for loading and unloading, the term OCR^{-n_O} provides a smooth switch between the two expressions in (14). Notice that states at which the current pressure P is larger than the preconsolidation pressure P_B, that means OCR < 1, are allowed.

Notice also that in (18) we introduced the function max() because the expression $1 - (e_i/e)^{n_O}\, K_i/K$ becomes negative for P smaller than

$$P_{neg} = \left[\frac{h_{si}}{n_{Bi}}\,\frac{c(n-2)(n-1)}{P_0^{1-n}}\left(\frac{\sqrt{3}}{h_{si}}\right)^{1-n_{Bi}}\right]^{\xi} \tag{21}$$

with $\xi = -1/(1 - n - n_{Bi})$.

6 Constitutive Model with Historiotropic Yield Surface

The notions developed for the 1-Dimensional model presented in Sect. 5 are now extended to represent more general stress states and strain paths. In particular, the 3-D model must be able to describe the asymptotic state reached upon shearing. The material behaviour observed in oedometric, isotropic, and triaxial tests should be also described by the model.

6.1 Historiotropic Yield Surface $g = 0$

For the 1D model (see Sect. 5), the overconsolidation ratio was defined in (19) as OCR $= p_B/p$. In this case, the state variable p_B could be related to the

[3] In clay, a family of parallel virgin compression lines (in the $e - \ln(p)$–diagram) can be obtained by deforming the samples at different straining rates.

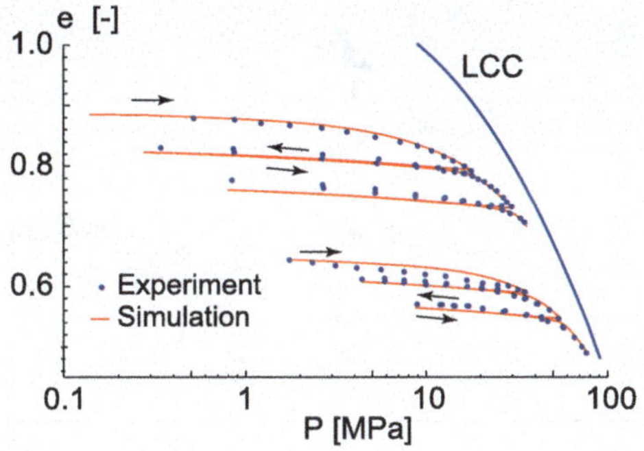

Fig. 2. Simulation of isotropic compression tests on Toyoura Sand [19] and [20] with the material constants of Table 1.

preconsolidation pressure, e.g. the maximum pressure at which the sample has been subjected before the actual test. However, the definition of OCR for the general case has to be modified in order to include the influence of the deviatoric portions of the stress $\boldsymbol{\sigma}$. Furthermore, it has to account for a more general definition of the preconsolidation pressure p_B, i.e. the preconsolidation stress tensor $\boldsymbol{\sigma}_B$

$$\boldsymbol{\sigma}_B \equiv p_B\,(-\,\mathbf{1} + \boldsymbol{\Omega}), \tag{22}$$

which is a new state variable with isotropic $-p_B\,\mathbf{1}$ and deviatoric $p_B\boldsymbol{\Omega}$ (i.e. $\operatorname{tr}\boldsymbol{\Omega} = 0$) parts. To achieve this goal, we propose the equation

$$g(\boldsymbol{\sigma}, \boldsymbol{\sigma}_B, e) \equiv \mathbf{w} : \mathbf{w} - \frac{2}{3} M_w^2\,(1-\alpha)(1-\beta)^2 = 0, \tag{23}$$

with

$$\alpha = (p/p_B)^{c_b} \quad \text{and} \quad \beta = \sqrt{\frac{3}{2}}\frac{\langle \vec{\mathbf{w}} : \boldsymbol{\Omega}\rangle}{M_o}, \tag{24}$$

to define a yield surface that can be used to compute the OCR in the general case, see Fig. 3. Stresses $\boldsymbol{\sigma}$ lying on the yield surface, i.e. all stresses that satisfy (23) for a given void ratio e and preconsolidation stress $\boldsymbol{\sigma}_B$, correspond to OCR = 1. The *size* p_B and *inclination* $\boldsymbol{\Omega}$ of the yield surface with respect to the isotropic axis in the stress space is determined by the new state variable $\boldsymbol{\sigma}_B$. In (24), c_b is a material constant and the Mc Cauley brackets operator is defined as $\langle \sqcup \rangle = \frac{1}{2}(\sqcup + |\sqcup|)$. The deviatoric tensor

$$\mathbf{w} = \hat{\boldsymbol{\sigma}}^* - \boldsymbol{\Omega} \quad \text{with} \quad \hat{\boldsymbol{\sigma}}^* = \frac{\boldsymbol{\sigma}^*}{p} \tag{25}$$

represents the inclination of the current stress with respect to *the major axis* of the yield surface. Near the origin of the p–axis, the maximum opening of the yield surface is controlled by peak friction angle φ_{peak} via

$$M_w = M\left(\varphi_{peak}, \theta_{\mathbf{w}}\right) \quad \text{and} \quad M_o = M\left(\varphi_{peak}, \theta_{\boldsymbol{\Omega}}\right) \tag{26}$$

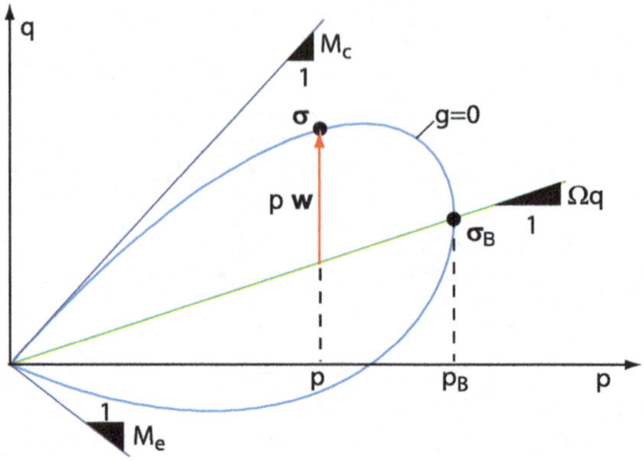

Fig. 3. Historiotropic yield surface $g = 0$ in the triaxial $p - q$–space

where M_w and M_o are the maximum stress ratios (q/p) in the direction of the tensors $\mathbf{w}$ and $\mathbf{\Omega}$, respectively. The peak friction angle φ_{peak} is a function of the void ratio and the pressure, see Sect. 6.4.2. For a given angle φ and a given tensor $\sqcup$ we use a general expression for the slope $M(\varphi, \theta_\sqcup)$ of the stress ratio on the $p - q$–plane. The slope M can be found as an interpolation between the two extreme values for triaxial compression M_c and triaxial extension M_e over the Lode's angle θ as $M = \frac{1}{2}\left[(M_c - M_e)\cos(3\theta) + (M_c + M_e)\right]$ or

$$M = M(\varphi, \theta) = \frac{6s\,(3 + sc)}{(3 + s)(3 - s)} \tag{27}$$

with $s = \sin\varphi$ and $c = \cos(3\theta)$. The Lode's angle $\theta_\sqcup$ of the tensor $\sqcup$ is defined as

$$\cos(3\theta_\sqcup) = -\sqrt{6}\vec{\sqcup}^*{}_{ij}\vec{\sqcup}^*{}_{jk}\vec{\sqcup}^*{}_{ki} \tag{28}$$

and $\vec{\sqcup}^* = (\sqcup^*)^{\rightarrow} \neq \left(\vec{\sqcup}\right)^*$. For isotropic tensors, i.e. $\|\sqcup^*\| = 0$, Eq. (28) cannot be used due to the division by zero in $\vec{\sqcup}^* = \sqcup^*/\|\sqcup^*\|$. In this case we arbitrary set $\cos(3\theta_\sqcup) = 1$. Triaxial compression and extension are especial cases of (27)

$$M_c = M(\varphi, \theta_{\boldsymbol{\sigma}} = 0^\circ) \quad \text{and} \quad M_e = M(\varphi, \theta_{\boldsymbol{\sigma}} = 60^\circ)$$

Conversely, for a given M and θ ($c = \cos(3\theta)$) we can find the mobilized angle φ from (27) as

$$\sin\varphi = \frac{-9 + 3\sqrt{9 + 6cM + M^2}}{6c + M} \tag{29}$$

Notice that the projection of the surface $q/p - M = 0$ (with M defined by (27)) onto the deviatoric plane may become concave for large friction angles. Although, alternative convex surfaces, like that from Matsuoka and Nakai [21], could be adopted, we use this concave surface because of its simplicity and because the flow rule is not associative (c.f. [22]).

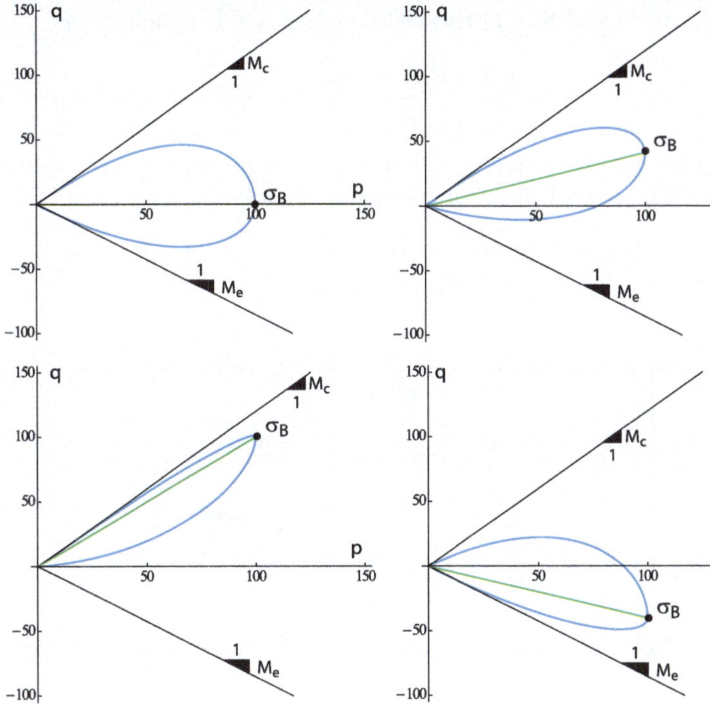

Fig. 4. Historiotropic yield surface $g = 0$ in the triaxial $p - q$–space for different backstresses $\boldsymbol{\sigma}_B$.

Similar to the yield surface of the SANISAND model [22], the surface (23) can be thought as a conical surface with sharp apex at $p = 0$. The opening of this cone around its axis $\boldsymbol{\Omega}$ varies from a maximal value M_w at $p = 0$ to zero at $p/p_B = 1$ (the cap of the surface), see α in (24) and Fig. 4. Furthermore, the opening of the cone reduces when the inclination $\boldsymbol{\Omega}$ reaches its maximum value in the direction of $\mathbf{w}$, see β in (24).

Notice that both the maximum opening M_w of the yield surface and the maximum slope M_o of the $\boldsymbol{\Omega}$ in the $p - q$–space are determined by the peak friction angle, φ_{peak}, which is a function of the void ratio and the pressure.

6.2 Overconsolidation Ratio OCR

In Sect. 5 we used the Overconsolidation Ratio in the multiplicator OCR^{-no} as a way to distinguish between first loading and unloading/reloading. In order to expand this idea to general stress states (and not only for the isotropic case), we combine the concept of overstress [23] with the yield surface $g = 0$. This generalization has been implemented in hypoplastic constitutive models for clay, e.g. [6] and [24]. Stress states lying on the yield surface $g = 0$ correspond to $\text{OCR} = 1$. Within ($\text{OCR} < 1$) the yield surface is the intensity of anelastic flow smaller than outside (OCR > 1) the yield surface. OCR is defined as

$$\mathrm{OCR} \equiv p_B / p_{B+}, \tag{30}$$

where p_{B+} is the "size" of a pseudo-yield surface on which the current stress $\boldsymbol{\sigma}$ lies. To compute p_{B+} we proceed as follows. For a given stress $\boldsymbol{\sigma}$, void ratio e, and inclination $\boldsymbol{\Omega}$, we find p_{B+} from a surface $g_+ = 0$

$$g_+(\boldsymbol{\sigma}, \boldsymbol{\sigma}_{B+}, e) = 0 \quad \text{with} \quad \boldsymbol{\sigma}_{B+} = p_{B+}(-\mathbf{1} + \boldsymbol{\Omega}) \tag{31}$$

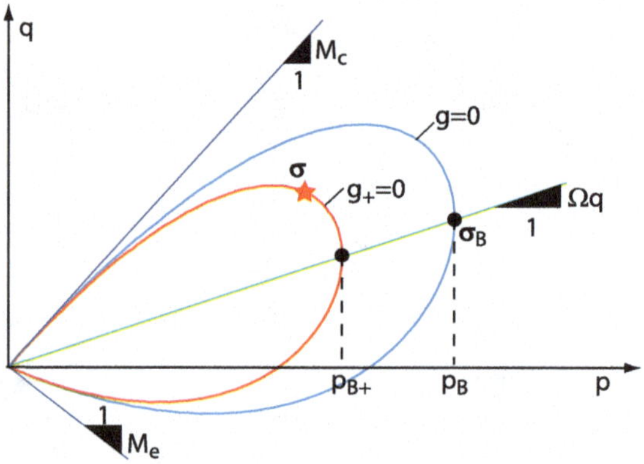

Fig. 5. Graphical interpretation of p_B and p_{B+} used in the definition of OCR (30)

The surface $g_+ = 0$ is affine to $g = 0$, i.e. both have the same inclination $\boldsymbol{\Omega}$, the same opening $\varphi_{peak}(e, p)$, but different back pressure p_{B+} (instead of p_B), see Fig. 5. Using (31) and (23) with the substitution (24) we obtain

$$p_{B+} = p \left[1 - \frac{3\mathbf{w} : \mathbf{w}}{2M_w^2 (1 - \beta)^2} \right]^{-1/c_b} \tag{32}$$

6.3 Generalized Taylor's Dilatancy M_d

The dilatancy M_d, which is the ratio of plastic volumetric $\dot{\epsilon}_v^p$ to the plastic deviatoric $\dot{\epsilon}_q^p$ strain rates, i.e. $M_d = \dot{\epsilon}_v^p / \left|\dot{\epsilon}_q^p\right|$, is a distinctive phenomenon of granular materials and is one of the most important aspects in the description of the mechanical behaviour of soils. The accurate modeling of dilatancy becomes even more relevant for cyclic loading. The experiments of Pradhan and Tatsuoka [25] and [26] show that the dilatancy M_d is a function not only of stress $\eta = q/p$ (like in Camclay model [10] $M_d(\boldsymbol{\sigma})$), or of the void ratio (like in [27] $M_d(\boldsymbol{\sigma}, e)$), but also a function of the direction of shearing $M_d(\boldsymbol{\sigma}, e, \vec{\dot{\boldsymbol{\epsilon}}}^*)$.

Consider a triaxial compression test on a dense sand sample at constant pressure. For stress ratios larger than the critical one $\eta = q/p > M_c$, the dilatancy is positive, i.e. the volume of the sample increases with shearing. At this state we may have two different dilatancies depending on the direction of shearing. Upon continued shearing, the sample dilates, i.e. $M_d > 0$. If the direction of shearing is reversed, say from triaxial compression to triaxial extension, then the sample contracts, i.e. $M_d < 0$. To describe this dependency of dilatancy on the strain direction, Pradhan and Tatsuoka [25] use two different equations: one for triaxial compression and other for triaxial extension.

In order to describe the relation between dilatancy, stress (ratio), void ratio and strain direction, we generalize of the Taylor's *interlocking* concept [7] by assuming that:

1. the division of strength into a portion due to internal friction and another due to dilatancy applies also for states different to the peak.
2. the dilatancy is zero at two stages: at PTL and at the critical state CS.
3. the Taylor's rule observed in simple shear test can be extended to multiaxial stress states.

We start from the concept of *interloking*, which was stated by Taylor [7] as follows. Consider a direct shear test on dense sample. The shear strength at the peak τ_{peak} consists of two portions:

$$\tau_{peak} = \tau_c + \tau_e \tag{33}$$

where $\tau_c = \sigma \tan \varphi_c$ is the *internal* shear strength, with φ_c as the critical friction angle, and τ_e is the portion of the shear strength related to interlocking. At the peak, the sample thickness h increases with the shearing displacement u, which represents a ratio of Δh_{peak} units of height per unit of Δu_{peak}. The expansion of the sample is resisted by the normal stress σ. For the occurrence of the expansion, energy must be supplied. The amount of energy E_e^{used} used during expansion is the product of the normal force σA applied on the top and bottom of the sample and the thickness increment Δh_{peak}, i.e. $E_e^{\text{used}} = \sigma A \Delta h_{peak}$. The portion of the shearing stress τ_e supplies the energy for expansion E_e^{supplied}, which is the product of the shear force $\tau_e A$ and the shearing displacement Δu_{peak}, i.e. $E_e^{\text{supplied}} = \tau_e A \Delta u_{peak}$. Setting these two energies equal, $E_e^{\text{used}} = E_e^{\text{supplied}}$, we obtain

$$\sigma A \Delta h_{peak} = \tau_e A \Delta u_{peak} \tag{34}$$

or $\tau_e = \sigma \Delta h_{peak} / \Delta u_{peak}$. We can therefore rewrite (33) as

$$\tau_{peak} = \sigma \tan \varphi_c + \sigma \Delta h_{peak} / \Delta u_{peak} \tag{35}$$

Dividing the last equation by σ and substituting $\tau_{peak}/\sigma = \tan \varphi_{peak}$ yields

$$\tan \varphi_{peak} = \tan \varphi_c + \tan \psi_{peak} \tag{36}$$

where

$$\tan \psi_{peak} \equiv \Delta h_{peak} / \Delta u_{peak} \tag{37}$$

and ψ_{peak} is defined as the dilatancy angle at peak state. If we assume that the relation between mobilized friction angle φ_m and dilatancy angle ψ given in (36) holds at any state (and not just at peak state where $\varphi_m = \varphi_{peak}$ and $\psi = \psi_{peak}$), (36) becomes

$$\tan\varphi_m = \tan\varphi_c + \tan\psi \tag{38}$$

Equation (38) can be interpreted as a constraint between stress and strain ratios during shearing. Notice that (38) has been obtained for constant σ. The advantage of direct shear or constant pressure triaxial tests is that volume changes induced by pure shearing can be easily identified. On the contrary, in drained triaxial tests, for example, a constitutive relation is required to identify which portion of the volume change is due to the change of applied pressure, and which portion is due just to interlocking.

We modify now the Taylor's rule given by (38) to incorporate two effects: the void ratio (and pressure) dependency and the strong contractancy upon reversals. Experiments of Ishihara [3] show that the dilatancy equals zero at two stages: at the critical state CS and at the Phase Transformation Line PTL. To reproduce this observation, we replace $\tan\varphi_c$ by $\tan\varphi_{PTL}$ in (38), where $\varphi_{PTL}(e, p)$ is a function of void ratio and pressure, see Sect. 6.4.3. For dense sand $e < e_c(p)$, the PTL lies below the critical state line CSL, $\varphi_{PTL} < \varphi_c$. For loose sand $e > e_c(p)$, the PTL lies above the CSL $\varphi_{PTL} > \varphi_c$ and may not be reached upon undrained shearing. After long monotonic shearing, the sample approaches the critical state, at which deviatoric straining produces neither stress nor volume change. Therefore, at the critical state (where $e = e_c$), the PTL coincides with CSL, i.e. $\varphi_{PTL} = \varphi_c$.

Experiments show also that the maximum contractancy is attained upon reversals of the strain path. We propose the factor $\vec{\boldsymbol{\sigma}}^* : \vec{\dot{\boldsymbol{\epsilon}}}^*$ as a way to reproduce this observation. Incorporating this two modifications into (38), we obtain

$$\tan\psi = \vec{\boldsymbol{\sigma}}^* : \vec{\dot{\boldsymbol{\epsilon}}}^* \tan\varphi_m - \tan\varphi_{PTL} \tag{39}$$

For a given mobilized friction angle $\varphi_m > \varphi_{PTL}$, we can obtain two different dilatancies from a single Eq. (39): a positive dilatancy $\tan\varphi_m - \tan\varphi_{PTL}$ if $\vec{\boldsymbol{\sigma}}^* : \vec{\dot{\boldsymbol{\epsilon}}}^* > 0$ or a stronger contractancy $-\tan\varphi_m - \tan\varphi_{PTL}$ if the strain rate is reversed, e.g. $\vec{\boldsymbol{\sigma}}^* : \vec{\dot{\boldsymbol{\epsilon}}}^* < 0$. Notice, however, that (39) relates three angles: ψ, φ_m, and φ_{PTL}. In order to generalize this equation for three dimensions, we must find a relation between the angle ψ and the volumetric and deviatoric invariants of the strain rate. Neglecting the difference between plastic and total deviatoric strain rates we can use (27), which express an angle in terms of stress invariants, to relate the dilatancy angle ψ with tensor invariants of the plastic strain rate

$$M_d = M(\psi, \theta_{\mathbf{W}}) = \frac{\dot{\epsilon}_v^p}{|\dot{\epsilon}_q|} = \frac{6s\,(3 + sc)}{(3 + s)(3 - s)} \tag{40}$$

where $s = \sin\psi$ and $c = \cos(3\theta_{\mathbf{W}})$. In hypoplasticity, the direction of the deviatoric anelastic strain has the same direction as the deviatoric stress. In (40), however, we have assumed that the plastic deviatoric shearing is parallel to $\mathbf{w}$. This assumption is based on the observation, that after long monotonic shearing, $\mathbf{w}$ and $\boldsymbol{\sigma}^*$ become parallel, see (53). Using (40) and writing $M_d = \operatorname{tr}\dot{\boldsymbol{\epsilon}}^p/\sqrt{\frac{2}{3}}\,\|\dot{\boldsymbol{\epsilon}}^*\|$, we can now formulate a flow rule for shearing as

$$\mathbf{m}_D = \overrightarrow{\left(\frac{1}{3}\sqrt{\frac{2}{3}}M_d\ \mathbf{1} + \mathbf{w}\right)} \tag{41}$$

Figure 6 compares the flow rule (41) with experimental results. In this simulations it has been assumed that the plastic strain rate is given by $\dot{\boldsymbol{\epsilon}}^p = \mathbf{m}_D\,\|\dot{\boldsymbol{\epsilon}}\|$ with $\mathbf{w} = \vec{\boldsymbol{\sigma}}^*$ and $\varphi_{PTL} = \varphi_c = 27°$.

6.4 Characteristic Curves

Besides the Limiting Compression Curve (LCC) defined in (11), the model requires explicit relations between the critical void ratio, the maximum friction angle, and the phase transformation line with the pressure.

6.4.1 Critical Void Ratio $e_c(p)$

After large monotonic shearing a sand sample approaches a state at which further isochoric shearing is possible at constant stress. This state is approached independently of the initial density or stress of the sample. At this state, the so called critical state, the material reaches a critical void e_c, which is a one-to-one function of the pressure p, and a critical stress ratio M (at which the mobilized friction angle $\varphi_m = \varphi_c$). The critical void ratio is described by the Bauer's formula [18]

$$e_c(p) = e_{c0}\exp\left[-(p/h_{sc})^{n_{Bc}}\right]$$

with the constants e_{c0}, h_{sc}, and n_{Bc}.

6.4.2 Maximum Friction Angle φ_{peak}

The maximum attainable stress obliquity depends on the void ratio and the pressure. Larger stress obliquities can be reached by dense sands than by looser ones, see Fig. 7. We describe the maximum stress obliquity in terms of the friction angle φ_{peak} via

$$\tan\varphi_{\text{peak}} = \tan\varphi_c\left(\frac{e_c}{e}\right)^{n_{peak}} \tag{42}$$

with the constant n_{peak}. Notice that at the critical state $e = e_c$ the maximum friction angle equals the critical one.

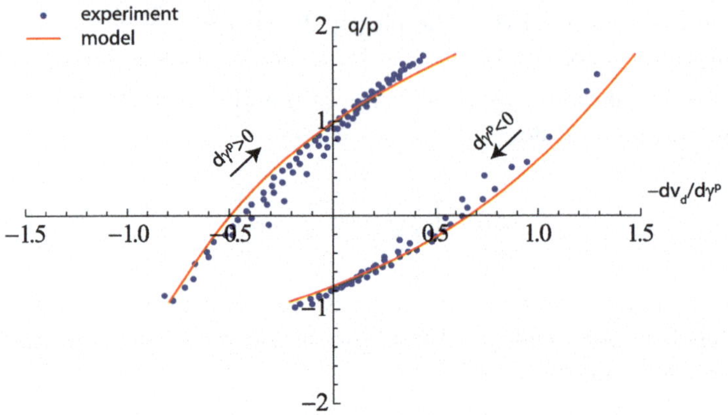

Fig. 6. Experimental results of constant pressure triaxial tests [25,26] compared with the dilatancy given by (40) and (41) with $\mathbf{w} = \vec{\sigma}^*$ and $\varphi_{PTL} = \varphi_c = 27°$. Conventions: $dv_d = \varepsilon_a + 2\varepsilon_r$, $d\gamma = \varepsilon_a - \varepsilon_r$ where ε_a and ε_r are the axial and radial strain components (compression positive), respectively. $d\gamma^d$ represents the plastic portion of $d\gamma$.

6.4.3 Friction Angle at Phase Transformation Line φ_{PTL}

The Phase Transformation Line (PTL) described by Ishihara et al. [3] indicates the stress ratio at which, upon monotonic shearing, the sand changes from contractive to dilative behaviour. The PTL (usually depicted in the $p-q$–space) is related to the friction angle φ_{PTL} and depends on the pressure and the void ratio

$$\tan\varphi_{PTL} = \tan\varphi_c \left(\frac{e}{e_c}\right)^{n_{PTL}} \tag{43}$$

with the constant n_{PTL}. Dense sands reach the PTL at lower stress ratios than loose ones, see Fig. 7. Notice that at the critical state $e = e_c$, where unbounded shear strain is possible with no volume changes, the PTL line becomes the CSL, i.e. $\tan\varphi_{PTL} = \tan\varphi_c$.

6.5 Intensity of Anelastic Flow

The intensity of anelastic flow Y is defined as an interpolation between two main loading cases: Y_D for shearing and Y_I for isotropic compression (see Sect. 5)

$$Y = [Y_I + (1 - Y_I)\,Y_D]\,\mathrm{OCR}^{-n_O} \tag{44}$$

Upon shearing, the degree of nonlinearity is given by (see Fig. 8)

$$Y_D = \left[\frac{a}{b}\right]^{n_{YD}} = \left[\frac{\mathbf{w} : (\hat{\sigma}^* - \mathbf{\Omega})}{\mathbf{w} : \left(\hat{\sigma}^*_f - \mathbf{\Omega}\right)}\right]^{n_{YD}} \tag{45}$$

with

$$\hat{\sigma}_f^* = \frac{\sigma_f^*}{p} \tag{46}$$

where σ_f^* is the deviatoric portion of the stress σ_f, σ_f is the image of the current stress σ projected onto the limiting surface $f = 0$ in the direction of $\mathbf{w}$, and the exponent n_{YD} is a material constant. The limiting surface $f = 0$ is an open cone in the stress space, whose opening is determined by the maximum friction angle φ_{peak}. This surface is given by the equation

$$f(\sigma, \varphi_{peak}) \equiv \frac{\|\sigma^*\|}{p} - \sqrt{\frac{2}{3}}M = 0 \quad \text{with} \quad M = M(\varphi_{peak}, \theta_\sigma). \tag{47}$$

Since σ_f^* is parallel to $\mathbf{w}$, we can find σ_f^* from (47) as

$$\sigma_f^* = p\sqrt{\frac{2}{3}}M_w\vec{\mathbf{w}} \quad \text{with} \quad M_w = M(\varphi_{peak}, \theta_{\mathbf{w}}) \tag{48}$$

6.6 Flow Rule m

The flow rule $\mathbf{m}$ is interpolated from the two special cases: the flow rule $\mathbf{m}_D$ for isochoric shearing and the flow rule $\mathbf{m}_I$ for radial compression

$$\mathbf{m} = [\mathbf{m}_D Y_D + \mathbf{m}_I \exp(-\xi Y_D)]^{\rightarrow} \tag{49}$$

where ξ is a large number, say $\xi \approx 1000$. During isochoric shearing, $Y_D > 0$ and the second term on the right-hand side of (49) vanishes. Upon radial compression, $\boldsymbol{\Omega}$ alienates asymptotically with the current stress ratio $\hat{\sigma}^*$ and $\mathbf{w}$ tends to zero. In this case Y_D cannot be determined from (45). For this special case we define

$$Y_D = 0 \quad \text{for} \quad \|\mathbf{w}\| < w_{\text{tol}} \tag{50}$$

where w_{tol} is a small number, say $w_{\text{tol}} = 10^{-4}$, and the first term on the right hand side of (49) vanishes. Notice that a similar interpolation of the flow rule has been used in the SANISAND model [22]. In order to reproduce the asymptotic behaviour upon radial compression, i.e. those tests with $\hat{\sigma}^* = \text{const}$ and $\text{tr}\,\dot{\epsilon} < 0$, the flow rule $\mathbf{m}$ should be given by

$$\mathbf{m}_I = \left(-\mathbf{1} + c_{mI}\hat{\sigma}^*\right)^{\rightarrow} \tag{51}$$

where c_{mI} is a material constant. The asymptotic value of the deviatoric portion $\boldsymbol{\Omega}$ of the back stress σ_B upon radial compression is $\boldsymbol{\Omega} = \hat{\sigma}^*$. At this stage $\dot{\boldsymbol{\Omega}} = \mathbf{0}$, the tensor $\mathbf{w}$ vanishes $\mathbf{w} = \mathbf{0}$, see Sect. 6.7, and $Y_D = 0$ (50). In addition the stress path approaches a constant stress ratio, at which the *directional homogeneity* $\vec{\sigma}(\mathbf{T}, \mathbf{D}) = \vec{\sigma}(\lambda^2\mathbf{T}, \mathbf{D})$ can be observed [6]. Using this property of the hypoplastic Eq. (1), the constant c_{mI} can be found by making the flow direction $\mathbf{m}_I$ compatible with the stress $\sigma_{K_0} \sim -\operatorname{diag}[1, K_0, K_0]$, which is reached after long oedometric compression. K_0 can be computed from the empirical "Jacky"

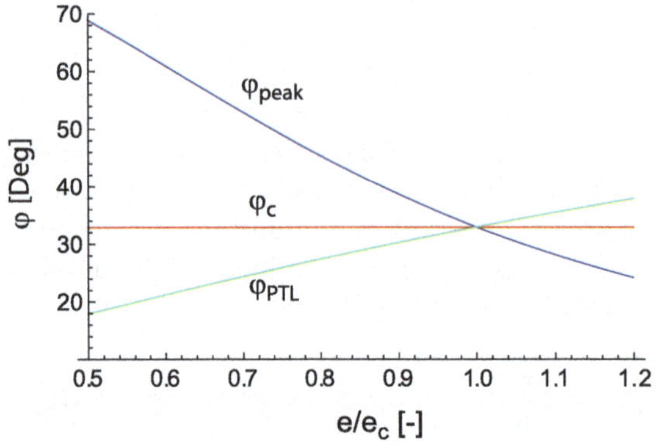

Fig. 7. Critical φ_c, maximum φ_{peak}, and phase transformation line φ_{PTL} angles as functions of the normalized void ratio e/e_c. Here: $\varphi_c = 33°$, $n_{peak} = 2$, and $n_{PTL} = 1$

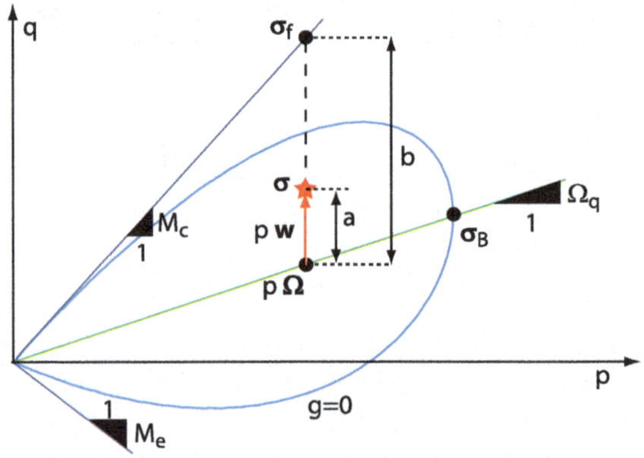

Fig. 8. Graphical interpretation of Y_D for triaxial compression.

formula $K_0 = 1 - \sin\varphi_c$. In the asymptotic case, the direction of flow must be uniaxial, i.e. $\mathbf{m}_I = -\operatorname{diag}[1, 0, 0]$. Therefore, c_{mI} is

$$c_{mI} = \frac{1 + 2K_0}{1 - K_0} = \frac{3}{\sin\varphi_c} - 2 \tag{52}$$

6.7 Evolution of the State Variable $\boldsymbol{\sigma}_B$

Similar as in the anisotropic viscohypoplastic model [24], the evolution of the state variable $\boldsymbol{\sigma}_B$ is given in separate equations for $\boldsymbol{\Omega}$ and p_B. For the evolution of $\boldsymbol{\Omega}$, i.e. the inclination of the yield surface $g = 0$, we propose

$$\dot{\boldsymbol{\Omega}} = C_2 \left(\hat{\boldsymbol{\sigma}}^* - \boldsymbol{\Omega}\right) \mathrm{OCR}^{-n_O} \, \|\dot{\boldsymbol{\epsilon}}\| \, . \tag{53}$$

According to this equation, $\boldsymbol{\Omega}$ evolves upon any kind of deformation (isotropic or deviatoric). The evolution rate is controlled by the material constant C_2 and is larger for stresses outside (OCR < 1) than for stresses inside (OCR > 1) the yield surface $g = 0$. $\boldsymbol{\Omega}$ evolves in the direction given by the difference between the current stress ratio $\hat{\boldsymbol{\sigma}}^*$ and $\boldsymbol{\Omega}$. To avoid numerical problems when the stress reaches the limit stress surface, and therefore $\|\boldsymbol{\Omega}\| > \left\|\hat{\boldsymbol{\sigma}}^*_f\right\|$, the stress $\hat{\boldsymbol{\sigma}}^*$ in (53) should be replaced by $\hat{\boldsymbol{\sigma}}^*_f$. After long radial compression, both $\boldsymbol{\Omega}$ and $\hat{\boldsymbol{\sigma}}^*$ approach an asymptotic stress ratio and the evolution of $\boldsymbol{\Omega}$ ceases.

The evolution of p_B, i.e. the size of the yield surface $g = 0$, consists of two portions

$$\dot{p}_B = \dot{p}_{BI} + \dot{p}_{BD}, \tag{54}$$

$\dot{p}_{BI}$ is related to volumetric and $\dot{p}_{BD}$ to deviatoric deformations. The evolution of p_B during isotropic compression has been discussed in Sect. 5 and is given by (20). Hence,

$$\dot{p}_{BI} = -K\,(1 - Y_I)\,\mathrm{OCR}^{-n_O}\,\mathrm{tr}\,\dot{\boldsymbol{\epsilon}} \tag{55}$$

With this equation, p_{BI} increases during volumetric compression and reduces during volumetric extension. However, during volumetric extension OCR increases making the reduction of p_{BI} negligible. Therefore, $\dot{p}_{BI}$ can be thought to be similar to the isotropic hardening of Camclay.

In case of shear deformations, we propose an evolution equation similar to (53)

$$\dot{p}_{BD} = C_2\,(p - p_B)\,\mathrm{OCR}^{-n_O}\,\|\dot{\boldsymbol{\epsilon}}^*\| \tag{56}$$

This equation allows for strong reductions of p_B during undrained cyclic loading, which resembles the material degradation observed in phenomena like cyclic mobility. Such reductions of p_B during pure deviatoric shearing can also describe the discrepancy between the compression curve and reconsolidation curves after some undrained shear cycles observed by Ishihara und Okada [2].

7 Comparison of the Model with Experimental Results

The proposed constitutive model has been implemented in a Fortran code (called UMAT – User MATerial subroutine) that is compatible with the commercial Finite Element program Abaqus. The UMAT is also compatible with the IncrementalDriver routine [28], which is an open-source code designed to verify the numerical implementation of the constitutive model and to simulate element tests.

To evaluate the performance of the model, experimental results of monotonic and cyclic triaxial tests on different sands were simulated. Furthermore, additional experiments on Karlsruhe fine sand, which were specially designed to study the influence of previous deformation histories on subsequent cyclic loading, were also simulated.

7.1 Monotonic Triaxial Tests

Figures 9, 10, and 11 compare the laboratory with the simulation results of triaxial compression tests on loose, medium dense, and dense Toyoura sand samples [29], respectively. These experiments are commonly used as benchmark for constitutive models because they account for the material response to monotonic undrained shearing over a wide range of pressures and densities. The material constants used in the simulations are listed in Table 1. Isotropic compression tests on Toyoura sand for different densities have been also simulated, Fig. 2. It can be seen that the model is able to reproduce the experimental results in an acceptable way for all pressures and densities.

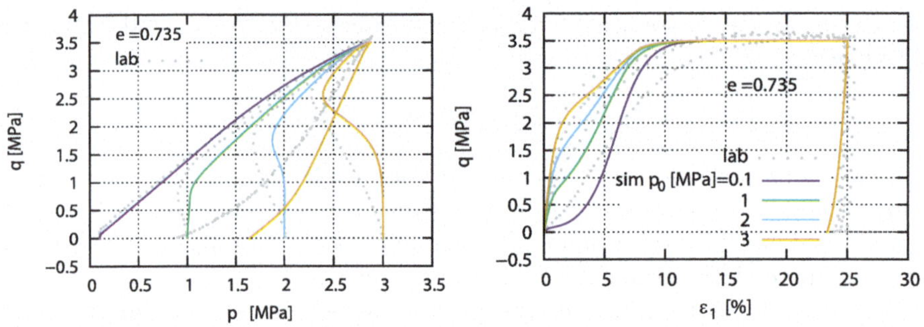

Fig. 9. Experiment results and simulation of triaxial compression tests on Toyoura sand for dense (e = 0.735) samples. Experiment data adapted from [29].

7.2 Drained and Undrained Triaxial Tests on Loose Sand

To show the influence of previous deformation history on the subsequent material response, Doanh [1] conducted a set of experiments on Hostun RF loose sand. Starting at an isotropic stress of p_0 = 100 kPa, the samples were subjected to drained triaxial compression up to different stress ratios denoted by the dots (1 to 10) in Fig. 12, upper part, left. Then the deviatoric stress was reduced in order to reach the initial isotropic stress with p = 100 kPa. At this point, all samples had similar void ratios, $e_0 \approx 0.94$. Then, undrained triaxial compression and extension tests were conducted. It can be observed that the maximum deviatoric stress upon undrained compression increases with the stress ratio attained during the preloading phase. Converserly, the maximum deviatoric stress attained in undrained extension reduces with the stress ratio reached during the preceding drained triaxial compression. This observation can be qualitatively reproduced by the proposed constitutive model, see Fig. 12, lower part, right.

7.3 Cyclic Triaxial Tests

The capabilities of the model were also tested under cyclic conditions. Figure 13 shows the simulation of undrained cyclic triaxial tests of small amplitude on

Karlsruhe fine sand [4]. Starting from an isotropic stress of p_0 = 200 kPa and e_0 = 0.952 (I_{D0} = 0.27) stress cycles of amplitude q^{ampl} = 40 kPa were applied. It can be observed that in the experiment as well as in the simulation, the accumulation of pore water pressure (or the reduction of effective pressure) after each cycle reduces with the number of cycles. The experiment was stopped when the accumulated axial strain reached 10%. This criterion was satisfied after 72 cycles in the experiment and after 120 cycles in the simulation. However, an acceptable agreement between experiment and simulation can be observed.

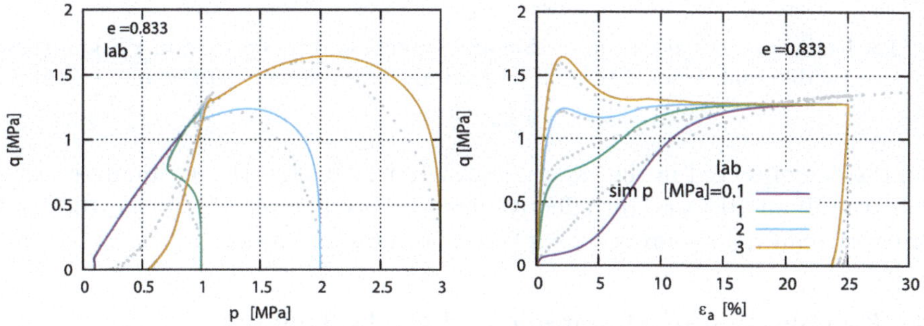

Fig. 10. Experiment results and simulation of triaxial compression tests on Toyoura sand for medium dense (e = 0.833) samples. Experiment data adapted from [29].

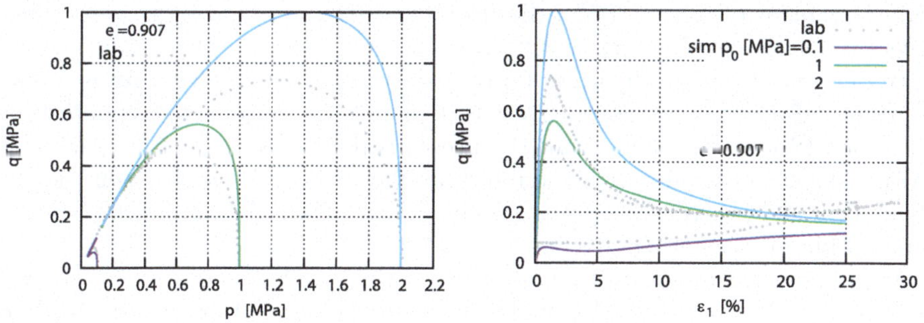

Fig. 11. Experiment results and simulation of triaxial compression tests on Toyoura sand for loose (e = 0.907) samples. Experiment data adapted from [29].

A cyclic undrained triaxial test with larger stress amplitude q^{ampl} = 60 kPa was also simulated, see Fig. 14. The cyclic loading started from an isotropic stress of p_0 = 200 kPa and an initial void ratio of e_0 = 0.726 (I_{D0} = 0.87). After about 80 cycles, the effective stress pressure reduces to nearly zero. Beyond that point, a cyclic attractor, a butterfly-shaped stress path in the $p - q$–space can be observed. The model simulates a stronger reduction of the effective stress

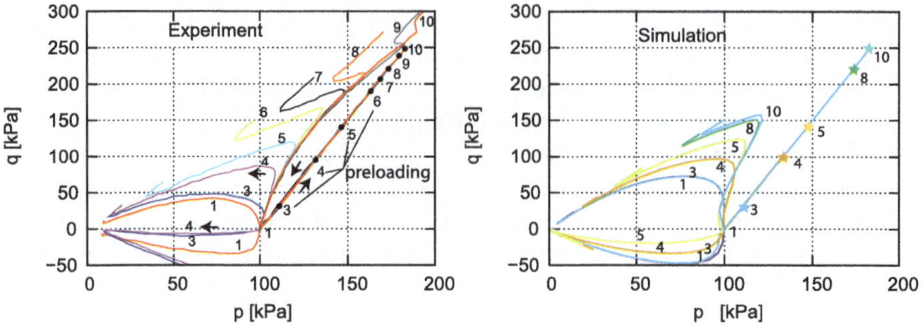

Fig. 12. Undrained triaxial compression and extension tests on Hostun RF loose sand preceded by different drained triaxial preloading. Experiment data adapted from [1].

than the one observed in the experiment. After 27 cycles, the accumulated axial strain was about 10% and the effective pressure became zero. However, the cyclic attractor could not be properly captured by the simulation.

7.4 Combination of Monotonic and Cyclic Loading

In a further validation attempt, experiments with cyclic loads of variable amplitude as well as combinations of monotonic and cyclic loading were simulated. Figure 15 shows the simulation of an undrained triaxial test with successive loading/unloading phases. The experiment starts from an isotropic stress of $p_0 = 200$ kPa and an initial density of $I_{D0} = 0.34$. After the application of a (compressive) strain increment of $\Delta\epsilon_1 = 0.05\%$, the deviatoric stress is reduced to zero. Then, successive cycles of reloading (with strain increment $\Delta\epsilon_1 = 0.05\%$) and unloading (reduction of deviatoric stress up to $q = 0$) phases took place. As reference, the results of a purely monotonic test is depicted. In contrast to hypoplasticity [13] or the ISA model [30], where the previous deformation path is tracked by a strain-like state variable (i.e. the intergranular strain), it can be seen that the proposed model can reproduce the cyclic loading of variable amplitude without overshooting or ratcheting effects.

A similar performance can be observed in the simulation of cyclic tests with small and large amplitudes. Figure 16 presents the simulation of a drained triaxial test with constant pressure and cycles of small and large amplitude on Toyoura sand. Figure 17 shows the simulation of an undrained triaxial test on dense Karlsruhe sand with cycles of small and large amplitude too. In both cases it can be observed that the model can "remember" the stress point from which the unloading (cycle of small amplitude) process started and that this stress point is reached again upon reloading. A stiffer and nearly elastic response during the small "unloading cycles" can be also reproduced by the model.

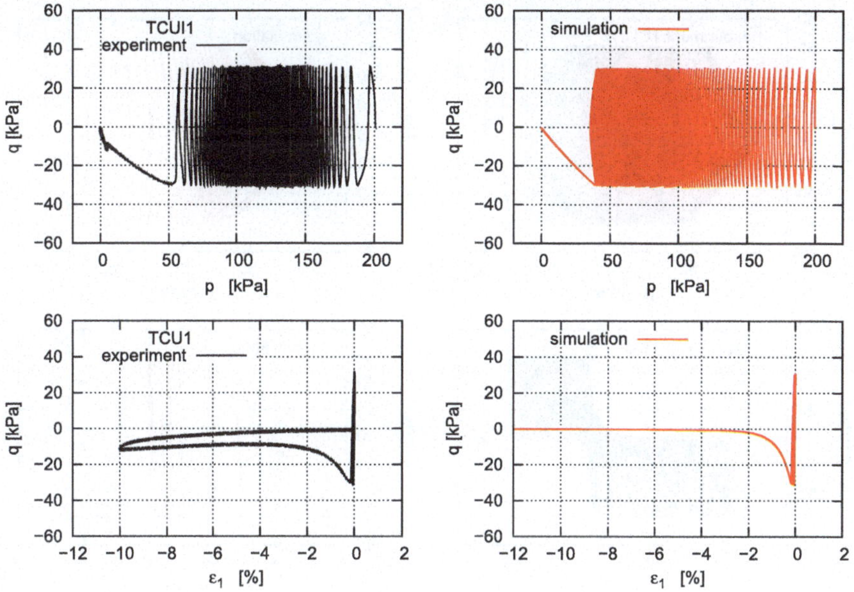

Fig. 13. Undrained cyclic triaxial tests of small stress amplitude on Karlsruhe fine sand. Experiment data adapted from [4].

7.5 Cyclic Loading with Different Deformation Histories

To investigate (and/or test) the evolution equation for the novel state variable $\boldsymbol{\sigma}_B$, we conducted special tests applying the same cyclic loads, but preceded by different deformation histories on dense sand samples. The Karlsruhe fine sand can be characterized by the following parameters: $d_{50} = 0.14$ mm, $C_u = 1.5$ (≤ 5, i.e. poorly graded sand), $e_{\min} = 0.677$, $e_{\max} = 1.054$ and $\varrho_s = 2.65\,\text{g/cm}^3$. The samples, with diameter $d = 100$ mm and height $h = 200$ mm, were prepared using either the dry pluviation or the moist tamping method. Then, the pores of the samples were fully saturated with water to allow a precise measurement of volume changes by means of a differential pressure transducer DPT that was connected to a pipette system.

All drained cyclic tests started from the same initial isotropic stress of $p_0 = 100$ kPa and from nearly the same density ($I_{D0} \approx 0.8$). The stress controlled cyclic loading consisted of stress increments of magnitude $l_{pq} = 40$ kPa applied successively upon 16 different stress ratios $\eta = \Delta q/\Delta p$ in the following sequence: $\eta = 1.125, 1.0, 0.875, \ldots, -0.625$, and -0.750. Each stress increment with stress ratio η_i (loading) was followed by another stress increment with $-\eta_i$ (unloading) until the initial stress $p_0 = 100$ kPa and $q_0 = 0$ was reached. Then, the next stress increment, with η_{i+1} was applied. The process was repeated for each of the 16 stress increments, see Fig. 18(a), left. The resulting strains are also shown in Fig. 18(a), right.

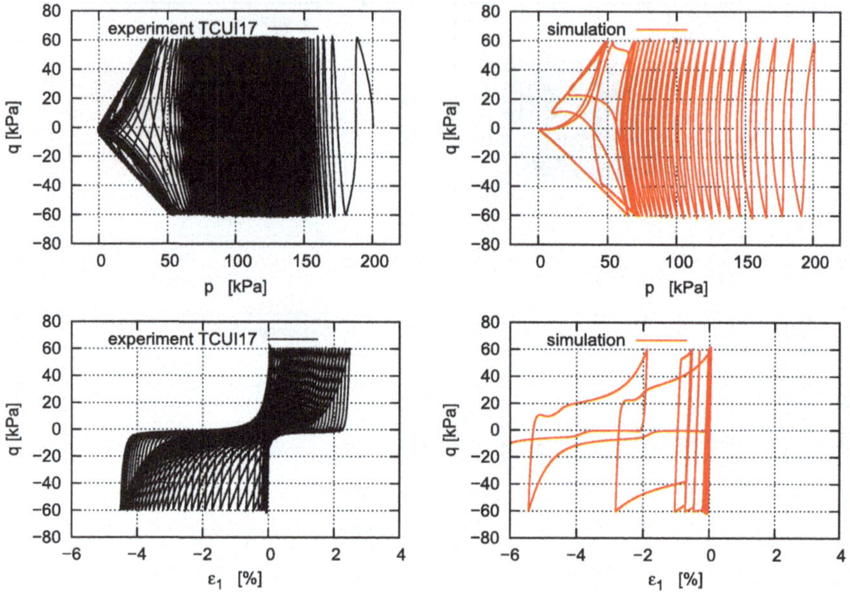

Fig. 14. Undrained cyclic triaxial tests of large stress amplitude on Karlsruhe fine sand. Experiment data adapted from [4].

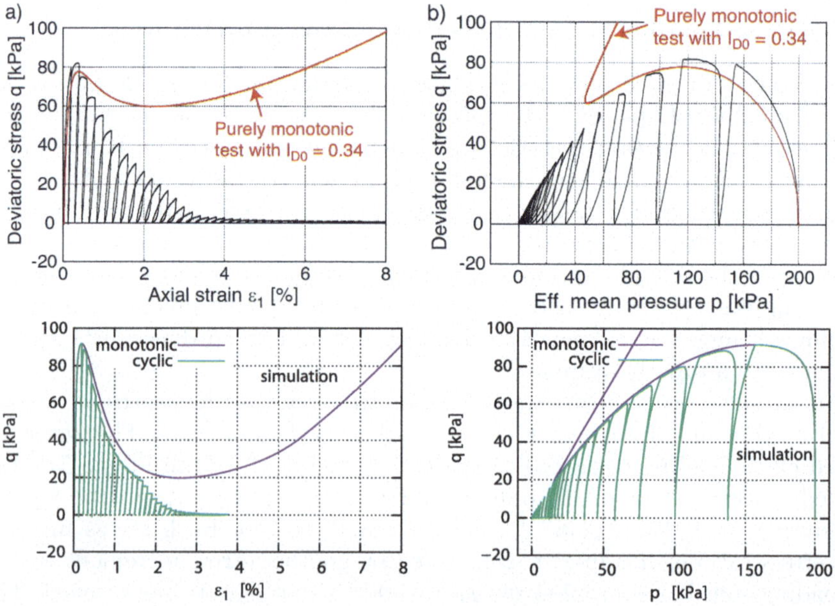

Fig. 15. Experiment results and simulation of an undrained triaxial compression test with successive loading/unloading phases on Karlsruhe fine sand. Experiment data adapted from [4].

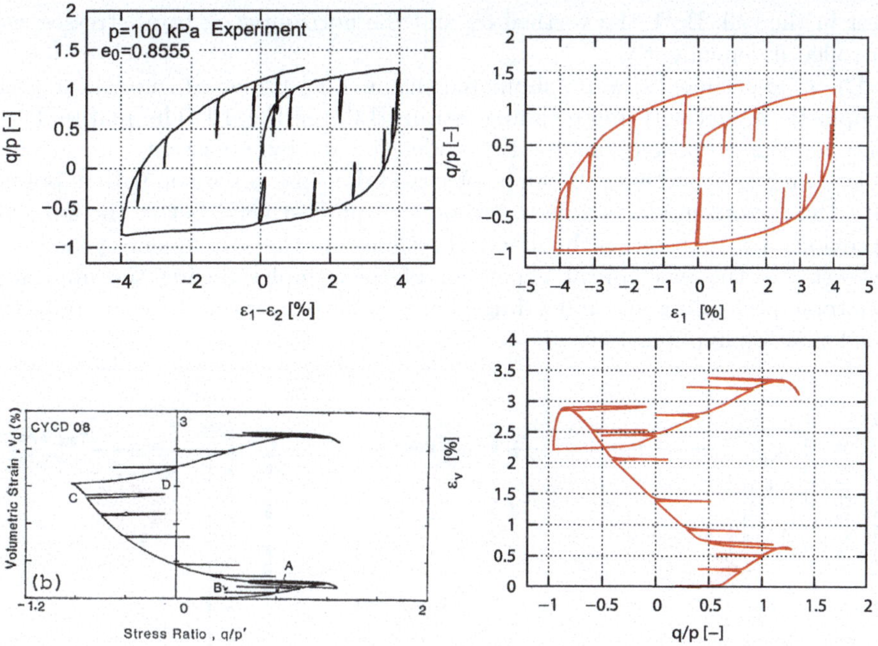

Fig. 16. Combination of cycles of large and small amplitude. Experiment data adapted from [25].

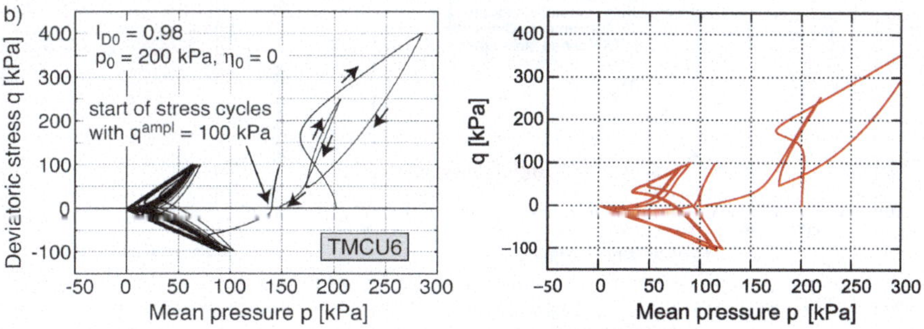

Fig. 17. Combination of cycles of large and small amplitude. Experiment data adapted from [4].

In order to investigate the influence of previous deformation history on the cyclic behaviour of the material, sand samples were subject to undrained triaxial deformations in compression as well as in extension before the drained cyclic tests described above were conducted, see Fig. 18(b) and (c). The monotonic and cyclic components of the axial loading were applied by means of a pneumatic cylinder located below the pressure cell. The cell pressure was applied pneumatically with water in the cell. Both the vertical σ_1 and the horizontal σ_3 total stresses were controlled independently.

These experiments were simulated using the proposed model and the hypoplastic model with intergranular strain [13], see Fig. 19. The material constants for the proposed constitutive model and for hypoplasticity are given in Tables 1 and 2, respectively. It can be observed that in comparison with hypoplasticity, the present model is able to reproduce (qualitatively) better the influence of previous deformation on the material response to the subsequent cyclic loads. Furthermore, the mechanical behaviour of the samples during the undrained monotonic preloading and unloading phases is better captured by the proposed model than by hypoplasticity.

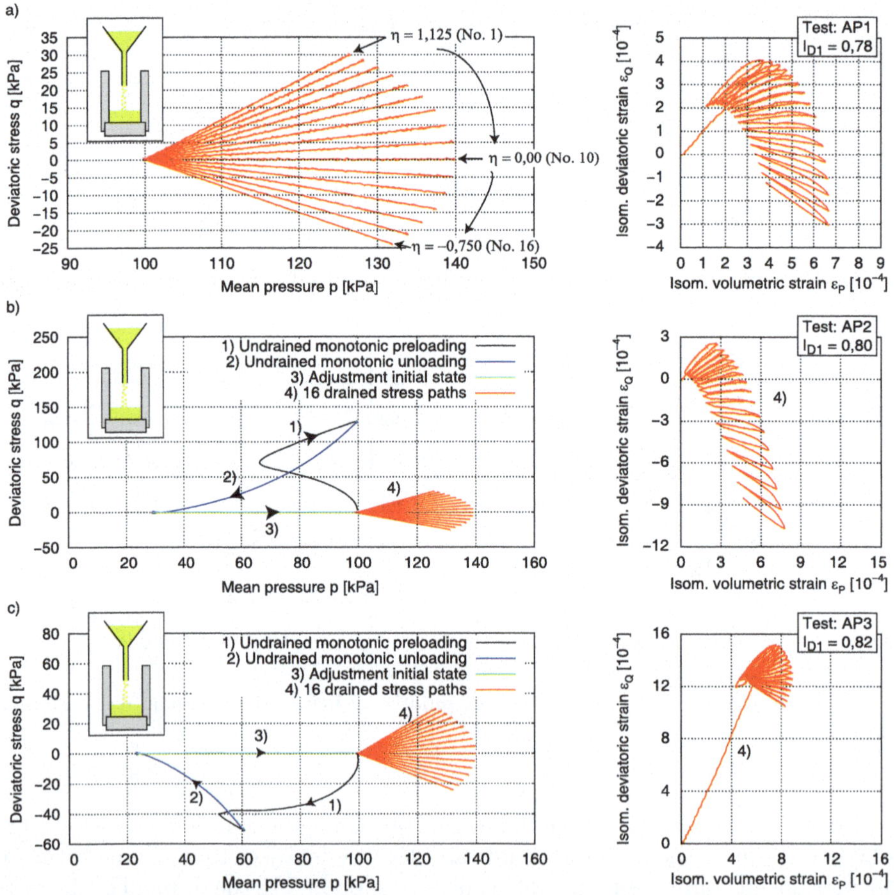

Fig. 18. Test results on Karlsruhe fine sand (a) without preloading history (Test AP1) as reference test, (b) preloading history in compression area (Test AP2) and (c) preloading history in extension area (Test AP3). I_{D0} after sample preparation (air pluviation) and I_{D1} at start of drained stress paths.

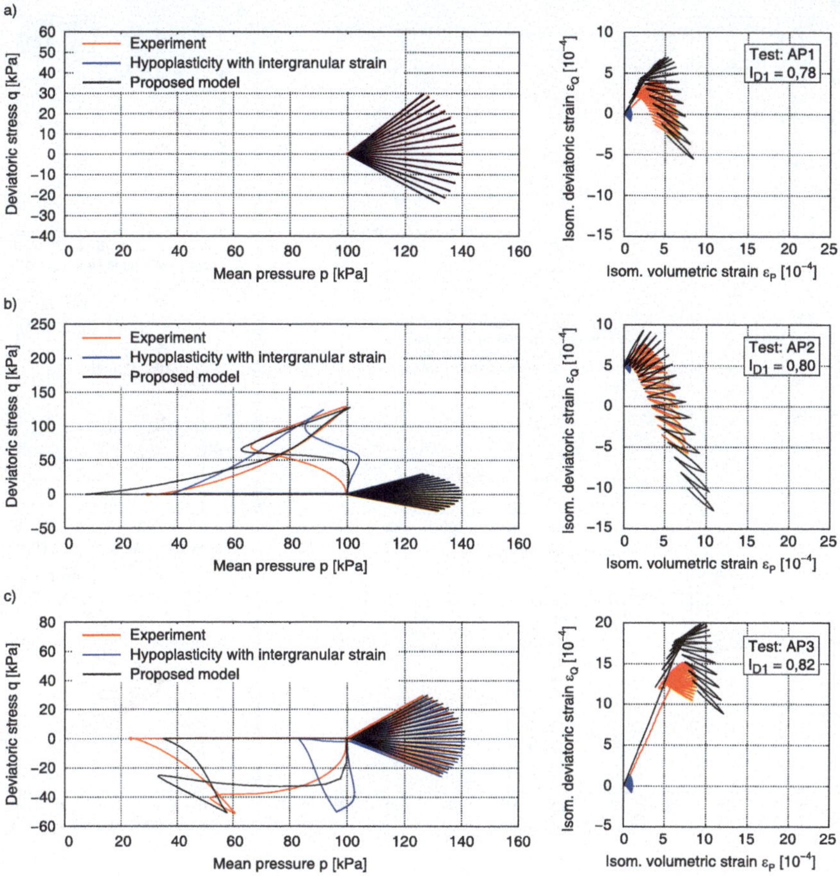

Fig. 19. Simulations of the experiments from Fig. 18 (red) with hypoplasticity with intergranular strain after Niemunis and Herle [13] (blue) and with the proposed model (black).

Figure 20 shows the same experiments as in Fig. 18, but on samples prepared by the moist tamping method. Compared to the air pluviation, the moist tamping method produces samples that show less contractancy during the undrained preloading and unloading phases. Furthermore, the overall response of the samples prepared by the moist tamping method is stiffer than that of the samples prepared by the air pluviation method. The influence of the preparation method on the mechanical behaviour of sand cannot be captured by the proposed constitutive model.

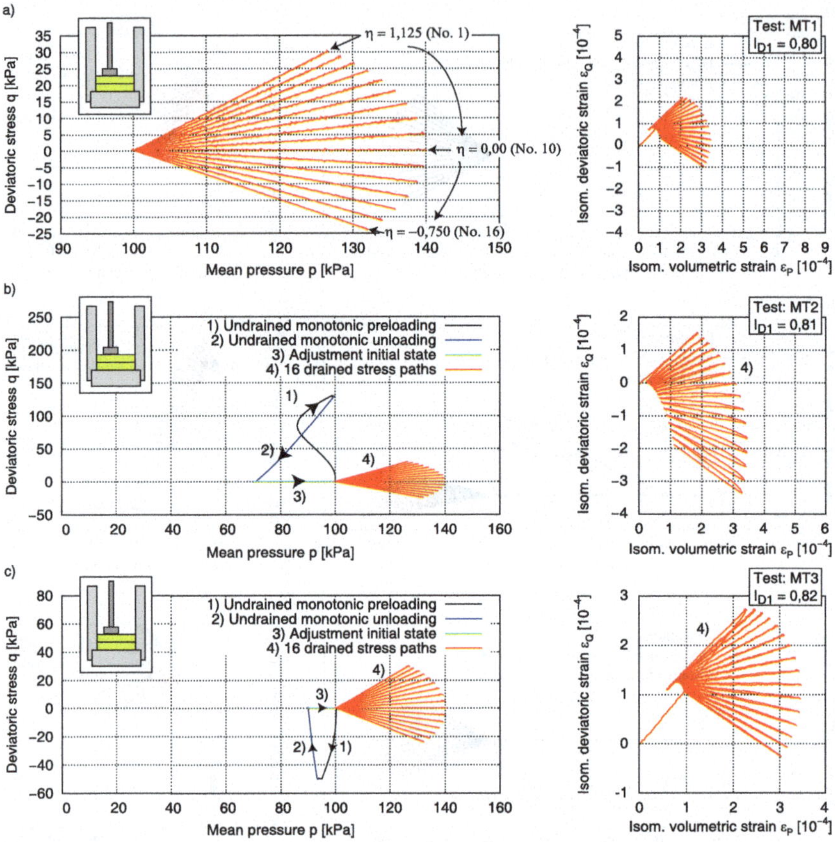

Fig. 20. Test results on Karlsruhe fine sand (a) without preloading history (Test MT1) as reference test, (b) preloading history in compression area (Test MT2) and (c) preloading history in extension area (Test MT3). I_{D0} after sample preparation (moist tamping) and I_{D1} at start of drained stress paths.

8 Conclusions

The present model introduces three main changes to the well known hypoplastic equation [5]: (1) a hyperelastic stiffness, (2) a yield surface to account for recent deformation history, and (3) a new state variable (a back-stress) $\boldsymbol{\sigma}_B$, which describes the size and inclination of the yield surface. These changes enable the proposed model to overcome some known shortcoming not only of hypoplastic, but also of elastoplastic constitutive models:

1. In contrast to yield surfaces with a constant elastic range defined in the strain space (like in hypoplasticity with intergranluar strain [13] or in the ISA model [30]), the *size* of the proposed yield surface $g = 0$, which is defined in the stress space, varies with the back stress $\boldsymbol{\sigma}_B$. Combined with the concept of OCR,

this yield surface allows the simulation of cyclic loads of different amplitudes without overshooting or excessive ratcheting.

2. The influence of recent deformation history on the mechanical behaviour of sand can be captured with the additional state variable $\boldsymbol{\sigma}_B$ and the yield surface.
3. The presented model is able to simulate the degradation of the material stiffness during undrained cyclic loading without additional state variables. Advanced constitutive models like SANISAND [22], ISA [30], or Neohypoplasticity [16] require an additional state variable (the so called **z** variable) to induce extra contractancy after reversals at large stress obliquities.
4. Radial compression paths with loading and unloading cycles can be also well reproduced by the proposed model.

Appendix

Table 1. Material constants for the proposed constitutive model

Symbol	Description	Toyoura	Karlsruhe
Hyperelasticity			
α	coefficient	0.1	0.1
n	barotropy exponent	0.6	0.677
c	coefficient	0.0015	0.001096
Critical state			
φ_c	critical friction angle (°)	32.5	33
e_{c0}	critical void ratio at $p = 0$	0.929	1.054
n_{Bc}	exponent	0.79	0.27
h_{sc}	reference pressure (kPa)	56000	$4 \cdot 10^6$
Limiting compression curve			
e_{i0}	max. void ratio at $p = 0$	1.115	1.1
n_{Bi}	exponent	0.79	0.48
h_{si}	reference pressure (kPa)	$1.9 \cdot 10^5$	$8.4 \cdot 10^6$
Friction angles			
n_{peak}	exponent peak	1	2
n_{ptl}	exponent PTL	1	1
Yield surface			
n_O	exponent of OCR	4	4
c_b	exponent	1	0.2
n_{YD}	exponent of Y_D	1	1
Evolution of $\boldsymbol{\sigma}_B$			
c_2	coefficient	30	50

Table 2. Hypoplastic material constants for Karlsruhe fine sand

φ_c (°)	e_{c0}	e_{i0}	e_{d0}	h_s (kPa)	n	α	β	m_T	m_R	$R_{\max}$	β_x	χ
33	1.054	1.212	0.677	$4 \cdot 10^6$	0.27	0.14	2.5	2	5	0.0001	0.5	6

References

1. Doanh, T., Dubujet, P., Touron, G.: Exploring the undrained induced anisotropy of Hostun RF loose sand. Acta Geotech. **5**, 239–256 (2010)
2. Ishihara, K., Okada, S.: Effects of stress history on cyclic behavior of sand. Soils Found. **18**(4), 31–45 (1978)
3. Ishihara, K., Tatsuoka, F., Yasuda, S.: Undrained deformation and liquefaction of sand under cyclic stresses. Soils Found. **15**(1), 29–44 (1975)
4. Wicthmann, T.: Soil behaviour under cyclic loading - experimental observations, constitutive description and applications. Heft 180, Institut für Boden und Felsmechanik, Karlsruhe Institut für Technologie (KIT) (2016)
5. Wolffersdorff, P.: A hypoplastic relation for granular materials with a predefined limit state surface. Mech. Cohesive-Frictional Mater. **1**, 251–271 (1996)
6. Niemunis, A.: Extended hypoplastic models for soils. Habilitation, monografia 34, Ruhr-University Bochum (2003)
7. Taylor, D.: Fundamentals of Soil Mechanics. Wiley, New York (1948)
8. Niemunis, A., Grandas-Tavera, C.E.: Computer aided calibration, benchmarking and check-up of constitutive models for soils. Some conclusions for neohypoplasticity. In: Triatafyllidis, T. (ed.) Holistic Simulation of Geotechnical Installation Processes, vol. 82, pp. 168–192. Springer, Heidelberg (2017)
9. Pestana, J., Whittle, A.: Compression model for cohesionless soils. GEOT **45**(4), 611–631 (1995)
10. Roscoe, K.H., Burland, J.B.: On the generalized stress-strain behaviour of 'wet' clay. In: Engineering Plasticity, pp. 535–609. Cambridge University Press, Cambridge (1968)
11. Schofield, A., Wroth, C.: Critical State Soil Mechanics. McGraw-Hill, London (1968)
12. Wu, W.: Hypoplastizität als mathematisches Modell zum mechanischen Verhalten granularer Stoffe. Veröffentlichungen des Institutes für Boden-und Felsmechanik der Universität Fridericiana in Karlsruhe, Heft Nr. 129 (1992)
13. Niemunis, A., Herle, I.: Hypoplastic model for cohesionless soils with elastic strain range. Mech. Cohesive-Frictional Mater. **2**, 279–299 (1997)
14. Niemunis, A., Prada-Sarmiento, L., Grandas-Tavera, C.: Paraelasticity. AG **6**(2), 67–80 (2011)
15. Prada Sarmiento, L.: Parelastic description of small-strain soil behaviour. Dissertation, Veröffentlichungen des Institutes für Bodenmechanik und Felsmechanik am Karlsruher Institut für Technologie, Heft 173 (2011)
16. Niemunis, A., Grandas Tavera, C.E., Wichtmann, T.: Peak stress obliquity in drained und undrained sands. Simulations with neohypoplasticity. In: Triatafyllidis, T. (ed.) Holistic Simulation of Geotechnical Installation Processes, pp. 85–114. Springer, Heidelberg (2015)
17. Gudehus, G.: A comprehensive constitutive equation for granular materials. Soils Found. **36**(1), 1–12 (1996)

18. Bauer, E.: Calibration of a comprehensive constitutive equation for granular materials. Soils Found. **36**, 13–26 (1996)
19. Miura, N., O-Hara, S.: Particle-crushing of a decomposed granite soil under shear stresses. Soils Found. **19**(3), 1–14 (1979)
20. Miura, N., Murata, H., Yasufuku, N.: Stress-strain characteristics of sand in a particle-crushing region. Soils Found. **24**(1), 77–89 (1984)
21. Matsuoka, H., Nakai, T.: Stress-strain relationship of soil based on the SMP, constitutive equations of soils. In: Specialty Session 9, Japanese Society of Soil Mechanics and Foundation Engineering, IX ICSMFE, Tokyo, pp. 153–162 (1977)
22. Taiebat, M., Dafalias, Y.F.: SANISAND: simple anisotropic sand plasticity model. IJNAMG **32**, 915–948 (2008)
23. Olszak, W., Perzyna, P.: The constitutive equations of the flow theory for a nonstationary yield condition. In: Applied Mechanics, Proceedings of the 11th International Congress, pp. 545–553 (1966)
24. Niemunis, A., Grandas-Tavera, C., Prada-Sarmiento, L.: Anisotropic viscohypoplasticity. Acta Geotech. **4**, 293–314 (2009)
25. Pradhan, T., Tatsuoka, F., Sato, Y.: Experimental stress-dilatancy relations of sand subjected to cyclic loading. Soils Found. **29**(1), 45–64 (1989)
26. Pradhan, T., Tatsuoka, F.: On stress-dilatancy equations of sand subjected to cyclic loading. Soils Found. **29**(1), 65–81 (1989)
27. Li, X., Dafalias, Y., Wang, Z.: State dependent dilatancy in critical state constitutive modelling of sand, vol. 36, no. 4, pp. 599–611 (1999)
28. Niemunis, A.: IncrementalDriver (2019). http://www.pg.gda.pl/~aniem/dyd.html
29. Verdugo, R., Ishihara, K.: The steady state of sandy soils. Soils Found. **36**(2), 81–91 (1996)
30. Fuentes Lacouture, W.M.: Contributions in mechanical modelling of fill materials. Dissertation, Veröffentlichungen des Institutes für Bodenmechanik und Felsmechanik am Karlsruher Institut für Technologie, Heft 179 (2014)

Geomechanical Influences of Interface Dilatancy

A. P. S. Selvadurai(✉)

Department of Civil Engineering and Applied Mechanics, McGill University, Montreal, QC H3A 0C3, Canada
patrick.selvadurai@mcgill.ca

Abstract. This paper examines the influence of dilatant processes that can occur at a discontinuity of finite dimensions located at a compressed elastic geological interface, due to relative shear movement in the plane of the discontinuity. The dilatant phenomena will result in displacements in a direction normal to the shear movement and these displacements will be influenced by the elasticity of the geological materials and the compression at the interface. The dilatant displacements will also create a zone where there is loss of contact at the unilaterally constrained interface. The resulting problem is examined by appeal to results of the mathematical theory of elasticity. The influence of dilatant processes on the development of shear at the elastically constrained interface is examined by considering the procedures proposed by D.W. Taylor to analyze dilatant phenomena. The mathematical developments illustrate the combined influence of elastic constraints and interface compression on the amplification of the shear stress generated at the finite region. In the absence of dilatancy, the developments reduce to the classical result involving only Coulomb friction.

1 Introduction

The mechanics of geologic interfaces, particularly faults and fractures, is important to the field of engineering geosciences dealing with stability of geologic strata, the development of tectonic motion, movements of pre-existing fractures and the interaction of constructed underground facilities. In the study of earthquake generating mechanisms during movement at a transform fault (Fig. 1), the limiting stresses and the movements necessary to rupture a locked-in region are important input parameters for developing earthquake models.

The literature dealing with contact mechanics is extensive (covering nearly six thousand references dealing with diverse areas of geomaterial interfaces, geologic fault zones, tribology, wear, biomechanics, contact mechanics, etc.) and a complete review is beyond the scope of this article. Historical studies related to the mechanics of contact between surfaces can be found in the volumes by Bowden and Tabor [1] and Hisano [2]. Other developments that emphasize engineering applications, mathematical modelling, computational modelling and experimental aspects of contact mechanics are given in several texts and review articles on contact mechanics and these developments are documented in articles by Selvadurai and Boulon [3], Selvadurai [4], Selvadurai and Yu [5], Selvadurai and Atluri [6] and Selvadurai et al. [7].

T. Triantafyllidis (Ed.): *Recent Developments of Soil Mechanics and Geotechnics in Theory and Practice*, LNACM 91, pp. 44–53, 2020.
https://doi.org/10.1007/978-3-030-28516-6_3

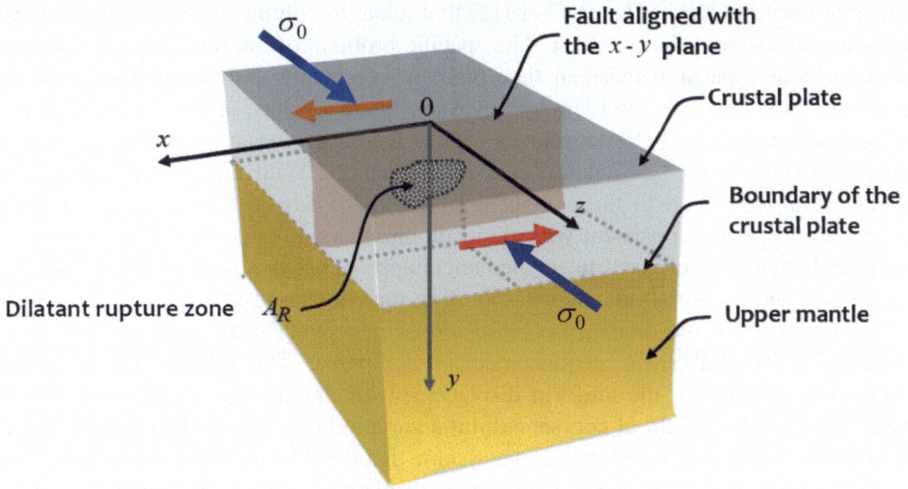

Fig. 1. Relative movement at a transform fault

In this article, we examine a rather idealized problem related to elastic isotropic geomaterial halfspace regions in *smooth contact*, pre-compressed by a normal stress σ_0 and containing a circular patch that possesses both *frictional* and *dilatant* mechanical properties. The Coulomb friction can be a result of the contact at the local scale of the idealized Euclidean surfaces and dilatant effects can occur due to surface irregularities. The dominant mechanism is asperity ride-up but the modelling can account for the deterioration of the dilatancy angle that can be identified with asperity breakage and damage, and indentation fracture (Selvadurai [8]). From a geo-environmental point of view, the dilatant movement at fault zones can lead to an increase in the aperture at contacts, which can enhance fluid flow through the fracture [5, 9–12]. The objective of this study is to develop a convenient analytical result that can be used to estimate the development of shear stresses at the *frictional-dilatant* circular patch during the relative shearing movement at the otherwise frictionless interface.

The problem posed here can be quite complicated if the influence of the frictional contact at the entire surfaces of the fracture and the dilatant effects of the circular patch are considered simultaneously. Frictional "asperity" patches surrounded by frictionless or stress free regions have been considered in the literature but these studies do not examine the possible development of dilatancy in the frictional patches. We have developed a theoretical approach for the study. However, it is unlikely that such a complete frictional interface contact problem will be amenable to analytical treatment, which is the basis for this study. The approach adopted here is to examine the mechanics of the pre-compressed *dilatant circular patch during shear*, while maintaining frictionless behaviour over the interface region exterior to the circular patch. The rationale for assuming frictionless behaviour exterior to the circular patch is to emphasize the constraint imposed by the dilatant patch and to evaluate the role that dilatancy at the patch has on limiting the shear capacity of the dilatant region. Observations of the property of dilatancy in geomaterials date back to the classical studies of Osborne Reynolds (1842–1912) that relate to volume expansion in the mass of a granular material during shear. The specific problem of interface dilatancy relates to the volume expansion that can take place at contacting surfaces due to the local geometric structure. The problem of interface dilatancy has several geomechanical applications, particularly those related to the load carrying capacity of embedded structural elements such as anchors and rock fractures. With interface dilatancy, the influences are restricted to the contact region and the adjacent domains can exhibit continuum properties consistent with the geologic material behaviour. The analysis of interface dilatancy lends itself to mathematical approaches that deal with the extended domain and the frictional-dilatancy response at the contacting regions.

We assume that the dilatant processes come into effect when the interface experiences a differential shear displacement in its plane. The dilatant movement is accommodated through the uniform displacement of the circular patch normal to its plane. The expansion can, of course, exhibit a variation over the circular region, but to preserve the simplicity of the model, we assume that the dilatant patch exerts a *uniform displacement* normal to its plane. This normal displacement can cause separation at the initially mated frictionless surfaces exterior to the circular patch, through its indentation into the deformable geologic media. The extent of separation will depend on (i) the elasticity characteristics of the geologic material, (ii) the magnitude of the dilatant displacement and (iii) the far-field normal stresses acting on the interacting surfaces. The paper presents an analysis of the elasticity problem associated with the dilatant patch and evaluates the normal stresses generated in the patch region during its dilatant expansion. This result is used in conjunction with a virtual work formulation to obtain a relationship for the shear stress at which the circular patch will experience failure or rupture.

2 Mathematical Modelling of the Contact Problem

We examine the problem of the smooth contact between two halfspace regions compressed by a normal stress σ_0 and containing a frictional-dilatant circular patch (Fig. 2). The dilatant behaviour at the circular patch can be caused by the relative shearing action between the halfspace regions and, in the process, the region can induce an indentation orthogonal to the shearing displacements.

We consider the elastostatic interaction between the frictionless interface containing the dilatant circular patch of radius a in a situation where the frictionless interface experiences a *total* relative movement $2\Delta u$ in its plane. (i) In general, the dilatancy-induced displacement normal to the circular patch can be variable within the dilatant region; for the purposes of developing a convenient analytical result, we assume that the induced normal displacement at the frictional interface has a constant value Δv imposed on both halfspace surfaces (Fig. 2). (ii) The dilatant behaviour of the circular patch will cause separation between the halfspace regions that are in smooth contact under the action of the compressive stress σ_0. (iii) The configuration of the boundary of the zone of separation can be elliptical in shape depending on the extent of shear. In this model, however, we assume that the boundary of separation can be approximated by a circular profile of radius b, which is an unknown (Fig. 2) and needs to be

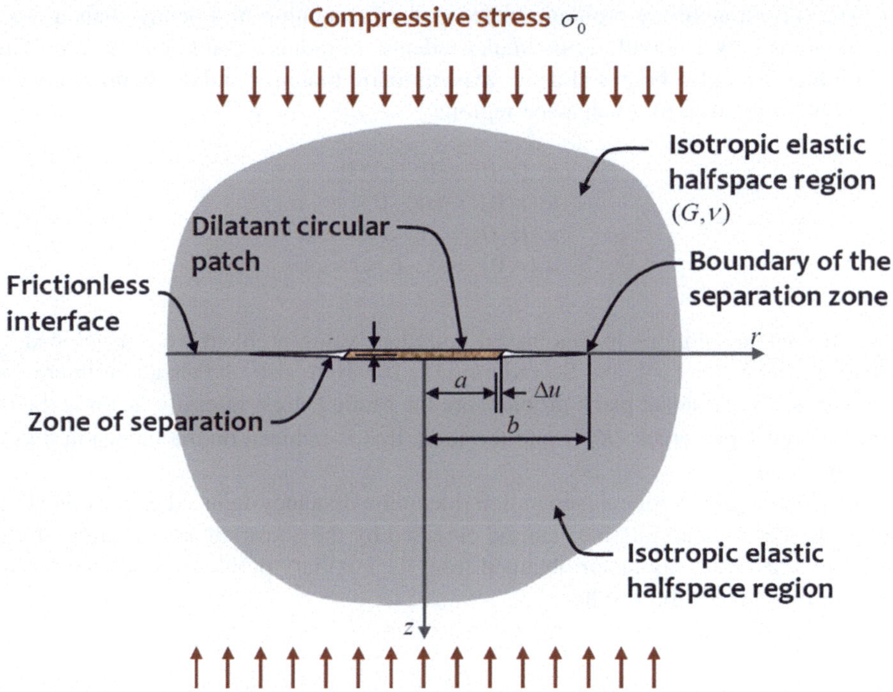

Fig. 2. Indentation of the pre-compressed halfspace regions by the dilatancy in the circular patch.

determined by solving a unilateral contact problem for the two halfspace regions [4]. To develop results for (a) the separation region during dilatancy-induced indentation at the circular region and (b) the force developed by the indentation displacement, we consider the following auxiliary problems:

(i) The internal pressurization of an annular crack of internal radius a and external radius b located in an elastic infinite space, by a pressure σ_0 and governed by the following three-part axisymmetric mixed boundary value problem referred to a halfspace region:

$$\begin{aligned} &\sigma_{rz}(r,0) = 0; \quad r \geq 0 \\ &u_z(r,0) = 0; \quad 0 \leq r \leq a \\ &\sigma_{zz}(r,0) = -\sigma_0; \quad a < r < b \\ &u_z(r,0) = 0; \quad b \leq r < \infty \end{aligned} \tag{1}$$

An approximate solution to this problem was developed by Selvadurai and Singh [13] and the important results relate to (a) the evaluation of the Mode I stress intensity factor at the outer boundary of the annular crack $(K_I^{\sigma_0})$, and (b) the resultant force developed in the ligament region $0 \leq r \leq a$.

(ii) The second auxiliary problem relates to the indentation of a penny-shaped crack of radius b by a smooth rigid circular indenter of radius a and thickness $2\Delta v$. This problem can also be posed as an axisymmetric three-part mixed boundary value problem referred to a halfspace region:

$$\begin{aligned} &\sigma_{rz}(r,0)=0; \quad r\geq 0\\ &u_z(r,0)=\Delta v; \quad 0\leq r\leq a\\ &\sigma_{zz}(r,0)=0; \quad a<r<b\\ &u_z(r,0)=0; \quad b\leq r<\infty \end{aligned} \tag{2}$$

Approximate solutions to this mixed boundary value problem were developed by Selvadurai and Singh [14], and Selvadurai [15, 16]. Here also, the results of interest to the analysis of the dilatant patch problem are the Mode I stress intensity factor at the tip of the penny-shaped crack ($K_I^{\Delta v}$) and the force that is induced on the inclusion due to the indentation.

The location of the zone of separation due to the dilatancy-induced expansion at the unilaterally constrained interface can be obtained by the constraint of vanishing of the combined stress intensity factor obtained from the auxiliary problems described above. This constraint gives the result

$$\left(\frac{G\Delta v}{2\sigma_0 a(1-v)}\right)cF_{\Delta v}(c) - F_{\sigma_0}(c) = 0 \tag{3}$$

where

$$F_{\Delta v}(c) = \begin{bmatrix} \frac{4c}{\pi} + \frac{16c^2}{\pi^3} + c^3\left(\frac{64}{\pi^5} + \frac{4}{3\pi}\right) \\ + c^4\left(\frac{80}{9\pi^3} + \frac{256}{\pi^7}\right) + c^5\left(\frac{448}{9\pi^5} + \frac{1024}{\pi^9} + \frac{4}{5\pi}\right) \end{bmatrix} \tag{4}$$

$$F_{\sigma_0}(c) = \begin{bmatrix} 1 - \frac{4c}{\pi^2} - \frac{16c^2}{\pi^4} - c^3\left(\frac{1}{8} + \frac{64}{\pi^6}\right) \\ - c^4\left\{\frac{16}{3\pi^4} + \frac{4}{\pi^2}\left(\frac{1}{24} - \frac{8}{9\pi^2} + \frac{64}{\pi^6} + \frac{4}{9\pi^3}\right)\right\} \\ - c^5\left\{\frac{16}{\pi^4}\left(\frac{1}{24} - \frac{8}{9\pi^3} + \frac{64}{\pi^6} + \frac{8}{9\pi^2}\right) + \frac{256}{9\pi^6} - \frac{4}{15\pi^2}\right\} + \mathrm{O}(c^6) \end{bmatrix} \tag{5}$$

and $c(=a/b)<1$. The lowest root of (3) gives the extent of the zone of separation. Omitting the details it can be shown that the force generated at the contact zone of the indenting region $0 \leq r \leq a$, with a separation region $a \leq r \leq b$ and a re-established contact zone $b \leq r \leq \infty$, can be evaluated in the form

$$P_N = \sigma_0\pi a^2 + \frac{4aG\Delta v}{(1-v)}P_N^{\Delta v} - \sigma_0\pi a^2 P_N^{\sigma_0} \tag{6}$$

where

$$P_N^{\Delta v} = \begin{bmatrix} \left(1 + \frac{4c}{\pi}\right) + \frac{16c^2}{\pi^4} + c^3\left(\frac{64}{\pi^6} + \frac{16}{9\pi^4} - \frac{8}{9\pi^2}\right) \\ + c^4\left(\frac{256}{\pi^8} + \frac{64}{9\pi^4}\right) + c^5\left(\frac{10240}{\pi^{10}} + \frac{9600}{225\pi^6} + \frac{92}{225\pi^2}\right) \end{bmatrix} \tag{7}$$

$$P_N^{\sigma_0} = \left(\begin{array}{l} -\frac{8}{\pi^2 c} + \left\{1 - \frac{32}{\pi^4}\right\} + 8c\left\{\frac{1}{\pi^2} - \frac{48}{\pi^6}\right\} \\ -\frac{c^2}{9\pi^8}\left\{4608 + \pi^3(32 - 64\pi + 3\pi^3)\right\} \\ -\frac{4c^3}{45\pi^{10}}\left\{23040 + \pi^3(-320 + 480\pi + 15\pi^3 + 6\pi^5)\right\} \\ -\frac{c^4}{675\pi^{12}}\left\{5529600 + \pi^3\left(-76800 + \pi\left[\begin{array}{l} 192000 \\ +\pi^2(3600 + \pi\{-320 + 3168\pi + 45\pi^3\}) \end{array}\right]\right)\right\} \end{array} \right) \tag{8}$$

3 The Shear of the Dilatant Circular Patch

In general, at the dilatant zone the response will be elasto-plastic. In a strict sense, the analysis of the shear rupture problem during the generation of frictional-dilatant phenomena should be examined by appeal to a theory of plasticity applicable for elasto-plastic phenomena with specified failure criteria and non-associated flow rules (Davis and Selvadurai [17]). Here, we focus on the evaluation of the peak rupture stress that can be generated at the circular patch when dilatancy is present. To estimate the limiting response, we utilize the procedure presented by D.W. Taylor [18] for the analysis of dilatancy processes in granular materials. In essence, when examining failure processes associated with dilatancy effects, Taylor proposed a criterion that neglects the elastic energy storage processes at the direct contact zone (see also Christian and Baecher [19]). (To a certain extent, this argument is consistent with the limit analysis concepts proposed by Drucker and Prager [20]; see also Davis and Selvadurai [17] and Ichikawa and Selvadurai [21].) The basic hypothesis involves the relationship between (i) the *work done by the shearing forces and normal forces* and (ii) the *energy dissipated at the frictional-dilatant region*, expressed in terms of force resultants rather than exact distributions over the contact zone. The work component W consists of the work of the shear force (P_T) acting at the onset of rupture and the work of the normal force (P_N) induced by dilatancy on the upper and lower surface: i.e.

$$W = 2P_T(\Delta u) + 2P_N(-\Delta v) \tag{9}$$

The energy dissipated at the dilatant circular patch is given by

$$D = 2P_N\,(\Delta u)\tan\varphi \tag{10}$$

where φ is the contact friction angle and Δv is the dilatant displacement. We note that the work of forces and the dissipation on both faces of the circular patch have to be included in the formulation. In Taylor's hypothesis, the conventional association between Δv and Δu is through the linear relationship $\Delta v = \Delta u \tan\alpha_0$, where α_0 is the constant dilatancy angle. The work of Selvadurai et al. [7], extends Taylor's definition of dilatant displacements to include dilatancy effects of the form

$$\Delta v = a\left(\frac{\Delta u}{a}\right)^2 \tan\alpha \tag{11}$$

and with variation in the dilatancy angle α is described by

$$\tan\alpha = \exp\left(-\lambda\left|\frac{\Delta u}{a}\right|\right)\tan\alpha_0 \tag{12}$$

where λ is a non-dimensional parameter and the modulus sign accounts for the invariance of the dilatancy angle on the sense of the shear displacement. Selvadurai et al. [7] show that the representation (12) correlates well with experimental observations of the mechanics of dilatant geologic interfaces. Omitting details, it can be shown that the shear stress developed on the dilatant circular patch can be expressed in the form

$$\begin{aligned}\tau_D = \sigma_0 &\left(1 + \frac{4G}{(1-\nu)\sigma_0\pi}\left(\frac{\Delta u}{a}\right)^2 \exp\left(-\lambda\left|\frac{\Delta u}{a}\right|\right)\tan\alpha_0 P_N^{\Delta v} - P_N^{\sigma_0}\right)\\ &\times\left(\tan\varphi + \frac{\Delta u}{a}\exp\left(-\lambda\left|\frac{\Delta u}{a}\right|\right)\tan\alpha_0\right)\end{aligned} \tag{13}$$

and the radius of the separation zone is now obtained from the smallest positive root of the characteristic equation

$$\left(\frac{G}{\sigma_0(1-\nu)}\right)\left(\frac{\Delta u}{a}\right)^2 \exp\left(-\lambda\left|\frac{\Delta u}{a}\right|\right)\tan\alpha_0\, cF_{\Delta v}(c) - 2F_{\sigma_0}(c) = 0 \tag{14}$$

The result (13) for the shear stress generated at failure at the dilatant circular patch can be compared with the analogous result for the non-dilatant case, which corresponds to the interface with purely Coulomb friction. The shear stress amplification factor (*SSAF*) is given by

$$\begin{aligned}SSAF = \frac{\tau_D}{\tau_C} = &\left(1 + \frac{4G}{(1-\nu)\sigma_0\pi}\left(\frac{\Delta u}{a}\right)^2 \exp\left(-\lambda\left|\frac{\Delta u}{a}\right|\right)\tan\alpha_0 P_N^{\Delta v} - P_N^{\sigma_0}\right)\\ &\times\left(1 + \frac{\Delta u}{a}\exp\left(-\lambda\left|\frac{\Delta u}{a}\right|\right)\frac{\tan\alpha_0}{\tan\varphi}\right)\end{aligned} \tag{15}$$

In the limiting case when $\alpha_0 \to 0$, the result (15) reduces to unity assuming that when $c \to 1$, $P_N^{\sigma_0} \to 0$. This assumption is invoked in view of the nature of the series approximations used in the solutions of the three-part mixed boundary value problems defined by (1) and (2). Figures 3 and 4 illustrate the typical variations in the shear stress amplification with (i) the shear displacement, (ii) the Coulomb friction angle, (iii) the dilatancy angle, (iv) the dilatancy degradation parameter and (v) the magnitude of the in situ compressive stress relative to the shear modulus. The basic approach can also be used to estimate the in-plane load carrying capacity of flat anchors [22] that are created by pressure grouting techniques (Fig. 5).

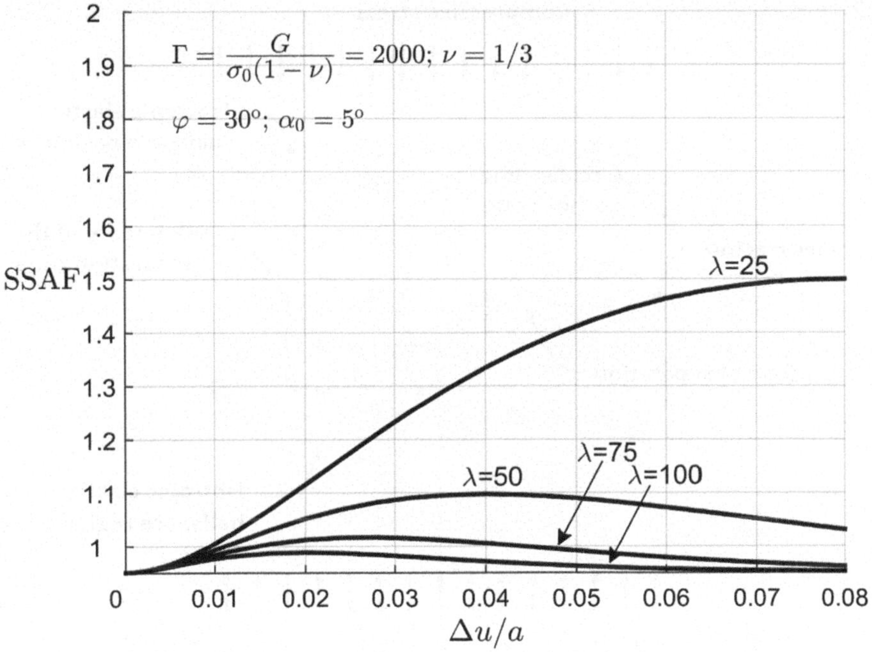

Fig. 3. The variation in the shear stress amplification factor with the relative shear movement

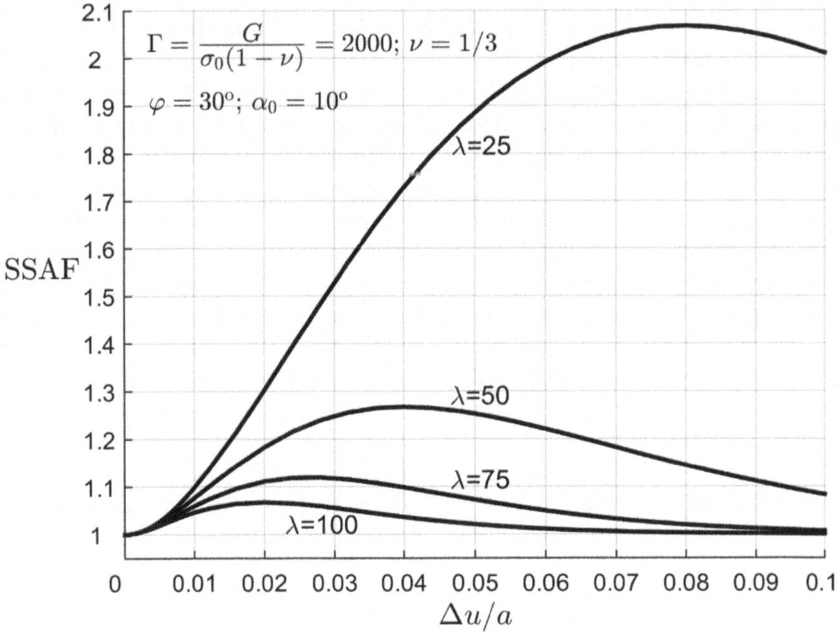

Fig. 4. The variation in the shear stress amplification factor with the relative shear movement

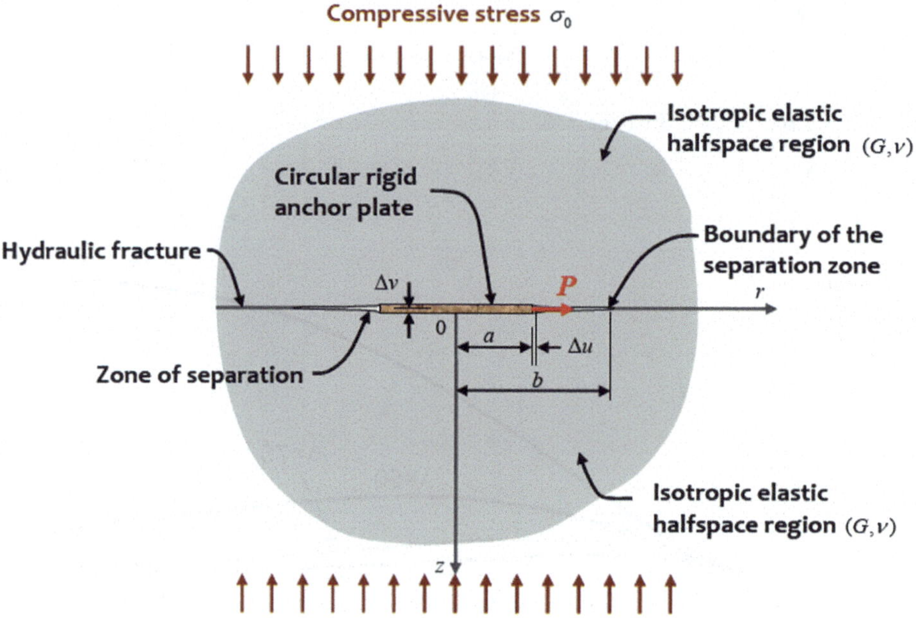

Fig. 5. In-plane loading for a flat anchor embedded in a fracture

4 Concluding Remarks

Dilatant processes at geological interfaces can have a strong influence on the shear stresses that are needed to rupture the region exhibiting dilatancy. The analysis of the shear rupture of an elastically and unilaterally constrained interface is non-routine since it requires the evaluation of the zone of separation induced by dilatancy effects with appeal to the solution of three-part mixed boundary value problems in elasticity theory applicable to a halfspace region. The paper demonstrates the efficient use of solutions available in the literature to evaluate the separation zone and the forces that are generated at the contact zone during dilatational displacement over a circular contact zone. The enhancement of the shear stresses needed to create shear rupture is demonstrated. This observation has implications for the interpretation of the seismic moment calculations, which currently accounts for only the relative shear displacements. It is shown that the analysis can also be extended to include interface dilatancy degradation that can result from asperity breakage during shear, resulting in a progressive decrease in the dilatancy angle with shear.

Acknowledgements. The author is grateful to the research support received by the Natural Sciences and Engineering Research Council of Canada and the James McGill Research Chairs program. The author would also like to acknowledge the contributions made by Dr. P.A. Selvadurai, Post-Doctoral Fellow at ETH Zurich and Dr. A.P. Suvorov, Assistant Professor, Moscow State University for Civil Engineering.

References

1. Bowden, F.P., Tabor, D.: Friction and Lubrication of Solids. Clarendon Press, Oxford (1986)
2. Hisano, M.: Friction at the Atomic Level - Atomistic Approaches in Tribology. Wiley-VCH, Weinheim (2018)
3. Selvadurai, A.P.S., Boulon, M.J. (eds.): Mechanics of Geomaterial Interfaces, Studies in Applied Mechanics, vol. 42. Elsevier, Amsterdam (1995)
4. Selvadurai, A.P.S.: On an invariance principle for unilateral contact at a bi-material elastic interface. Int. J. Eng. Sci. **41**, 721–739 (2003)
5. Selvadurai, A.P.S., Yu, Q.: Mechanics of a discontinuity in a geomaterial. Comput. Geotech. **32**, 92–106 (2005)
6. Selvadurai, A.P.S., Atluri, S.N. (eds.): Contact Mechanics in the Engineering Sciences - Material Characterization, Micromechanical Processes and Modelling, Geosciences. Tech Science Press, Forsyth (2010)
7. Selvadurai, A.P.S., Selvadurai, P.A., Suvorov, A.P.: Contact mechanics of a dilatant region located at a pre-compressed elastic interface. Int. J. Eng. Sci. **133**, 144–168 (2018)
8. Selvadurai, A.P.S.: Fracture evolution during indentation of a brittle elastic solid. Mech. Cohes.-Frict. Mater. **5**, 325–339 (2000)
9. Boulon, M.J., Selvadurai, A.P.S., Benjelloun, H., Fuega, B.: Influence of rock joint degradation on hydraulic conductivity. Int. J. Rock Mech. Min. Sci. **30**, 1311–1317 (1993)
10. Nguyen, T.S., Selvadurai, A.P.S.: A model for coupled mechanical and hydraulic behaviour of a rock joint. Int. J. Numer. Anal. Methods Geomech. **22**, 29–48 (1998)
11. Selvadurai, A.P.S., Nguyen, T.S.: Mechanics and fluid transport in a degradable discontinuity. Eng. Geol. **53**, 243–249 (1999)
12. Selvadurai, A.P.S.: Normal stress-induced permeability hysteresis of a fracture in a granite cylinder. Geofluids **15**, 37–47 (2015)
13. Selvadurai, A.P.S., Singh, B.M.: The annular crack problem for an isotropic elastic solid. Q. J. Mech. Appl. Math. **38**, 233–243 (1984)
14. Selvadurai, A.P.S., Singh, B.M.: On the expansion of a penny-shaped crack by a rigid disc inclusion. Int. J. Fract. **25**, 69–77 (1984)
15. Selvadurai, A.P.S.: On an integral equation governing an internally indented penny-shaped crack. Mech. Res. Commun. **12**, 347–351 (1985)
16. Selvadurai, A.P.S.: Separation at a pre-fractured bi-material geological interface. Mech. Res. Commun. **21**, 83–88 (1994)
17. Davis, R.O., Selvadurai, A.P.S.: Plasticity and Geomechanics. Cambridge University Press, Cambridge (2003)
18. Taylor, D.W.: Fundamentals of Soil Mechanics. Wiley, New York (1948)
19. Christian, J.T., Baecher, G.P.: D.W. Taylor and the foundations of modern soil mechanics. J. Geotech. Geoenvironm. Eng. **141**, 02514001 (2015)
20. Drucker, D.C., Prager, W.: Soil mechanics and plastic analysis or limit design. Q. Appl. Math. **10**, 157–165 (1952)
21. Ichikawa, Y., Selvadurai, A.P.S.: Transport Phenomena in Porous Media. Springer, Berlin (2012)
22. Selvadurai, A.P.S.: In-plane loading of a rigid disc inclusion embedded in a crack. Int. J. Solids Struct. **36**, 1701–1714 (1999)

State-Dependent Dilatancy of Soils: Experimental Evidence and Constitutive Modeling

Merita Tafili[1(✉)] and Theodoros Triantafyllidis[2]

[1] Institute of Soil Mechanics and Rock Mechanics (IBF), Karlsruhe, Germany
merita.tafili@kit.edu
[2] IBF, Karlsruhe, Germany
theodoros.triantafyllidis@kit.edu

Abstract. This work provides a new evaluation method of the *dilatancy* for cohesive soils from monotonic and cyclic undrained triaxial tests. Herein it is implemented for experiments performed on Kaolin. Leastwise for this soft soil the *dilatancy* turns out to be a function of the stress ratio η and the void ratio e along with the intrinsic material parameters. Furthermore, an OCR-definition, which includes the influence of both the stress ratio and void ratio such that $d = f(\text{OCR}, C)$ with C being a set of inherent parameters is proposed. In addition based on the experimental observations it is suggested that there is an overconsolidation ratio OCR_{ci} at which the soft soil behaviour changes from contractant in case of $\text{OCR} < \text{OCR}_{ci}$ to dilatant (the material can both contract and dilate depending on η) in case of $\text{OCR} > \text{OCR}_{ci}$ with the PTL lying below the CSL in this case. Finally, a constitutive relation describing the behaviour of soft soils including the *dilatancy* and viscosity is proposed. Some simulations of monotonic as well as cyclic tests are shown to prove the accurate performance of the model.

1 Introduction

The design of geotechnical installations in offshore constructions for example gravity platforms, piled installations, suction anchors, drilling rigs, wind turbines or constructions along the coast for example harbours, breakwaters, dams, storm-surge barriers is governed by the bearing capacity and the serviceability of the structures under cyclic loading [1]. In most of these cases, the underground consists of soft and normalconsolidated clays, sand-clay or sand-silt mixtures. A soft normalconsolidated clay subjected to cyclic loading passes through different overconsolidation ratios $\text{OCR} \geq 1$. Though if the cyclic behaviour of a normalconsolidated clay with a moderately number of cycles is described, the behaviour of an overconsolidated clay is captured by the way. Various effects of clays become increasingly important at different stages of these cyclic tests as for example at the beginning $\text{OCR} = 1$ the time-dependent behaviour of a plastic clay achieves its maximum. For $1 < \text{OCR} \leq 2$ the rate-dependency of

T. Triantafyllidis (Ed.): *Recent Developments of Soil Mechanics and Geotechnics in Theory and Practice*, LNACM 91, pp. 54–84, 2020.
https://doi.org/10.1007/978-3-030-28516-6_4

the clay decays sparsely and depending on the subjected loading magnitude the response of the material is contractant. This leads to the generation of excess pore water pressure resulting in a reduction of the bearing capacity and decay of the barotropic soil stiffness. Towards higher OCR > 2 a significant reduction of the viscous effects is evident, whereby now the materials' phase transformation line (PTL) lies below the critical state line, thus besides contractancy also dilatant behaviour can be observed when reaching the PTL.

Dilatancy and contractancy, hereafter referred to collectively as *dilatancy*, are highly relevant for geotechnical issues dealing with volumetric change due to shearing. Under undrained conditions the prevented dilatancy leads to an increase of the mean pressure and the prevented contractancy can lead to a considerable decay of the same pressure. The consequences are then an increasing accumulation of the pore water preassure, a decay of the barotropic stiffness as well as a considerable reduction of the shear strength q_{max}. Accurate and reliable predictions of post-seismic irreversible displacements in retaining structures supporting fine-grained and low-permeable soils can therefore be obtained only if the adopted constitutive model is capable of capturing correctly the soil *dilatancy* as a function of the current stress and loading history [2]. Furthermore, the *dilatancy* is shown to depend also on other state variables as the void ratio e [3–8] as well as on the sample preparation method [9].

The *dilatancy* of sands has been the research target of many authors [3–8] in the last decades. Among others, Taylor documented in [3]: "The shearing strength" $\tau_m = \sigma \tan \varphi_m$ "in sand may be said to consist of two parts, the internal, frictional resistance between grains, which is a combination of rolling and sliding friction," $\tau_c = \sigma \tan \varphi_c$ "and a second factor for which the most common name is *interlocking*." $\tau_d = \sigma \tan \varphi_d = \sigma \tan \psi$. "Interlocking contributes a large portion of the strength in dense sands; this phenomenon does not occur in very loose sands. The gradual loss of strength after the peak point is passed, ..., may be attributed to a gradual descrease in interlocking which takes place because the sample is decreasing in density. The angle of internal friction, in spite of its name, does not depend solely on internal friction, since a portion of the shearing stress on a plane of failure is utilized in overcoming interlocking."[1] which mathematically can be expressed through:

$$\tau_m = \tau_c + \tau_d \tag{1}$$

whereby the subscripts m, c and d denote the mobilized, critical and *dilatant* state, respectively. Further Taylor [3] suggests: "Interlocking can best be explained by considerations of strain energy. Sands generally are undergoing increase in volume" (expansion) "when the ϕ-obliquity condition is reached," $E_e^d = \sigma A \Delta h$ (superscript d stays for dissipated, subscript e stays for expansion) "and the part of the shearing stress that is acting to overcome interlocking may also be said to be supplying the energy that is being expended in volume increase." (See footnote 1) $E_e^s = \tau A \Delta u$ (superscript s stays for supplied). Assuming $\sigma = const.$ as defined in Fig. 1 we can establish the equilibrium between these two energies:

[1] Taylor in 1954 [3], pp. 345–347.

$$E_e^d = E_e^s \Rightarrow \sigma A \Delta h = \tau A \Delta u \Rightarrow \tau = \sigma \Delta h / \Delta u \tag{2}$$

Due to the equality between the shear strength during expansion τ with the shear strength resulting from interlocking τ_d, Eq. 1 can be rewritten to:

$$\sigma \tan \varphi_m = \sigma \tan \varphi_c + \sigma \Delta h / \Delta u \tag{3}$$

$$\Rightarrow \quad \tan \varphi_m = \tan \varphi_c + \tan \psi. \tag{4}$$

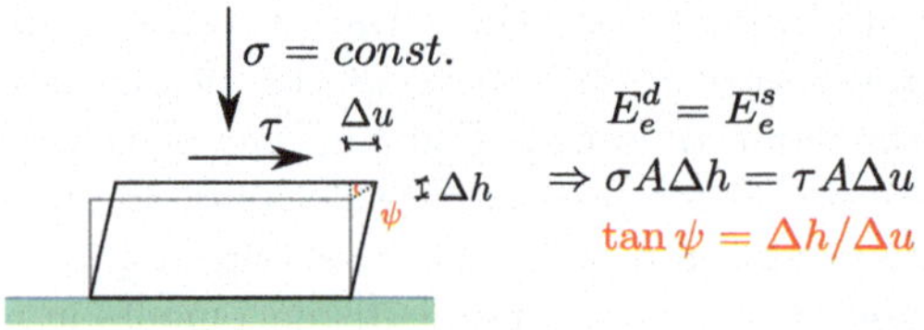

Fig. 1. Shearing of a dense sand with interlocking.

Note that the assumption $\sigma = const.$ guarantees that the volume change dv_d is induced by pure shearing. Otherwise, for example in an undrained or drained triaxial test ($p \neq const.$) the knowledge about the elasticity of the material is required in order to identify which portion of the volumetric and shear change is due to the variation of the applied stresses and which one due to interlocking. In this case instead of the normal σ and shear τ stresses, the stress tensor $\boldsymbol{\sigma}$ is used and instead of the height change Δh and the horizontal displacement Δu, the strain tensor $\boldsymbol{\varepsilon}$ is used. Then, the *dilatancy* is expressed through the ratio between the increment of the plastic volumetric change and the plastic shear strain increment:

$$d = dv^p / |d\gamma^p|. \tag{5}$$

Although the particular forms proposed by Taylor [3] and by Rowe [4] were different, both of them recommended a unique function dependent on the stress ratio $\eta = q/p$ for the *dilatancy*, as depicted by Li and Dafalias [6] $d = d(\eta, C)$ whereby C is a set of inherent material properties. However, experimental evidence postulates [5,9–11] a density dependence of the *dilatancy* as well. For example dense sand dilates after minor initial contractancy and loose sand contracts during shearing. For this purpose researchers proposed various expressions of state-dependent *dilatancy* for sands, which will be reviewed in Sect. 3 after the formal extension from direct shear to triaxial conditions, which are provided in Sect. 2.

On the other side, too little attention has been paid to the dilatant behaviour of clays. Li and Dafalias [6] indicate that $d = d(\eta, M)$ with the slope of the critical state line denoted as M works quite well for clays. This may correspond for normalconsolidated clays (compare the behaviour of loose sands). However, experimental evidence has shown that overconsolidated clays dilate [9,12–15] showing a similar response as medium dense and dense sand. Section 4 provides an overview and the evaluation of experimental data on normal- and overconsolidated clays regarding their *dilatancy*. Then, in Sect. 5 a state dependent *dilatancy* for clays is formulated. Finally, the reference model [15] is extended to account for the *dilatancy* of clays. The models' performance is evaluated through simulations of experimental data of Kaolin with variation of initial OCR_0.

2 From Direct Shear to General Conditions

In a general case the material state is described through the stress tensor $\boldsymbol{\sigma}$ replacing the normal stress σ and the shear stress τ and through the strain rate (increment tensor) $\dot{\boldsymbol{\varepsilon}}$ ($d\boldsymbol{\varepsilon}$). The dilatancy is then defined as the negative ratio between the irreversible strain invariants: plastic volumetric strain increment and the plastic deviatoric strain increment:

$$d = -d\varepsilon_v^p / |d\varepsilon_q^p|. \tag{6}$$

In order to generalize the dilatancy rule given in Eq. 4 we start with the thermodynamic equilibrium of a direct shear test whereby the introduced power density $\tau\dot{\gamma} + \sigma\dot{\varepsilon}$ is partly stored as elastic power $\tau\dot{\gamma}^e + \sigma\dot{\varepsilon}^e$ and partly converted into heat due to the friction between the grain contacts $\tan\varphi_c\,\sigma|\dot{\gamma}^p|$. Thus, a constraint on the plastic strain rates (i.e. on the flow rule) can be established from the relation:

$$\tau\dot{\gamma}^p + \sigma\dot{\varepsilon}^p = \tan\varphi_c\sigma|\dot{\gamma}^p|. \tag{7}$$

Due to the vanishing plastic strains at reversals in elasto-plasticity, it is more convenient to rewrite Eq. 7 with total strain rates:

$$\tau\dot{\gamma} + \sigma\dot{\varepsilon} = \tan\varphi_c\sigma|\dot{\gamma}|. \tag{8}$$

Note that the relation 8 is deduced considering $\sigma = const.$ For triaxial conditions Eq. 8 reads $q\dot{\varepsilon}_q + p\dot{\varepsilon}_v = M\,p\,|\dot{\varepsilon}_q|$ and using the energetically conjugated tensorial values $\boldsymbol{\sigma} : \dot{\boldsymbol{\varepsilon}}$ for the left hand side and $\dot{\boldsymbol{\varepsilon}}^*$ for the last therm on the right hand side the generalization of the *dilatancy* rule given in Eq. 8 is straightforward and reads:

$$\boldsymbol{\sigma} : \dot{\boldsymbol{\varepsilon}} = \sqrt{\frac{2}{3}}\frac{1}{3}M\,\mathbf{1} : \boldsymbol{\sigma}|\dot{\boldsymbol{\varepsilon}}^*| \tag{9}$$

with $p = \mathbf{1} : \boldsymbol{\sigma}/3 = const.$ Relations 7, 8 and 9 operate as *dilatancy* constraints for both loading $\boldsymbol{\varepsilon}_q > 0$ and for unloading $\boldsymbol{\varepsilon}_q < 0$ conditions. In order to impose this constraint in an for example isotropic elastic model an anisotropic term E_{pq} of the stiffness tensor for the consideration of dilatancy is necessary. Without loss of generality we consider triaxial conditions:

$$\begin{bmatrix}\Delta p\\ \Delta q\end{bmatrix} = \begin{bmatrix}E_{pp} & E_{pq}\\ 0 & E_{qq}\end{bmatrix}\begin{bmatrix}\Delta\varepsilon_v\\ \Delta\varepsilon_q\end{bmatrix} \tag{10}$$

with $E_{pp} = K$ and $E_{qq} = 3G$. Imposing the *dilatancy* constraint according to Eq. 9 for $\Delta p = 0$ the coupling stiffness therm between the increment of p and of ε_q reads:

$$E_{pq} = E_{pp}\,(M - \eta). \tag{11}$$

Obviously, an undrained shearing $\dot{\varepsilon}_v = 0$ simulated with the model given in Eq. 10 produces contractancy resulting in mean pressure reduction $\Delta p < 0$. To impose the condition given in Eq. 9 in a sophisticated constitutive model is however almost impossible. For this reason the generalization of other variables that define the *dilatancy* is indispensable.

Pradhan and Tatsuoka reported in [10] the following two relations between the *dilatancy* d (herein defined in strain invariants, see Eq. 6) and the dilatancy angle ψ for isobaric triaxial conditions:

$$d = \begin{cases} \dfrac{3}{2}\dfrac{4\sin\psi}{3-\sin\psi} = \dfrac{6\sin\psi}{3-\sin\psi} & \text{for axial compression } (d\varepsilon_q > 0) \\ -\dfrac{3}{2}\dfrac{4\sin\psi}{3+\sin\psi} = -\dfrac{6\sin\psi}{3+\sin\psi} & \text{for axial extension } (d\varepsilon_q < 0). \end{cases} \tag{12}$$

Note that the factor 3/2 on the RHS is introduced because Pradhan & Tatsuoka used $d\gamma^p$ instead of the invariant $d\varepsilon_q^p = 2/3\, d\gamma^p$ appearing in Eq. 6. Taking into consideration for example the Matsuoka-Nakai failure criterion [16] and the Mohr-Coulomb criterion as well, the slope of the critical state line in triaxial compression and triaxial extension with respect to the critical friction angle φ_c reads:

$$M = \begin{cases} \dfrac{6\sin\varphi_c}{3-\sin\varphi_c} & \text{for compression } (q > 0) \\ -\dfrac{6\sin\varphi_c}{3+\sin\varphi_c} & \text{for extension } (q < 0) \end{cases} \tag{13}$$

The analogy between d given in Eq. 12 and M presented in Eq. 13 is obvious. Yet, it is mandatory to realize the difference in the definition of compression and extension whether using the stress or the strain. The definition for the critical state (CS) is realised in the **stress space** with $q > 0$ for compression and $q < 0$ for extension. Thus, the generalization of M to multiaxial space is executed by means of the Lode angle of the **stress** θ_σ:

$$M = \frac{6\sin\varphi_c}{3-\sin\varphi_c\cos(3\theta_\sigma)} \tag{14}$$

A similar relation leading to same results was proposed by Argyris et al. [17] and was used for other constitutive models e.g. [18–20]. The link between the stress and the strain in CS is established through the critical void ratio, to be defined in Eq. 23.

Yet, when we consider the same stress ratio η (viz. same stress state), two different values for d depending on the direction of straining increment are possible as documented by Pradhan et al. [5] for sands (experiments from [5] are illustrated exemplarily in Fig. 3). Figure 2b shows two different stress states (but three strain states Fig. 2a, c) with different loading and straining directions: 1 = loading in axial compression leading to the blue point, whereby the deviatoric strain tensor and its rate have the same direction as illustrated in Fig. 2c; 2 = loading reversal from axial compression to axial extension or unloading leading to stress and strain reversal and to the red point, the deviatoric strain tensor and its rate have opposite directions as illustrated in Fig. 2c; 3 = continued loading in axial extension leading to the green square. Eventhough the stress state of the green square coincides with the one of the blue point, the deviatoric strain tensor of the green square and its rate have certainly the same direction but nonetheless opposite to the ones of the blue point. For all states the deviatoric stress direction shows in the same direction but the stress increment direction is unknown. Thus, for the generalization of the *dilatancy* to multiaxial space, neither the stress tensor nor its unknown increment can be used. However, given the definition in **strain space** through Eq. 12, the most natural is to use the direction of the deviatoric strain rate, which is also in accordance with the observations presented in Fig. 2 and with the experimental behavior of sand reported for example in [5,10,21]. Thus, we propose a generalization of d considering the direction of the **strain rate** through its Lode angle $\theta_{\dot{\varepsilon}} = \theta_{\dot{\varepsilon}^*}$ (note that for the Lode angle calculation only the deviatoric component of the tensor is needed):

$$d = \frac{6 \sin \psi}{3 - \sin \psi \cos(3\theta_{\dot{\varepsilon}})} \tag{15}$$

and thus the stress ratio at the phase transformation line η_{PTL} which we will denote in general with M_d reads:

$$\eta_{PTL} = \frac{6 \sin \varphi_{PTL}}{3 - \sin \varphi_{PTL} \cos(3\theta_{\dot{\varepsilon}})} \overset{!}{=} M_d. \tag{16}$$

Note that Eq. 16 establishes the interrelation between the strain space and the stress space for the *dilatancy*. The phase transformation angle of friction φ_{PTL} will be discussed and defined in the next Section.

Furthermore, Pradhan et al. [5,10,21] studied the relationship between the dilatancy d and the stress ratio η of sand subjected to cyclic loading, whereby they postulate a unique relationship between the stress ratio and the *dilatancy* irrespective of void ratio and pressure level. During for example undrained cyclic loading the void ratio remains the same but the mean pressure changes resulting in a change of the critical void ratio. Hence, after a certain reduction of p the soil state is dense $e < e_c$ and when the stress state goes towards the phase transformation state $\eta \overset{!}{=} \eta_{PTL} = q_{PTL}/p_{PTL}$ both the mean stress and the deviatoric stress increase reaching a peak value. At the reversal point the stress path goes towards lower p and q viz. lower η with the greatest contractancy. Consequently, the *dilatancy* relation $d - \eta$ of each cycle may be alike in shape, but not the same because until $e > e_c$ $d \leq 0$ holds and when $e < e_c$ $d \lesseqqgtr 0$ holds depending on η. Yet, neither Pradhan et al. [5,10,21] nor the relations 7–9 consider the void ratio dependence of the *dilatancy*.

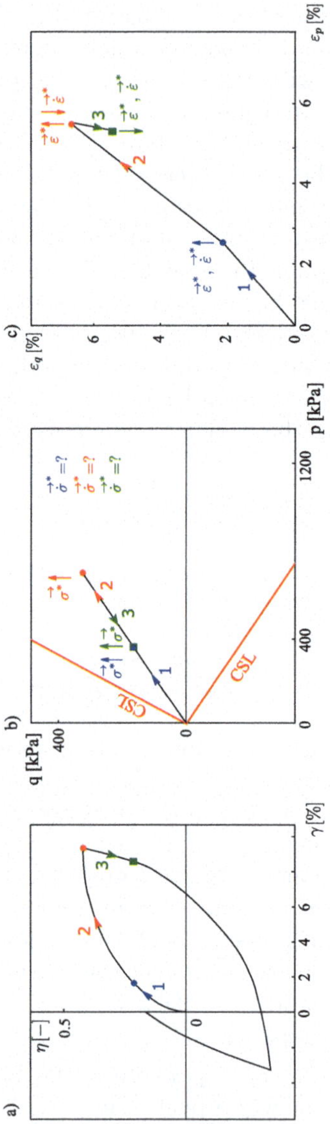

Fig. 2. Example of two different stress states (but three strain states) with different loading and straining directions: 1 = loading in compression, the deviatoric strain tensor and its rate have the same direction; 2 = loading in axial extension or unloading leading to stress and strain reversal, the deviatoric strain tensor and its rate have opposite directions; 3 = loading in extension leading to the green square. Eventhough the stress state of the green square coincides with the one of the blue point, the deviatoric strain tensor of the green square and its rate have certainly the same direction but nonetheless opposite to the one of the blue point. For all states the deviatoric stress direction shows in the same direction but the stress increment direction is unknown. (a) $\eta - \gamma$ space, (b) effective stress in $p - q$ plane, (c) $\varepsilon_p - \varepsilon_q$ space.

3 Overview of Expressions Describing the State-Dependent *Dilatancy* of Cohesionless Soils

Except of the state dependence, the *dilatancy* has to fulfill certain requirements arising from experimental evidence. On one side, at the phase transformation line (PTL) i.e. at $\eta = M_d$ the dilatancy vanishes $d(\eta = M_d) = 0$ [22]. On the other side, the critical state represents an ultimate state of failure at which $\eta = M$ and $e = const.$ thus no volumetric strain occurs. The constant volumetric strain $dv_d = 0$ implies $d(\eta = M, e = e_c) = 0$ at the critical state. A unique relationship between d and η would render $M = M_d$, which can represent an exceptional case for a certain value of e_0. However, in general laboratory data show that dense sand dilates $M_d < M$ and loose sand contracts $M_d > M$.

In the following we will review two approaches for the constitutive description of the state-dependent *dilatancy* of cohesionless soils.

3.1 State-Dependent *Dilatancy* for Monotonic Loading Described by Li and Dafalias [6]

In order to account for the density dependence of the *Dilatancy* for sands, Li and Dafalias [6] proposed a state-dependent expression for d:

$$d = \frac{d_0}{M} \left(M \exp(m\psi) - \eta \right), \quad \psi = e - e_c(p'). \tag{17}$$

The fulfillment of the critical state requirements for $e = e_c \Rightarrow \psi = 0$ and $\eta = M$ is evident. The second requirement $d(\eta = M_d) = 0$ implies a void ratio dependence of the phase transformation stress ratio $M_d = M \exp(m\psi)$.

The relationship proposed in Eq. 17 is illustrated in Fig. 3, whereby blue dots represent the experimental results of Pradhan et al. [5]. It is obvious that the relation given in Eq. 17 (orange lines in Fig. 3) is not in accordance with experimental evidence for axial extension. These disadvantages of Eq. 17 resulted in the introduction of the so-called fabric-dilatancy internal tensor-valued variable $\mathbf{z}$ by Dafalias and Manzari [23].

3.2 State-Dependent *Dilatancy* Described by Grandas and Triantafyllidis [24]

The dilatancy relation derived in Eq. 4 is void ratio independent and from Fig. 3 (purple lines) it can be observed that the Taylor's rule neither captures the strong contractancy upon reversals. To describe a state-dependent *Dilatancy* Grandas and Triantafyllidis [24] proposed a modification of the term describing the critical state $\tan \varphi_c$ analogous to the idea proposed by [6]. For this purpose the tangens of the critical state angle of friction is replaced by $\tan \varphi_{PTL} = \tan \varphi_c / f_e$, whereby f_e is a scalar function of void ratio and mean pressure. Some requirements has to be imposed on the formulation of the void ratio function f_e:

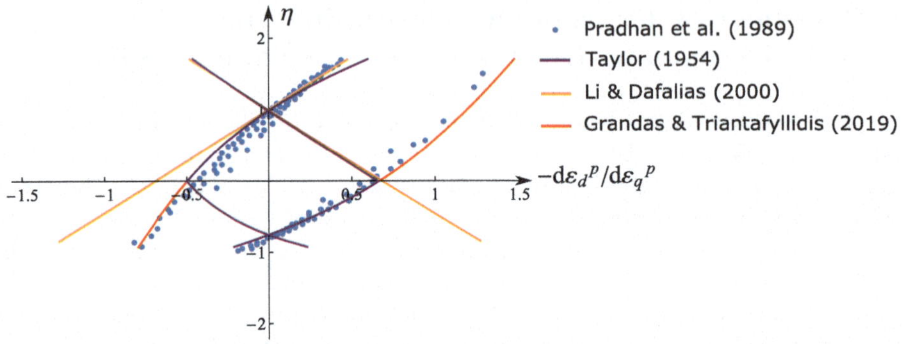

Fig. 3. Experimental results of constant pressure triaxial tests of Pradhan et al. [5] compared with the *dilatancy* relations proposed by Li and Dafalias [6] and Grandas and Triantafyllidis [24].

- if the mobilized state of the material is at failure i.e. the critical state is reached then $\varphi_m = \varphi_c$ and the dilatancy vanishes viz. $\psi = 0$. Thus, the void ratio function should render $f_e = 1$.
- when subjected to shearing, dense sand dilates which renders $M_d < M$. Hence, $\tan\varphi_{PTL} < \tan\varphi_c$ resulting in $f_e > 1$.
- loose sand upon shearing contracts and the phase transformation line lies then above the critical state line $M_d > M$ and may not be reached under monotonic loading. Thus, $\tan\varphi_{PTL} > \tan\varphi_c$ results in $f_e < 1$.

Equation 4 now reads:

$$\tan\psi = \tan\varphi_m - \tan\varphi_c / f_e. \tag{18}$$

Experiments evidence the maximum contractancy upon reversals of the strain path, see Fig. 3 and the reflections made in Sect. 2. This effect can be described by extending the dilatancy rule from Eq. 18 to account for the direction of shearing. A thermodynamically conform extension is proposed through the relation:

$$\tan\psi = \vec{\boldsymbol{\sigma}}^* : \vec{\dot{\boldsymbol{\varepsilon}}}^* \tan\varphi_m - \tan\varphi_c / f_e. \tag{19}$$

This relation is depicted in Fig. 3 with red lines and shows a perfect agreement with the experimental data of Pradhan et al. [5]. Thus, the *dilatancy* expressed by d or ψ is not only a function of stress as in Cam Clay theory $d = f(\boldsymbol{\sigma}, C)$ or from stress and void ratio as assumed by Li and Dafalias [6] $d = f(\boldsymbol{\sigma}, e, C)$, but also a function of the direction of shearing $\psi = f(\boldsymbol{\sigma}, e, \vec{\dot{\boldsymbol{\varepsilon}}}^*, C)$. Intrinsic material parameters denoted with C are of course included.

4 Experimental Findings Regarding the *Dilatancy* of Cohesive Soils

It is tentatively assumed that the *dilatancy* of fine-grained soils is only dependent on the stress ratio and on the intrinsic material parameters [25–29]. Yet, experimental evidence gained for example in [9] shows that with increasing OCR resulting in a lower void ratio e_0 prior to the undrained monotonic shearing, the material response is rendered more dilative and an increase of the undrained shear strength is evidenced. For normalconsolidated samples the PTL may not be reached during shearing $M_d \geq M$ rendering the material response contractant. However, the PTL is observed to lie below the CSL for overconsolidated samples $M_d < M$. Similar dependencies of the *dilatancy* on the overconsolidation ratio can be concluded from the experiments presented in [12,30–40]. Nevertheless, the *dilatancy* effects and their dependence on the overconsolidation ratio, which is not trivial, was mostly ignored in the constitutive modeling of soft soils.

4.1 Basic Postulates

In order to proceed with some findings the therms normalconsolidated, lightly overconsolidated and heavily overconsolidated need to be defined. Consider isotropic states ($q = 0$) and isotropic loading directions ($\dot{q} = 0$) then the state of the material is defined by a combination of the void ratio e and the mean pressure p, whereby the normal consolidated state (OCR = 1) is given by the maximum void ratio e_i, see also Fig. 4:

$$e_i = e_{i0} - \lambda \ln(p/p_{ref}). \tag{20}$$

The slope of the normal consolidated line (NCL) in e vs. p diagram is determined through the compression index λ and its position is defined through the material constant e_{i0} specified at a reference mean pressure p_{ref}. Equation 20 can be solved for the void ratio equivalent normalconsolidated pressure imposing the condition $e = e_i$. Then, the relation proposed by Hvorslev [41] is obtained:

$$p_{ei} = \exp\left(\frac{e_{i0} - e}{\lambda}\right). \tag{21}$$

p_{ei} can be termed also as preconsolidation pressure (similarly to p_c in Cam Clay) and the formulation of its evolution rate is straightforward:

$$\dot{p}_{ei} = -\frac{\dot{e}}{\lambda} \exp\left(\frac{e_{i0} - e}{\lambda}\right) \tag{22}$$

Similarly to the NCL, the critical void ratio line in e vs. p diagram can be defined through:

$$e_c = e_{c0} - \lambda \ln(p/p_{ref,c}). \tag{23}$$

In analogy to sands and gravels, which are defined as loose or dense depending on the position of the state with respect to the critical void ratio line, also clays can be defined as lightly overconsolidated or heavily oversoncolidated. The analogy between the behaviour of grained and fine-grained soils has been depicted also in the pioneer work of Atkinson [42], but has been almost ignored in subsequent research. We define a material state lying between the NCL (OCR = 1)

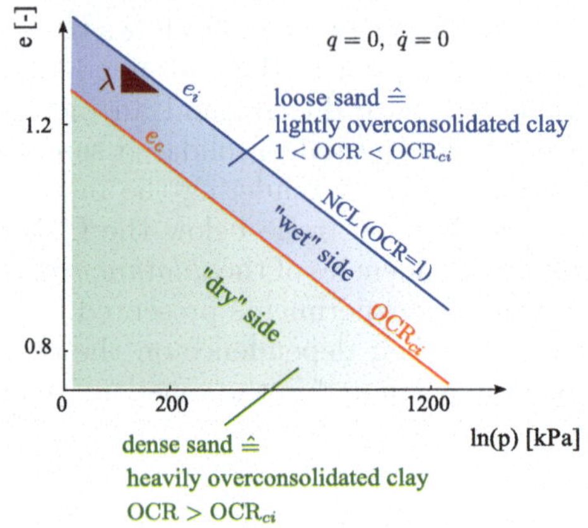

Fig. 4. State of soils for isotropic initial $q_0 = 0$ and isotropic loading conditions $\dot{q} = 0$.

and the critical void ratio line (OCR $=$ OCR$_{ci} > 1$) (blue area in Fig. 4) as lightly overconsolidated ($1 <$ OCR $<$ OCR$_{ci}$), similar to the definitions in [42]. The subscript *ci* stays for the *c*ritical *i*sotropic line. The necessity for this definition will be clear after the generalization to three dimensional states in the next pages. However, the aforementioned states of soils are likely to be called on the "wet" side of the critical, not to be mixed up with the saturation of the soil. The soil is either saturated or dry and "wet" means that the void ratio e or the water content w at a mean pressure p is higher than the critical one e_c or more wet than the critical one w_c, respectively. For a heavily overconsolidated clay OCR $>$ OCR$_{ci}$ (green area in Fig. 4) the void ratio is lower than the critical one. Therefore, the soil is termed on the "dry" side i.e. drier than at the critical line. These classifications correspond to the analogous characterisations of loose and dense sands. Note that only isotropic states $q = 0$ and isotropic loading directions $\dot{q} = 0$ are discussed above and illustrated in Fig. 4.

Now let us consider general stress states with $q \neq 0$. Then the soil is at its critical state when both the critical state line (CSL) $\eta = q_c/p_c \overset{!}{=} M$ and the critical void ratio line $e \overset{!}{=} e_c$ are reached. Eventhough the definition of the CS is "ordinary", the classification lightly or heavily overconsolidated is not straightforward. We first consider the state A in Fig. 5a on top, which is normalconsolidated. The most natural is to think that point A which is in e vs. p space uniquely defined, possesses infinite projections on the p vs. q plane along the line $p_A = const$. Three marking states along this line are illustrated in Fig. 5a at the bottom: A1 at $q = 0$, A2 at $q < q_c$ and A3 at the critical state $q = q_c$. Each of these states are normalconsolidated if they are passed in the numbered order A1–A2–A3. In order to ensure uniqueness for general states we introduce the widely used concept of the loading surface. The (pre)loading surface represents a constant void ratio $e = const.$ and constant overconsolidation ratio surface which intersects the p-axis at $p = p_{ei}$ if OCR $= 1$. The critical state line and the loading surface intersect each other at $p = p_{ei}/\mathrm{OCR}_{ci}$ and $q = \eta_c p$. Consequently, the points A1, A2 and A3 correspond to three different loading surfaces intersecting the p-axis at different mean preassures $p_{ei,A1} = p_A < p_{ei,A2} < p_{ei,A3}$. Figure 5(a and b) shows exemplary the loading surface for A1, which expands for the loading sequence A1–A2–A3 keeping the same shape. If they were projections of point A then $p_{ei,A1} = p_A = p_{ei,A2} = p_{ei,A3}$ would hold, which obviosly is in contradiction with the loading surface theory. Therefore, the projection of the NCL defined in $e-p$ space can be only realised for $q = 0$, hence in the isotropic axis of $p-q$ diagram and then A = A1 is uniquely defined in both spaces. If the initial state of the material corresponds to A2 and the sample is subjected to deviatoric unloading until the isotropic axis is reached, then the final state of the material is overconsolidated eventhough the void ratio and the mean stress remain unchanged $p_{ei,A2} > p_{ei,A1} = p_A$.

Secondly we will analyze states lying at the critical void ratio line denoted with point B in e vs. p diagram (Fig. 5a on top) having the same void ratio as A but lower mean pressure. Once more, one may think that infinite projections of B are possible in the $p-q$ space. In Fig. 5a on bottom, exemplary three of these projections are illustrated. Along the same lines of thoughts as for the NCL, also the critical void ratio line is to be projected at the isotropic axis. Thus, the unique projection of B corresponds to B1 ($q = 0$) and is equivalent to the states which separate the "wet" side from the "dry" side corresponding to a critical overconsolidation ratio OCR_{ci} (the subindex **ci** = **c**ritical **i**sotropic) in Fig. 4. For this purpose, we propose to define the critical void ratio as a shift by the OCR_{ci} of the maximum void ratio:

$$e_c = e_{i0} - \lambda \ln(\mathrm{OCR}_{ci}) - \lambda \ln(p/p_{ref}). \tag{24}$$

Thus, the loading surface can be constructed by knowing the current stress and void ratio of the material, current OCR and OCR_{ci} solely. The state B2 lies between $q = 0$ and $q = q_c$ and at a loading surface, which intersects the p-axis at $p^+_{ei,B2} > p^+_{ei,B1}$. Thus, B2 is characterised with a lower $\mathrm{OCR}_{B2} < \mathrm{OCR}_{B1} = \mathrm{OCR}_{ci}$ as on the "wet" side of the critical state. Point B3 lies at the critical state in the void ratio - pressure diagramm as well as in the effective stress plane. Because the critical state represents the "constant" state of the material and is independent of the initial state then it defines also a normalconsolidated state $\mathrm{OCR}_c = 1$. Considering these observations the isotropic critical overconsolidation ratio OCR_{ci} can be calibrated through an undrained shearing test of a normalconsolidated sample following the path A1–B3 $\Rightarrow \mathrm{OCR}_{ci} = p_{A1}/p_{B3}$. Note, that it is mandatory for the state B3 to fulfill both critical state requirements $e = e_c$ and $\eta = \eta_c$. Thus, in order to reach the critical state strains greater than 10% are needed, in some soils even greater than 50% [42].

Finally, we will consider the state C lying at the "dry" side of the critical void ratio line in Fig. 5(a). Its projection C1 on the $p-q$ plane, Fig. 5a is straightforward and corresponds to $\mathrm{OCR}_{C1} > \mathrm{OCR}_{ci}$ characterizing a heavily overconsolidated soft soil. For a state C2 ($q \neq 0$) lying inside the loading surface

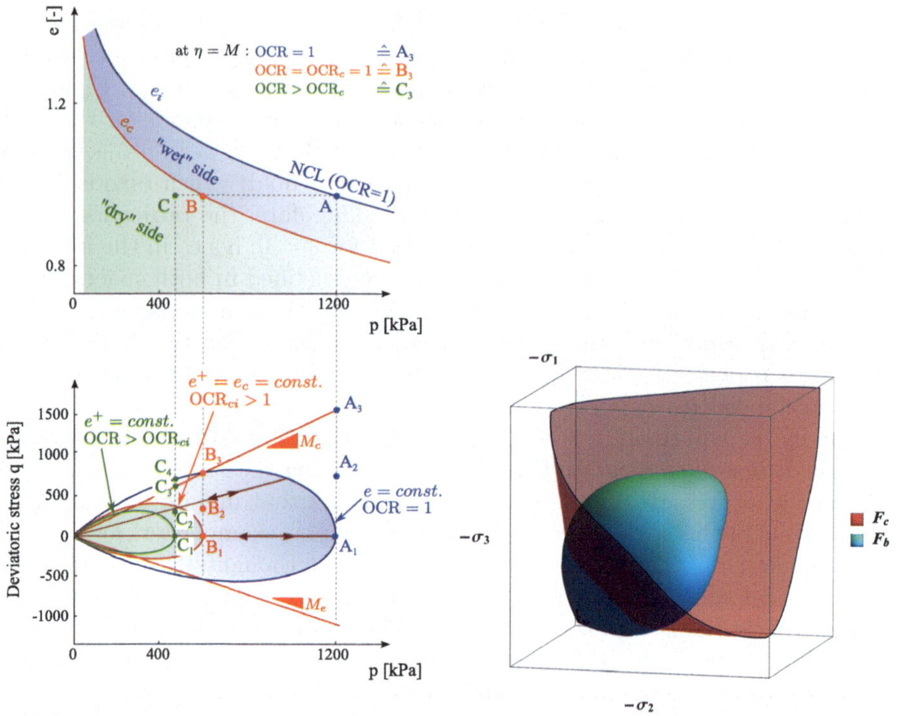

(a) State of soils for general initial and loading conditions $\dot{q} \neq 0$.

(b) Critical state surface F_c and bounding surface F_b in the principal stress space.

Fig. 5. Interrelation between the stress space and the void ratio space.

of B1 the overconsolidation ratio is less than at C1 but greater than the isotropic critical oversondolidation ratio $\mathrm{OCR}_{C1} > \mathrm{OCR}_{C2} > \mathrm{OCR}_{ci}$. Similarly, a state C3 lying outside the isotropic critical loading surface possesses $\mathrm{OCR}_{C3} < \mathrm{OCR}_{ci}$ characterizing a lightly overconsolidated soft soil. Eventhough C3 lies at the CSL $\eta = \eta_c$ the critical state is not reached as $e \neq e_c$. At long last, the state C4 lies at the initial or preloading surface reaching a peak stress q_p and $\mathrm{OCR}_{C4} = 1$. Thus state C, depending on the deviatoric stress q, can be characterized as all of them: heavily overconsolidated at C1 $p = 0$ and at C2 $0 < q < q(\mathrm{OCR} = \mathrm{OCR}_{ci})$, lightly overconsolidated at C3 $q(\mathrm{OCR} = \mathrm{OCR}_{ci}) < q < q(\mathrm{OCR} = 1)$ and normalconsolidated at C4 $q = q(\mathrm{OCR} = 1) = q_p$. Thus, for isotropic states Fig. 4 or Fig. 5a (e vs. p relationship) is sufficient in order to determine whether the soil is normal-, lightly- or highly consolidated. At anisotropic states, both the e vs. p relation and its generalization in $p - q$ space is necessary.

A lightly overconsolidated clay when subjected to shearing is expected to contract. We suggest that there is an overconsolidation ratio OCR_{ci} at which the soft soil behaviour changes from contractant $\mathrm{OCR} < \mathrm{OCR}_{ci}$ to dilatant (the material can both contract and dilate depending on η) $\mathrm{OCR} > \mathrm{OCR}_{ci}$ with the

PTL lying below the CSL, as will be shown through experiments in the next Sects. 4.2 and 4.3.

4.2 Triaxial Tests Under Monotonic Loading

In the last decades, many stress-*dilatancy* relations for sand have been proposed. An example therefore presents Fig. 3. For these studies, mostly constant pressure triaxial tests have been used, because then the volumetric change viz. the volumetric strain rate ${\dot{\varepsilon}_v}^2$ corresponds to the plastic one $\dot{\varepsilon}_v^p$. However, drained tests for clays are highly time-consuming and therefore very rare to find. For this reason, in this work we used undrained triaxial tests (monotonic and cyclic) of Kaolin Clay [14]. For the evaluation of its *dilatancy* behaviour some assumptions and relations need to be established.

In general, the volumetric strain rate consists of a *dilatancy* component $\dot{\varepsilon}_v^d$ and of an another component denoted with $\dot{\varepsilon}_v^c$:

$$\dot{\varepsilon}_v = \dot{\varepsilon}_v^c + \dot{\varepsilon}_v^d \tag{25}$$

$\dot{\varepsilon}_v^d$ is totally plastic and is induced by a plastic shear strain rate $\dot{\varepsilon}_q^p$, whereas $\dot{\varepsilon}_v^c$ contains both elastic and plastic components induced by the change in the effective mean stress p. Thus, the volumetric strain during tests with $p = const.$ is equally to the *dilatant* one:

$$\dot{\varepsilon}_v = \dot{\varepsilon}_v^p = \dot{\varepsilon}_v^d, \quad \text{for } p = const. \tag{26}$$

Similar relations are documented also by Shibata in [43]. Hence, a rearrangement of Eq. 6 is necessary for $p \neq const.$ The dilatancy is then defined as the relation between the dilatant volumetric strain increment (instead of total plastic one) and the plastic deviatoric strain increment:

$$d = -d\varepsilon_v^d / |d\varepsilon_q^p| \tag{27}$$

Assuming that clays possess a unique normal consolidation line (NCL), then reloading occurs at nearly the same line (with (strain)hysthereticai behaviour at the reverse direction) as the foregoing unloading did, see Fig. 6(a) and (b). For undrained conditions the total volumetric change vanishes $d\varepsilon_v = 0$ and hence the *dilatant* strain increment results from the prevention of the volumetric change and can be termed also prevented *dilatancy*. If the loading was not undrained, then the volumetric change would follow the unloading or unloading-reloading line for OCR $\leq$ OCR$_{ci}$ or OCR $>$ OCR$_{ci}$, respectively (marked with dark-green dottet line in Fig. 6(b)). On the other side, Fig. 6(c) and (d) show that sands do not behave the same way. Its compaction during for example isotropic unloading as presented in Fig. 6(c) leads to a new normal consolidation line lying below the first one (marked with dark-green dottet line in Fig. 6(d). Thus, the volumetric change upon subsequent undrained shearing goes along with the slope λ in the

2 Note the equality $\dot{\sqcup} = d\sqcup/dt$.

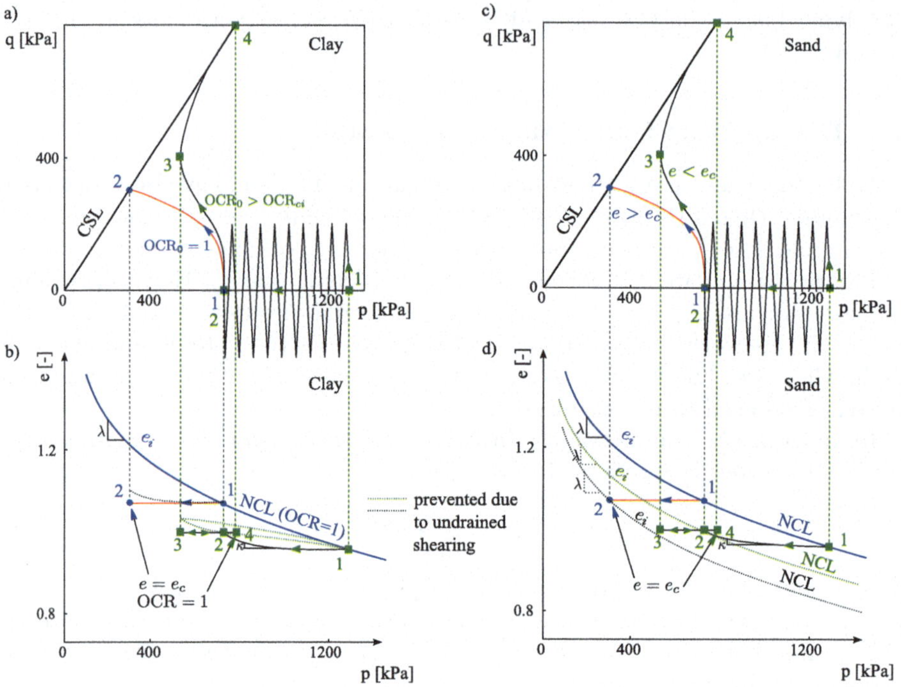

Fig. 6. Illustrative examples of isochoric shearing of a normalconsolidated ($\mathrm{OCR}_0 = 1$) and highly overconsolidated ($\mathrm{OCR}_0 > \mathrm{OCR}_{ci}$) clay sample (left) compared to a loose ($e_0 > e_{c0}$) and dense ($e_0 < e_{c0}$) sand sample (right). The overconsolidated clay sample or the dense sand sample are obtained by drained isotropic unloading until either $\mathrm{OCR}_0 > \mathrm{OCR}_{ci}$ or e_{c0}, in order to satisfy the requirement $e < e_{c0}$, are reached. In a) it is shown that when subjected to undrained shearing the normalconsolidated sample (blue point 1) is expected to contract (reaching blue point 2 at CSL) and a highly overconsolidated sample (dark-green square 2) to dilate after an initial contraction (reaching dark-green square 4 at CSL). These effects result due to the prevented *dilatancy* viz. if drained shearing occured then the unloading in the $e - \ln p$ space in (b) would follow the line with the slope κ. Therefore, also the dilatant behaviour starting from the dark-green square 3 in (a) and (b) would follow this line (marked with a dotted line in (b)) in the reverse direction rendering a totally elastic response at reloading until the NCL is reached. Thus for undrained shearing of clays holds: $\dot{\varepsilon}_v = \dot{\varepsilon}_c + \dot{\varepsilon}_d = \dot{\varepsilon}_{el} + \dot{\varepsilon}_d = 0 \Rightarrow \dot{\varepsilon}_d = -\dot{\varepsilon}_{el} = -\dfrac{\kappa \dot{p}}{p(1+e)}$. Sands however possess more than one NCL (depending on the initial void ratio e_0 and the loading history), hence after a compaction they always follow the line with the slope λ (new NCL) in the $e - \ln p$ space as demonstrated in (d) and thus $\dot{\varepsilon}_d \neq -\dot{\varepsilon}_{el}$ but $\dot{\varepsilon}_c = \dfrac{\lambda \dot{p}}{p(1+e)} = -\dot{\varepsilon}_d$.

$e - \ln p$ space consisting of a reversible part $d\varepsilon_v^{el}$ (belonging to the slope κ) and a plastic one $d\varepsilon_v^c + d\varepsilon_v^d$ (with the slope $\lambda - \kappa$). Following these realizations, for clays we suggest that the volumetric change due to the change in p is solely elastic $\dot{\varepsilon}_v^c = \dot{\varepsilon}_v^{el}$. Hence, the *dilatant* strain increment follows from:

$$d\varepsilon_v = d\varepsilon_v^{el} + d\varepsilon_v^p \quad \Rightarrow \quad d\varepsilon_v^d = d\varepsilon_v^p = -d\varepsilon_v^{el} \tag{28}$$

The elastic strain increment can be calculated from the constitutive equation:

$$\begin{bmatrix} dp^{el} \\ dq^{el} \end{bmatrix} = \begin{bmatrix} E_{pp} & E_{pq} \\ E_{qp} & E_{qq} \end{bmatrix} \begin{bmatrix} d\varepsilon_v^{el} \\ d\varepsilon_q^{el} \end{bmatrix} \tag{29}$$

$$\Rightarrow \begin{bmatrix} d\varepsilon_v^{el} \\ d\varepsilon_q^{el} \end{bmatrix} = \begin{bmatrix} C_{pp} & C_{pq} \\ C_{qp} & C_{qq} \end{bmatrix} \begin{bmatrix} dp^{el} \\ dq^{el} \end{bmatrix}. \tag{30}$$

Omitting the non-diagonal terms for an isotropic material and using the hypoelastic stiffness:

$$\mathsf{E}_{iso} = 3K \,\vec{\mathbf{1}}\,\vec{\mathbf{1}} + 2G\,\mathsf{I}^{dev} \tag{31}$$

one obtains:

$$d\varepsilon_v^{el} = dp/E_{pp} = \frac{2\lambda\kappa}{\lambda + \kappa}\frac{dp}{(1+e)p}. \tag{32}$$

Note that the void ratio during undrained shearing corresponds to the initial void ratio $e = e_0$.

For a transversal isotropic material the scaled hypoelastic stiffness can be used instead as proposed in [15]:

$$\mathsf{E}_{abcd} = \mathsf{Q}_{abij} : \mathsf{E}_{ijkl} : \mathsf{Q}_{klcd} \tag{33}$$

$$\mathsf{Q}_{abcd} = \mu_{ac}\mu_{bd} \tag{34}$$

$$\boldsymbol{\mu} = \sqrt{\alpha}\mathbf{1} + (1 - \sqrt{\alpha})\mathbf{m}_s \otimes \mathbf{m}_s, \tag{35}$$

with the vector along the sedimentation axis $\mathbf{m}_s$ and the material parameter α denoted as anisotropic coefficient [15]. For a vertical sedimentation axis viz. $\mathbf{m}_s = \{0, 0, 1\}$ corresponding to a vertically cutted Kaolin sample from [14] then the hypoelastic compliance from Eq. 30 in conjunction with the transversal isotropic hypoelastic stiffness given through Eqs. 33–35 reads:

$$C_{pp} = \frac{3K(\alpha-1)^2 + G(\alpha+2)^2}{27GK\alpha^2}$$

$$C_{pq} = \frac{(\alpha-1)(6\alpha K + 3K + 2G(\alpha+2))}{27\sqrt{2}GK\alpha^2}$$

$$C_{qp} = \frac{(\alpha-1)(6\alpha K + 3K + 2G(\alpha+2))}{27\sqrt{2}GK\alpha^2}$$

$$C_{qq} = \frac{4G(\alpha-1)^2 + 3K(2\alpha+1)^2}{54GK\alpha^2}$$

This set of equations present the basis for the evaluation of the experimental data published by Wichtmann and Triantafyllidis in [14]. Hereby we will discuss both the evaluation procedure and the results of the monotonic tests on normal and overconsolidated samples OCR = {1, 2, 4, 8}. When trying to evaluate the data, the change of the mean pressure dp underlied a huge data scattet such that for example during some time points the same mean pressure was evaluated. This led to a vanishing *dilatancy* $d = 0$ for unrealistic stress ratios η and not even for a specific OCR_0 a unique relationship between $d - \eta$ could be established. For this reason, the development of the mean pressure was approximated through polynomial functions depending on the only equidistantly captured quantity i.e. the time t as follows:

$$p_{approx} = \sum_{i=0}^{n} a_i t^i, \quad \text{with} \quad 3 \leq n \leq 9. \tag{36}$$

The value of n was varied aiming the minimum residual of $(p_{approx} - p) \to 0.1$ kPa. The affinity of the curves represented another important criterium. For a transversal hypoelastic stiffness also the deviatoric stress q need to be approximated because its increment dq is used for the calculation of the irreversible components $d\varepsilon_v^d$ and $d\varepsilon_q^p$.

Finally, we will now discuss the results of the $d - \eta - \mathrm{OCR}_0$ relation. Figure 7a shows the relationship between the stress ratio and the dilatancy for different initial overconsolidation ratios OCR_0. It is evident that the normal consolidated sample $\mathrm{OCR}_0 = 1$ as well as the lightly overconsolidated sample $\mathrm{OCR}_0 = 2$ contracts $d < 0$ upon undrained compressive shearing $d\varepsilon_q > 0$, whereas the heavily overconsolidated samples with $\mathrm{OCR}_0 = 4$ and $\mathrm{OCR}_0 = 8$ dilate $d > 0$ for $d\varepsilon_q > 0$ after an initial contraction $d < 0$ for $d\varepsilon_q > 0$. The initial contraction decays with an increasing initial overconsolidation ratio. Hence the subsequent dilation increases and the phase transformation stress ratio $\eta = \eta_{PTL}$ at $d = 0$ is reduced. These observations form the basis for the classification of clays' *dilatancy* behaviour very similar to that of sands. The question arises which value can be assigned to OCR_{ci} (see Eq. 24) and whether it represents the same number for each clay. The Modified Cam Clay theory renders the value $\mathrm{OCR}_{ci} = \exp(1) = 2.71$ as a constant value. However, for the first we will let this value as a material parameter to be calibrated until we evaluate some cyclic undrained triaxial tests in the next Section.

Sands are classified as dense or loose in therms of their relative density value which would in this work coincide with the overconsolidation ratio for clays. The classification dependent on the void ratio itself is not the most appropriate for clays due to the experimental data scatter of the initial void ratio. On the other side, knowing the consolidation process and the preloading of the soil, the overconsolidation ratio can be determined by:

$$\mathrm{OCR} = \frac{p_{ei}}{p_{ei}^{+}} \tag{37}$$

with the Hvorslev stress p_{ei} defined in Eq. 21. To eleminate the data scatter of the initial void ratio one may use for $p_{ei} = \max(p)$ the maximum mean pressure the soil was subjected to. Here we of course assume that the loading history

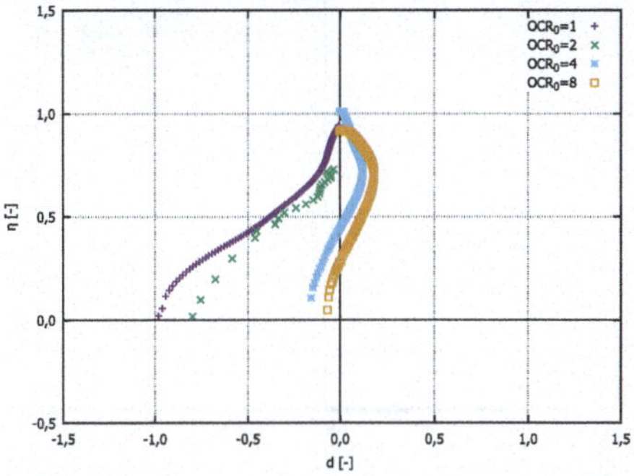

(a) Stress ratio η vs. dilatancy d relationship for different initial overconsolidation ratios OCR_0 evaluated from monotonic undrained triaxial tests performed on Kaolin. (Experiments borrowed from [14])

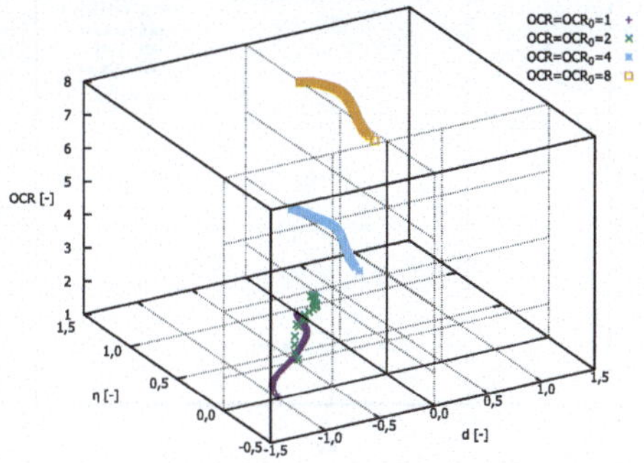

(b) $d - \eta - \mathrm{OCR}$ for the first approach $\mathrm{OCR} = \mathrm{OCR}_0 = const.$

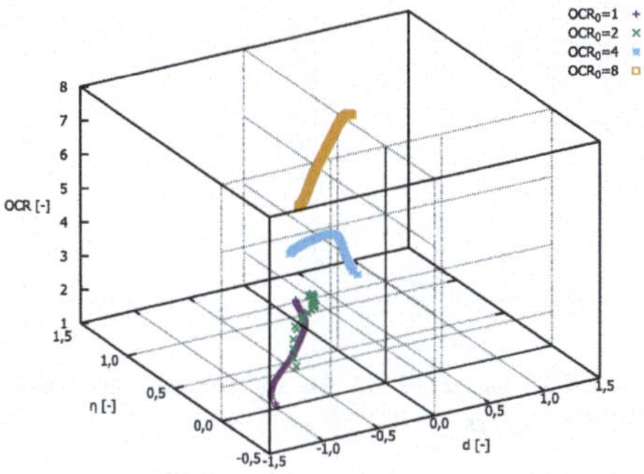

(c) $d - \eta - \mathrm{OCR}$ for the second approach $\mathrm{OCR} = p_e/p$.

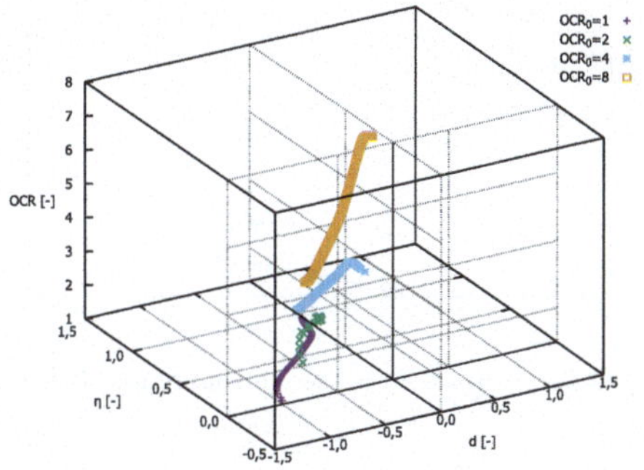

(d) $d - \eta - \mathrm{OCR}$ for the third approach $\mathrm{OCR} = p_e/p_{ei}^+$.

Fig. 7. Two- and three-dimensional relation between $d - \eta - \mathrm{OCR}$ evaluated from monotonic undrained triaxial tests performed on Kaolin. (Experiments borrowed from citewt17)

of the soil is known, which is mostly the case with reconstituted samples. Now three different approaches for the definition of the mean pressure corresponding to the actual void ratio of the soil p_{ei}^+ are presented. Firstly, as a very simple but also a very rough approximation can serve $\mathrm{OCR} = \mathrm{OCR}_0$ thus $p_{ei}^+ = p_{ei}$ and $\mathrm{OCR} = const.$ during the undrained shearing. The evaluated data corresponding to this (absurde) assumption is presented in Fig. 8a and b. Thus, the critical state $\mathrm{OCR} = 1$ and $\eta = \eta_c$ for initially overconsolidated samples $\mathrm{OCR}_0 \neq 1$ will never be reached in therms of OCR.

Secondly, as a more natural, appropriate and still simple but as will be explained not ideal approach one can approximate the mean pressure corresponding to the actual isotropic stress state ignoring the influence of the

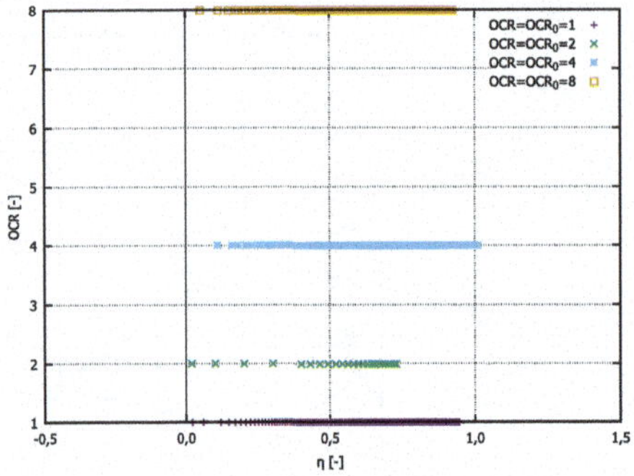

(a) OCR $-\ \eta$ space for the first approach OCR = OCR$_0$ = *const*.

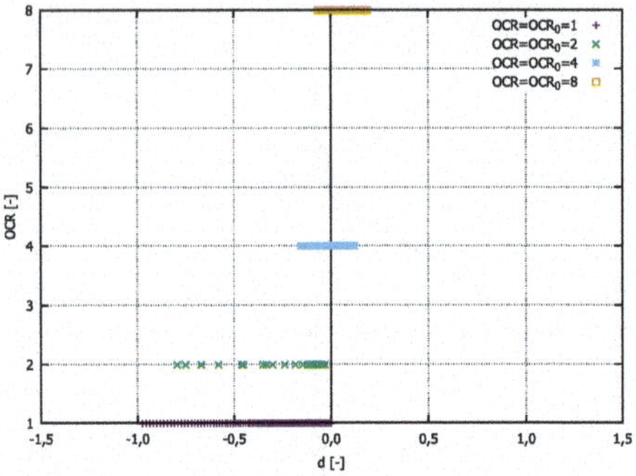

(b) OCR vs. d plane for the first approach OCR = OCR$_0$ = *const*.

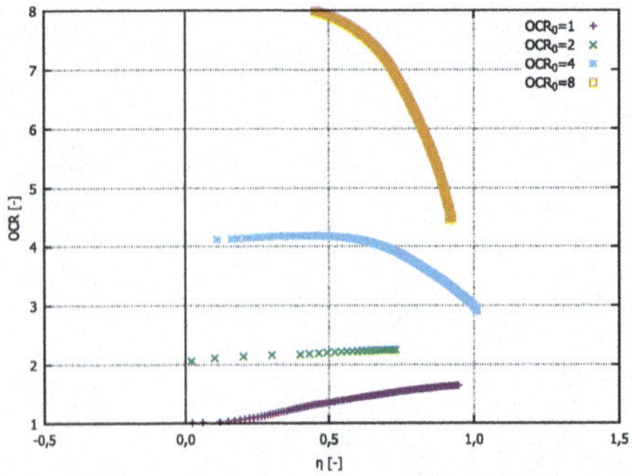

(c) OCR$-\eta$ space for the second approach OCR = p_e/p.

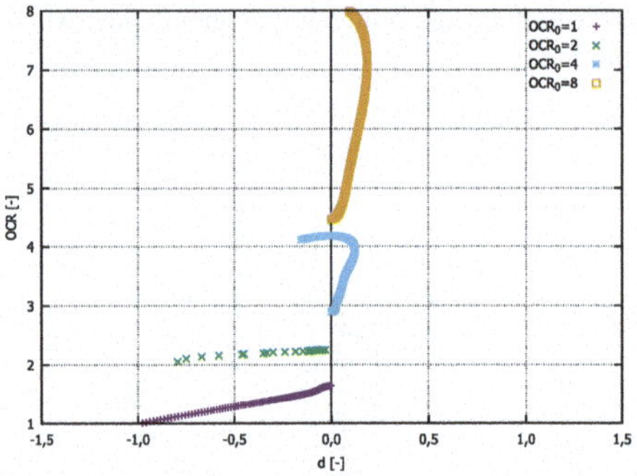

(d) OCR vs. d plane for the second approach OCR = p_e/p.

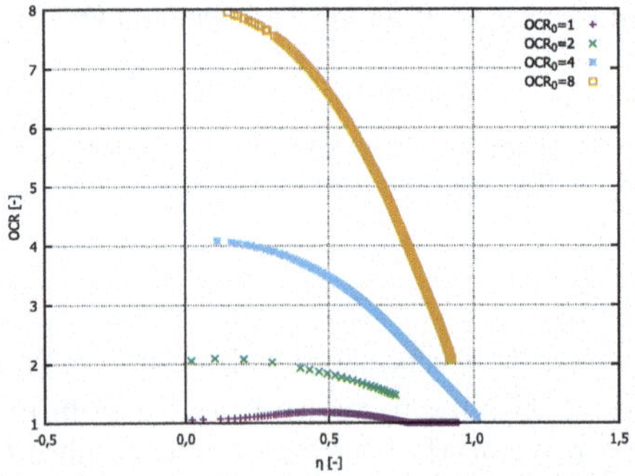

(e) OCR $-\ \eta$ space for the third approach OCR = p_e/p^+_{ei}.

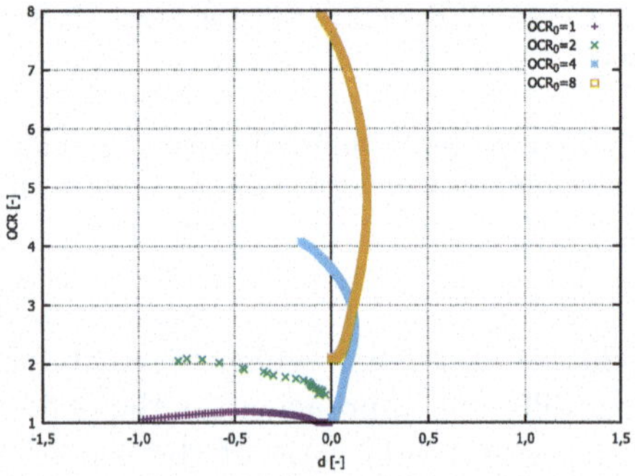

(f) OCR vs. d plane for the third approach OCR = p_e/p^+_{ei}.

Fig. 8. OCR$-\eta$ and OCR$-d$ relationships evaluated from monotonic undrained triaxial tests performed on Kaolin with different OCR definitions. (Experiments borrowed from [14]) deviatoric stress viz. $p_{ei}^{+} = p$. Figure 8c and d present the results for this approach in $\mathrm{OCR} = p_e/p$ vs. η space and in OCR $-\,d$ plane, respectively. An increase of OCR for the normalconsolidated sample can be observed during the shearing. If the postulate that normalconsolidated samples contract when subjected to undrained shearing holds, then with this approach the critical state line implying $\mathrm{OCR} = 1$ and $\eta = \eta_c$ will never be reached for initially normalconsolidated samples. More controversial is the fact that an initially normal consolidated sample becomes overconsolidated during undrained shearing. As one might expect, the initially overconsolidated samples are not able to reach the CSL for this evaluating approach neither. One may build a conceptual case when $\mathrm{OCR} = 1$ can be reached then the dilation must be so large that the mean pressure reachs its initial value $p \geq p_0$. To the authors knowledge this behaviour is not supported by any experimental evidence.

Finally, we suggest for the definition of p_{ei}^{+} the same relation we already proposed in [15,44]:

$$p_{ei}^{+} = \exp\left(\frac{e_{i0} - e^{+}}{\lambda}\right) \tag{38}$$

Therefore the loading surface relation from [15,44] is solved for the void ratio e^{+} corresponding to the actual stress state:

$$e^{+} = A^{1/n_f}\, e_i, \quad A = 1 - \left(\frac{|\eta|}{M f_{b0}}\right) \tag{39}$$

$$n_f = \frac{\ln(f_{b0}^2 - 1)/f_{b0}^2}{\ln(e_c/e_i)} \tag{40}$$

with the bounding surface material parameter f_{b0}. This formulation of the bounding surface and of OCR is also used for the explanations presented in Fig. 5b. The dilatancy relationships for Kaolin are once more for this approach illustrated in Fig. 8e and f. Hereby the consistency with the critical state surface is evident, hence the critical state is reached when $\mathrm{OCR} = 1$ is obtained. Note, that this limit state is reached only when $\eta = \eta_c$ (which arises also from the limit calculation $\lim\limits_{\eta \to \eta_c} \mathrm{OCR} = 1$). Thus $\mathrm{OCR} = 1$ represents a sufficient condition for the CSL! Fig. 8e and f indicate that for the sample with $\mathrm{OCR}_0 = 8$ the critical state is not reached yet which is in accordance with the documentation in [14]. In [42] Atkinson explained: "The critical state will normally be reached after strains greater than 10%."... At this stage "the movements of grains are essentially turbulent, involving relative movements and rotations of both clay and sand grains. At larger displacements, however, the strains become localized into distinct zones of intense shearing and the shear stresses applied to the clay soil decrease." Thus, it is likely that for this sample the critical state was not reached, but a sort of "laminar flow" of flat clay grains which became parallely oriented to a very thin shearing zone occured. Furthermore, the limits of the application of continuum theory are reached in this case.

Figures 7b, c and d present the above discussed relations in the three-dimensional space $d - \eta - \mathrm{OCR}$. Once more it is evident that the third approach presented in Fig. 7d renders both $d = 0$ and OCR $= 1$ at the critical state, following the first approach illustrated in Fig. 7b the CSL will never be reached and according to the second approach in Fig. 7c both the CSL is reached under controversial conditions and the development of OCR for example for initially normal consolidated samples is very unrealistic. Note that these observations along with Eqs. 37, 38, 38, 40 and $p_{ei} = p_0$ (for initially isotropically consolidated samples) represent a sufficient condition for the CSL, avoiding the experimental void ratio scatter. Thus, the OCR value calculated through Eq. 37 is a very good indicator whether the ultimate state is reached or not.

4.3 Triaxial Tests Under Cyclic Loading

A cyclic undrained triaxial test named C08 and presented in [14] is evaluated for the *dilatancy* relation in the following. With a constant rate of axial strain $\dot{\varepsilon}_1 = 0.12/\mathrm{h}$ and a deviatoric amplitude of $q_{ampl} = 70$ kPa the sample could withstand nearly 7 cycles until the failure criterion of $|\varepsilon_1| = 10\%$ is reached. For purposes of the *dilatancy* evaluation, the cycles are separated according to the direction of the shear strain increment $\Delta\gamma$ (separation points correspond to the loading direction reversals) in 0.5 units beginning with 0.25. An essential point (for all time-depentent materials) is that the shearing velocity was held constant during the test to $||\Delta\gamma|| = 0.0113 = const.$ corresponding to $\Delta\gamma = 0.0113$ at loading in axial compression (the sample is axially compressed) and $\Delta\gamma = -0.0113$ at unloading in axial compression (the sample is axially extended).

Two different approaches are used for the calculation of elasticity. First the isotropic hypoelasticity as described in [20] is adopted. The compression index λ and the swelling index κ are calibrated to fit the oedometric tests: λ as the slope of the virgin compression line and κ the average slope of the unloading-reloading hystheresis. The initial void ratio in the evaluation was adjusted to account for the initial overconsolidation ratio $e_0 = e_{i0} - \lambda \ln(p_0 \cdot \mathrm{OCR}_0)$. The authors of the experiments [14] reported normalconsolidated sample for C08, thus $\mathrm{OCR}_0 = 1$. The maximum void ratio e_{i0} was calibrated at a reference pressure of $p_{ref} = 1$ kPa for the OCR $= 1$ isotache, as described in [15,20,45]. Secondly, the transverse isotropic hypoelasticity as introduced by the authors in [15,44] is utilized. Therefore, the calibration of the anisotropic coefficient α is required. A calibration scheme of α in conjunction with the Poisson ratio ν is also given in [15,44]. Table 1 lists the parameters' values used for the evaluation of the elasticity in C08.

Finally, Fig. 9 shows the resulting *dilatancy* relations. Note that throughout all these figures $d = -d\varepsilon_v^d/d\gamma^{tot}$ has been used in order to provide a more comparative study with the results of Pradhan et al. for sand [5]. To obtain the relation described in Eq. 6 a factor of 3/2 ($\varepsilon_q = 2/3\gamma$ for triaxial conditions) should be introduced which would solely elongate the *dilatancy* curves. In absence of $p' = const.$ tests with intermediate small loops of unloading and reloading, γ^{tot} presents a good approximation for γ^p. Further research in this point is required. The figures on the left side (Figs. 9a, c and e) accounted for

Table 1. Parameters used for the elasticity evaluation in C08

Elasticity approach	ν [–]	α [–]	λ [–]	κ [–]	e_{i0} [–]
Hypoelasticity	0.3	–	0.13	0.05	1.76
Transverse isotropic hypoelasticity	0.3	1.8	0.13	0.05	1.76

hypoelasticity, while the figures on the right side (Figs. 9b, d and f) for the transverse isotropic hypoelasticity. It is to be noted, that when the loading direction was reversed the different approaches did not yield the same behaviour. By comparing the Figs. 9a, b, c, d, e and f the following points may be noticed.

- Considering all cycles Figs. 9a and c (hypoelasticity used), b and d (transverse isotropic hypoelasticity used) the $d - \eta$ relation is different among different loading cycles, especially at loading reversals in axial extension. Differences between the elasticity approaches may be observed intensively comparing Figs. 9a or c with b or d, respectively. While using the hypoelasticity the points at loading reversals from axial compression to axial extension jump all to the dilatant area. Yet, Kaolin turned out to be transverse isotropic as showed in [15,20,44], thus the usage of transverse isotrope hypoelasticity would provide a more representative approach for Kaolin. Then, it can be observed that among some data scatter the loading reversals from axial compression to axial extension jump to the contractant area as observed also for sands. Note that no data points has been omitted during these evaluations.
- Figures 9e and f account only for regular cycles. The scattering in the relation is no longer present. As discussed also before it is now in fact more obvious that accounting for hypoelasticity the loading reversal $\Delta\gamma > 0 \rightarrow \Delta\gamma < 0$ renders dilatancy, whereas using transverse isotropic hypoelasticity it results in dilatancy, which is conform with the findings for sand in [5]. The shape of the curves is also different comparing both elasticity approaches. It may be concluded once more that the transverse isotropic hypoelasticity corresponds better to the behaviour of Kaolin.
- Pradhan et al. [5,10,21] claimed that using $d\gamma^p$ instead of $d\gamma^{tot}$ would reduce the scattering in the relations for sand. We however observe a scattering in the first (non regular cycles), see Figs. 9e and f. The scattering in the first and intermediate cycles can possibly be reduced by using $d\gamma^p$.
- The "elastic dilatancy" in reversals evident for example in Fig. 9c, which is eliminated using the transverse isotropic hypoelasticity, see for example Fig. 9d, provokes an increase of the effective mean stress resulting in a slope to the upper left of the effective stress path [14,15,44]. It is well known that clays possess a greater elastic locus compared to sands. If the elasticity was isotropic than the described $p - q$ path would be vertical with respect to $p-$axis. Thus, we can call the inherent anisotropy responsible for an even lower excessive pore water pressure after loading reversal. Sands in contrast react with the "highest" contraction after loading reversal. Hence, the greater

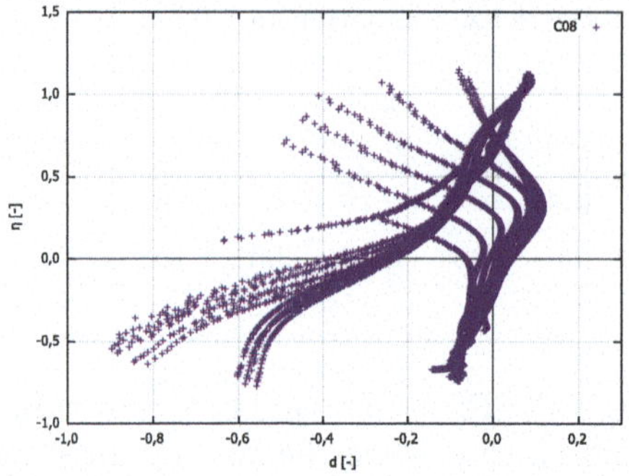

(a) All cycles. It was accounted for the isotropic hypoelasticity.

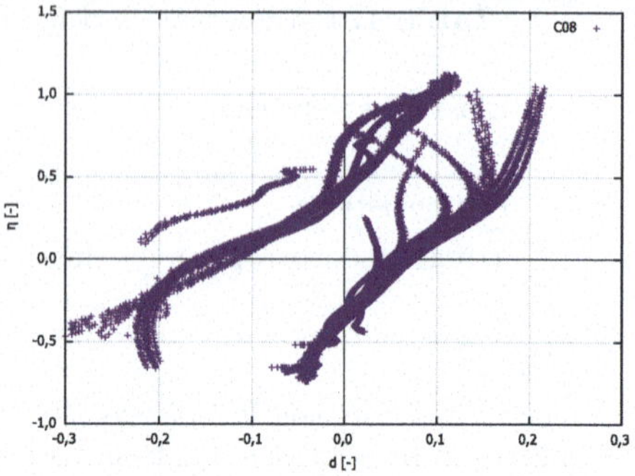

(b) All cycles. It was accounted for the transverse hypoelasticity.

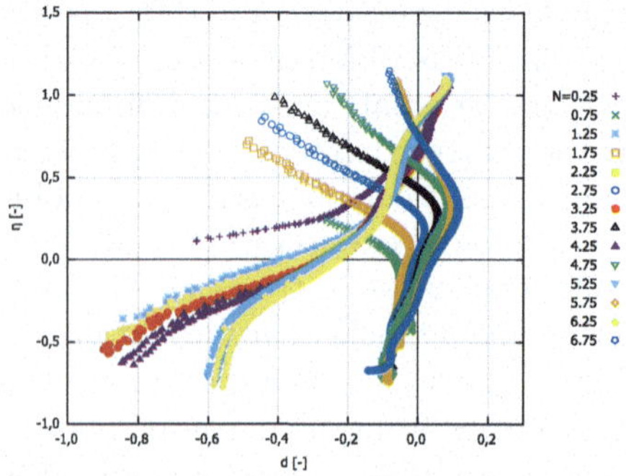

(c) Cycles separated in 0.5 cycles. It was accounted for the isotropic hypoelasticity.

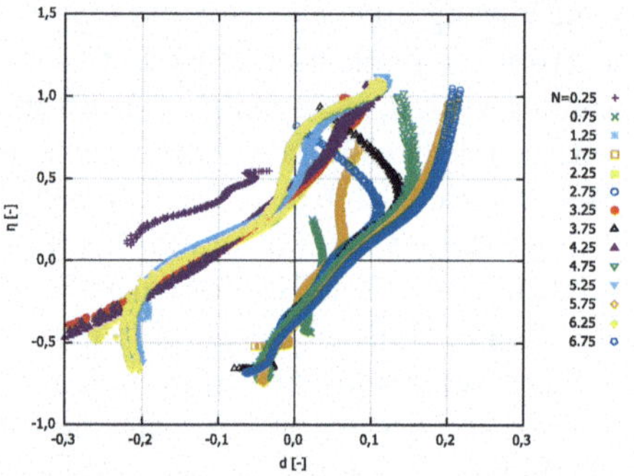

(d) Cycles separated in 0.5 cycles. It was accounted for the transverse hypoelasticity.

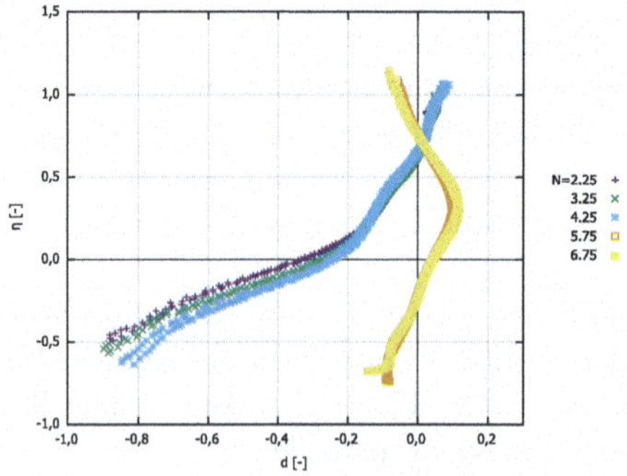

(e) Only regular cycles are plotted. It was accounted for the isotropic hypoelasticity.

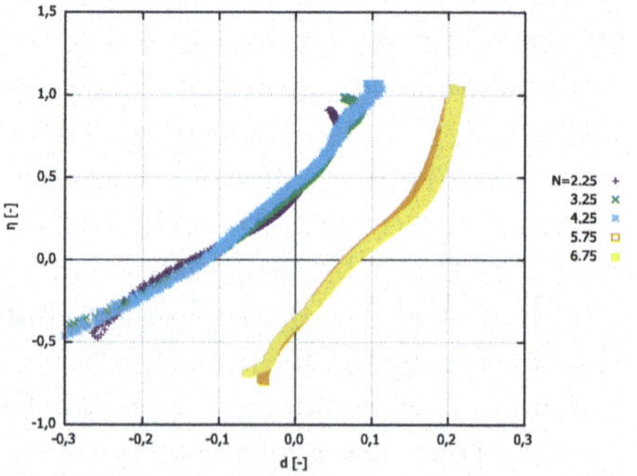

(f) Only regular cycles are plotted. It was accounted for the transverse hypoelasticity.

Fig. 9. d-η relationship evaluated for C08. (Experiment borrowed from [14])

elastic regime cannot be overcomed through the contractancy and thus results in a non vanishing mean pressure at cyclic mobility $p' \neq 0$. This effect can be implified with the inherent anisotropy according to the sedimentation axis. Thus, a cutting direction of the sample may be found at which the soft soil would also liquify (weak axis). The last assumption requires further research. Some authors explain the non-liquefaction behaviour of clays with the viscosity and its cohesive effect. If this was supposed to be truth, than the intensity of creep would not vanish with higher over consolidation ratio as has been experimentally documented in [28,46]. On the other side, it implies the existence of a special strain rate (loading velocity) with which the cyclic undrained shearing of a clay sample would result in liquefaction. Following our theory this would be the case for the greatest velocity $\dot{\varepsilon}_1 \to \infty$, hence the viscous effects wouldn't have time to develope. The experimentes presented in [14] and the discussions made in [44] give a hint to this phenomenon. However, further research work is required in order to bring more light and explain this phenomenon.
- With the undrained cyclic loading of normal consolidated samples the overconsolidation ratio increases and the stress ratios at the phase transformation line M_{dc} and M_{de} from Eq. 16 appears to be not affected as illustrated in Figs. 9e and f. Considering the non-regular cycles, Figs. 9a–d, η_{PTL} seems to increase with the number of cycles in axial extension. Such effects has not been reported in the literature, moreover the monotonic tests illustrated in Figs. 8 and 7 show opposite behaviour (but only for axial compression, no monotonic tests were available for axial extension). Further research works are required at this point.

In general, we can conclude that following the experimental evidence gained on Kaolin in [14] and evaluated in this work the *dilatancy* is a function of the stress ratio η and the void ratio e along with the intrinsic material parameters summarized under C, hence $d = f(\eta, e, C)$. Furthermore, we proposed an OCR-definition in Eq. 37, which includes the influence of both the stress ratio and void ratio such that $d = f(\text{OCR}, C)$.

If the assumption $d = f(\eta, C)$ would hold, the direction of plastic flow influencing the direction of the undrained stress path would be able to render dilatant direction only for $\eta = M \Rightarrow M_d = M$ independent of the overconsolidation ratio of the material. Thus, the same behaviour would be obtained for each OCR contradicting the findings gained in experiments and explained in this Section as illustrated in Figs. 7, 8 and 9.

5 Constitutive Description of the Dilatancy of Cohesive Soils

Perusal of actual research works regarding constitutive modeling of soft soils suggest that for clays a unique relation between the *dilatancy* d and the stress ratio η exist. This relation was used by the Cam Clay theory $d = M - \eta$ in [25] or $d = (M^2 - \eta^2)/\eta$ [26] resulting in a dilative response only when the stress ratio reaches the critical state line thus $M_d = M$. Similarly, the hypoplastic model developed by Masin [27] is able to reproduce dilatant behaviour only for states lying above the critical state. The usage of a unique relation between d and η is reported as satisfactorily for cohesive soils by some authors [6,27]. This assumption may be consistent for normalconsolidated clays subjected to monotonic loading. However, we showed in Sects. 4.2 and 4.3 that the overcritical dilation is not sufficiently described for general initial states of cohesive soils and the less for cyclic loading.

5.1 Incorporation in a General Hypoplastic Model

We start with the *dilatancy* relation given in Eq. 6. Note the difference between this definition of d and the one determined by Pradhan et al. in [5,10] $d = -d\varepsilon_v^p/|d\gamma^p|$ with $d\gamma = 3/2 d\varepsilon_q$. For multiaxial generalization purposes we will however stick to our formulation, Eq. 6, which can be rearranged to:

$$d = -\frac{\mathrm{tr}(\dot{\varepsilon}^p)}{\sqrt{2/3}\,||\dot{\varepsilon}^{*p}||} \quad \Rightarrow \quad -\mathbf{1} : \dot{\varepsilon}^p = d\sqrt{2/3}\,||\dot{\varepsilon}^{*p}|| \tag{41}$$

with the deviator of the plastic strain rate $\dot{\varepsilon}^{*p}$. The *dilatacy* evaluated on sands rendered a similar relationship when using $d\varepsilon_q$ instead of $d\varepsilon_q^p$. In particular, eventhough the scattering of the data is reduced when using $d\varepsilon_q^p$, it was very difficult to evaluate d after the loading reversal, because when both values $d\varepsilon_v^p$ and $d\varepsilon_q^p$ are very small $||\dot{\varepsilon}|| < 10^{-5}$ it can be defined also as completely elastic regime [5]. Similar effects are shown in Sect. 4.3 for the behaviour of clays. Moreover, for the evaluations presented in Figs. 7, 8 and 9 we used the total deviatoric strain rate instead of the plastic one, thus in Eq. 41 we can substitute $||\dot{\varepsilon}^{*p}|| \approx ||\dot{\varepsilon}^{*}||$ to obtain:

$$-\mathbf{1} : \dot{\varepsilon}^p = d\sqrt{2/3}\,||\dot{\varepsilon}^{*}||. \tag{42}$$

From the point of view of a constitutive equation, the plastic strain rate has to be identified from Eq. 41. We start by expressing $\dot{\varepsilon}^p$ by its isotropic and deviatoric components with unknown magnitudes a and b respectively:

$$\dot{\varepsilon}^p = a\mathbf{1} + b\,\overrightarrow{\dot{\varepsilon}^{*p}} \tag{43}$$

By multiplying both sides from left of Eq. 43 with 1 and inserting Eq. 42 we obtain for the isotropic magnitude a:

$$\mathbf{1} : \dot{\varepsilon}^p = 3a \overset{42}{=} -d\sqrt{2/3}\,||\dot{\varepsilon}^{*}|| \tag{44}$$

$$\Rightarrow a = -\frac{1}{3}\sqrt{\frac{2}{3}}\,d\,||\dot{\varepsilon}^{*}||. \tag{45}$$

Inserting Eq. 45 and $b = ||\dot{\varepsilon}^{*p}|| \approx ||\dot{\varepsilon}^{*}||$ in Eq. 43 and dividing both sides of Eq. 43 with $||\dot{\varepsilon}^{*}||$ renders the following relation for $\dot{\varepsilon}^p$:

$$\frac{\dot{\varepsilon}^p}{||\dot{\varepsilon}^{*}||} = -\frac{1}{3}\sqrt{\frac{2}{3}}\,d\,\mathbf{1} + \overrightarrow{\dot{\varepsilon}^{*}}\,. \tag{46}$$

For the derivation of Eq. 46 the assumption $\overrightarrow{\dot{\varepsilon}^{*p}} \approx \overrightarrow{\dot{\varepsilon}^{*}}$ was involved as well. Finally the well known hypoplastic constitutive equation:

$$\dot{\boldsymbol{\sigma}} = \mathsf{E} : \left(\dot{\varepsilon} - \dot{\varepsilon}^{hp}\right) = \mathsf{E}\left(\dot{\varepsilon} - Y\mathbf{m}||\dot{\varepsilon}||\right) \tag{47}$$

is considered. The viscous strain rate is without loss of generality disregarded in respect thereof. Drawing parallels between the plastic (Eq. 46) and the hypoplastic strain rate (Eq. 47) explicit relations for the flow rule and the degree of nonlinearity are obtained:

$$\mathbf{m} = \overrightarrow{\left(-\frac{1}{3}\sqrt{\frac{2}{3}}\,d\,\mathbf{1} + \overrightarrow{\dot{\varepsilon}^{*}}\right)} \tag{48}$$

$$Y_D = \frac{||\dot{\varepsilon}^{*}||}{||\dot{\varepsilon}||}\left|\left|-\frac{1}{3}\sqrt{\frac{2}{3}}\,d\,\mathbf{1} + \overrightarrow{\dot{\varepsilon}^{*}}\right|\right|. \tag{49}$$

The last relation defining Y can be simplified to read:

$$Y_D = \sqrt{1 - \frac{1}{3}\frac{\mathrm{I}_1^2(\dot{\varepsilon})}{\mathrm{I}_1(\dot{\varepsilon}^2)}}\left|\left|-\frac{1}{3}\sqrt{\frac{2}{3}}\,d\,\mathbf{1} + \overrightarrow{\dot{\varepsilon}^{*}}\right|\right| \tag{50}$$

Note that the relation 50 serves only to define the degree of nonlinearity for shearing and should not be mixed up or used for the isotropic or radial compression. For radial compression Y_I the relations proposed in [15,20,44,47] or in [24] can be used for clays or sands, respectively. An interpolation function between Y_I for isotropic states and radial loading directions and Y_D for shearing is convenient.

A fundamental drawback of the flow rule presented in Eq. 48 is the deviatoric direction. For hypoplastic models a deviatoric direction of the flow rule equally to the deviatoric direction of the strain rate would mean that at loading reversals from stress states lying at the critical state the flow rule would try to get the stress state outside the critical state (or the bounding surface if introduced). The FE-simulation would loose the controllability and abort the calculation. Moreover, experimental observations show indeed deviatoric flow direction at the critical state but corresponding to the direction of the deviatoric stress. For this purpose a modified flow rule for hypoplastic models will be used:

$$\mathbf{m} = \overrightarrow{\left(-\frac{1}{3}\sqrt{\frac{2}{3}}\,d\,\mathbf{1} + \left(\frac{\overrightarrow{\boldsymbol{\sigma}^{*}}}{p}\right)\right)} \tag{51}$$

For the stiffness tensor E in Eq. 47 the hypoelastic stiffness from [15,44,47] is adopted. The equations of the reference model presented in [44] which are ommitted in this section are therefore summarized in the Appendix.

5.2 Simulations

Now the simulations compared with the experimental results (borrowed from [14]) which are reviewed, evaluated and discussed in Sects. 4.2 and 4.3 are shown. For the different tests the same set of material parameters, listed in Table 2, has been used.

Table 2. Material parameters

ν	α	λ	κ	e_{i0}	M_c	f_{b0}	n_{PTL}	n_{OCR}	χ	I_v
0.3	1.8	0.13	0.05	1.76	1.0	1.5	3	2	20	0.025

Figure 10 presents the simulations of the monotonic undrained triaxial tests with different initial overconsolidation ratios $\text{OCR}_0 = \{1, 2, 4, 8\}$, whereby the dilatant behaviour is evident for higher $\text{OCR}_0 > 2$. These tests are also used for the evaluations in Sect. 4.2. The simulations show very good agreement with the experimental results in both effective stress paths (right figure) and deviatoric stress vs. axial strain space (left figure). The shearing at strains greater than approximately 12% can provoke inhomogenity of the sample and localization of shear strains, thus it was not considered by the model. In the $q - \varepsilon_1$ space it corresponds to the softening of the deviatoric stress evident in the experimental curves. Furthermore, the inherent anisotropy of Kaolin is captured well by the model as can be seen from the simulation for $\text{OCR}_0 = 1$.

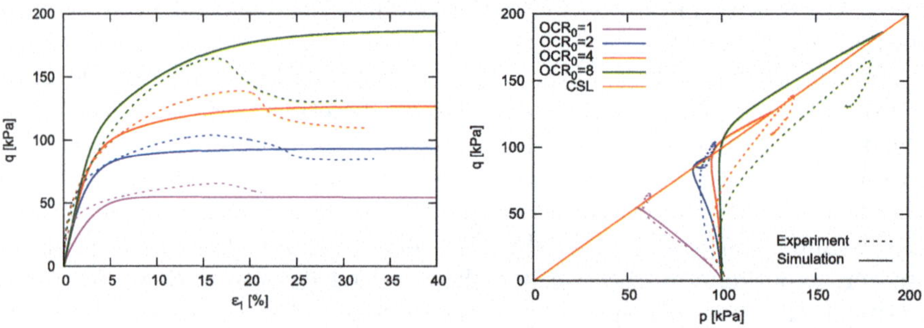

Fig. 10. Experimental results from [14] and simulations with the proposed model of undrained triaxial tests with variation of the initial overconsolidation ratio $\text{OCR}_0 = \{1, 2, 4, 8\}$. The displacement rate was held constant in each test to $\dot{s} = 0.025\,\text{mm/min}$. The laboratory tests are presented with dashed lines, whereas the simulations with solid lines.

Figure 11 presents an undrained triaxial cyclic test of a normalconsolidated sample $\text{OCR}_0 = 1$ with deviatoric amplitude of $q_{ampl} = 70\,\text{kPa}$. The failure criterion for the experiment was defined as an axial strain of 10%. The simulation in $q-\varepsilon_1$ space is in good accordance with the experimental findings. Furthermore, the hysteresis when reaching the cyclic mobility is satisfactorily described. The same conclusions can be drawn for the effective stress path whereby 7 cycles has been surpassed. The 8-shaped cyclic mobility of the $p-q$ path is in good agreement with the experiment.

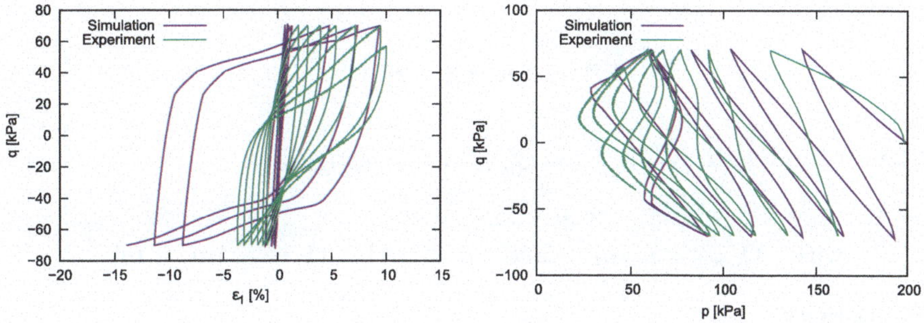

Fig. 11. Experimental result from [14] and simulation with the proposed model of a cyclic undrained triaxial test with constant displacement rate $\dot{s} = 0.1\,\text{mm/min}$ and deviatoric amplitude of $q_{ampl} = 70\,\text{kPa}$.

6 Concluding Remarks

This work provides a new evaluation method of the *dilatancy* from monotonic and cyclic undrained triaxial tests. Herein it has been realized for the experiments performed on Kaolin [14]. In general we can conclude that leastwise for this soft soil the *dilatancy* is a function of the stress ratio η and the void ratio e along with the intrinsic material parameters (as observed for sands). Furthermore, we proposed an OCR-definition, which includes the influence of both the stress ratio and void ratio such that $d = f(\text{OCR}, C)$. We suggest that there is an overconsolidation ratio OCR_{ci} at which the soft soil behaviour changes from contractant $\text{OCR} < \text{OCR}_{ci}$ to dilatant (the material can both contract and dilate depending on η) $\text{OCR} > \text{OCR}_{ci}$ with the PTL lying below the CSL, as was also shown in the present study. We further proposed a constitutive relation describing the behaviour of soft soils including the *dilatancy* and viscosity. Some simulations of monotonic as well as cyclic tests are shown.

In order to verify these results for plastic clays, soft soils with higher plasticity should be evaluated in a subsequent work. In addition laboratory tests with constant mean pressure are required for the improvement of the *dilatancy* relation for cohesive soils.

Acknowledgements. The financial support from the German Research Community (DFG TR 218/27-1) is herewith gratefully acknowledged.

Appendix

Constitutive model

Constitutive equation:

$$\dot{\boldsymbol{\sigma}} = \mathsf{E} : \left(\dot{\boldsymbol{\varepsilon}} - \dot{\boldsymbol{\varepsilon}}^{hp} - \dot{\boldsymbol{\varepsilon}}^{vis}\right) \tag{52}$$

Elasticity:

$$\mathsf{E}^{iso} = 3K \overrightarrow{\mathbf{1}}\overrightarrow{\mathbf{1}} + 2G\, \mathsf{I}^{dev} \tag{53}$$

$$\mathsf{E}_{abcd} = \mathsf{Q}_{abij} : \mathsf{E}_{ijkl} : \mathsf{Q}_{klcd} \tag{54}$$

$$\text{with} \quad \mathsf{Q}_{abcd} = \mu_{ac}\mu_{bd} \quad \text{and} \quad \mu = \sqrt{\alpha}\mathbf{1} + (1 - \sqrt{\alpha})\mathbf{m}_{\text{s}} \otimes \mathbf{m}_{\text{s}}. \tag{55}$$

Hypoplasticity:

$$\dot{\boldsymbol{\varepsilon}}^{hp} = Y\, \mathbf{m}\, ||\dot{\boldsymbol{\varepsilon}}||, \tag{56}$$

$$Y = [Y_0 + (1 - Y_0)\, Y_D]\ \text{OCR}^{-n_{OCR}}, \quad Y_0 = \frac{\lambda - \kappa}{\lambda + \kappa}\left(\frac{p_{ei}}{p}\right)^2, \tag{57}$$

$$\mathbf{m} = \overrightarrow{\left(-\frac{1}{3}\sqrt{\frac{2}{3}}\, d\, \mathbf{1} + \overrightarrow{\boldsymbol{\sigma}^*}\right)} \tag{58}$$

Viscosity:

$$\dot{\boldsymbol{\varepsilon}}^{vis} = I_v\, \lambda \left(\frac{1}{\text{OCR}}\right)^{1/I_v} \mathbf{m}, \tag{59}$$

$$\text{OCR} = \frac{p_{ei}}{p_{ei}^+}, \quad p_{ei}^+ = \exp\left(\frac{e_{i0} - e^+}{\lambda}\right) \tag{60}$$

$$e^+ = A^{1/n_f}\, e_i, \quad A = 1 - \left(\frac{|\eta|}{M f_{b0}}\right), \quad n_f = \frac{\ln(f_{b0}^2 - 1)/f_{b0}^2}{\ln(e_c/e_i)} \tag{61}$$

References

1. Andersen, K.H.: Bearing capacity under cyclic loading - offshore, along the coast, and on land. Can. Geotech. J. **46**(5), 513–535 (2009)
2. Cattoni, E., Tamagnini, C.: On the seismic response of a propped r.c. diaphragm wall in a saturated clay. Acta Geotech. (2019). https://doi.org/10.1007/s11440-019-00771-4
3. Taylor, D.W.: Fundamentals of Soil Mechanics. Wiley, New York (1954)
4. Rowe, P., Taylor, G.I.: The stress-dilatancy relation for static equilibrium of an assembly of particles in contact. Proc. Roy. Soc. Lond. Math. Phys. Sci. **269**, 500–527 (1962)
5. Pradhan, T.B.S., Tatsuoka, F., Sato, Y.: Experimental stress-dilatancy relations of sand subjected to cyclic loading. Soils Found. **29**(1), 45–64 (1989)

6. Li, X.S., Dafalias, Y.F.: Dilatancy for cohesionless soils. Géotechnique **50**(4), 449–460 (2000)
7. Yang, Z., Xu, T.: J2-deformation type model coupled with state dependent dilatancy. Comput. Geotech. **105**, 129–141 (2019)
8. Wan, R.G., Guo, P.J.: Stress dilatancy and fabric dependence on sand behavior. J. Eng. Mech. **160**(6), 635–645 (2004)
9. Wichtmann, T.: Soil behaviour under cyclic loading - experimental observations, constitutive description and applications. Veröffentlichungen des Instituts für Bodenmechanik und Felsmechanik am KIT, Habilitation, no. 181 (2016)
10. Pradhan, T.B.S., Tatsuoka, F.: On stress-dilatancy equations of sand subjected to cyclic loading. Soils Found. **29**(1), 65–81 (1989)
11. Verdugo, R., Ishihara, K.: The steady state of sandy soils. Soils Found. **36**(2), 65–81 (1996)
12. Coop, M.R., Atkinson, J.H., Taylor, R.N.: Strength and stiffness of structured and unstructured soils. In: Proceedings of the 11th European Conference on Soil Mechanics and Foundation Engineering, Copenhagen, vol. 1, pp. 55–62 (1995)
13. Amorosi, A., Rampello, S.: An experimental investigation into the mechanical behaviour of a structured stiff clay. Géotechnique **57**(2), 153–166 (2007)
14. Wichtmann, T., Triantafyllidis, T.: Monotonic and cyclic tests on kaolin: a database for the development, calibration and verification of constitutive models for cohesive soils with focus to cyclic loading. Acta Geotech. **13**(5), 1103–1128 (2017)
15. Tafili, M., Triantafyllidis, T.: A simple hypoplastic model with loading surface accounting for viscous and fabric effects of clays. Submitted for IJNAMG (2017)
16. Matsuoka, H., Nakai, T.: A new failure for soils in three-dimensional stresses. In: Proceedings of the IUTAM Symposium on Deformation and Failure of Granular Materials, Delft, pp. 253–263 (1982)
17. Argyris, J.H., Faust, G., Szimmat, J., Warnke, E.P., Willam, K.J.: Recent developments in the finite element analysis of prestressed concrete reactor vessels. Nucl. Eng. Des. **282**(1), 42–75 (1974)
18. Taiebat, M., Dafalias, Y.F.: SANISAND: simple anisotropic sand plasticity model. Int. J. Numer. Anal. Meth. Geomech. **32**, 915–948 (2008)
19. Fuentes, W., Triantafyllidis, T.: ISA model: a constitutive model for soils with yield surface in the intergranular strain space. Int. J. Numer. Anal. Meth. Geomech. **39**(11), 1235–1254 (2015)
20. Fuentes, W., Tafili, M., Triantafyllidis, T.: An ISA-plasticity-based model for viscous and non-viscous clays. Acta Geotech. **13**(2), 367–386 (2017)
21. Pradhan, T.B.S., Tatsuoka, F., Mohri, Y., Sato, Y.: An automated triaxial testing system using a simple triaxial cell for soils. Soils Found. **29**(1), 151–160 (1989)
22. Ishihara, K., Tatsuoka, F., Yasuda, S.: Undrained deformation and liquefaction of sand under cyclic stresses. Soils Found. **15**(1), 29–44 (1975)
23. Dafalias, Y.F., Manzari, M.T.: Simple plasticity sand model accounting for fabric change effects. J. Eng. Mech. **130**(6), 622–634 (2004)
24. Grandas, C., Triantafyllidis, T.: Personal communication. Institute of Soil Mechanics and Rock Mechanics (IBF, KIT) (2019)
25. Roscoe, K.H., Schofield, A.N.: Mechanical behaviour of an idealized 'wet' clay. In: Proceedings of the European Conference on Soil Mechanics and Foundation Engineering, Wiesbaden, vol. 1, pp. 47–54 (1963)
26. Roscoe, K.H., Burland, J.B.: On the generalized stress-strain behaviour of 'wet' clay. Engineering plasticity, pp. 535–609 (1968)
27. Masin, D.: Hypoplastic models for fine-grained soils. Ph.D. thesis, Charles University, Prague (2006)

28. Niemunis, A.: Extended hypoplastic models for soils. Habilitation, Monografia 34, Ruhr-University Bochum (2003)
29. Niemunis, A., Grandas, C., Prada, L.: Anisotropic Visco-hypoplasticity. Acta Geotech. **4**(4), 293–314 (2009)
30. Finno, R.J., Chung, C.K.: Stress-strain-strength responses of compressible Chicago glacial clays. J. Geotech. Eng. ASCE **118**(10), 1607–1625 (1992)
31. Yasuhara, K., Hirao, K., Hyde, A.: Effects of cyclic loading on undrained strength and compressibility of clay. Soils Found. **32**(1), 100–116 (1992)
32. Bohac, J., Herle, I., Feda, J., Klablena, P.: Properties of fissured Brno clay. In: Proceedings of the 11th European Conference on Soil Mechanics and Foundation Engineering, Copenhagen, vol. 3, pp. 19–24 (1995)
33. Sheahan, T.C., Ladd, C.C., Germaine, J.T.: Rate dependentundrained shear behavior of saturated clay. J. Geotech. Eng. ASCE **122**(2), 99–108 (1996)
34. Burland, J.B., Rampello, S., Georgiannou, V.N., Calabresi, G.: A laboratory study of the strength of four stiff clays. Géotechnique **46**(3), 491–514 (1996)
35. Rampello, S., Callisto, L.: A study on the subsoil of the tower of Pisa based on results from standard and high-quality samples. Can. Geotech. J. **35**(6), 1074–1092 (1998)
36. Sivakumar, V., Doran, I.G., Graham, J.: Particle orientation and its influence on the mechanical behavior of isotropically consolidated reconstituted clay. Eng. Geol. **66**, 197–209 (2002)
37. Gasparre, A., Nishimura, S., Coop, M.R., Jardine, R.J.: The influence of structure on the behavior of London clay. Géotechnique **57**(1), 19–31 (2007)
38. Chu, D.B., Stewart, J.P., Boulanger, R.W., Lin, P.S.: Cyclic softening of low-plasticity clay and its effect on seismic foundation performance. J. Geotech. Geoenviron. Eng. ASCE **134**(11), 1595–1608 (2008)
39. Abdulhadi, N.O., Germaine, J.T., Whittle, A.J.: Stress-dependent behavior of saturated clay. Can. Geotech. J. **49**, 907–916 (2012)
40. Duong, N.T., Suzuki, M., Hai, N.V.: Rate and acceleration effects on residual strength of kaolin and kaolin-bentonite mixtures in ring shearing. Soils Found. **58**, 1153–1172 (2018)
41. Hvorslev, M.: Physical components of the shear strength of saturated clays. In: ASCE Research Conference, Shear Strength of Cohesive Soils, Boulder Colorado (1960)
42. Atkinson, J.: The Mechanics of Soils and Foundations. McGraw-Hill, New York (1993)
43. Shibata, T.: On the volume changes of normally-consolidated clays. In: Annuals, Disaster Prevention Research Institute, Kyoto (1963). (in Chinese)
44. Tafili, M., Triantafyllidis, T.: AVISA: anisotropic visco ISA model and its performance at cyclic loading. Submitted for Acta Geotech. (2019)
45. Tafili, M., Triantafyllidis, T.: Constitutive model for viscous clays under the ISA framework. In: Triantafyllidis, T. (ed.) Holistic Simulation of Geotechnical Installation Processes - Theoretical Results and Applications, vol. 82, pp. 324–340. Springer, Heidelberg (2017)
46. Niemunis, A., Krieg, S.: Viscous behaviour of soils under oedometric conditions. Can. Geotech. J. **33**, 159–168 (1996)
47. Fuentes, W., Hadzibeti, M., Triantafyllidis, T.: Constitutive model for clays under the ISA framework. In: Triantafyllidis, T. (ed.) Holistic Simulation of Geotechnical Installation Processes - Benchmarks and Simulations, vol. 80, pp. 115–130. Springer, Heidelberg (2015)

On FE Modeling of Two-Phase Media Using the *up*-Discretization

Andrzej Niemunis(✉)

Institute of Soil Mechanics and Rock Mechanics (IBF), Karlsruhe Institute of Technology (KIT), Karlsruhe, Germany
andrzej.niemunis@kit.edu

Abstract. In saturated soils the difference between the acceleration of soil skeleton and water is usually disregarded. In this paper it is reintroduced in a simplified manner into the balance equation and into the conservation of mass. This leads to a symmetric element matrix. Also the inertial interaction force associated with this difference are taken into account. It turns out, however, that numerical errors due to the standard Galerkin spatial discretization and due to the usual trapezoidal rule of Newmark time integration scheme blur the physically motivated extension. The conventional FEM is insufficient to calculate boundary value problems of dynamic nature even if displacements is better interpolated than pressure. Stabilization techniques and discrete Galerkin formulation seem to be indispensable to judge about effects of the proposed extension.

1 Introduction

We consider water-saturated granular media, $S \equiv 1$, and assume a simple constitutive law $T_{ij} = T^0_{ij} + E_{ijkl}u_{k,l} + C_{ijkl}\dot{u}_{k,l}$ with a homogeneous stiffness E_{ijkl} of skeleton and its viscosity tensor C_{ijkl}. The bulk densities of quartz and water are $\rho_s(x_i, t) = \text{const}$ and $\rho_w(x_i, t) = \text{const}$. Grains and water are almost incompressible (as a substance). The soil skeleton (arrangement of grains) is compressible i.e. $E_{iikk} < \infty$. For such medium we formulate the condition of equilibrium. Due to coupling between to the pore water pressure $p_0(x_i) + p(x_i, t)$ and volumetric deformation rate $\dot{\epsilon}_{ii} = \dot{u}_{i,i}$ of the skeleton, the equilibrium and water transport must be solved jointly. According to the simplified Biot [1] theory, we have the following system of PDE

$$\begin{cases} T_{ij,j} - (p_0 + p),_{\,i} + \rho g_i &= \rho \ddot{u}_i \\ w_{i,i} + \frac{1}{Q}\,\dot{p} + \dot{u}_{i,i} &= 0 \end{cases} \tag{1}$$

wherein $Q = (n/K_f + (1-n)/K_s)^{-1} \approx \infty$ denotes the compressibility of the substance. We call T^0_{ij} the geostatic stress and p_0 the hydrostatic pore pressure. The seepage velocity w_i is defined as the difference between the average velocities of water $\dot{U}_i$ and of skeleton $\dot{u}_i$

T. Triantafyllidis (Ed.): *Recent Developments of Soil Mechanics and Geotechnics in Theory and Practice*, LNACM 91, pp. 85–96, 2020.
https://doi.org/10.1007/978-3-030-28516-6_5

$$w_i = n(\dot{U}_i - \dot{u}_i) \tag{2}$$

where n denotes the porosity. The classical *up*-formulation (1) uses the displacement $u_i(x_j, t)$ of the skeleton and the pore water overpressure[1] $p(x_i, t)$ as the primary unknown fields. The simplification $\ddot{U}_i \approx \ddot{u}_i$ equating the acceleration of skeleton $\ddot{u}_i$ and the acceleration of water $\ddot{U}_i$ is attributed to Zienkiewicz [5]. Hence we could have written $\rho\ddot{u}_i$ instead of $\rho_d\ddot{u}_i + n\rho_w\ddot{U}_i$ in $(1)_1$. Moreover, $(1)_1$ disregards the so-called inertial coupling originally postulated by Biot. Both $\ddot{U}_i \neq \ddot{u}_i$ and the inertial coupling will be taken into account in (3).

1.1 Notation

A fixed orthogonal Cartesian coordinate system x_1, x_2, x_3 is used. A repeated (dummy) indices, Greek = nodal or Latin = Cartesian, imply summation. Frequently used variables are listed below

$\dot{\sqcup}, \ddot{\sqcup}$	first and second time derivatives of $\sqcup$
$\sqcup_n \sqcup_{n+1}$	value of $\sqcup$ at the beggining/end of an increment
$\sqcup_{,i} \sqcup_{,x}$	spaetial derivatives of $\sqcup$
$\breve{\sqcup}$	prescribed value of $\sqcup$,usually a boundary condition
β_1, β_2	Newmark integration parameters
Δt	time increment
ρ_s, ρ_w	intrinsic density of quarz and water
ρ_d	dry bulk density of soil
ρ_a	additional mass (density) for inertial coupling
τ	tortuosity
ψ	test function in the weak form of mass conservation
$\Delta t_i^{\text{tot}}, \Delta t^{\text{tot}}$	traction of a boundary
$b_i^{\diamond}$	buoyancy force due to total pressure gradient
C_{ijkl}, C	viscosity of soil skeleton
E_{ijkl}, E	elastic stiffness
g_i	specific gravity force
$j_i^{\diamond}$	seepage force due to viscous flow
K_f, K_s	bulk modulae of fluid and skeleton
$\mathbf{K}_{UU}, \mathbf{K}_{UP} \ldots$	blocks of element matrix der FEM
k_{ij}, k	specific permeability
n	porosity
$N_\beta, \bar{N}_\gamma$	interpolation functions for displacement or overpressure
p, p_0	overpressure and hydrostatic portion of pore pressure
p_γ	nodal value of overpressure
Q	bulk stiffness of the (undrained) substance
$\mathbf{R}_U, \mathbf{R}_P \ldots$	portions of load in FEM

[1] In p compression is positive.

S	degree of saturation
T_{ij}, T	stress (tension positive)
T^0_{ij}, T^0	geostatic stress
u_i, U_i	displacement of soil and water
u_α	nodal value of soil displacement
v	test function in the weak form of equilibrium
w_i	seepage velocity

1.2 Strong Form of the Equilibrium and the Mass Conservation

In the following we allow for $\ddot{u}_i \neq \ddot{U}_i$. The equilibrium equations are written for individual phases including three interaction forces: static $b^\diamond$, viscous $j^\diamond$and inertial $\rho_a(\ddot{u}_i - \ddot{U}_i)$.

$$\begin{cases} T_{ij,j} + (1-n)\rho_s g_i + b^\diamond + j^\diamond & = \rho_d \ddot{u}_i + \rho_a(\ddot{u}_i - \ddot{U}_i) & \text{equil. skeleton} \\ -(p_0 + p)_{,i} + n\rho_w g_i - b^\diamond_i - j^\diamond_i & = n\rho_w \ddot{U}_i - \rho_a(\ddot{u}_i - \ddot{U}_i) & \text{equil. water} \\ w_i & = k_{ij} j^\diamond_j / n & \text{Darcy} \\ w_{i,i} + {}^1/_Q\, \dot{p} + \dot{u}_{i,i} & = 0 & \text{conservation of mass} \end{cases} \tag{3}$$

The static interaction force $b^\diamond_i = -(1-n)(p_0 + p)_{,i}$ corresponds to the buoyancy $b^\diamond_{0i} = -(1-n)p_{0,i} = -(1-n)\rho_w g_i$ in the hydrostatic case. Note, however, that $b^\diamond_i$ is caused by the *total* pore pressure gradient (and not by its gravitational part only). Moreover, due to the presence of static interaction $b^\diamond$, the effective stress T_{ij} may be used and not the "partial stress", as it is often the case in the literature. The viscous interaction is described by the density $j^\diamond_i$ of drag forces and the inertial interaction is described by the fictitious density ρ_a appearing at the difference $\ddot{u}_i - \ddot{U}_i$,

$$\rho_a = (\tau - 1)n\rho_w \qquad \text{with} \qquad \tau \approx n^{-1/2} \tag{4}$$

wherein τ is the so-called tortuosity[2] [2]. The geostatic effective stress T^0_{ij} and the hydrostatic pore pressure p_0 satisfy (3) in the special case of $\ddot{U}_i = \ddot{u}_i = \dot{u}_i = u_i = w_i = j^\diamond_i = \dot{p} = p = 0$ and the well known geotechnical relations kann be recovered

$$-p_{0,i} + n\rho_w g_i - b^\diamond_{0i} = \boxed{0 = p_{0,i} - \rho_w g_i} \quad \text{and} \tag{5}$$

$$T^0_{ij,j} + \rho_d g_i + b^\diamond_{0i} = \boxed{0 = T^0_{ij,j} + (\rho - \rho_w) g_i = 0} \tag{6}$$

Let us first resolve the second equation in (3) for the viscous interaction[3],

$$j^\diamond_i = -(p_0 + p)_{,i} + n\rho_w g_i + \overbrace{(1-n)(p_0+p)_{,i}}^{=-b^\diamond_i} - \left[n\rho_w \ddot{U}_i - \rho_a(\ddot{u}_i - \ddot{U}_i) \right] \tag{7}$$

[2] For low porosity $n \to 0$ the "added mass" increases, $\rho_a \to \infty$, which results in $\ddot{U}_i = \ddot{u}_i$.

[3] Between pore water and skeleton.

or dividing by n

$$k_{ij}^{-1} w_j = j_i^{\diamond}/n = -(p_0 + p)_{,i} + \rho_w g_i - \left[\rho_w \ddot{U}_i - (\tau - 1)\rho_w(\ddot{u}_i - \ddot{U}_i)\right] \quad (8)$$

$$= -p_{,i} - \left[\rho_w \ddot{U}_i - (\tau - 1)\rho_w(\ddot{u}_i - \ddot{U}_i)\right] \quad (9)$$

In (8) we have prepended the Darcy law[4] as the proportionality between w_i and $j_i^{\diamond}$,

$$w_i = \frac{1}{n} k_{ij} j_j^{\diamond} \quad (10)$$

We have also canceled $-p_{0,i} + \rho_w g_i = 0$ basing on $(6)_1$ and assumed $k_{ij} = \text{const}$. Let us add the first two lines of the system (3) in order to cancel all interaction forces. We obtain the equilibrium condition of the mixture

$$T_{ij,j} - (p_0 + p)_{,i} + (1-n)\rho_s g_i + n\rho_w g_i = (1-n)\rho_s \ddot{u}_i + n\rho_w \ddot{U}_i \quad (11)$$

Next we introduce $\ddot{U}_i = \ddot{u}_i + {}^1/n\ \dot{w}_i$ from the time derivative of the definition (2) of w_i (with $n \approx$ const). The inertial term in the balance of momentum (11) is

$$n\rho_w \ddot{U}_i + (1-n)\rho_s \ddot{u}_i = n\rho_w(\ddot{u}_i + {}^1/n\ \dot{w}_i) + (1-n)\rho_s \ddot{u}_i = \rho \ddot{u}_i + \rho_w \dot{w}_i \quad (12)$$

We substitute $\ddot{U}_i = \ddot{u}_i + {}^1/n\ \dot{w}_i$ also into the Darcy equation in the form (8). The expression in square brackets is

$$\rho_w \left[\ddot{U}_i - (\tau - 1)(\ddot{u}_i - \ddot{U}_i)\right] = \rho_w \left[(\ddot{u}_i + {}^1/n\ \dot{w}_i) - (\tau - 1)(-{}^1/n\ \dot{w}_i)\right]$$

$$= \rho_w(\ddot{u}_i + {}^\tau/n\ \dot{w}_i) \quad (13)$$

and hence the Darcy law takes the form

$$w_i = -k_{ij}(p_{,j} + \rho_w(\ddot{u}_j + {}^\tau/n\ \dot{w}_j)) \quad (14)$$

Equations (3) take the form[5]

$$\begin{cases} T_{ij,j} - (p_0 + p)_{,i} + \rho g_i & = \rho \ddot{u}_i + \rho_w \dot{w}_i \\ w_{i,i} + {}^1/Q\ \dot{p} + \dot{u}_{i,i} & = 0 \end{cases} \quad (15)$$

Now we need to approximate $\dot{w}_i$ in (15) as a function of primary variables p and u. For this purpose we differentiate the Darcy's law (14) wrt time and make the simplifying assumption $\dddot{u} + {}^\tau/n\ \ddot{w} \approx 0$. We obtain[6]

$$\dot{w}_i \approx -k_{ij}\ \dot{p}_{,j} \quad (16)$$

---

4 Note that the specific permeability k_{ij} $[(\text{m/s})/(\text{kN/m}^3)]$ is used and not the hydraulic conductivity $k_{ij}^w = k_{ij}\gamma_w$ [m/s].

5 Note the impending numerical instability at small permeabilities, e.g. in 1D $\dot{w} = -k^{-1}w + \ldots$ with $k^{-1} \approx 10^{10}$ m/s.

6 Note that the third time derivatives are ignored in the GN22 time integration anyway. Alternatively one can try out $\dddot{U}_j \approx \dddot{u}_j$ i.e. $\ddot{U}_i - \ddot{u}_i = -{}^\tau/n\ k_{ij}\ (\dot{p}_{,j} + \dddot{u}_j \rho_w)$.

instead of neglecting the portion $^{\tau}/_{n}\ \dot{w}$ completely, as proposed by Zienkiewicz [5] and preserving the inertial term in the Darcy law unlike Li et al. [4]. Next, we substitute (16) into (15) and use a simple visco-elastic constitutive model in the form $T_{ij} = T^0_{ij} + \Delta T_{ij}$ with $\Delta T_{ij} = E_{ijkl} u_{k,l} + C_{ijkl} \dot{u}_{k,l}$ obtaining

$$\overbrace{E_{ijkl} u_{k,lj} + C_{ijkl} \dot{u}_{k,lj}}^{=\Delta T_{ij}} - p_{,i} = \overbrace{-T^0_{ij,j} + p_{0,i} - \rho g_i}^{=0} + \rho \ddot{u}_i - \rho_w k_{ij} \dot{p}_{,j} \tag{17}$$

and the system has the form

$$\begin{cases} \overbrace{E_{ijkl} u_{k,lj} + C_{ijkl} \dot{u}_{k,lj}}^{=\Delta T_{ij}} - p_{,i} & = \rho \ddot{u}_i - \rho_w k_{ij} \dot{p}_{,j} \\ -[k_{ij}(p_{,j} + (\ddot{u}_i - {}^{\tau}/_{n}\ k_{ij}\ \dot{p}_{,j})\rho_w)]_{,i} + {}^{1}/_{Q}\ \dot{p} + \dot{u}_{i,i} & = 0 \end{cases} \tag{18}$$

Moreover we have used the equilibrium of geostatic stress (6) to cancel a part of the r.h.s. in our final strong form. In the next section we formulate the weak form starting from (15) rather than from (18) and simplifying everything to 1D case.

1.3 Strong and Weak Forms of Equilibrium and Mass Conservation in 1D

In the 1D case $i = j = k = l = 1$ and hence $\delta_{ij} \rightarrow 1$, $E_{ijkl} \rightarrow E$ und $k_{ij} \rightarrow k$. We obtain the strong form with displacement $u(x, t)$ and pore water overpressure $p(x, t)$ as primary unknowns

$$\begin{cases} \Delta T^{\text{tot}}_{,x} - \rho \ddot{u} + \rho_w k \dot{p}_{,x} & = 0 \\ w_{,x} + \frac{1}{Q}\ \dot{p} + \dot{u}_{,x} & = 0 \end{cases} \tag{19}$$

wherein

$$\begin{aligned} T^{\text{tot}} &= T_0 - p_0 + \Delta T^{\text{tot}} \\ \Delta T^{\text{tot}} &= E u_{,x} + C \dot{u}_{,x} - p \\ w &= -k(p_{,x} - g\rho_w + (\ddot{u} - \frac{\tau}{n}\ k\ \dot{p}_{,x})\rho_w) \end{aligned} \tag{20}$$

Having noticed the 2-nd derivatives of the primary unknowns wrt x in the system we may defer the substitution of T^{tot} and w until (25). In order to restrict the formulation to 1-st x derivatives we build the weak form of (19) with test functions $v(x)$ and $\psi(x)$ for $x \in [0, L]$

$$\begin{cases} \int v \Delta T^{\text{tot}}_{,x} \mathrm{d}x - \int v \rho \ddot{u} \mathrm{d}x + \int v \rho_w k \dot{p}_{,x} \mathrm{d}x & = 0 \\ \int \psi w_{,x} \mathrm{d}x + \int \psi \frac{1}{Q}\ \dot{p} \mathrm{d}x + \int \psi \dot{u}_{,x} \mathrm{d}x & = 0 \qquad \text{with} \qquad \int = \int_0^L \end{cases} \tag{21}$$

The first terms in both equations will be integrated by parts obtaining $\int v \Delta T^{\text{tot}}_{,x} \mathrm{d}x = (v \Delta T^{\text{tot}})|_0^L - \int v_{,x} \Delta T^{\text{tot}} \mathrm{d}x$ and $\int \psi w_{,x} \mathrm{d}x = (\psi w)|_0^L - \int \psi_{,x} w \mathrm{d}x$.

This partial integration lowers the degree of differentiation of the primary unknowns u (hidden in ΔT^{tot}) and p, viz.

$$\begin{cases} (v\Delta T^{\text{tot}})|_0^L - \int v_{,x}\Delta T^{\text{tot}}\mathrm{d}x - \int v\rho\ddot{u}\,\mathrm{d}x + \int v\rho_w k\dot{p}_{,x}\mathrm{d}x & = 0 \\ (\psi w)|_0^L - \int \psi_{,x} w\,\mathrm{d}x + \int \psi\frac{1}{Q}\,\dot{p}\mathrm{d}x + \int \psi\dot{u}_{,x}\mathrm{d}x & = 0 \end{cases} \tag{22}$$

Moreover, this form facilitates the formulation of the von Neumann BC: mechanical ones with total tractions $t_i^{\text{tot}} = (T_{ij} - \delta_{ij}p)n_j$ and hydraulic ones with influxes $q = -w_j n_j$ with outer normal n_j. In our 1D case the indices can be omitted and von Neumann BC can be defined as

$$\begin{cases} q = -w, & \Delta t^{\text{tot}} = \Delta T^{\text{tot}} & \text{for} & x = L \\ q = w, & \Delta t^{\text{tot}} = -\Delta T^{\text{tot}} & \text{for} & x = 0 \end{cases} \tag{23}$$

because we have just two boundary points $x = 0$ and $x = L$ with $n = -1$ and $n = 1$, respectively. Using prescribed $\Delta\breve{t}^{\text{tot}}$ and $\breve{q}$ we can write

$$\sum \psi\breve{q} = -\,(\psi w)|_0^L \quad \text{and} \quad \sum v\Delta\breve{t}^{\text{tot}} = (v\Delta T^{\text{tot}})|_0^L \tag{24}$$

In the usual FE assembling procedure we need to consider just the nodes on the boundary of the region. With these boundary conditions we have

$$\begin{cases} \int v_{,x}\Delta T^{\text{tot}}\mathrm{d}x + \int v\rho\ddot{u}\,\mathrm{d}x - \int v\rho_w k\dot{p}_{,x}\mathrm{d}x & = \sum v\Delta\breve{t}^{\text{tot}} \\ -\int \psi_{,x} w\,\mathrm{d}x + \int \psi\frac{1}{Q}\,\dot{p}\mathrm{d}x + \int \psi\dot{u}_{,x}\mathrm{d}x & = \sum \psi\breve{q} \end{cases} \tag{25}$$

Using (25) in a single element we may set $\breve{q} = 0$ and $\Delta\breve{t}^{\text{tot}} = 0$ on all internal nodes because the contributions from two adjacent elements must cancel. It is so due to identical stresses ΔT^{tot} and velocities w but opposite outer normal vectors n. Finally let us substitute w and ΔT^{tot} from (20) into the weak form (25)

$$\begin{cases} \int v_{,x}Eu_{,x}\mathrm{d}x + \int v_{,x}C\dot{u}_{,x}\mathrm{d}x - \int v_{,x}p\,\mathrm{d}x + \int v\rho\ddot{u}\,\mathrm{d}x \\ \qquad\qquad - \int v\rho_w k\dot{p}_{,x}\mathrm{d}x = \sum v\Delta\breve{t}^{\text{tot}} \\ k\int \psi_{,x}\left[p_{,x} + (\ddot{u} - \frac{T}{n}\,k\,\dot{p}_{,x})\rho_w\right]\mathrm{d}x + \int \psi\frac{1}{Q}\,\dot{p}\mathrm{d}x \\ \qquad\qquad + \int \psi\dot{u}_{,x}\mathrm{d}x = \sum \psi\breve{q} \end{cases} \tag{26}$$

1.4 Spatial Galerkin Discretization

Our the system of equations contains convective terms. For dominant convection these equation may need stabilization similar to techniques used in computational fluid dynamics. Nevertheless we implement the conventional Bubnov-Galerkin FE discretization, i.e. the test functions are the same as the trial functions. Different basis functions are used for pore pressure field and for displacement field (like in Taylor-Hood elements). Quadratic approximation of displacement with $N_\alpha(x)$ compared with linear approximation of pore pressure should mollify the impending sattle-point oscillations in the case of incompressible substance combined with low permeability.

The weak form (26) contains only first spatial derivatives and hence it is suitable for FE discretization. For this purpose we introduce the following interpolations $u(x) = N_\alpha(x)u_\alpha$, $v(x) = N_\beta(x)v_\beta$, $p(x) = \bar{N}_\eta(x)p_\eta$, $\psi(x) = \bar{N}_\gamma(x)\psi_\gamma$, with nodal values $u_\alpha, v_\beta, p_\eta, \psi_\gamma$. The displacement indices $\{\alpha, \beta\}$ take values $1, 2, 3$ and the pressure indices $\{\gamma, \eta\}$ take values $1, 2$. Integrals from (26) are evaluated for a single 3-nodal finite element[7] with $x \in [x_1, x_3]$,

$$\begin{cases} \int v_\beta N_{\beta,x} E N_{\alpha,x} u_\alpha \mathrm{d}x + \int v_\beta N_{\beta,x} C N_{\alpha,x} \dot{u}_\alpha \mathrm{d}x - \int v_\beta N_{\beta,x} \bar{N}_\eta p_\eta \,\mathrm{d}x \\ \qquad + \int v_\beta N_\beta \rho N_\alpha \ddot{u}_\alpha \,\mathrm{d}x - \int v_\beta N_\beta \rho_w k \bar{N}_{\eta,x} \dot{p}_\eta \mathrm{d}x = \sum v_\beta N_\beta \Delta \breve{t}^{\text{tot}} \\ k \int \left[\psi_\gamma \bar{N}_{\gamma,x} \bar{N}_{\eta,x} p_\eta + \psi_\gamma \bar{N}_{\gamma,x} \rho_w N_\alpha \ddot{u}_\alpha - \psi_\gamma \bar{N}_{\gamma,x} \rho_w \frac{\tau}{n}\, k\, \bar{N}_{\eta,x} \dot{p}_\eta \right] \mathrm{d}x \\ \qquad + \int \psi_\gamma \bar{N}_\gamma \frac{1}{Q}\, \bar{N}_\eta \dot{p}_\eta \mathrm{d}x + \int \psi_\gamma \bar{N}_\gamma N_{\alpha,x} \dot{u}_\alpha \mathrm{d}x = \sum \psi_\gamma \bar{N}_\gamma \breve{q} \end{cases} \tag{27}$$

We introduce the following submatrices for the future element matrix

$$\begin{cases} v_\beta (K_{\beta\alpha} u_\alpha + C_{\beta\alpha} \dot{u}_\alpha - Q_{\beta\eta} p_\eta + M_{\beta\alpha} \ddot{u}_\alpha - P_{\beta\eta} \dot{p}_\eta) & = \sum v_\beta \Delta \breve{t}_\beta^{\text{tot}} \\ \psi_\gamma (H_{\gamma\eta} p_\eta + P^*_{\gamma\alpha} \ddot{u}_\alpha - J_{\gamma\eta} \dot{p}_\eta + S_{\gamma\eta} \dot{p}_\eta + Q^*_{\gamma\alpha} \dot{u}_\alpha) & = \sum \psi_\gamma \breve{q}_\gamma \end{cases} \tag{28}$$

wherein $\breve{t}_\beta^{\text{tot}}$ values of traction at boundary node b (zero on internal nodes) and $\breve{q}_\gamma$ is the influx through the boundary node g. The submatrices are in the first equation are

$$\begin{aligned} K_{\beta\alpha} &= \int N_{\beta,x} E N_{\alpha,x} \mathrm{d}x \\ C_{\beta\alpha} &= \int N_{\beta,x} C N_{\alpha,x} \mathrm{d}x \\ Q_{\beta\eta} &= \int N_{\beta,x} \bar{N}_\eta \,\mathrm{d}x \\ M_{\beta\alpha} &= \int N_\beta \rho N_\alpha \mathrm{d}x \\ P_{\beta\eta} &= \int N_\beta \rho_w k \bar{N}_{\eta,x} \mathrm{d}x \end{aligned} \tag{29}$$

and the submatrices are in the second equation are

[7] Within a single element one works with the dimensionless coordinate $\xi \in [-1, 1]$ with the operations $\int_{x_1}^{x_3} F(x)\mathrm{d}x = \frac{L_e}{2} \int_{-1}^{1} f(\xi)\mathrm{d}\xi$ and $F_{,x} = \frac{2}{L_e} f_{,\xi}$ wherein $F(x(\xi)) = f(\xi)$ and $L_e = x_3 - x_1$ is the length of the element. Moreover expressions $f|_{-1}^{1}$ are of importance only for nodes on the boundary of the whole region. For internal nodes the sum of contributions $f|_{-1}^{1}$ from neighbouring elements is zero.

$$
\begin{aligned}
H_{\gamma\eta} &= \int k\bar{N}_{\gamma,x}\bar{N}_{\eta,x}\mathrm{d}x \\
P^*_{\gamma\alpha} &= \int k\bar{N}_{\gamma,x}\rho_w N_\alpha \mathrm{d}x \\
J_{\gamma\eta} &= \int \bar{N}_{\gamma,x}\rho_w \frac{\mathcal{T}}{n}\, k^2\, \bar{N}_{\eta,x}\mathrm{d}x \\
S_{\gamma\eta} &= \int \bar{N}_\gamma \frac{1}{Q}\, \bar{N}_\eta \mathrm{d}x \\
Q^*_{\gamma\alpha} &= \int \bar{N}_\gamma N_{\alpha,x}\mathrm{d}x
\end{aligned} \tag{30}
$$

and (31) holds for any combination of test function coefficients v_β and ψ_γ so these coefficients can be omitted and (31) generates a system of N equations, one eq. pro DOF. Moreover we change the sign of the second equation so that the system becomes symmetric

$$
\begin{cases}
K_{\beta\alpha}u_\alpha + C_{\beta\alpha}\dot{u}_\alpha - Q_{\beta\eta}p_\eta + M_{\beta\alpha}\ddot{u}_\alpha - P_{\beta\eta}\dot{p}_\eta &= \Delta\breve{t}^{\text{tot}}_\beta \\
-H_{\gamma\eta}p_\eta - P^*_{\gamma\alpha}\ddot{u}_\alpha + J_{\gamma\eta}\dot{p}_\eta - S_{\gamma\eta}\dot{p}_\eta - Q^*_{\gamma\alpha}\dot{u}_\alpha &= \breve{q}_\gamma
\end{cases} \tag{31}
$$

because the starred matrices can be obtained by transposition $\mathbf{P}^* = \mathbf{P}^T$ and $\mathbf{Q}^* = \mathbf{Q}^T$.

1.5 Temporal Discretization with GN22 for u and GN11 for p

The temporal discretization is performed with finite differences after the spatial discretization with finite elements is completed. More freedom in modelling and some stabilizing mechanisms offer the discrete Galerkin FEM [3], however we prefer to use the traditional Newmark time integration schemes first. In this section we prefer to use the boldface notation for matrices and vectors, for example we write $\mathbf{u}, \mathbf{p}$ rather than u_α, p_η. Moreover we notice $\mathbf{Q}^* = \mathbf{Q}^T$ and $\mathbf{P}^* = \mathbf{P}^T$ so multiplying the second equation in (27) by -1 we may expect the symmetry of the future element matrix. Our system (31) takes the form

$$
\begin{cases}
\mathbf{K}\cdot\mathbf{u} + \mathbf{C}\cdot\dot{\mathbf{u}} + \mathbf{M}\cdot\ddot{\mathbf{u}} - \mathbf{Q}\cdot\mathbf{p} - \mathbf{P}\cdot\dot{\mathbf{p}} &= \Delta\breve{\mathbf{t}}^{\text{tot}} \\
-\mathbf{Q}^*\cdot\dot{\mathbf{u}} - \mathbf{P}^*\cdot\ddot{\mathbf{u}} + (\mathbf{J}-\mathbf{S})\cdot\dot{\mathbf{p}} - \mathbf{H}\cdot\mathbf{p} &= -\breve{\mathbf{q}}
\end{cases} \tag{32}
$$

This makes room for time indices n and $n+1$ corresponding to the beginning and to the end of the current increment. We use the Newmark time integration scheme in which all end-of-increment values $\mathbf{u}_{n+1}, \dot{\mathbf{u}}_{n+1}, \ddot{\mathbf{u}}_{n+1}, \mathbf{p}_{n+1}, \dot{\mathbf{p}}_{n+1}$ are expressed using just two of them: acceleration $\ddot{\mathbf{u}}_{n+1}$ and pore pressure rate $\dot{\mathbf{p}}_{n+1}$.

$$
\begin{aligned}
\mathbf{u}_{n+1} &= \mathbf{u}^E + \frac{1}{2}\Delta t^2\beta_2\ddot{\mathbf{u}}_{n+1} \\
\dot{\mathbf{u}}_{n+1} &= \dot{\mathbf{u}}^E + \Delta t\beta_1\ddot{\mathbf{u}}_{n+1} \\
\mathbf{p}_{n+1} &= \mathbf{p}^E + \Delta t\beta_1\dot{\mathbf{p}}_{n+1}
\end{aligned} \tag{33}
$$

with integration parameters β_1, β_2 and time increment Δt. The extrapolators calculated solely from the IC or from the end-values of the preceding increment[8].

$$
\begin{aligned}
\mathbf{u}^E &\approx \mathbf{u}_n + \Delta t \dot{\mathbf{u}}_n + \frac{\Delta t^2}{2}(1-\beta_2)\ddot{\mathbf{u}}_n \\
\dot{\mathbf{u}}^E &\approx \dot{\mathbf{u}}_n + \Delta t(1-\beta_1)\ddot{\mathbf{u}}_n \\
\mathbf{p}^E &\approx \mathbf{p}_n + \Delta t(1-\beta_1)\dot{\mathbf{p}}_n
\end{aligned} \tag{34}
$$

Our system (32) must be written for $t+\Delta t$ and hence we obtain the form

$$
\begin{cases}
\mathbf{K}\cdot(\mathbf{u}^E + \frac{1}{2}\Delta t^2\beta_2\ddot{\mathbf{u}}_{n+1}) + \mathbf{C}\cdot(\dot{\mathbf{u}}^E + \Delta t\beta_1\ddot{\mathbf{u}}_{n+1}) + \mathbf{M}\cdot\ddot{\mathbf{u}}_{n+1} \\
\qquad -\mathbf{Q}\cdot(\mathbf{p}^E + \Delta t\beta_1\dot{\mathbf{p}}_{n+1}) - \mathbf{P}\cdot\dot{\mathbf{p}}_{n+1} = \Delta\check{\mathbf{t}}^{\text{tot}}_{n+1} \\
-\mathbf{Q}^*\cdot(\dot{\mathbf{u}}^E + \Delta t\beta_1\ddot{\mathbf{u}}_{n+1}) - \mathbf{P}^*\cdot\ddot{\mathbf{u}}_{n+1} \\
\qquad +\mathbf{J}\cdot\dot{\mathbf{p}}_{n+1} - \mathbf{H}\cdot(\mathbf{p}^E + \Delta t\beta_1\dot{\mathbf{p}}_{n+1}) - \mathbf{S}\cdot\dot{\mathbf{p}}_{n+1} = -\check{\mathbf{q}}_{n+1}
\end{cases} \tag{35}
$$

All extrapolators are known so we may send them to the r.h.s. and the contribution from a single element is

$$
\begin{cases}
\overbrace{[\mathbf{M} + \Delta t\beta_1\mathbf{C} + 1/2\Delta t^2\beta_2\mathbf{K}]}^{=\mathbf{K}_{UU}}\cdot\ddot{\mathbf{u}}_{n+1} \overbrace{-[\mathbf{P} + \Delta t\beta_1\mathbf{Q}]}^{=\mathbf{K}_{UP}}\cdot\dot{\mathbf{p}}_{n+1} \\
\qquad = \overbrace{\Delta\check{\mathbf{t}}^{\text{tot}}_{n+1} - \mathbf{C}\cdot\dot{\mathbf{u}}^E - \mathbf{K}\cdot\mathbf{u}^E + \mathbf{Q}\cdot\mathbf{p}^E}^{=\mathbf{R}_U} \\
\overbrace{-[\mathbf{P}^* + \Delta t\beta_1\mathbf{Q}^*]}^{=\mathbf{K}_{PU}}\cdot\ddot{\mathbf{u}}_{n+1} + \overbrace{[(\mathbf{J}-\mathbf{S}) - \Delta t\beta_1\mathbf{H}]}^{=\mathbf{K}_{PP}}\cdot\dot{\mathbf{p}}_{n+1} \\
\qquad = \overbrace{-\check{\mathbf{q}}_{n+1} + \mathbf{Q}^*\cdot\dot{\mathbf{u}}^E + \mathbf{H}\cdot\mathbf{p}^E}^{=\mathbf{R}_P}
\end{cases} \tag{36}
$$

The vector of unknowns (degrees of freedom) of an element may consist of different number of nodal accelerations $\ddot{\mathbf{u}}$ and pore pressure rates $\dot{\mathbf{p}}$ so the blocks $\mathbf{K}_{UP}$ and $\mathbf{K}_{PU}$ need not be quadratic. However the element matrix (here in block form)

$$
\begin{bmatrix} \mathbf{K}_{UU} & \mathbf{K}_{UP} \\ \mathbf{K}_{PU} & \mathbf{K}_{PP} \end{bmatrix} \cdot \begin{Bmatrix} \ddot{\mathbf{u}}_{n+1} \\ \dot{\mathbf{p}}_{n+1} \end{Bmatrix} = \begin{Bmatrix} \mathbf{R}_U \\ \mathbf{R}_P \end{Bmatrix} \tag{37}
$$

is symmetric.

1.6 Temporal Discretization: GN22 for u and GN21 for p

Contrarily to GN11 used in Sect. 1.5 we try out the GN21 time scheme for the pore pressure. This means that all end-of-increment values $\mathbf{u}_{n+1}, \dot{\mathbf{u}}_{n+1}, \ddot{\mathbf{u}}_{n+1}, \mathbf{p}_{n+1}, \dot{\mathbf{p}}_{n+1}, \ddot{\mathbf{p}}_{n+1}$ are expressed using the *second* derivatives $\ddot{\mathbf{u}}_{n+1}$ and $\ddot{\mathbf{p}}_{n+1}$ as the unknowns. We obtain

$$
\mathbf{u}_{n+1} = \mathbf{u}^E + \frac{1}{2}\Delta t^2\beta_2\ddot{\mathbf{u}}_{n+1} \qquad \dot{\mathbf{u}}_{n+1} = \dot{\mathbf{u}}^E + \Delta t\beta_1\ddot{\mathbf{u}}_{n+1} \tag{38}
$$

[8] Extrapolators can be calculated using the global vectors.

$$\mathbf{p}_{n+1} = \mathbf{p}^E + \frac{1}{2}\Delta t^2 \beta_2 \ddot{\mathbf{p}}_{n+1} \qquad \dot{\mathbf{p}}_{n+1} = \dot{\mathbf{p}}^E + \Delta t \beta_1 \ddot{\mathbf{p}}_{n+1} \tag{39}$$

with extrapolators $\sqcup^E$ calculated as follows

$$\begin{aligned}
\mathbf{u}^E &\approx \mathbf{u}_n + \Delta t \dot{\mathbf{u}}_n + \frac{\Delta t^2}{2}(1-\beta_2)\ddot{\mathbf{u}}_n \\
\dot{\mathbf{u}}^E &\approx \dot{\mathbf{u}}_n + \Delta t(1-\beta_1)\ddot{\mathbf{u}}_n \\
\mathbf{p}^E &\approx \mathbf{p}_n + \Delta t \dot{\mathbf{p}}_n + \frac{\Delta t^2}{2}(1-\beta_2)\ddot{\mathbf{p}}_n \\
\dot{\mathbf{p}}^E &\approx \dot{\mathbf{p}}_n + \Delta t(1-\beta_1)\ddot{\mathbf{p}}_n
\end{aligned} \tag{40}$$

Except for the additional initial conditions the new system is very similar to (35)

$$\begin{cases}
\mathbf{K}(\mathbf{u}^E + \frac{1}{2}\Delta t^2 \beta_2 \ddot{\mathbf{u}}_{n+1}) + \mathbf{C}(\dot{\mathbf{u}}^E + \Delta t \beta_1 \ddot{\mathbf{u}}_{n+1}) + \mathbf{M}\ddot{\mathbf{u}}_{n+1} \\
\qquad -\mathbf{Q}(\mathbf{p}^E + \frac{1}{2}\Delta t^2 \beta_2 \ddot{\mathbf{p}}_{n+1}) - \mathbf{P}(\dot{\mathbf{p}}^E + \Delta t \beta_1 \ddot{\mathbf{p}}_{n+1}) = \Delta \check{\mathbf{t}}^{\text{tot}}_{n+1} \\
-\mathbf{Q}^*(\dot{\mathbf{u}}^E + \Delta t \beta_1 \ddot{\mathbf{u}}_{n+1}) - \mathbf{P}^* \ddot{\mathbf{u}}_{n+1} \\
\qquad -\mathbf{H}(\mathbf{p}^E + \frac{1}{2}\Delta t^2 \beta_2 \ddot{\mathbf{p}}_{n+1}) + (\mathbf{J}-\mathbf{S})(\dot{\mathbf{p}}^E + \Delta t \beta_1 \ddot{\mathbf{p}}_{n+1}) = -\check{\mathbf{q}}_{n+1}
\end{cases} \tag{41}$$

As before, all extrapolators are sent them to the r.h.s. and the block Eq. (37) has the following form

$$\begin{bmatrix} \mathbf{K}_{UU} & \mathbf{K}_{UP} \\ \mathbf{K}_{PU} & \mathbf{K}_{PP} \end{bmatrix} \cdot \begin{Bmatrix} \ddot{\mathbf{u}}_{n+1} \\ \ddot{\mathbf{p}}_{n+1} \end{Bmatrix} = \begin{Bmatrix} \mathbf{R}_U \\ \mathbf{R}_P \end{Bmatrix} \tag{42}$$

wherein

$$\mathbf{K}_{UU} = \mathbf{M} + \Delta t \beta_1 \mathbf{C} + {}^1\!/_2 \Delta t^2 \beta_2 \mathbf{K} \tag{43}$$

$$\mathbf{K}_{UP} = -\left[\Delta t \beta_1 \mathbf{P} + \frac{1}{2}\Delta t^2 \beta_2 \mathbf{Q}\right] \tag{44}$$

$$\mathbf{K}_{PU} = -\left[\mathbf{P}^* + \Delta t \beta_1 \mathbf{Q}^*\right] \tag{45}$$

$$\mathbf{K}_{PP} = (\mathbf{J}-\mathbf{S})\Delta t \beta_1 - \frac{1}{2}\Delta t^2 \beta_2 \mathbf{H} \tag{46}$$

$$\mathbf{R}_P = -\check{\mathbf{q}}_{n+1} + \mathbf{Q}^* \cdot \dot{\mathbf{u}}^E - (\mathbf{J}-\mathbf{S}) \cdot \dot{\mathbf{p}}^E + \mathbf{H} \cdot \mathbf{p}^E \tag{47}$$

$$\mathbf{R}_U = \Delta \check{t}^{\text{tot}}_{n+1} - \mathbf{C} \cdot \dot{\mathbf{u}}^E - \mathbf{K} \cdot \mathbf{u}^E + \mathbf{Q} \cdot \mathbf{p}^E + \mathbf{P} \cdot \dot{\mathbf{p}}^E \tag{48}$$

It turns out that the GN22/GN21 time integration does not improve the accuracy. In coarse time stepping one cannot reproduce the desired Dirichlet BC $\check{\mathbf{q}}(t)$ function exactly. Using its second derivative $\check{\mathbf{q}}''(t)$ we need smaller time increments than using $\check{\mathbf{q}}'(t)$. The reason for this is the accumulation of error. For example, some inaccuracies in $\dot{\mathbf{p}} \neq 0$ may remain after prescribed accelerations vanish, $\ddot{\mathbf{p}}_{n+1} = 0$. These inaccuracies lead to cumulative change in $\mathbf{p}$ with time. The conventional time discretization GN22/GN11 is better, in the sense that its inaccuracies do not grow with time causing some small oscillations only.

1.7 Initial Static Equilibrium

Our differential equations contain $\mathbf{u}, \dot{\mathbf{u}}, \ddot{\mathbf{u}}, \dot{\mathbf{p}}$ and $\mathbf{p}$ so the usual time integration needs the initial conditions (IC) imposed on $\mathbf{u}, \dot{\mathbf{u}}, \dot{\mathbf{p}}$. However, the Newmark time integration method requires also the IC for the highest time derivatives $\mathbf{p}$ and $\ddot{\mathbf{u}}$ appearing in the problem. We determine them from the equilibrium condition and mass conservation

$$\begin{cases} \mathbf{K}\cdot\mathbf{u} + \mathbf{C}\cdot\dot{\mathbf{u}} - \mathbf{Q}\cdot\mathbf{p} + \mathbf{M}\cdot\ddot{\mathbf{u}} - \mathbf{P}\cdot\dot{\mathbf{p}} & = \Delta\check{\mathbf{t}}^{\mathrm{tot}} \\ -\mathbf{Q}^*\cdot\dot{\mathbf{u}} - \mathbf{P}^*\cdot\ddot{\mathbf{u}} - \mathbf{H}\cdot\mathbf{p} + (\mathbf{J}-\mathbf{S})\cdot\dot{\mathbf{p}} & = -\check{\mathbf{q}} \end{cases} \tag{49}$$

which should be satisfied for the initial time $t = 0$. After the usual setting of the initial vectors $\mathbf{u}, \dot{\mathbf{u}}, \mathbf{p}$ we may find the initial distributions $\ddot{\mathbf{u}}$ and $\dot{\mathbf{p}}$ from (49). It is not necessary to begin with the geostatic stress and hydrostatic pore pressure.

1.8 FEM Implementation and Conclusion

The 1D implementation is based on 1D 3-nodal element with linear interpolation of pore pressure and quadratic interpolation of displacement. The classical consolidation problem of a soil layer of height L drained on the upper side only was calculated:

$L = 10$ m, $k = 10^{-3}\,(\mathrm{m/s})/(\mathrm{kPa/m})$, $n = 0.5$, $K_f = 2.2\cdot 10^6\,\mathrm{kPa}$, $K_s = \infty$, $E = 10^4$ kPa, $C = 0$ $\rho_w = 1, \rho = 2\ \mathrm{t/m^3}$.

mechanical IC: $u(x,0) = \dot{u}(x,0) = \ddot{u}(x,0) = 0$
hydraulic IC: $p(x,0) = 100$ kPa
mechanical BC: $\ddot{u}(0,t) = 0$ and $t^{\mathrm{tot}}(L,t) = 100$ kPa
hydraulic BC: $q(0,t) = 0$ and $p(L,t) = 100 - 50\cos(\pi t/0.01)$

FEM: $n_{\mathrm{el}} = 20$, $\Delta t = 0.001$, $\beta_1 = \beta_2 = 0.55$, GN2211

Individual portions of the element matrix are optional and could be switched on or off as desired. At first only the portions $\mathbf{K}, \mathbf{H}, \mathbf{Q}, \mathbf{Q}^*$ were on and the calculation was stable giving the expected results. Portions $\mathbf{M}$ and $\mathbf{S}$ could be added without strong changes in the result. After the matrices $\mathbf{P}$ and $\mathbf{P}^*$ were switched on the pore pressure diagram (smoothly distributed in space) showed dynamic effects in form of small waves running in time. As soon as the portion $\mathbf{J}$ was switched on strong numerical oscillations occur.

Using similar parameters but with $L = 400$ m and $n_{\mathrm{el}} = 800$, a short disturbance was applied at $x = L$. It could propagate in the form of a smooth sine wave with $P-$wave velocity of locally undrained soil. Unfortunately the calculation was fragile and wave propagation could not be simulated for some combinations of material parameters. Of course, $\mathbf{M}$ was switched on for this calculation. Activation of matrices $\mathbf{P}$ and $\mathbf{P}^*$ did not affect the solution visibly.

We may conclude that the effect of the difference $\ddot{U} - \ddot{u}$ between accelerations cannot be investigated unless stabilizing algorithms are used in the solution.

Numerical implementation with the conventional Galerkin discretization leads to numerical instabilities although the element matrix is symmetric and different discretizations are used for displacement and pore pressure, as it is known from Taylor-Hood elements.

References

1. Biot, M.A.: Theory of propagation of elastic waves in a fluid saturated porous solid. I. Low-frequency range. J. Acoust. Soc. Am. **28**(2), 168–178 (1956)
2. Gajo, A.: The effects of inertial couplng in the interpretation of dynamic soil tests. Géotechnique **46**(2), 245–257 (1996)
3. Hesthaven, J.S., Warburton, T.: Nodal Discontinuous Galerkin Methods. Springer, New York (2008)
4. Li, X., Han, X., Pastor, M.: An iterative stabilized fractional step algorithm for finite element analysis in saturated soil dynamics. Comput. Methods Appl. Mech. Eng. **192**, 3845–3859 (2003)
5. Zienkiewicz, O.C., Chan, A.H.C., Pastor, M., Schrefler, B.A., Shiomi, T.: Computational Geomechanics with Special Reference to Earthquake Engineering. Wiley, Hoboken (1998)

Possibilities and Limitations of ALE Large Deformations Analyses in Geotechnical Engineering

Frank Rackwitz(✉)

Chair of Soil Mechanics and Geotechnical Engineering, Technische Universität Berlin, Gustav-Meyer-Allee 25, 13355 Berlin, Germany
frank.rackwitz@tu-berlin.de

Abstract. A number of geotechnical engineering problems involve large deformations in the soil. While small deformation geotechnical problems can be adequately analyzed by means of conventional Lagrangian FEM, such an approach exhibits considerable shortcomings when the soil undergoes significant deformation. Hence methods have been developed which overcome these shortcomings. Among the methods using a computational mesh, the most promising approaches include the ALE methods: Coupled Eulerian-Lagrangian (CEL) method, the Simplified or Single-Material Arbitrary Lagrangian-Eulerian (SALE) method, and the Multi-Material Arbitrary Lagrangian-Eulerian (MMALE) method. In this contribution the possible advantages and limitations of ALE in comparison to other numerical approaches are presented. The performance of ALE methods is evaluated by means of two application examples. First the vibratory driven installation of open-ended tubular steel piles in sand is modeled and local large deformation due to initial pile imperfections and soil heterogeneity is analyzed. Second the sand column collapse problem is simulated using different modeling approaches and soil models as well. Simulation results assess the feasibility of ALE methods in geotechnical large deformation problems. It can be concluded that the ALE method could be considered as a promising framework for solving complex large deformation problems in geotechnical engineering.

1 Introduction

The Finite Element Method (FEM) is probably the most widely used numerical approach for solving boundary value problems in geotechnical engineering. The traditional formulations of FEM are based on either the Lagrangian or the Eulerian viewpoint. In the Lagrangian formulation the mesh deformation coincides with the material deformation, where the material can be the soil or a structure. In the Eulerian formulation the mesh is fixed and the movement of the material is independent of the mesh. The conventional Lagrangian FEM approach exhibits considerable limitations when the soil undergoes significant, large deformation. Methods that overcome the limitations of classical FEM can be divided into two groups: mesh-based and mesh-free or point (particle)-based (such as the material point methods (MPM) [1] and smoothed

T. Triantafyllidis (Ed.): *Recent Developments of Soil Mechanics and Geotechnics in Theory and Practice*, LNACM 91, pp. 97–112, 2020.
https://doi.org/10.1007/978-3-030-28516-6_6

particle hydrodynamics (SPH) [2]) approaches as well. Figure 1 gives an overview of the continuum based approaches in use.

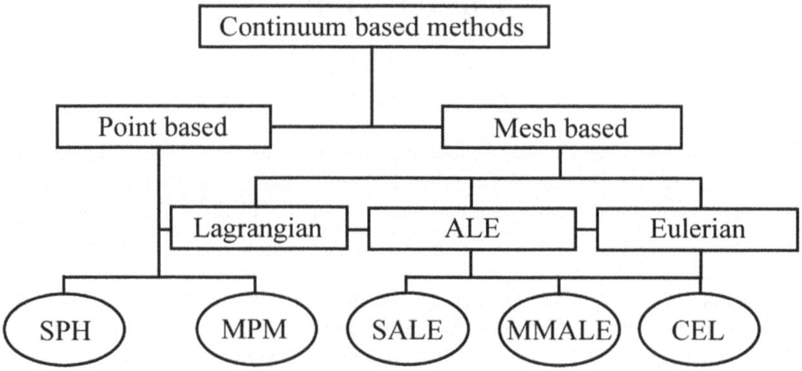

Fig. 1. Overview of the continuum based approaches for the analysis of large deformation problems

The most promising mesh-based approaches include the Coupled Eulerian-Lagrangian (CEL) method, the Simplified or Single-Material Arbitrary Lagrangian-Eulerian (SALE) method, and the Multi-Material Arbitrary Lagrangian-Eulerian (MMALE) method. Irrespective of the method used, the material behavior of the soil shall be described by a suitable non-linear constitutive model. Also the behavior of the interface between the soil and the structure, if it exists in the problem, has to be modeled adequately. Both mentioned requirements complicate the formulation and implementation of the chosen approach.

There are commercial as well as freely available programs, the latter primarily developed at universities for use in research. Not every program contains all the methods listed above. In addition to that there are two well known general strategies to solve the algebraic systems of discretized finite element equations: implicit and explicit solution procedures. Implicit solution procedures are based on the control of the equilibrium of forces and of the error of the solution variables. This makes it possible to control and achieve the convergence of the solution. But at the same time this has a disadvantage due to the need for a convergent solution in any case. In explicit solution procedures there is no such control, instead a minimal time step size has to be used to achieve a reliable solution. It leads to a high number of time steps compared to the same problem solved by an implicit scheme, but the need for a convergent solution is not an issue. A comprehensive comparison of different approaches has to consider the mentioned details of modeling aspects and solution procedures. In this way, the possibilities and limits of the approaches can be demonstrated.

2 ALE Methods

Due to the difficulties and restrictions occurring from the simply Lagrangian and Eulerian approaches when simulating large deformation problems, a more general approach was called for. The arbitrary Lagrangian–Eulerian (ALE) method has been developed in order to overcome these limitations and to combine the advantages of the purely Lagrangian and Eulerian approaches [3]. In an ALE method the computational mesh is regarded as an independent reference domain. The ALE mesh can be continuously smoothed so that the element quality is maintained until the end of the calculation. Mesh connectivity is kept unchanged hence the solution variables can be mapped to the improved mesh by means of advection algorithms. The most general approach within the ALE framework is referred to as MMALE – Multi-Material ALE – which means, that there can be more than one material within a finite element. In addition to the general method, some simplifications have been developed. One simplification allows only one material in a finite element and it is named SALE – Single-Material or Simplified ALE. Another specialization resigns any mesh update, i.e. a fixed Eulerian mesh is used throughout the analysis and this method is called CEL. The CEL method links overlapping but otherwise independent Lagrange and Euler meshes. An edge of the structure discretized with a Lagrange mesh usually forms the contact surface for the connection of the two independent meshes.

All above mentioned approaches are schematically drawn in Fig. 2 illustrating a Lagrangian modeled structure, such as a pile or foundation, penetrating into a deforming material, such as soil.

In recent years the CEL approach has been often used to solve large deformation geotechnical problems [5–8]. Qui et al. [5] analyze a strip footing, pile penetration, and ship grounding problem by means of CEL. Penetration of a jack-up spudcan foundation into seabed was investigated by Qui and Henke [6] and Tho et al. [7]. Hamann et al. [8] investigated the pile installation process into fully saturated soil under partially drained conditions.

After the ALE method had reached a certain level of development and application in the engineering sciences [3], first applications in geomechanics were made by Di et al. [9]. They developed an ALE method where the soil is modeled as a saturated porous material with solid–fluid coupling and strong material non-linearity. Later Aubram et al. [10] investigated a strip footing as well as a monotonic pile penetration problem into sandy soil by means of a SALE model. Recently Aubram et al. [11] provided the theoretical background and basis of development of ALE methods for modeling of vibro-pile penetration problems into sandy soils by means of the MMALE approach.

3 Practical Applications

The possibilities and limitations of the methods are best revealed in the analysis of real applications where large soil deformations occur. From the variety of possible applications, two are selected and analysed in the following sections. First, a current problem is investigated in connection with the dynamic installation of open tubular

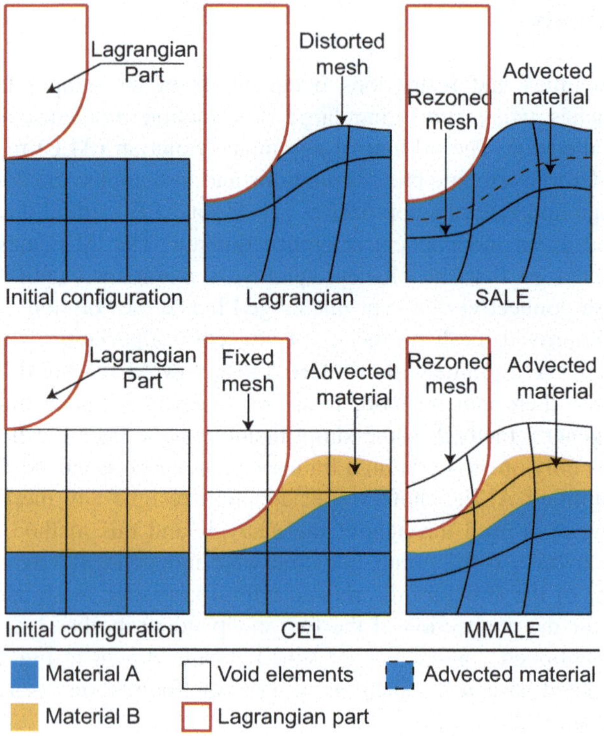

Fig. 2. Illustration of different FEM modeling approaches with respect to mesh and material deformation [4]

steel piles into soil. The second application example is the numerical simulation of a laboratory experiment in which a sand column flows out of a hollow cylinder opened at the bottom. The comparison between numerical and experimental results allows a validation of the calculation model. All numerical analyses have been performed using the commercial code LS-DYNA by means of explicit solution techniques.

3.1 Steel Pipe Buckling During Installation

A numerical model has been developed simulating a pile installation into soil by means of vibratory forces. The finite element model, dimensions and vibratory loading with time are shown in Fig. 3. The model pile has dimensions height/outer diameter/wall thickness = 1.5/0.2/0.005 m. The pile is modeled using the conventional Lagrangian element formulation with single point reduced integration and a uniform element size of 2 cm (3,000 elements). An elastic-perfectly plastic material model based on the von Mises failure criterion is used for the pile with the material properties density 7,850 kg/m^3, Young's modulus 2.1E5, Poisson's ratio 0.3 and Yield stress 250 MPa. The pile head is fixed against horizontal movements.

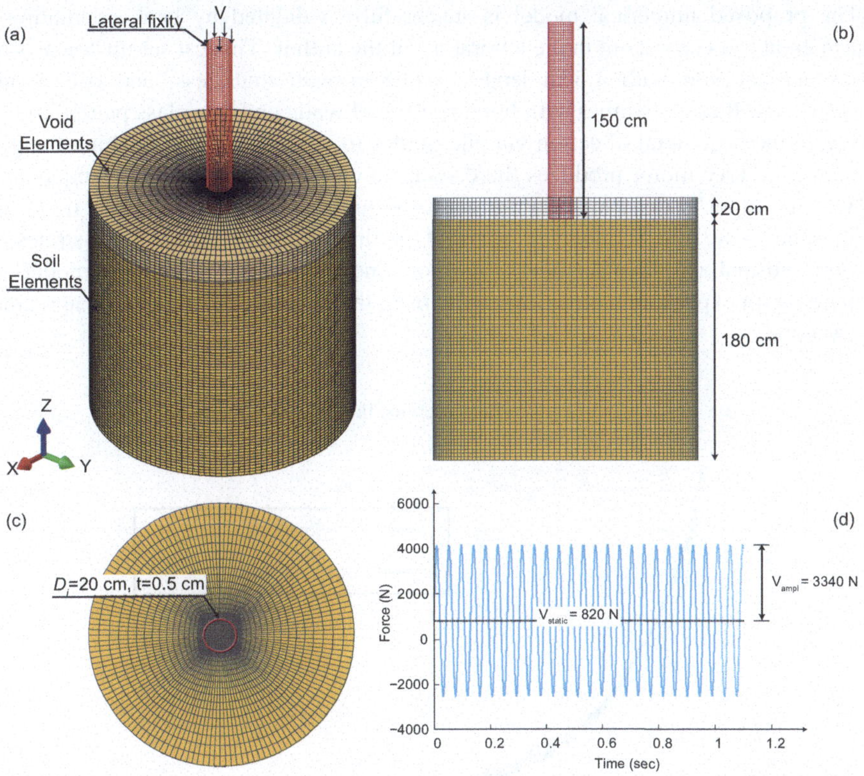

Fig. 3. FE model of pipe pile installation: (a) isometric, (b) side, and (c) planar view, and (d) vibratory loading time function [12]

The overall soil and void volume with 2 m height and 1 m radius is discretized by means of MMALE element formulation using one-point integration with in total 367,200 elements. The void domain with 0.2 m height, which has neither mass nor strength, was defined above the soil material to enable the soil to move to this domain during penetration of the pile. The lateral sides of the soil are constrained against movements in a direction perpendicular to their faces, while fixity in all directions is applied to the bottom of the soil.

A simple Mohr-Coulomb constitutive equation is used to capture the Berlin sand behavior in the model with parameters density 1,900 kg/m^3, friction angle 35°, dilatancy angle 1°, cohesion 1 kPa, Young's modulus 20 MPa and Poisson's ratio 0.2. The initial stress state in the soil is prescribed using the gravity acceleration of 10 m/s^2. Penalty contact between the Lagrangian pile elements and the MMALE soil elements is defined with a tangential friction coefficient of 0.1. The equipotential smoothing technique is applied where the computational grids are rearranged to maintain the mesh quality [13]. For the ALE advection step, the 2nd-order accurate van Leer method is chosen [14].

The proposed numerical model is successfully validated by back-calculating an experimental test carried out in the laboratory of the author. The test set-up consists of a half-cylindrical pile with 1.5 m length, 0.005 m wall thickness, and 0.2 m outer diameter as well as a chamber with three rigid steel walls and one glass panel. The pile is fixed in the horizontal direction via pile guides to ensure penetration along the glass panel. A vibratory motor produces the driving force of 1,670 N with the frequency of 23 Hz. The imposed dead load on the pile is about 410 N. The chamber is filled with Berlin sand. The pile is modeled as rigid in this analysis. Figure 4 illustrates the resulting normalized displacement curve obtained from the numerical model and compared with experimental measurements from the model test. The overall agreement is very good.

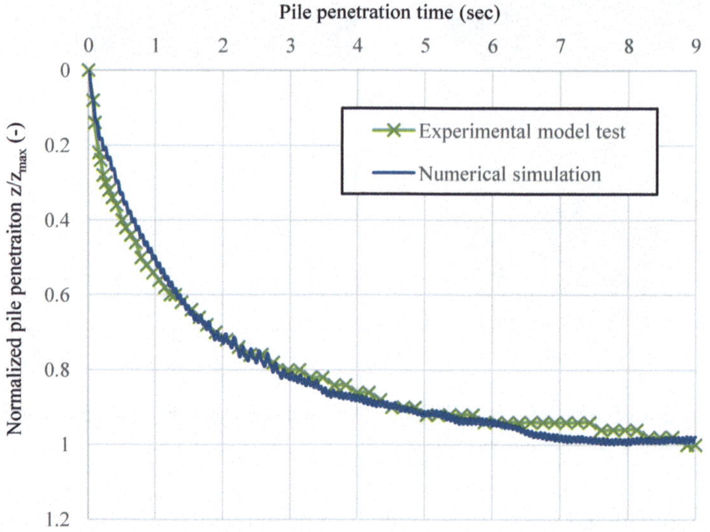

Fig. 4. Normalized model pile penetration z/z_{max} vs. time: experimental model test measurements and numerical simulation results

To investigate the pile buckling behavior during vibratory installation with 23 Hz under different conditions the elastic-plastic pile was first numerically penetrated into the homogeneous model (Fig. 3) for about 8 s, which corresponds to 0.65 m penetration. No pile buckling or large deformations occurred until this penetration depth. The vertical stress contours in the soil during vibratory penetration are shown in Fig. 5. Six different time steps from 6.29 to 6.34 s were selected, corresponding to slightly more than one cycle of sinusoidal frequency loading (Fig. 5a).

For results interpretation the evaluation criteria are mean strain, internal energy, pile vertical displacement, and pile lateral (horizontal) displacement. The mean strain is calculated as one-third of the strain tensor trace and defined based on the infinitesimal theory. The internal energy is the work done to induce strain in a unit volume of the solid part which can be used here to evaluate the accumulated strain in the pile during

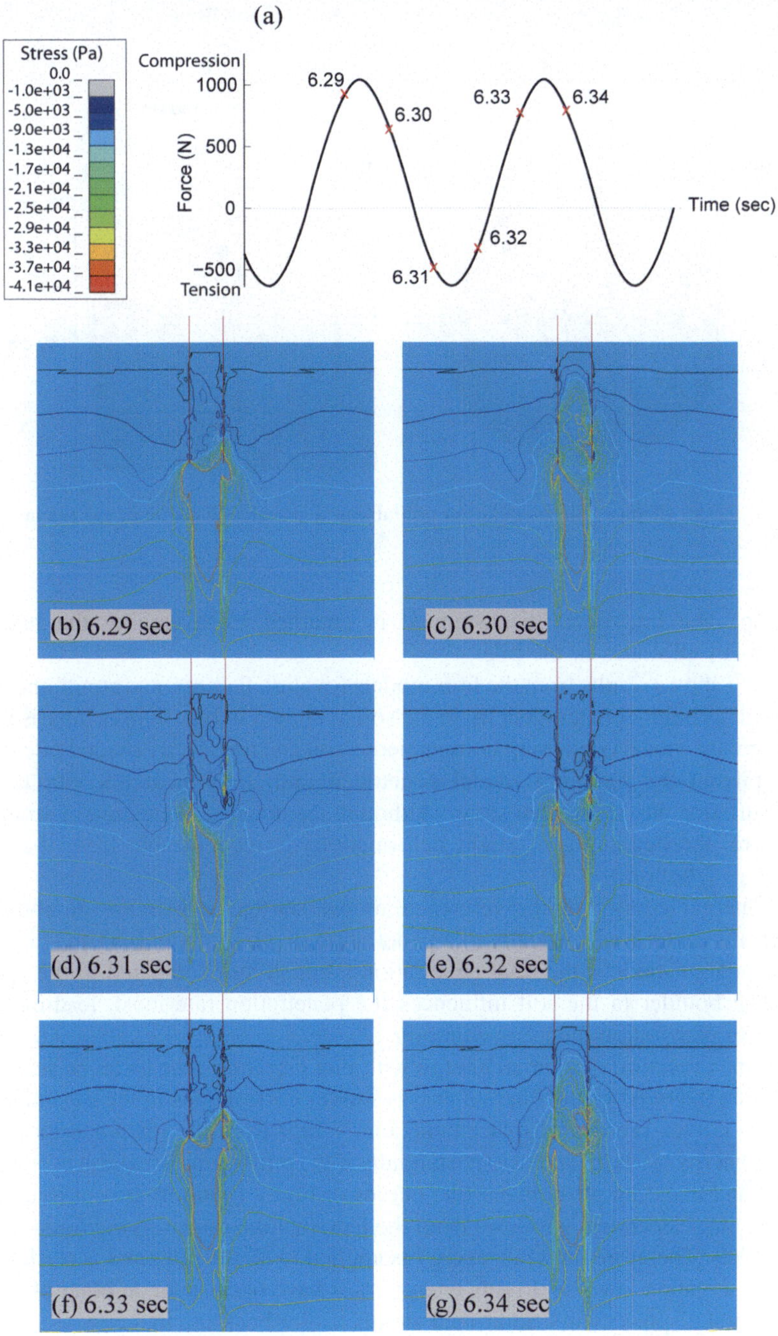

Fig. 5. Vibratory loading function (a) and isolines of induced vertical stresses in the soil at different times during one vibration cycle (b) to (g)

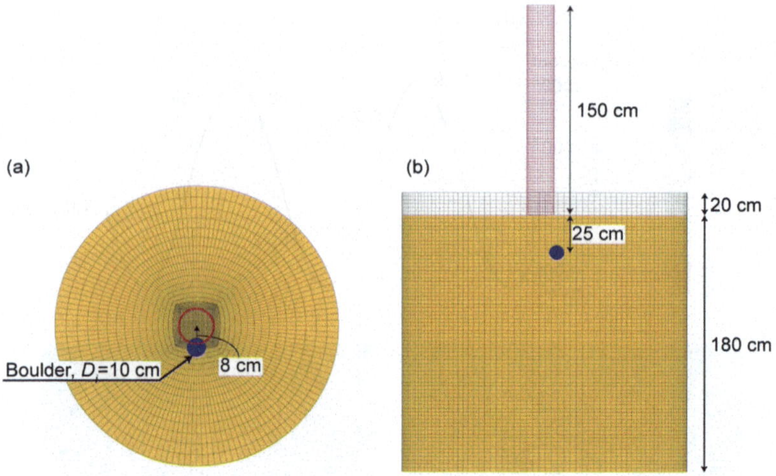

Fig. 6. FE model of pipe pile installation containing a boulder in the soil: (a) planar, (b) side view [12]

installation. The horizontal displacement is obtained by averaging the nodal displacements of all nodes in the pile.

Figure 7 shows results from the four models (ovality, flatness, and straightness of the pile, as well as boulder heterogeneity in the soil as shown in Fig. 6) in comparison to the initial reference model containing homogeneous soil and a perfectly shaped pile (Fig. 3).

The pile in the reference model penetrated until 0.65 m at 8 s vibration time without suffering any significant strain. Although the detailed evaluation of strain is not shown here, the beginning of slight deformations is recognizable from the internal energy (Fig. 7 bottom).

The internal energy in the reference model started to increase at about 7.2 s vibration time, which corresponds to a penetration depth of about 0.63 m (Fig. 7 top). In the other four models, the final penetration is much less compared to the reference model. The boulder in the soil influences the penetration rate most, leading to only about 0.3 m penetration after 8 s vibration.

From the evolution of internal energy with pile vibration time it can be seen (Fig. 7 bottom), that the trend of the internal energy curves corresponds to the deviation of the penetration curves (vertical displacement, Fig. 7 top) from the reference model. That means the energy starts to increase substantially if the pile begins to exhibit accumulation of strains and buckling phenomena (the curves in Fig. 7 (bottom) are cut to a value of 42 J). This interdependency follows from the fact, that the same driving force was used for all models. Therefore the same energy is applied to all the piles and according to the energy conservation law, the driving energy must have been spent on other phenomena such as lateral displacements, additional strains and/or buckling in a pile.

Consequently the evolution of the horizontal pile displacement with time (Fig. 7 middle) correlates in general also with the before mentioned behavior. But the lateral

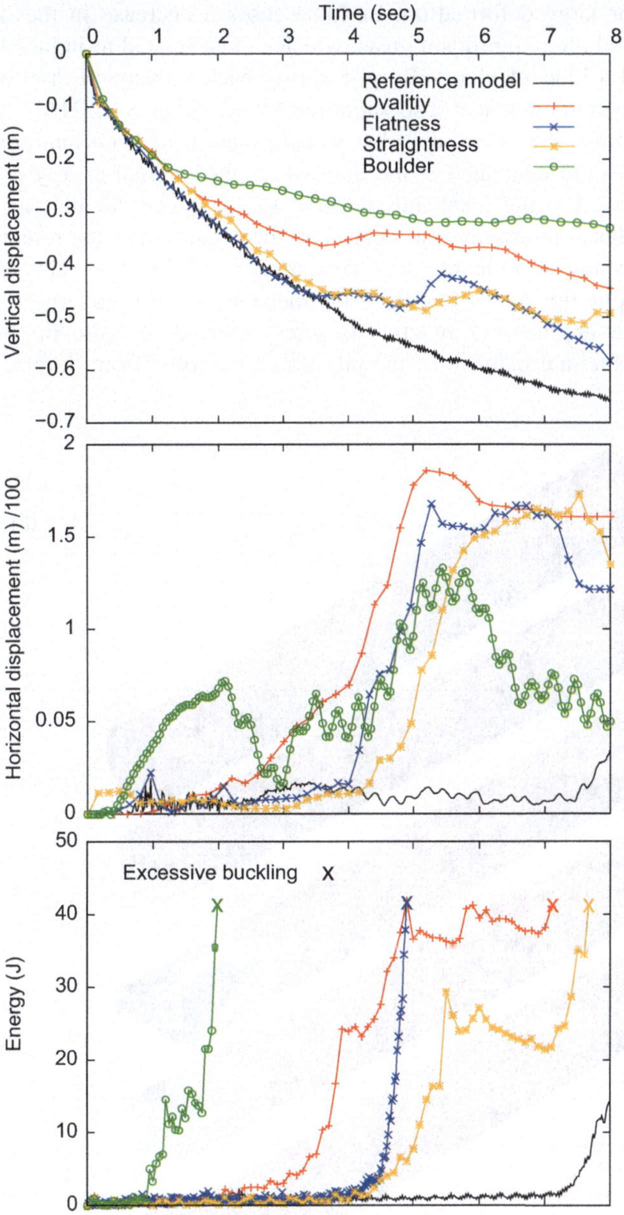

Fig. 7. Influence of pile imperfections and soil heterogeneity on pile penetration behavior: (top) vertical pile displacement, (middle) horizontal pile displacement, and (bottom) internal energy [12]

pile displacement is limited and maintained after some penetration due to the resistance of the surrounding soil. Thus, the remaining driving energy must have been spent on buckling and/or large deformation. In some cases a decrease in the internal energy value is observed after significant grow, which can be related to induced elastic strains in the pile and a kind of elastically pile spring back movement due to further penetration. This phenomenon was also supposed by Aldridge et al. [15].

Figure 8 shows the pile shape due to deformation of each analyzed model. The deformed shapes are evaluated at the time where the internal energy reached a maximum. Also plotted is the mean infinitesimal strain as contour plot on the deformed shape of the piles at the same time step. With the exception of the reference model, all other models show significant large deformation of the pile. This deformation is greatest at the pile tip. At the pile tip is the mean strain also the largest. Therefore this phenomenon in engineering practice is often referred to “pile tip buckling”. The buckling and potential damage of the pile starts to evolve from the tip.

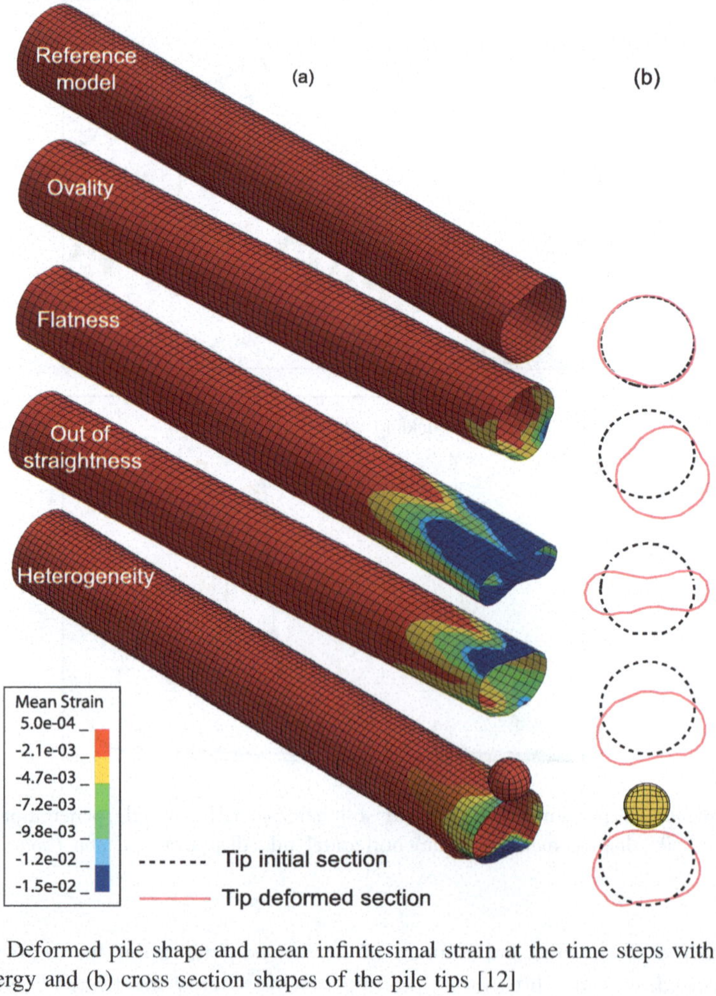

Fig. 8. (a) Deformed pile shape and mean infinitesimal strain at the time steps with maximum internal energy and (b) cross section shapes of the pile tips [12]

The amount of large deformation depends on the initial imperfection of the pile and the heterogeneity of the soil respectively. Each pile shows a different buckling mode due to the different initial imperfection. The largest pile tip deformation occurs with an initially flat pile. It is also interesting to note, that the deformed pile shape becomes non-symmetric with further penetration. It can be observed from the numerical results (Fig. 8b), that the cross section shapes of the tips of the imperfect piles tend to take the forms of the so-called "peanut-shape" as reported also in the literature [15].

It can be assumed that the proposed numerical model covers the complex pile-soil interaction behavior during dynamic installation of open-ended tubular piles quite well. Phenomena and boundary conditions like soil resistance, pile imperfection, and heterogeneity are included therein.

3.2 Sand Column Collapse

The controlled outflow of a defined volume of sand out of a hollow cylinder or container is a common laboratory experiment to investigate fundamental granular flow phenomena and behavior. This experiment is also called sand column collapse in the literature [16, 17].

The general experimental set-up is shown in Fig. 9 left. A column of sand is held in rest inside an rectangular container. The holding gate or wall is suddenly uplifted allowing the sand to collapse and flow out by its own weight. The investigations performed by Lube et al. [16] are used as an experimental reference to analyze and to validate own numerical ALE models.

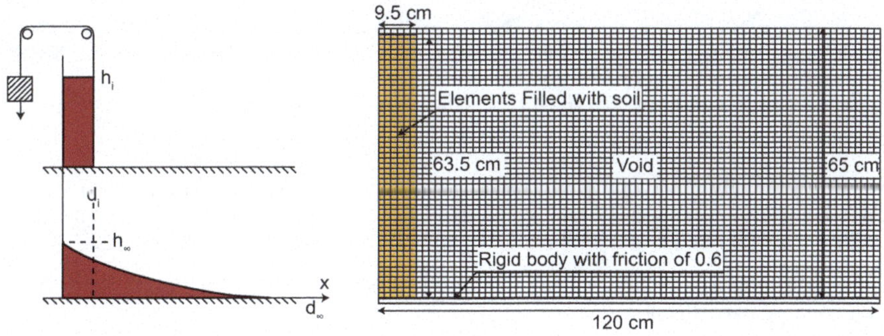

Fig. 9. Experimental set-up (left) and numerical ALE model in LS-Dyna (right) of the sand collapse problem [18]

The left hand side wall of the experimental set-up is considered to be a symmetry condition in the numerical model. Initial dimensions of the soil column are: width 9.5 cm, height 65 cm and depth 20 cm in direction normal to flow.

Two different constitutive models are used to represent the soil behavior. First a simple Mohr-Coulomb relation and second an advanced hypoplastic model. The hypoplastic model parameters are estimated to fit the Mohr-Coulomb parameters using numerical simulations of triaxial tests. This procedure and the resulting model parameters are given in Bakroon et al. [18].

In the 3D finite element model single point reduced integration elements with 15 mm element size are used. To achieve a plane strain condition one element in depth direction was modeled. A void region is defined to let the soil material flow to these void elements after the sand solumn collapse initiates. The container is modeled as a frictionless rigid body part which is removed as the initial stresses are initialized. Gravity acceleration is assigned with 9.806 m/s^2 in order to calculate the initial stresses. The modeled surface friction (penalty contact) of the rigid body at the bottom where the sand runs along is taken equal to the internal friction of the sand. Sand column collapse is initiated by abruptly lifting up one side of the rectangular container. The total calculation time of the problem is 2 s only.

In Fig. 10 the sand surface evolution and run-out distances with time are shown. Although all simulations give acceptable results in comparison with the experiment, the best fit is obtained when using a hypoplastic soil model in connection with the MMALE numerical model. Both the run-out distances and the heights of the sand slope with time are best predicted by hypoplasticity compared to the simple Mohr-Coulomb model. The worst result compared to the experiment is achieved using the CEL approach in conjunction with the Mohr-Coulomb model.

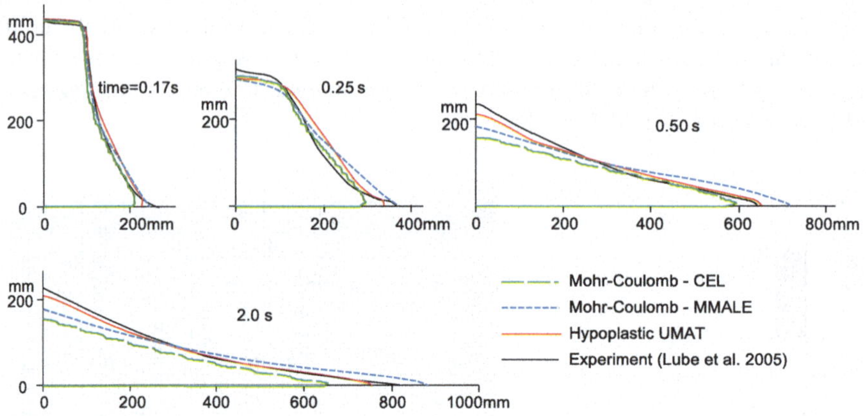

Fig. 10. Sand surface evolution with time: comparison of numerical results with the sand solumn collapse experiment

A fair judgment between the different ALE models is to compare CEL with MMALE by using the same simple Mohr-Coulomb model. The CEL simulation underpredicts the sand surface at any time step, whereas the MMALE simulation is closer to the surface measured in the experiment, but after some distance from the container it predicts a higher surface and longer run-out distance than in the experiment. The difference in the predicted run-out distance between CEL and MMALE is about 200 mm with the experiment being closer to the MMALE result.

Another interesting results evaluation concerns the void ratio distribution calculated with the hypoplastic MMALE numerical model, which is presented in Fig. 11 at various times. The initial void ratio was 0.695 before the sand column collapse started.

The minimum and maximum void ratios of the modeled sand are 0.55 and 0.95 respectively. Right after the collapse was initiated the void ratio tends to increase, only at the bottom and inside the core of the sand column there is the initial void ratio kept. This core of the column is compacted in the following time but its expansion size remains almost constant during the entire analysis.

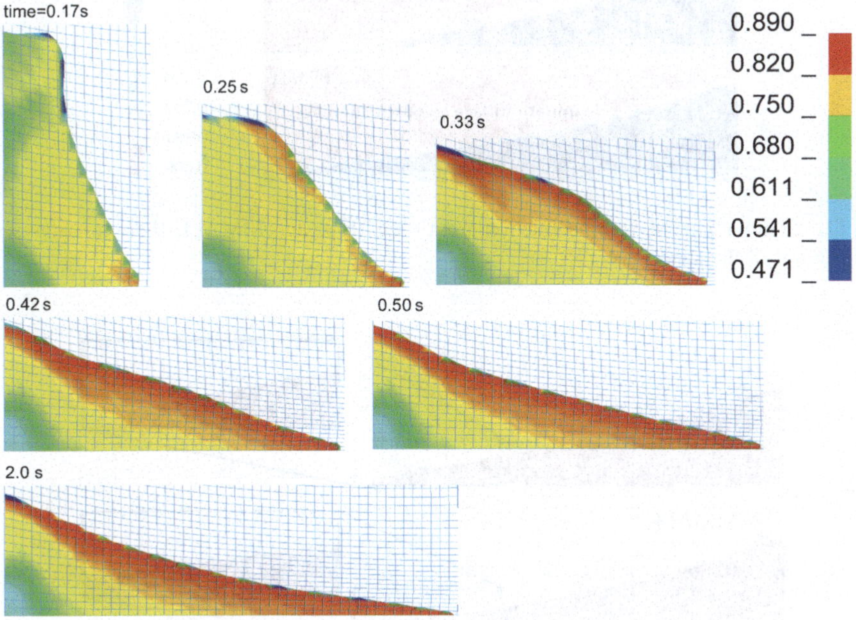

Fig. 11. MMALE with hypoplastic soil model: void ratio distribution at various times of the simulation.

The top sand layers exhibit loosening behavior (red color in Fig. 11), but do not reach the maximum void ratio of the sand. Finally at the end of the flow process a clearly visible three layered structure of the sand mass is achieved, with a dense core, a medium dense middle part (yellow color in Fig. 11) and a loose top layer. The void ratio distribution in the results must be further evaluated to demonstrate whether the change of void ratio over the height is continuous or not.

The interpretation of the void ratio at the surface has to be done with special care. Due to the presence of the void elements outside the sand volume the void ratio contour values are averaged void ratio values for mixed elements containing soil and void. Since void ratio of the void elements is always zero, the resulting averaged void ratio is not the real void ratio of the soil.

A detailed comparison of the CEL and the MMALE models is shown in Figs. 12 and 13. The final shape of the sand surface as well as the interface details at the end of the flow process highlight the difference between CEL and MMALE models (Fig. 12). The fixed Eulerian mesh in the CEL model leads to a rough surface reconstruction, illustrated by means of the plotted volume fraction contours. It results in a stronger numerical diffusion in the CEL simulation, which can also explain the shorter run-out distance from this analysis compared to MMALE.

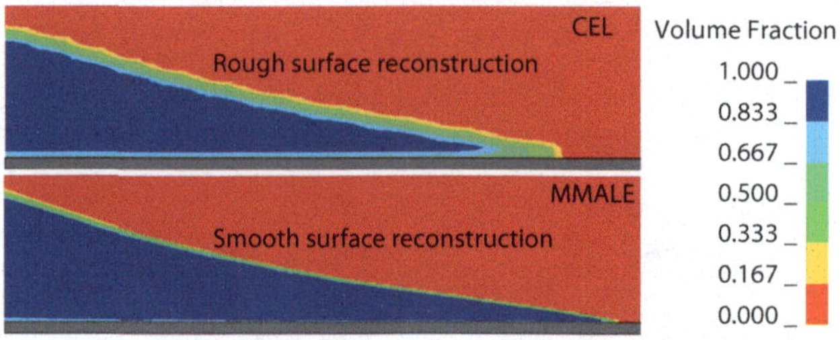

Fig. 12. Contour plot of volume fraction from CEL (top) and MMALE (bottom) analyses at final run-out of the sand mass [4]

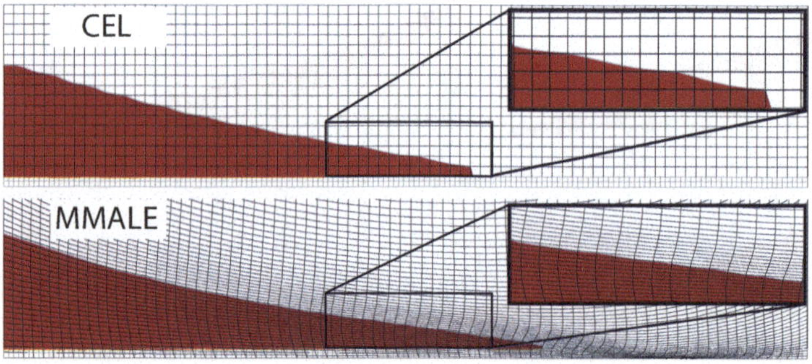

Fig. 13. Shape of material and mesh from CEL (top) and MMALE (bottom) analyses at final run-out of the sand mass [4]

The smoother surface reconstruction is also clearly visible in Fig. 13 (bottom). MMALE is capable of doing a much smoother surface reconstruction compared to CEL due to the mesh update algorithms. Of course one can achieve a smoother surface with CEL too, but that requires a finer mesh in the whole discretized domain, leading to much higher computational costs.

4 Summary and Conclusions

The performance of ALE methods is evaluated by means of two application examples: the vibratory driven installation of open-ended tubular steel piles in sand and the sand column collapse problem. The MMALE soil-pile interaction model provides reliable measures to assess pile buckling. As a result from this pile buckling analyses it can be argued that the driving energy for the pile installation is reflected in the model mainly in three different forms: vertical and horizontal pile displacement, and pile buckling. Overall it can be concluded that the proposed numerical model covers the complex soil-pile interaction behavior during dynamic installation of open-ended tubular piles quite well. Phenomena and boundary conditions like soil resistance, pile imperfections, and heterogeneity are included therein.

From the second application, the sand column collapse, different results from two constitutive models as well as two ALE approaches are gained. The latter comparison proofs the advantage of MMALE with much smoother surface reconstruction compared to CEL due to the mesh update algorithms in MMALE with less computational costs.

All simulation results assess the feasibility of ALE methods in geotechnical large deformation problems, concluding that the ALE method is a promising framework for the solution of complex large deformation problems in geotechnical engineering.

Acknowledgments. The author is grateful for carrying out the numerical analysis and preparation of figures by Montaser Bakroon and Reza Daryaei. Special thanks to Daniel Aubram for countless discussions over the years, not only about the ALE methods. Last but not least, the author would like to acknowledge his highly esteemed colleague Prof. Dr.-Ing. habil. Theodorus Triantafyllidis for numerous inspiring and creative discussions in joint research projects of the past years - thank you very much!

References

1. Beuth, L., Benz, T., Vermeer, P.A., Coetzee, C.J., Bonnier, P., van den Berg, P.: Formulation and validation of a quasi-static Material Point Method. In: Numerical Models in Geomechanics: Proceedings of the 10th International Symposium on Numerical Models in Geomechanics (NUMOG X), Rhodes, Greece, 25–27 April 2007 (2007)
2. Bui, H.H., Fukagawa, R., Sako, K.: Smoothed particle hydrodynamics for soil mechanics. In: Numerical Methods in Geotechnical Engineering, pp. 275–281 Taylor & Francis Group, London (2006)
3. Benson, D.J.: An efficient, accurate, simple ALE method for nonlinear finite element programs. Comput. Methods Appl. Mech. Eng. **72**(3), 305–350 (1989)
4. Bakroon, M., Daryaei, R., Aubram, D., Rackwitz, F.: Investigation of mesh improvement in multi-material ALE formulations for large deformation problems. Int. J. Geomech. ASCE (2019, submitted)
5. Qiu, G., Henke, S., Grabe, J.: Application of a Coupled Eulerian-Lagrangian approach on geomechanical problems involving large deformations. Comput. Geotech. (2011). https://doi.org/10.1016/j.compgeo.2010.09.002
6. Qiu, G., Henke, S.: Controlled installation of Spudcan foundations on loose sand overlying weak clay. Marine Structures (2011). https://doi.org/10.1016/j.marstruc.2011.06.005
7. Tho, K.K., Leung, C.F., Chow, Y.K., Swaddiwudhipong, S.: Eulerian finite-element technique for analysis of Jack-Up Spudcan Penetration. Int. J. Geomech. ASCE (2012). https://doi.org/10.1061/(ASCE)GM.1943-5622.0000111
8. Hamann, T., Qiu, G., Grabe, J.: Application of a Coupled Eulerian-Lagrangian approach on pile installation problems under partially drained conditions. Comput. Geotech. (2014). https://doi.org/10.1016/j.compgeo.2014.10.006

9. Di, Y., Yang, J., Sato, T.: An operator-split ALE method for large deformation analysis of geomaterials. Int. J. Numer. Anal. Meth. Geomech. (2007). https://doi.org/10.1002/nag.601
10. Aubram, D., Rackwitz, F., Wriggers, P., Savidis, S.A.: An ALE method for penetration into sand utilizing optimization-based mesh motion. Comput. Geotech. (2015). https://doi.org/10.1016/j.compgeo.2014.12.012
11. Aubram, D., Rackwitz, F., Savidis, S.A.: Contribution to the Non-Lagrangian formulation of geotechnical and geomechanical processes. In: Triantafyllidis, T. (ed.) Holistic Simulation of Geotechnical Installation Processes – Theoretical Results and Applications. LNACM, vol. 82. Springer, Heidelberg (2017). https://doi.org/10.1007/978-3-319-52590-7_3
12. Bakroon, M., Daryaei, R., Aubram, D., Rackwitz, F.: Numerical evaluation of buckling in steel pipe piles during vibratory installation. Soil Dyn. Earth. Eng. (2019). https://doi.org/10.1016/j.soildyn.2018.08.003
13. Winslow, A.M.: Equipotential zoning of two-dimensional meshes (UCRL-7312). United States (1963)
14. van Leer, B.: Towards the ultimate conservative difference scheme. J. Comput. Phys. (1997). https://doi.org/10.1006/jcph.1997.5704
15. Aldridge, T.R., Carrington, T.M., Kee, N.R.: Propagation of pile tip damage during installation. Frontiers in offshore geotechnics. In: Proceedings of the 1st International Symposium on Frontiers in Offshore Geotechnics, ISFOG 2005, p. 823-7 (2005)
16. Lube, G., Huppert, H.E., Sparks, R.S.J., Freundt, A.: Collapses of two-dimensional granular columns. Phys. Rev. E (2005). https://doi.org/10.1103/PhysRevE.72.041301
17. Thompson, E.L., Huppert, H.E.: Granular column collapses: further experimental results. J. Fluid Mech. (2007) https://doi.org/10.1017/S0022112006004563
18. Bakroon, M., Daryaei, R., Aubram, D., Rackwitz, F.: Implementation and validation of an advanced hypoplastic model for granular material behavior. In: Proceedings of the 15th International LS-DYNA Users Conference (2018). https://www.dynalook.com/conferences/15th-international-ls-dyna-conference

Approaches for the Design of Foundations for Offshore Wind Turbines

A Review Based on Comparisons with HCA-Based Models

Hauke Zachert[1](✉) and Torsten Wichtmann[2]

[1] Institute and Laboratory of Geotechnics, Technische Universität Darmstadt, Franziska-Braun-Straße 7, 64287 Darmstadt, Germany
zachert@geotechnik.tu-darmstadt.de
[2] Chair for Soil Mechanics, Foundation Engineering and Environmental Geotechnics, Ruhr-University Bochum, Universitätsstraße 150, 44789 Bochum, Germany
torsten.wichtmann@rub.de

Abstract. The design of foundations for offshore wind turbines is still a challenging engineering task. Design rules are not always settled in national standards and especially tasks like the design against cyclic loading remain a field where results of scientific research are immediately implemented in practical design approaches. The PISA project demonstrated, that even the static design of monopiles can be improved by validating new design approaches and implementing them in design tools. This paper compares different approaches for the static and cyclic design of foundations for offshore wind turbines, including new approaches like PISA, SOLDYN and TANDEM, as far as available. Focus is laid on the validation process of a cyclic approach, based on the high cycle accumulation (HCA) model. The main results of this validation on monopiles and shallow foundations in different scales are discussed. The HCA model showed to be very well suited to simulate and predict the cyclic soil structure interaction processes of foundations for offshore wind turbines. Effects like the accumulation of deformations or the redistribution of stresses and internal forces induced by a cyclic loading on monopiles or shallow foundations are well reproduced by the HCA model. This is demonstrated based on a comparison of results from model tests and in situ tests with numerical predictions. Furthermore, the results of these studies on the HCA model are set in relation to the approaches from other research groups. The outcome of the PISA project for the static design of monopiles is shown and reviewed based on the own results of numerical simulations.

1 Introduction

Wind energy has a key role in the worldwide transition from fossil energy sources to renewable energies. While onshore wind energy can be considered a matured industry, the offshore wind market still exhibits a vast growth. By 2018, a total capacity of 23.1 GW was installed offshore worldwide and forecasts predict a growth to about 100 GW

T. Triantafyllidis (Ed.): *Recent Developments of Soil Mechanics and Geotechnics in Theory and Practice*, LNACM 91, pp. 113–135, 2020.
https://doi.org/10.1007/978-3-030-28516-6_7

by 2025 [18]. Europe has the highest installed offshore wind capacity, but Asia is catching up and shows yearly additions of about 5–7 GW [18]. New markets like Taiwan are developing and existing markets like Europe experience a boost by the first auctions which were closed by so called "zero bids", where the development, construction and operation of the offshore wind farms is offered subsidy free [45]. This encourages the public, like the Dutch government to increase their targets for the offshore wind development.

These achievements were enabled by a huge professionalisation of the whole supply chain but also by a continuous increase of turbine size and efficiency [18]. A turbine capacity of about 8 MW is already reached and the trend leads towards 12 MW [13]. This goes along with an increase in the diameter of wind turbine blades, which already reached 167 m for a modern 8 MW turbine [44].

Increasing rotor diameters will transfer increasingly higher loads to the foundations. The dominant foundation type for offshore wind turbines (OWT) is the monopile: a single pile with about 6 m to 8 m diameter and an embedment length of about 20 m to 30 m below seabed. It represents about 80% of the installed OWT foundations [46]. Beside monopiles, jacket or tripod foundations and gravity base foundations have been applied for the secure transfer of offshore loads to the ground. OWT foundations have to withstand different types of loading: intense horizontal loads due to wind and large single waves, vertical loads caused by the self-weight of the turbine and a high-cyclic loading due to wind and water waves. The latter may cause an accumulation of permanent deformations in the soil. The serviceability of the OWTs may get lost due to an excessive tilting of the tower, considering the strict operational requirements of the turbines (a maximum tilting of 0.25° is often stated, see e.g. [15] or [31]). In addition, the change of the bedding reaction due to an accumulation of deformation or changes of lateral stress along the pile shaft may cause a shift in the eigenfrequencies of the turbine. Experience from conventional offshore foundations (e.g. oil rigs) cannot be easily adapted since the ratio of the horizontal cyclic loads and the self-weight of the structure is significantly larger for the OWTs. Early offshore wind parks had smaller turbines (2 to 3 MW) and were installed in shallow waters up to 20 m [46]. New offshore windfarm areas are predominantly assigned further offshore with increasing water depths up to 40 m [46]. Thus, bending moments in monopile foundations are much larger for the new OWT foundations. Furthermore, the increased rotor diameter and the increased height of the turbines lead to higher loads transferred to the foundations. Practical long-term experience regarding such large pile diameters in combination with high ratios of cyclic horizontal load to the self-weight of the structure is still limited.

Only few design codes like API RP2GEO [2], DIN EN ISO 19901-4 [12], DIN EN ISO 19902 [11] or DNV-GL [15] are considering cyclic loading at all and their approaches are based on in situ tests on flexible piles with high slenderness (length L to diameter D ratio), diameters below 1 m and subjected to only about 100 load cycles. Since the discrepancy to the current industry practice for monopiles applying small slenderness and large pile diameters, and considering high numbers of load cycles is obvious, several research projects on horizontally loaded monopiles were conducted in the past decade.

Besides the few mentioned design standards, some technical recommendations for the design of offshore wind turbine foundations were established based on these research projects to support the safe design of such foundations, like BSH-standard [9],

SOLCYP [15] or EA-Pfähle [16]. Still, most of these recommendations are based on small-scale 1 g-model tests or centrifuge tests. Large-scale in situ tests on horizontally loaded monopiles have been performed recently to support the validation of such approaches but are mostly still under evaluation or have only recently been published [7, 16, 33].

The most common approaches for the design of offshore monopiles will be briefly discussed in Sect. 2.

2 Industry Standards for the Design of Foundations for Offshore Wind Turbines

The general design of offshore wind turbine foundations has to comply with the applicable standards and regulations, e.g. defined in the BSH-Standard [9], DIN EN 1997-1 [14] in combination with EA-Pfähle [16], API [2] or DNVGL [15]. However, as stated before, most of these regulations do not refer to the design of cyclically loaded foundations on a validated basis or they have not been validated for the current standard of large diameter monopiles. For monopiles, most designs follow the API-approach [2] with p-y curves defining the load-deflection relationships for the horizontal soil reaction. Cyclic loading is only referred to by a moderate reduction of the stiffness.

Several dominant effects for the safe and economic design of OWT foundations are neglected or not accurately covered by the API-approach. The horizontal bearing behaviour of very stiff monopiles like the modern XXL-monopiles has not been studied in detail yet (see e.g. [13, 21]), the accumulation of deformations due to the high-cyclic loading is not covered accurately, the potential accumulation of excess pore water pressure is often neglected [21] and furthermore, the soil stiffness during dynamic loading is not covered accurately although it heavily affects the dynamic response of the wind turbine [32]. These uncertainties lead to the development of a variety of design approaches on national and international level. One approach for the design of monopiles which is addressing the mentioned inaccuracies is described in [21] and [31], another one is proposed in [4].

The behaviour of monopiles with large diameters is often studied by means of 3D finite element simulations, which is also recommended by DNVGL [15] to overcome the uncertainties regarding the large pile diameters. However, a validation of these numerical approaches can only be achieved by large-scale tests as they are currently performed in several studies, like the PISA [7] or the TANDEM [33] projects.

In the TANDEM project, two series of piles were tested in the "Testzentrum Tragstrukturen Hannover" by Fraunhofer IWES. With reference to a monopile with Diameter $D = 8.0$ m, embedment length $L_{pen} = 32.0$ m and a maximum horizontal loading $H_{ULS} = 25.4$ MN, three different test piles with different scales were tested in sand compacted in layers: D1220 ($D = 1.220$ m, $L_{pen} = 7.50$ m, $H_{ULS} = 217$ kN), D914 ($D = 0.914$ m, $L_{pen} = 6.15$ m, $H_{ULS} = 108$ kN) and D610 ($D = 0.610$ m, $L_{pen} = 4.70$ m, $H_{ULS} = 45$ kN). The deflection curves of a first series of these three test piles in the TANDEM project are shown in Fig. 1, together with the deflection obtained from 3D finite element simulations applying a modified hypoplastic constitutive model for the sand [13]. In principle, the agreement between the test data and the model prediction seems to be very good, but especially the deformation at the pile tip, which

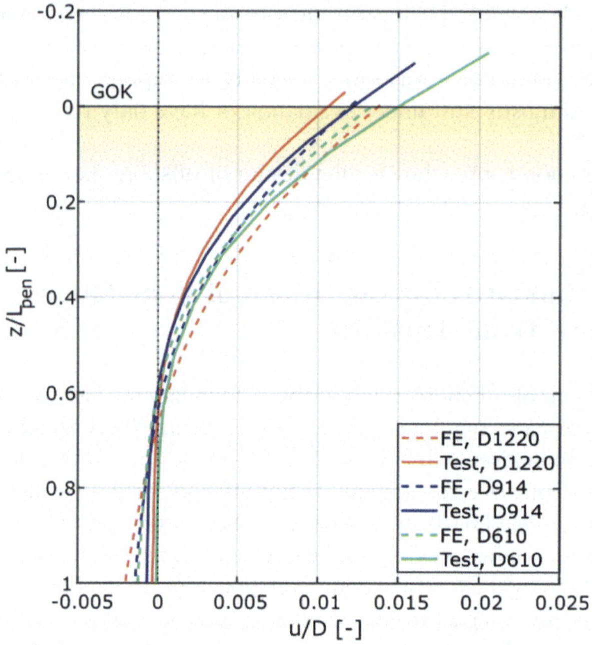

Fig. 1. Pile deflection curve for three tests in the TANDEM project ("Test") and their numerical approximation ("FE"), D1220 ($D = 1.220$ m, $L_{pen} = 7.50$ m, $H_{ULS} = 217$ kN), D914 ($D = 0.914$ m, $L_{pen} = 6.15$ m, $H_{ULS} = 108$ kN) and D610 ($D = 0.610$ m, $L_{pen} = 4.70$ m, $H_{ULS} = 45$ kN) [13].

is critical for the development of the base shear (see explanations below) shows some deviations from the tests.

Especially the PISA project is attracting a lot of attention. In [8] a new approach for the static design of monopiles was proposed which is based on the results of the large-scale monopile tests in [7] and [5], where monopiles were tested in sand and in clay. The monopiles in sand had dimensions of $D = 0.273$ m ($L/D = 5.25$, 8 and 10), $D = 0.762$ m ($L/D = 3$, 5.25 and 8) and $D = 2.00$ m ($L/D = 4$ and 5.25) and were subjected to horizontal static and cyclic loads. In contrast to the majority of the common approaches, it is explicitly considering large-scale monopiles and its almost rigid behaviour. This led to the introduction of a base shear force at the pile tip [6]. While the regular monopile design was requiring an almost vertical tangent at the pile tip or at least very small deflections (5% of the pile head deflections, see [21]), the PISA approach is acknowledging the more rigid pile behaviour by introducing this base shear. This general approach is also followed in the TANDEM project, see [13].

The PISA approach has already been introduced into a commercial PLAXIS tool, called MoDeTo (Monopile Design Tool), which enables designers to apply the PISA approach for their monopile design, see e.g. [27]. Up to now, the PISA approach is only referring to the design of monopiles subjected to static loading. An extension for cyclic loading has not been published yet. The PISA approach will be discussed in more detail in Sect. 4.1.

The cyclic behaviour of piles was the main subject of a French research group which gathered their results in the SOLCYP recommendations [29]. The recommendations concentrate on piles under vertical cyclic loading and are based on results of small-scale model tests, large-scale field tests and centrifuge tests on piles. Horizontally loaded piles have been studied in detail by means of centrifuge model tests in non-cohesive soils [30]. Large-scale tests on horizontally loaded piles were not part of the program. Furthermore, the horizontally loaded piles investigated in the SOLCYP were rather flexible than rigid, which puts the applicability of the results to modern monopiles into question. Besides these limitations, the model tests showed some relevant results for the cyclic pile-soil-interaction. Figure 2 shows the cyclic *P*-*y*-curves for a model pile in dry sand at different depths below ground with *P* defined in [29] as the soil reaction of the pile at a certain depth. It is obvious, that the stiffness increases with depth and that the residual deformations grow with increasing number of cycles, which is usually considered by a decrease of the *p*-*y*-stiffness.

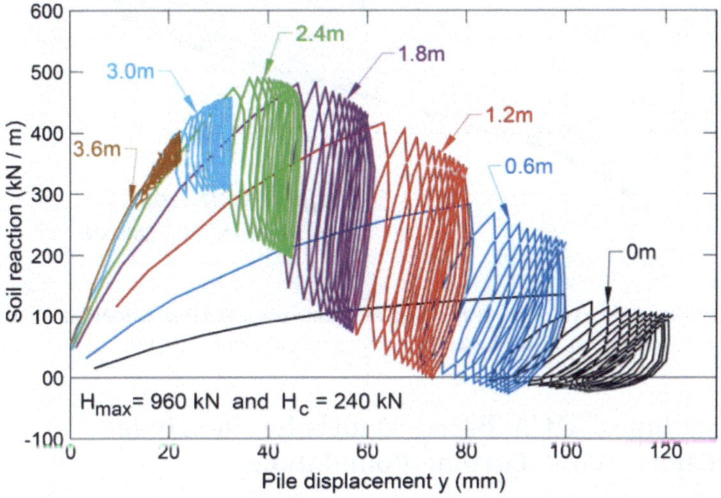

Fig. 2. Example of cyclic *P*-*y*-curves at different depths for a model monopile in dense (I_D = 86%) dry sand tested in a geotechnical centrifuge, representing a pile with L = 12.0 m and D = 0.72 m, [30] according to [29].

Furthermore, an increase of the bending moment of the pile was reported as shown in Fig. 3. This increase demonstrates the complex soil-structure-interaction occurring during cyclic loading. Such a change in the bending moment should not be neglected in a monopile design. Similar evolutions of bending moments have been reported in [47].

For the design of monopiles, most of the described shortcomings could also be overcome by a numerical model which is capable to capture all effects of a cyclic loading, in particular the accumulation of deformations or excess pore water pressures and changes of stresses in the soil and the structure, and the scale effects. For that purpose, the authors of the present paper propose the application of the High Cycle Accumulation model (HCA) developed at the Institute of Soil Mechanics and Rock Mechanics (IBF) in Karlsruhe. The huge advantage of such a numerical approach is the flexibility with regard to the foundation type and size investigated. Furthermore, it allows insights into the effects of cyclic soil-structure-interaction which are difficult to identify in model tests. The HCA model was validated by several comparisons with boundary value problems with high-cyclic loading [48]. Section 3 will show some examples for this validation.

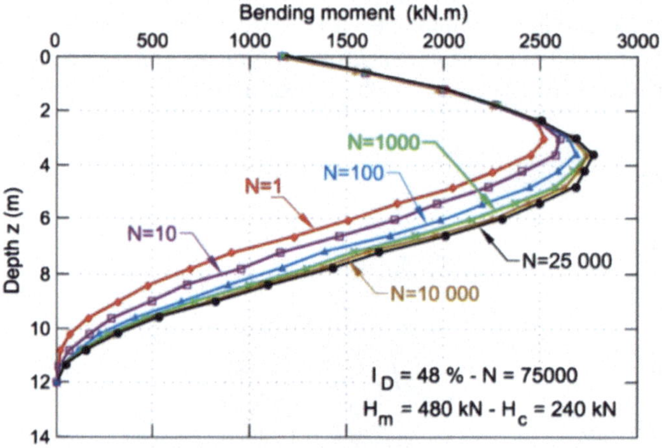

Fig. 3. Example of the evolution of bending moments for a flexible model pile (representing a pile with L = 12 m, D = 0.72 m) tested in a geotechnical centrifuge [30] according to [29].

3 Application of HCA Based Models for the Design of Offshore Wind Turbine Foundations

3.1 HCA Model

The procedure for the proof of serviceability for offshore wind turbine foundations is based on FE calculations with the HCA model. The HCA model of Niemunis et al. [25] is only applicable to a high-cyclic loading. Monotonic loading and unloading phases and a few number of cycles, especially at the beginning of a load package, have to be calculated with conventional soil models. Here, Hypoplasticity with Intergranular Strain [26] is used for these so-called *implicit* steps. Steps 1–5 in Fig. 4 represent for example the application of the self-weight and of the first two load cycles. Step 4 is called *irregular cycle*, because it includes the first loading to the maximum load and can produce a much higher residual strain than the following (*regular*) cycles. The strain path during the first regular cycle (step 5) is used to evaluate the strain amplitude $\varepsilon^{\mathrm{ampl}}$. One basic assumption of the HCA model is, that strain and stress paths resulting from cyclic loading can be divided into an oscillating part and a trend. The oscillating part is captured by ε^{ampl}, which is assumed to remain constant over a certain number of cycles with constant load amplitude. ε^{ampl} is one of the main input parameters of the HCA model, which is used for the calculation of the accumulation trend, i.e. the development of permanent deformations with increasing number of cycles, see step 6 in Fig. 4. This trend due to an increment ΔN of the number of cycles is calculated in analogy to a creep process within the so-called *explicit* step. During the explicit parts of the calculation the external loads are kept constant on their average values. The strain amplitude can be updated in a so-called control cycle calculated implicitly (step 7 in Fig. 4).

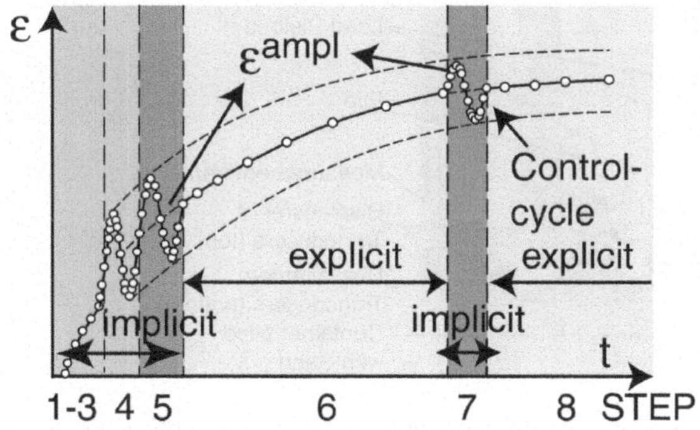

Fig. 4. Implicit and explicit steps in FE calculations with the HCA model.

The predition of the development of permanent deformations with the HCA model (step 8 in Fig. 4) is then continued with this updated value of ε^{ampl}. The constitutive equations are described in detail by Niemunis et al. [25]. The HCA model captures the main influencing parameters on the accumulated strain, such as the strain amplitude ε^{ampl}, the number of cycles N, the void ratio e, the average mean pressure p^{av}, the average stress ratio η^{av} and polarization changes. The equations of the HCA model are based on numerous cyclic triaxial tests on different sands [39].

3.2 Validation of the HCA – Based Approach for Monopiles

Developed in a theoretical framework and calibrated with numerous cyclic triaxial tests on sand, the HCA model has also been applied in several simulations of offshore wind turbine foundations subjected to cyclic loading (e.g. [43, 48, 50]).

The validation of the HCA model for the application on monopiles was based on small-scale 1 g-model tests in two different scales, see e.g. [50] and [43]. In the following mainly the small-scale model tests performed by the first author at the Institute of Soil Mechanics and Rock Mechanics (IBF) at Karlsruhe Institute of Technology (KIT) will be discussed.

The test setup, applied for the model tests on monopiles consists of a steel container (diameter: 95 cm, depth: 145 cm) which was filled with Karlsruhe fine sand (clean quartz sand, mean grain size $d_{50} = 0.14$ mm, uniformity coefficient $Cu = d_{60}/$

d_{10} = 1.5) by pluviation. The model pile was then driven by impact into the dry sand until it reached the target embedment length L which was chosen between 40 cm and 100 cm. Two separate frames, the measurement frame and the loading frame were mounted on top of the steel container. The test setup is presented in Fig. 5 together with a scheme showing the positions of sensors for displacement and strain measurements at the model pile.

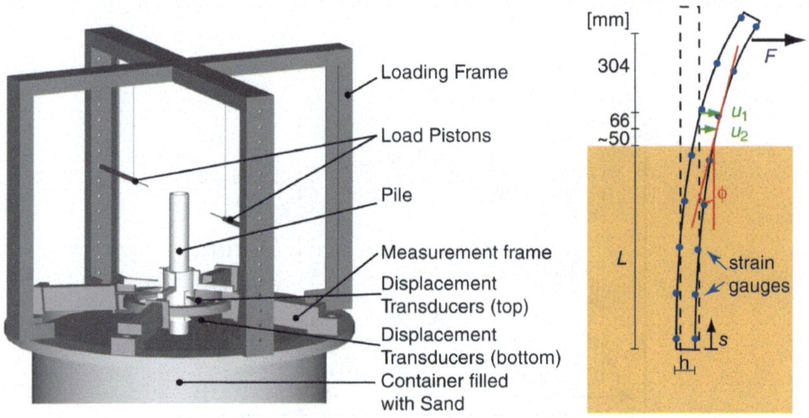

Fig. 5. Small-scale model tests on monopiles as a basis for the validation of the HCA-model approach.

The model pile consists of a PVC pipe (outer diameter: 61 mm, wall thickness: 3 mm) with an elastic modulus E = 3650 N/mm^2 leading to a bending stiffness EI = 0.8412 kNm. EI has been confirmed in calibration tests with static loading on a statically determined system. The pile was equipped with eight layers of strain gauges, equally distributed over the pile length. From the strain difference $\Delta\varepsilon$ measured for each layer the bending moments and the whole deflection curve of the model pile can be derived for every load cycle. This enables a deeper insight into the cyclic soil-pile interaction compared to the common practice where only the pile displacements are measured close above the sand surface, e.g. in [17] or [22]. For details of the model tests at KIT the interested reader is referred to [48, 50] and [47].

The model tests formed the basis for numerical simulations applying the HCA-based approach as described in Sect. 3.1. Static load steps were calculated with Hypoplasticity and intergranular strain while the high-cyclic loading was captured by the HCA model. All FE simulations were performed using the commercial FE software Abaqus Standard and applying material parameters derived independently from laboratory tests. These material parameters for the Karlsruhe fine sand (KFS) are shown in Tables 1 and 2.

All material parameters were completely obtained from element or index tests in the laboratory and were not fitted afterwards to reach a better prediction of the model test results. This independent determination of the parameters of the constitutive models used in the FE simulations represents an important requirement in order to validate the

Table 1. Material parameters of hypoplasticity and intergranular strain for KFS

φ_c	e_{d0}	e_{c0}	e_{i0}	h_s	n	α	β	R	m_R	m_T	β_R	χ
33.1	0.677	1.054	1.212	2000	0.32	0.12	2.0	10^{-4}	2.6	1.3	0.1	6.0

φ_c in [°] and h_s in [MPa]; all other parameters dimensionless

Table 2. Material parameters of the HCA model for KFS (all parameters dimensionless)

C_{ampl}	C_e	C_p	C_Y	C_{N1}	C_{N2}	C_{N3}
1.33	0.60	0.23	1.68	$2.95 \cdot 10^{-4}$	0.41	$1.90 \cdot 10^{-5}$

numerical approach. The FE model of the model tests is shown in Fig. 6. It utilizes the symmetry of the system by representing only one half of it. The pile installation process was not considered in the numerical simulations, that means the pile was modelled *wished in place*. Further details on the FE model can be found in [48] and [47].

As a first step of the validation, the pile behaviour under static loading in the model test and the FE simulations was compared, as shown in Fig. 6. The distribution of the bending moment along depth for different magnitudes of the horizontal load F applied at the pile head was very well captured by the corresponding numerical simulations. The same applied for the static deformations (see [48] for detailed results). The accurate simulation of the static load steps (*implicit steps*) is essential for a successful prediction with the HCA model since the strain amplitude ε^{ampl} is recorded in these steps, which is one of the decisive input parameters of the HCA model.

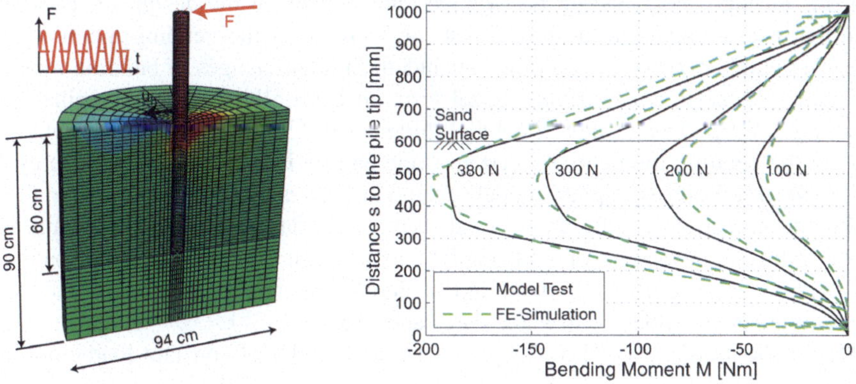

Fig. 6. FE model using C3D8R elements to simulate the model tests on monopiles and distributions of bending moments along depth measured in the model tests in comparison with the results from the numerical simulations for different loading stages.

Figure 7 presents the total horizontal displacements measured slightly above the sand surface in four tests with different constant load amplitudes $F^{ampl} = F^{av}$. Each test was conducted in a freshly pluviated sand. Only the average deformation u^{av} is displayed in Fig. 7. The measured total deformations (static and accumulated part, solid curve in Fig. 7) show again a very good agreement with the FE predictions (dashed curves in Fig. 7). However, the test results for F^{ampl} = 7.5 N reveal a certain scatter which is inevitable in geotechnical model tests, see e.g. [10].

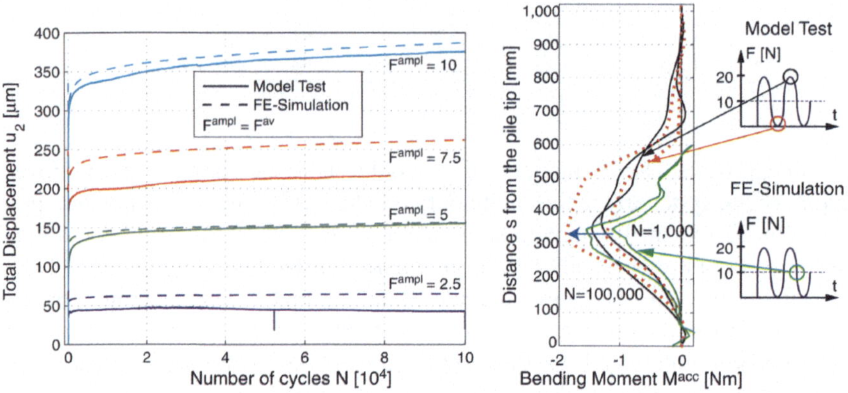

Fig. 7. Total horizontal displacement measured at the model piles in tests with different load levels and with about 100,000 cycles compared to the numerical prediction and change of the bending moment due to the cyclic loading in the model test and the numerical simulation.

Beside the total and accumulated deformations, the development of bending moments during cyclic loading has a governing impact on the design of monopiles. From the measurements in the model tests an increase of the bending moments with increasing number of cycles could be concluded. Similar changes of bending moments have been observed in centrifuge model tests on cyclically loaded monopiles ([30], according to [29]). The change of bending moment M^{acc} is calculated as the difference between the bending moment at a certain cycle and its initial value after three cycles. M^{acc} is shown for one exemplary model test and the corresponding simulation in Fig. 7. Again, the numerical results show a satisfying accordance with the test data. This demonstrates the advantage of a FE model capturing all effects of a cyclic loading at once without relying on phenomenological descriptive relations between deflections and the number of cycles as in some early approaches.

Beside the model tests with a single package of cycles of constant load amplitude, tests with several such packages of different amplitudes applied in succession have been performed. They were also simulated with the FE model. Figure 8 demonstrates the good agreement between horizontal displacements measured in such a model test with 13 different load packages and the corresponding numerical prediction, applying the HCA approach.

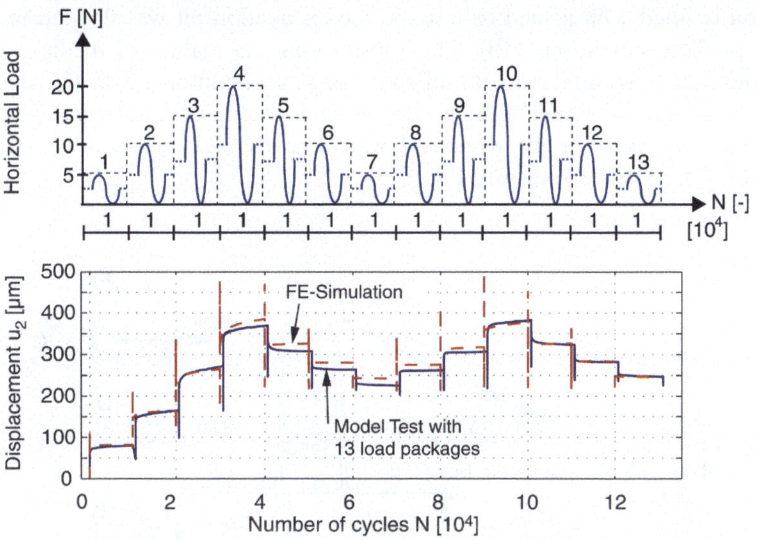

Fig. 8. Total horizontal displacement of a monopile subjected to 13 load packages with different amplitudes and 10,000 cycles each. Comparison of the data from the model test with the corresponding numerical prediction.

Beside these small-scale model tests performed at the KIT, [48] and [43] demonstrate the application of the HCA-based approach to model tests with a larger pile diameter, performed at TU Berlin [35]. These simulations showed as well a satisfying agreement with the model tests.

Several experimental studies on the horizontal bearing capacity of large-scale monopiles and on the effect of cyclic loading on monopiles are still in execution or have only recently been published. Some of them apply industry dimensions like [24], others use a scale of 1:3 to 1:10, which still results in test pile diameters of 1 to 2 m like in [33] and [6]. These tests have not been simulated with the HCA-based approach yet. Successful recalculations of these tests would mean a validation of the HCA-based approach for monopiles in a larger scale and would thus further improve the confidence in the approach.

3.3 Validation of the HCA – Based Approach for Gravity-Based Foundations

Some of the first offshore windfarms were founded on shallow foundations. In the meanwhile, monopiles represent about 80% of all installed offshore wind foundations [46]. With the aim to provide a new foundation structure for OWT, the Ed. Züblin AG developed a gravity-based foundation, designed for water depths up to 40 m. In order to demonstrate the bearing capacity and serviceability of this innovative foundation under static and cyclic loading, a large-scale test was performed on a prototype foundation. Details of this test are presented in [19] and [20]. The foundation consists of a concrete box girder with a shaft and four precast concrete plates (A–D). The foundation is filled with a sand ballast and the excavation pit was flooded in order to simulate offshore conditions [19]. The subsoil consists mainly of medium dense to dense sand which is similar to the offshore in situ conditions. The foundation was subjected to an unidirectional high-cyclic loading with about 1.5 million cycles in order to simulate the loading from wind and water waves. The foundation and the test loading arrangement are shown in Fig. 9.

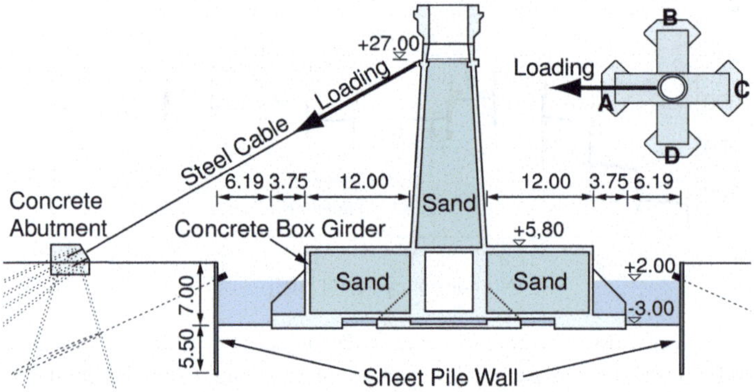

Fig. 9. Test foundation subjected to 1.5 million load cycles, after [19].

Not only the test foundation but also the subsoil was extensively instrumented in order to capture for example the settlements of the foundation itself (e.g. vertical extensometers V1 and V2 in Fig. 10) or in different depths below the plates. Furthermore, pore water pressures and total pressures (soil pressures) were measured beneath each of the plates at different locations and depths.

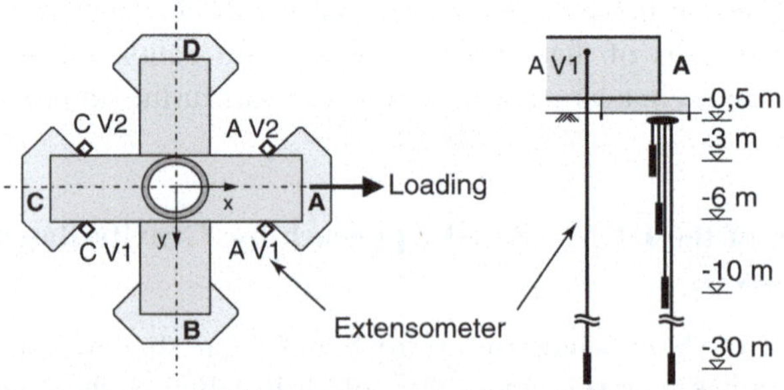

Fig. 10. Top view on the plates of the test foundation and the installed extensometers below and aside the plates, after [19].

One of the main concerns in terms of the application of new design approaches is the missing validation on prototype tests. This still holds true for monopiles. The shown test foundation however can serve as a basis for the evaluation of numerical models for shallow foundations for OWT. The test foundation is a unique example of an in-situ test, with a huge amount of measurement equipment and thoroughly performed. It was utilized in [48–50] to validate the HCA-approach for OWT shallow foundations. The finite element mesh applied for the numerical simulations is shown in Fig. 11.

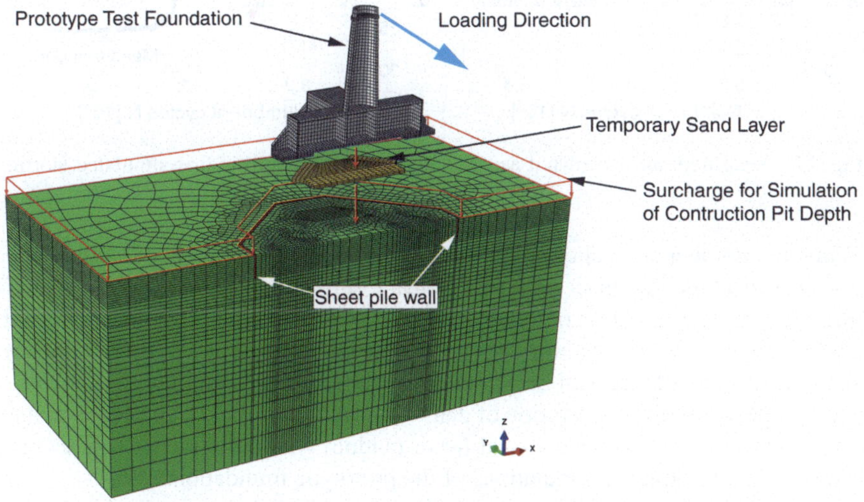

Fig. 11. FE mesh for the simulation of the test foundation.

The subsoil conditions were evaluated based on CPT profiles determined prior to the construction of the foundation and on a set of static and cyclic laboratory tests on the fine and medium sands gained from liner drillings exactly below the plates of the foundation. Based on these laboratory tests, three different approaches according to [40] and [41] were applied to determine the material parameters of the constitutive models. The approaches apply different numbers of cyclic triaxial tests and thus can reach different levels of accuracy regarding the HCA model parameters.

Figure 12 shows the comparison between measured and simulated settlements in the first tested storm event, with a satisfying agreement of both curves, in particular for foundation plate C. The simulation has been performed with the best estimate material parameters based on all available cyclic laboratory tests and with the most accurate evaluation of the in-situ conditions with respect to the relative density. It is important to note that these simulations were treated like a prediction, not a back-calculation. Thus, no results of the test itself were used to calibrate the material parameters in [48]. Applying different parameter sets, it was confirmed that a more detailed determination of material parameters also leads to a higher level of accuracy with respect to the accumulation of foundation settlements under cyclic loading. However, already the approach which purely relies on correlations and does not require any cyclic triaxial test [41] was giving a reasonably good agreement with the measurements at the prototype foundation, cf. [49].

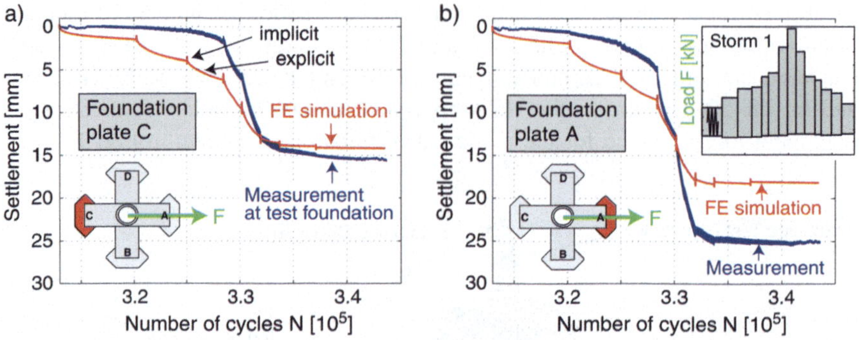

Fig. 12. Measured and calculated accumulated settlements in the first simulated storm.

While the number of cyclic triaxial tests used for the determination of the material parameters had a limited impact on the numerical results in this case, it was proven in [49] that the accurate determination of the relative density I_D in the field has a huge impact on the predicted accumulation of deformation. However, a sufficiently accurate prediction of the settlement accumulation can be achieved by applying mean values of I_D across soil layers. A more detailed consideration of density fluctuations, e.g. by means of sublayers (e.g. 20 cm layers, each with a constant I_D) or random void ratio fluctuations seems not necessary, at least not in the simulations of the prototype foundation.

Not only the settlements were compared between simulation and test, also the excess pore water pressures and the contact pressures between the plates and the subsoil were assessed in detail, cf. [50] and [43]. The contact pressures were exhibiting a strong evolution caused by the cyclic loading. The pressures below the mainly loaded plates (Fig. 10: plates A and C) were decreasing with every storm event while the load plates perpendicular to the loading direction experienced a significant increase of contact pressures. This redistribution of stresses was well predicted by the numerical models applying the HCA-based approach. Detailed results can be found in [48] and [50].

Considering the overall good reproduction of the behaviour of the test foundation by the HCA-based approach and the good agreement on small-scale model tests and on other applications [43], the HCA-based approach can be regarded as validated for cyclically loaded structure in non-cohesive soils. Section 4 will compare the results of HCA-based simulations with the findings from other research groups working on the behaviour of OWT foundations under static and cyclic loading, with a focus on monopiles.

4 Comparison of HCA-Based Simulations with Other Standards

4.1 PISA Approach for Static Loading of Monopiles

The geotechnical design of monopiles is governed by static loading. The PISA approach has gained a lot of attention in the last years and is starting to be implemented into current wind farm designs. Thus, it will be shown more in detail in this Section and will be set into relation to numerical simulations using a hypoplastic model.

The PISA model relies on the data of field tests on scaled monopiles with diameters D between 0.27 m and 2.00 m and *L*/*D*-ratios between 3 and 10 (majority at L/D = 5.25) cf. [5] and [7]. Based on the test results with static loading, a new approach for the design of monopiles was developed [8]. In principal it is following the well-established approach of defining the bedding reaction of the pile as function of the deflections by *p*-*y*-curves (horizontal deformation *y* is referred to as *v* in the PISA approach). Instead of relying on the *p*-*v* curves only, the PISA approach introduces further relations for additional soil reactions applicable to less slender monopiles [8], see Fig. 13:

- Distributed lateral load p vs. horizontal deformation v
- Distributed bending moment of the pile m vs. local cross section rotation ψ
- Base shear force at pile tip H_B vs. horizontal displacement at pile tip v_B
- Base moment M_B vs. rotation of the pile tip ψ_B

All these additional reaction forces were determined in 3D finite element simulations and are simplifications of the shear forces which mainly occur at the inside and the outside of the large diameter piles. The base shear H_B is a result of the stiff pile behaviour which enables the so-called toe kick and a reaction force in the soil-soil

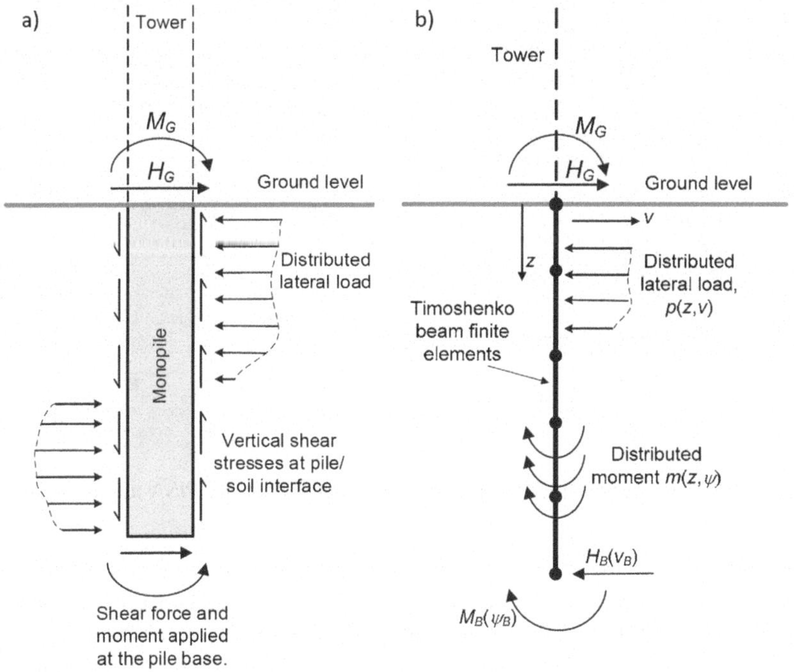

Fig. 13. (a) Soil reactions implemented in the PISA design model and (b) their simplification as Timoshenko beam for the 1D FE approach [8].

interface at the pile tip. These general forces are applied in the design approach [13] based on the pile tests in the TANDEM project.

Two approaches for the design of monopiles are proposed in [8], the so-called "Rule-Based Method" and the "Numerical-Based Method", see Fig. 14. Both methods aim to predict the lateral response of the monopiles based on soil reaction curves. The numerical-based approach combines FE simulations on a 3D model with simplified FE simulations of a 1D Timoshenko beam representing the monopile and capturing the soil reaction curves. 3D simulations have to be performed to determine the normalized soil reaction curves for a first draft pile geometry and the exact soil conditions, represented by a suitable soil model. [8] applied the bounding surface model after [23] modified by [34] for simulations in sand. These calibrated soil reaction curves are then applied in a 1D FE simulation covering the Timoshenko beam shown in Fig. 13. This 1D FE model is applied to allow efficient parameter studies on the pile geometry which finally lead to the optimized pile design. One of the ambitions of the PISA project was to develop design tools which are as simple as possible but consider the relevant soil-pile interaction mechanisms and the relevant soil behaviour. This shall enable a wide group of design engineers to apply these techniques in their offshore design practice.

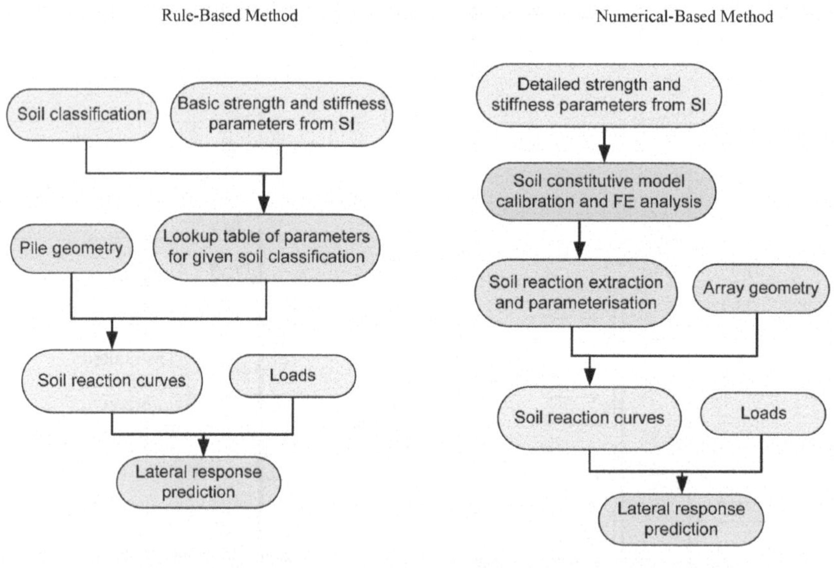

Fig. 14. Rule-Based and Numerical-Based Method for the PISA approach [8].

The application of the PISA model in engineering practice is even further improved by a specific Plaxis design tool: MoDeTo (Monopile Design Tool), which enables the user to perform the 3D and 1D FE simulations automized and even more important, to extract the soil reaction curves fully automatized from the 3D model and transfer them into the 1D simulations [27].

At the TU Darmstadt, studies are ongoing on detailed simulations with the PISA model as well based on an own developed program covering the 1D approach and the soil reaction curves but also using the Plaxis MoDeTo environment. The results of these simulations are set in relation to 3D FE simulations with Hypoplasticity and Intergranular Strain performed in Abaqus. Monopile dimensions and soil properties are chosen in line with the structure presented in [38]. The monopile with a diameter $D = 5$ m, embedment length $L = 30$ m and a wall thickness of $t = 6$ cm is loaded at a height of 20 m above ground level with different horizontal loads (0.5 MN, 1.5 MN, 2.5 MN). Soil conditions are chosen to be a homogenous fine quartz sand with three different relative densities.

Figure 15 shows the monopile deformations and the bending moments obtained from the simulations with the three different models: the PISA approach following the numerical-based method in the MoDeTo environment, the p-y curves according to API and a 3D FE model using Hypoplasticity and Intergranular Strain in Abaqus. The latter simulations are referred to has HIA (Hypoplasticity, Intergranular Strain, Accumulation Model), because generally they could be extended by an explicit phase using the HCA model (which has not been done here). The soil parameters for API and HIA are given in [37]. Material parameters for the PISA approach were determined according to the recommendations of the MoDeTo user manual [28]. However, these PISA parameters are still under investigation since they preferably should be determined from laboratory tests to reach the same level of accuracy as the HIA parameters.

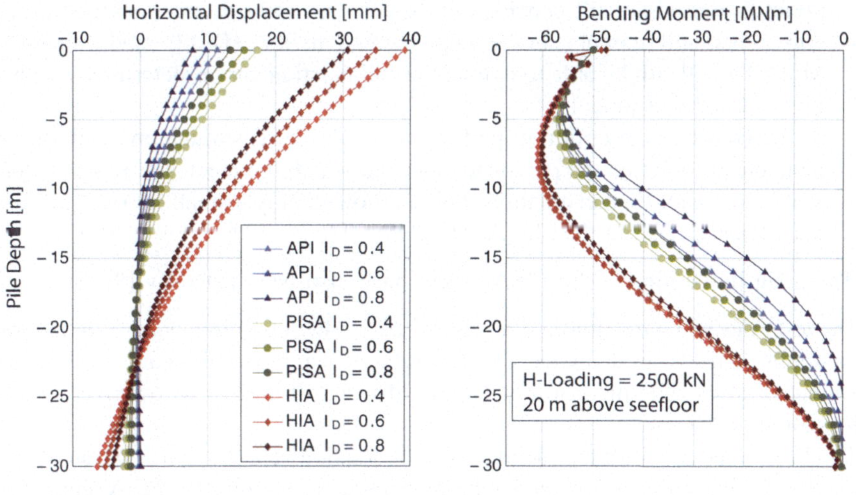

Fig. 15. Pile displacement and bending moment for three relative densities ($I_D = 0.4$, 0.6 and 0.8) calculated with the PISA model following the numerical-based method using the MoDeTo environment in comparison to results applying the API – approach and using Hypoplasticity with Intergranular Strain in an Abaqus model.

The comparison in Fig. 15 shows a good agreement between API and PISA with the main difference in the pile tip displacement. The HIA simulations were giving much higher displacements and the influence of relative density is more pronounced. The larger deformations go along with higher bending moments in HIA compared to the other approaches. Further studies at TU Darmstadt are currently dealing with the comparison between the analytical approach of PISA in comparison to the MoDeTo as well as testing the sensitivity of the model. Since these studies are still ongoing, results can only be shared in future publications.

Preliminary conclusions applying the PISA model are the following:

- Although the theoretical framework was intended to be easily adoptable for individual applications, the coding of the model turned out to be very sophisticated and not straight forward at some points.
- The material parameters for the PISA model need to be chosen carefully (as for every other model as well).
- Extracting the soil reaction curves, bending moment m, base shear H_B and base moment m_B is a very sophisticated task if not performed via the MoDeTo environment.
- Working in the MoDeTo environment leads to quick results without requiring intense preparation. It somehow has the structure of a black box since some steps are not easy to trace back. However, the easy handling includes the chance that also unexperienced users can successfully apply this environment.
- MoDeTo has some further features which efficiently support the design of monopiles and open the system for future developments
  - The efficient combination of a 3D model with the 1D approach enables fast optimization while still considering the relevant soil-pile interaction effects.
  - Any other soil model can be implemented in the 3D FE environment and MoDeTo will still be able to extract the soil reaction curves determined with this user-defined soil model.
  - MoDeTo offers the option to perform a set of 3D FE simulations with different geometries to define the parameter space in which the mean soil reaction curves will be defined. First evaluations on this showed only a small impact of different parameter spaces on the pile deformations.

The main limitations of the PISA model have also been listed by [8] as:

- The test piles had a maximum diameter of 2 m, which is still a factor 5 to upcoming monopiles with 10 m diameter. The *L*/*D* ratio of real monopiles can still differ from validated ratios. The inclusion of the 3D Model will help to reduce/overcome this limitation.
- The PISA model is only applicable to static loading. The implementation of cyclic loading is currently a matter of further research within the PISA consortium. Accumulated excess pore water pressure can not be accounted for in the current version.
- The approach is only considered to be validated for two soil types, sand and clay. All other types of soil have to be dealt with engineering judgement.
- Layered soils can not be described by the PISA model for the time being.
- Pile installation effects (impact driven or vibrated) are not covered.

Future studies and the application in commercial offshore projects will reveal if the PISA approach can overcome the limitations and be established as a new standard. In principal, the approach can be seen as a very successful step towards the improvement of the monopile design and will contribute to the cost reduction for the production of wind energy.

4.2 Comparison with Models for Cyclic Monopile Design

As noted before, the PISA approach is not applicable yet on the design for cyclic loading. Also the SOLDYN [29] studies did not conclude with a new approach for horizontally loaded monopiles. A detailed study on the comparison of different approaches for the design of cyclically loaded monopiles was performed in [37] and [38]. By comparing the horizontal deformations of a monopile subjected to 100,000 load cycles the notable differences in the assessed approaches were shown. The following approaches have been studied: the *p-y* model from API [3], the stiffness degradation model according to Achmus [1], the enhanced strain wedge model according to Taşan [36], the modified *p-y* approach after Dührkop [17] and the HCA based approach HIA. The results for different relative densities are shown in Fig. 16. Beside considerable differences in the total displacements between the various approaches, also the sensitivity on changes in the void ratio or the load regime were found model-dependent.

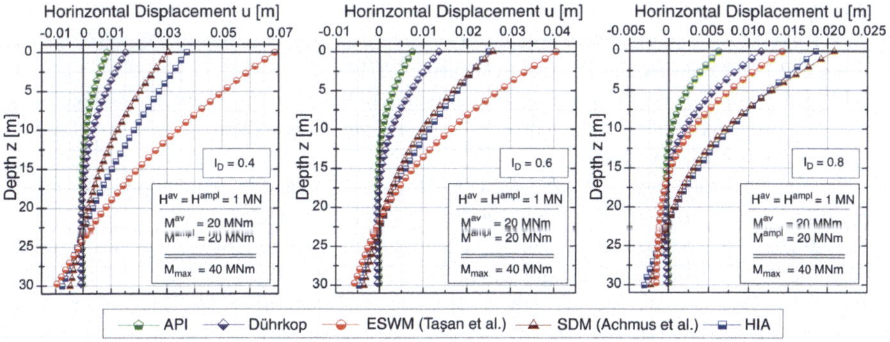

Fig. 16. Pile displacement after 100,000 load cycles obtained with different approaches covering the cyclic monopile design, for three different soil densities and the same loading.

These huge differences in the results demonstrated the need for validated and flexible approaches for the design of foundations for offshore wind turbines. By applying the HCA-based approach to the large-scale test foundation results and to the model tests on monopiles, such a validation was aimed for. Given the satisfying results for the HCA approach, the development of HCA-based engineering models for monopiles and shallow foundations was started [42]. However, the new tests on monopiles ([24, 32, 33]) with a more realistic *L*/*D* ratio give room for further validation not only with respect to the cyclic loading phase but also to static loading.

Summary
The design of foundations for offshore wind turbines remains a challenging task, especially because of the rapid changes and developments in turbine sizes and thus foundation sizes. The variety of studies on static and cyclic loading on monopiles was shown and set in relation to the HCA-based approach. The HCA-based approach showed good results while being applied to small-scale model tests on monopiles and to the prototype test of the gravity based Züblin foundation. Finally, the new PISA approach for the design of monopiles was discussed. First results of a comparison with established soil models like Hypoplasticity with Intergranular Strain showed some differences to the PISA approach which need further investigation. In principal, a combination of HCA-based engineering approaches like those proposed in [42] with popular approaches for the static design, like PISA, should be aimed for to further improve the design of foundations for offshore wind turbines.

Acknowledgments. Parts of the research presented in this paper have been funded by the Federal Ministry for the Environment, Nature Conversation and Nuclear Safety (BMU) under research grants nos. 0325175 and 0325405A/B.

References

1. Achmus, M., Abdel-Rahmann, K.: Zur Bemessung von Monopiles für zyklische Lasten. In: 3rd Workshop des DGGT Arbeitskreises 1.6 Numerik in der Geotechnik, Karlsruhe, pp. 61–74 (2009)
2. API RP2GEO: Recommended Practice, Geotechnical and Foundation Design Considerations, RP 2GEO, American Petroleum Institute (2014)
3. API American Petroleum Institute: Recommended Practice for Planning, Designing and Constructing Fixed Offshore Platforms – Working Stress Design. API RP 2A-WSD, Washington DC, USA (2000)
4. Arany, L., Bhattacharya, S., Macdonald, J., Hogan, S.J.: Design of monopiles for offshore wind turbines in 10 steps. Soil Dyn. Earthq. Eng. **92**, 126–152 (2017)
5. Beuckelaers, W.: Numerical modelling of laterally loaded piles for offshore wind turbines. PhD-thesis. St Catherine's College, University of Oxford (2017)
6. Burd, H.J., Byrne, B.W., McAdam, R.A., Houlsby, G.T., Martin, C.M., Beuckelaers, W.J.A. P., Zdravkovic, L., Taborda, D.M.G., Potts, D.M., Jardine, R.J., Gavin, K., Doherty, P., Igoe, D., Skov Gretlund, J., Pacheco Andrade, M., Muir Wood, A.: Design aspects for monopile foundations. In: Proceedings of TC 209 Workshop on Foundation Design for Offshore Wind Structures, 19th ICSMGE, Seoul, September 2017, pp. 35–44 (2017)
7. Byrne, B., Mcadam, R., Burd, H., Houlsby, G., Martin, C., Gavin, K., Doherty, P., Igoe, D., Taborda, D., Potts, D., Jardine, R., Sideri, M., Schroeder, F., Muir-Wood, A., Kallehave, D., Skov-Gretlund, J.: Field testing of large diameter piles under lateral loading for offshore wind applications. In: Proceedings of the European Conference of Soil Mechanics and Geotechnical Engineering (2015)
8. Byrne, B., Mcadam, R., Burd, H., Houlsby, G., Martin, C., Beuckelaers, W., Zdravković, L., Taborda, D., Potts, D., Jardine, R., Ushev, E., Liu, T., Abadias, D., Gavin, K., Igoe, D., Doherty, P., Skov Gretlund, J., Pacheco Andrade, M., Muir Wood, A., Schroeder, F., Turner, S., Plummer, M.: PISA: new design methods for offshore wind turbine monopiles. In: Proceedings of the Offshore Site Investigation and Geotechnics, SUT, vol. 1, pp. 142–161, London, United Kingdom (2017)

9. BSH Standard Konstruktion: Mindestanforderungen an die konstruktive Ausführung von Offshore-Bauwerken in der ausschließlichen Wirtschaftszone (AWZ) (BSH-Standard 7005-12). Bundesamt für Seeschifffahrt und Hydrographie (2015)
10. Cuéllar, P.: Pile foundations for offshore wind turbines: numerical and experimental investigations on the behaviour under short-term and long-term cyclic loading. PhD-thesis. Veröffentlichungen des Grundbauinstituts der TU Berlin, 52 (2011)
11. DIN EN ISO 19902:2018-05 – Draft. Petroleum and natural gas industries - Fixed steel offshore structures
12. DIN EN ISO 19901-4:2017-01. Petroleum and natural gas industries - Specific requirements for offshore structures - Part 4: Geotechnical and foundation design considerations
13. Dührkop, J., Siegl, K., Heins, E., Pucker, T.: Bemessung von XXL-Monopiles – Aktuelle Erfahrungen und Herausforderungen im Geotechnischen Design. In: Pfahlsymposium 2019, Mitteilungen des Institutes für Grundbau und Bodenmechanik der TU Braunschweig, Heft Nr. 107, pp. 89–112 (2019)
14. EC 7-1/DIN 1054:2011: Normenhandbuch Eurocode 7, Geotechnische Bemessung, Band 1: Allgemeine Regeln. Beuth Verlag, 1. Auflage
15. DNVGL: DNVGL-ST-0126, Support structures for wind turbines, April 2016 (2016)
16. EA-Pfähle: Empfehlungen des Arbeitskreises 2.1 "Pfähle". DGGT Deutsche Gesellschaft für Geotechnik, 2. Aufl., Ernst & Sohn, Berlin (2012)
17. Dührkop, J.: Zum Einfluss von Aufweitungen und zyklischen Lasten auf das Verformungsverhalten lateral beanspruchter Pfähle in Sand [Dissertation]. Veröffentlichungen des Instituts für Geotechnik und Baubetrieb der Technischen Universität Hamburg-Harburg, Heft Nr. 20 (2009)
18. GWEC: Global Wind Report 2018, Global Wind Energy Council, April 2019 (2019)
19. Hartwig, U.: Planung der versuchstechnischen Simulation des Verhaltens einer Flachgründung für Offshore-Windenergieanlagen im Originalmaßstab. In: Workshop "Gründung von Offshore-Windenergieanlagen", Karlsruhe 2010, Publications of the Institute of Soil Mechanics and Rock Mechanics, KIT, Issue No. 172, pp. 81–97 (2010)
20. Hartwig, U., Mayer, T.: Entwurfsaspekte bei Gründungen für Offshore-Windenergieanlagen. Bautechnik **89**(3), 153–161 (2012)
21. Kirsch, F., Richter, T., Coronel, M.: Geotechnische Aspekte bei der Gründungsbemessung von Offshore-Windenergieanlagen auf Monopfählen mit sehr großen Durchmessern. In: Stahlbau Spezial, pp. 61–67 (2014)
22. Leblanc, C., Houlsby, G.T., Byrne, B.W.: Response of stiff piles in sand to long-term cyclic lateral loading. Géotechnique **60**(2), 79–90 (2010)
23. Manzari, M., Dafalias, Y.: A critical state two-surface plasticity model for sands. Géotechnique **47**(2), 255–272 (1997)
24. Moormann, C., Kirsch, F., Herwig, V.: Vergleich des axialen und lateralen Tragverhaltens von vibrierten und gerammten Stahlrohrpfählen. In: Proceedings of the 34. Baugrundtagung, pp. 73–81, DGGT, Bielefeld (2016)
25. Niemunis, A., Wichtmann, T., Triantafyllidis, Th.: A high-cycle accumulation model for sand. Comput. Geotech. **32**(4), 245–263 (2005)
26. Niemunis, A.: Extended hypoplastic models for soils. Habilitation, Veröffentlichungen des Institutes für Grundbau und Bodenmechanik, Ruhr-Universität Bochum, Heft Nr. 34 (2003)
27. Panagoulias, S., Brinkgreve, R., Minga, E., Burd, H., McAdam, R.: Application of the PISA framework to the design of offshore wind turbine monopile foundations. In: Proceedings of the WindEurope 2018 Conference, Hamburg, Germany, 25–28 September 2018 (2018)
28. Panagoulias, S., Brinkgreve, R., Zambich, L.: Plaxis MoDeTo Manual 2018. Delft, Netherlands (2018)

29. Puech, A., Garnier, J.: Design of Piles Under Cyclic Loading - SOLCYP Recommendations (2017)
30. Rakotonindriana, J.: Comportement des pieux et des groupes de pieux sous chargement latéral cyclique, PhD thesis, ENPC (2009)
31. Savidis, S., Richter, T., Kirsch, F., Rücker, W.: Entwurf und Bemessung von Gründungen für Offshore-Windenergieanlagen. In: Grundbau-Taschenbuch Teil 3, Gründungen und geotechnische Bauwerke, 8. Auflage, Ernst & Sohn, Berlin 2018 (2018)
32. Seidel, M., Voormeeren, S., van der Steen, J.-B.: State-of-the-art design process for offshore wind turbine support structures. Stahlbau, vol. 85, issue 9 (2016)
33. Spill, S., Kohlmeier, M., Maretzki, S., Dührkop, J., Wefer, M.: Auslegung und Umsetzung großmaßstäblicher Versuche zum Tragverhalten von Monopiles. In: Mitteilung des Instituts für Grundbau und Bodenmechanik der Technischen Universität Braunschweig, Pfahlsymposium 2017, Heft Nr. 102, pp. 309–325 (2017)
34. Taborda, D., Zdravković, L., Kontoe, S., Potts, D.: Computational study on the modification of a bounding surface plasticity model for sands. Comput. Geotech. **59**, 145–160 (2014)
35. Taşan, H.E., Rackwitz, F., Glasenapp, R.: Experimentelle Untersuchungen zum Verhalten von zyklisch horizontal belasteten Monopiles. Bautechnik **88**(2), 102–112 (2011)
36. Taşan, E., Rackwitz, F., Glasenapp, R.: Ein Bemessungsmodell für Monopilegründungen unter zyklischen Horizontallasten. In: Bautechnik 88, Heft 5, pp. 301–318 (2011)
37. Westermann, K., Zachert, H., Wichtmann, T.: Vergleich von Ansätzen zur Prognose der Langzeitverformungen von OWEA Monopilegründungen in Sand. Teil 1: Grundlagen der Ansätze und Parameterkalibration. Bautechnik **91**(5), 309–323 (2014)
38. Westermann, K., Zachert, H., Wichtmann, T.: Vergleich von Ansätzen zur Prognose der Langzeitverformungen von OWEA Monopilegründungen in Sand. Teil 2: Simulationen und Schlussfolgerungen. Bautechnik **91**(5), 324–332 (2014)
39. Wichtmann, T., Niemunis, A., Triantafyllidis, T.: Validation and calibration of a high-cycle accumulation model based on cyclic triaxial tests on eight sands. Soils Found. **49**(5), 711–728 (2009)
40. Wichtmann, T., Niemunis, A., Triantafyllidis, T.: On the determination of a set of material constants for a highcycle accumulation model for non-cohesive soils. Int. J. Numer. Anal. Meth. Geomech. **34**(4), 409–440 (2010)
41. Wichtmann, T., Niemunis, A., Triantafyllidis, T.: Simplified calibration procedure for a high-cycle accumulation model based on cyclic triaxial tests on 22 sands. In: ISFOG, Perth, Australia (2010)
42. Wichtmann, T., Triantafyllidis, Th., Chrisopoulos, S., Zachert, H.: Prediction of long-term deformations of offshore wind power plant foundations using engineer-oriented models based on HCA. Int. J. Offshore Polar Eng. **27**(4), 346–356 (2017)
43. Wichtmann, T., Machaček, J., Zachert, H., Günther, H.: Validierung eines hochzyklischen Akkumulationsmodells anhand von Modellversuchen und Messungen an realen Bauwerken. Bautechnik **96**(2), 160–175 (2019)
44. Windenergie auf See Pressekonferenz; Berlin, 21 January 2019
45. Windeurope: Wind Energy in Europe in 2018 – Trends and Statistics published in February 2019 (2019)
46. Windeurope: Offshore Wind in Europe - Key trends and statistics 2018, published in February 2019 (2019)
47. Zachert, H.: Zur Gebrauchstauglichkeit von Monopile-Gründungen für Offshore-Windenergieanlagen. In: Mitteilung des Instituts für Grundbau und Bodenmechanik der Technischen Universität Braunschweig, Pfahlsymposium 2015, Heft Nr. 99, pp. 2–15 (2015)

48. Zachert, H.: Zur Gebrauchstauglichkeit von Gründungen für Offshore-Windenergieanlagen. Dissertation, Veröffentlichungen des Institutes für Bodenmechanik und Felsmechanik am Karlsruher Institut für Technologie, Heft Nr. 180 (2015)
49. Zachert, H., Wichtmann, T., Triantafyllidis, Th., Hartwig, U.: Simulation of a full-scale test on a gravity base foundation for offshore wind turbines using a high cycle accumulation model. In: 3rd International Symposium on Frontiers in Offshore Geotechnics (ISFOG), Oslo (2015)
50. Zachert, H., Wichtmann, T., Triantafyllidis, Th.: Soil structure interaction of foundations for offshore wind turbines. In: 26th International Ocean and Polar Engineering Conference (ISOPE-2016), Rhodos (2016)

Behaviour of Granular Soils Under Uni- and Multidimensional Drained High-Cyclic Loading

Torsten Wichtmann[1(✉)] and Lukas Knittel[2]

[1] Chair for Soil Mechanics, Foundation Engineering and Environmental Geotechnics, Ruhr-University Bochum, Universitätsstraße 150, 44789 Bochum, Germany
torsten.wichtmann@rub.de

[2] Karlsruhe Institute of Technology, Institute of Soil Mechanics and Rock Mechanics, Engler-Bunte-Ring 14, 76131 Karlsruhe, Germany
lukas.knittel@kit.edu

Abstract. The paper summarizes a long-term experimental research program on the behaviour of granular soils under high-cyclic loading, that means a loading with many cycles ($N \geq 10^3$) and relatively small strain amplitudes ($\varepsilon^{\text{ampl}} \leq 10^{-3}$). Numerous drained cyclic triaxial and hollow cylinder triaxial tests have been performed, most of them on Karlsruhe fine sand. The effects of the stress or strain amplitude, average stress, relative density, grain size distribution curve, particle shape and a content of non-plastic fines or shell fragments on the cumulative deformations are discussed. The important role of initial fabric (sample preparation method) is demonstrated, while the geometry and dimensions of the samples seem to be of secondary importance. The effect of multiple changes of the polarization (i.e. direction) of the cycles was found less important than previously thought. The bundling of a cyclic loading with continuously changing amplitude into bundles of cycles each with a constant amplitude is demonstrated to be conservative. An interesting effect has been observed in tests with bundles of cycles interrupted by monotonic loading phases: The monotonic loading can lead to a loss of the cyclic preloading memory, and thus to an increase of the rate of strain accumulation in the next bundle of cycles. Results from tests with 1D, 2D, 3D and 4D stress and strain paths confirm the amplitude definition for multi-dimensional strain loops incorporated into a high-cycle accumulation (HCA) model. The description of the strain accumulation rates by the HCA model is discussed based on the results from the various test series.

1 Introduction

A high-cyclic loading may be caused by traffic (e.g. high-speed trains, magnetic levitation trains), industrial sources (e.g. crane rails, machine foundations), wind and waves (e.g. on-shore and off-shore wind turbines, coastal structures),

T. Triantafyllidis (Ed.): *Recent Developments of Soil Mechanics and Geotechnics in Theory and Practice*, LNACM 91, pp. 136–165, 2020.
https://doi.org/10.1007/978-3-030-28516-6_8

repeated filling and emptying processes (e.g. locks, tanks and silos), construction processes (e.g. vibration of sheet piles) or mechanical compaction (e.g. vibratory compaction). Permanent deformations caused by such high-cyclic loading may endanger the serviceability of foundations and thus must be accurately predicted in the design stage. For that purpose a high-cycle accumulation (HCA) model has been developed by Niemunis et al. [10]. It is based on an extensive laboratory study with drained cyclic tests on various sands [18]. The most important findings of that and subsequent experimental work [19] are summarized herein. The consideration of the various parameters in the HCA model is discussed.

If not otherwise stated in the following sections, the tests were performed in a cyclic triaxial device using samples of 10 cm diameter and 10 cm height. A scheme of the devices at IBF, KIT is shown in Fig. 1. The samples were prepared by air pluviation and tested in the water-saturated condition using a back pressure of 200 or 500 kPa. The effective lateral stress σ_3' was kept constant, while the cyclic loading was applied in the vertical direction, with an average value of effective axial stress $\sigma_1'^{\text{av}}$ and a stress amplitude $\sigma_1'^{\text{ampl}}$. The cyclic loading was realized by means of a pneumatic loading system. In this paper the stress conditions in a test are described by the average mean pressure p^{av} (with $p = (\sigma_1' + 2\sigma_3')/3$), the average deviatoric stress q^{av} (with $q = \sigma_1' - \sigma_3'$), average stress ratio $\eta^{\text{av}} = q^{\text{av}}/p^{\text{av}}$ and deviatoric stress amplitude $q^{\text{ampl}} = \sigma_1'^{\text{ampl}}$.

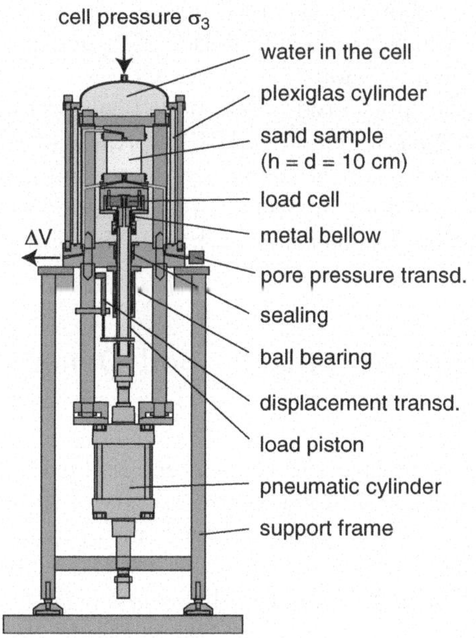

Fig. 1. Scheme of the cyclic triaxial devices at IBF, KIT

Axial deformations were measured with a displacement transducer attached to the load piston, while volume changes were obtained from a burette system connected to the fully water-saturated pore space, using a differential pressure transducer. The axial strain ε_1 and the volumetric strain $\varepsilon_v = \varepsilon_1 + 2\varepsilon_3$ derived from these measurements are used to calculate the lateral strain $\varepsilon_3 = (\varepsilon_v - \varepsilon_1)/2$ and the deviatoric strain $\varepsilon_q = 2/3(\varepsilon_1 - \varepsilon_3)$. The total strain is defined as $\varepsilon = \|\boldsymbol{\varepsilon}\| = \sqrt{(\varepsilon_1)^2 + 2(\varepsilon_3)^2}$, i.e. as the norm of the strain tensor. All these strain quantities can be splitted into a resilient/elastic/amplitude portion $\sqcup^{\text{ampl}}$ and a residual/plastic/accumulated portion $\sqcup^{\text{acc}}$. In the context of a HCA model a rate is understood as the derivative of a variable with respect to the number of cycles, e.g. $\dot{\sqcup} = \partial \sqcup / \partial N$.

The first *irregular* cycle usually generates larger deformations and is thus applied with a relative low frequency (0.01 Hz). The subsequent 10^5 regular cycles are applied with a higher frequency of 0.2 Hz. Since the HCA model predicts the strain accumulation rates during the regular cycles only, the first irregular cycle is not discussed in this paper. In the following, $N = 1$ thus refers to the end of the first regular cycle.

In the HCA model the tensor of the rate of strain accumulation is described by $\dot{\boldsymbol{\varepsilon}}^{\text{acc}} = \dot{\varepsilon}^{\text{acc}}\mathbf{m}$, i.e. as the product of the scalar *intensity* of strain accumulation $\dot{\varepsilon}^{\text{acc}} = \|\dot{\boldsymbol{\varepsilon}}^{\text{acc}}\|$ (norm of strain rate tensor) and the *direction* of accumulation $\mathbf{m} = \dot{\boldsymbol{\varepsilon}}^{\text{acc}}/\|\dot{\boldsymbol{\varepsilon}}^{\text{acc}}\|$ (unit tensor). In the triaxial case, the intensity of accumulation is calculated as $\dot{\varepsilon}^{\text{acc}} = \partial\varepsilon^{\text{acc}}/\partial N$ and the direction of accumulation can be expressed by the ratio of the volumetric and deviatoric strain accumulation rates $\dot{\varepsilon}_v^{\text{acc}}/\dot{\varepsilon}_q^{\text{acc}}$. In the following the test results are analyzed with respect to both, the measured intensity of accumulation $\dot{\varepsilon}^{\text{acc}}$ or their integral value ε^{acc}, and the direction of accumulation $\dot{\varepsilon}_v^{\text{acc}}/\dot{\varepsilon}_q^{\text{acc}}$.

Most of the test series presented herein have been performed at the Institute of Soil Mechanics and Rock Mechanics (IBF) at Karlsruhe Institute of Technology (KIT). The only exception is the study on the influence of the grain size distribution curve documented in Sect. 3. Those tests have been conducted at the Chair for Foundation Engineering and Soil Mechanics at Ruhr-University Bochum (RUB).

2 Stress or Strain Amplitude, Soil Density and Average Stress

The influence of the soil density and the average and cyclic stresses is inspected based on a series with 17 drained cyclic triaxial tests performed on Karlsruhe fine sand (KFS), a clean fine sand with mean grain size $d_{50} = 0.14$ mm and uniformity coefficient $C_u = d_{60}/d_{10} = 1.5$. Four different stress amplitudes ($20\,\text{kPa} \leq q^{\text{ampl}} \leq 80\,\text{kPa}$), seven different initial relative densities ($0.25 \leq I_{D0} = (e_{\max} - e)/(e_{\max} - e_{\min}) \leq 0.90$, measured before the start of the regular cycles), five different average mean pressures ($50\,\text{kPa} \leq p^{\text{av}} \leq 300\,\text{kPa}$) and four different average stress ratios ($0.5 \leq \eta^{\text{av}} = q^{\text{av}}/p^{\text{av}} \leq 1.25$) have been tested. In four test series, one of these parameters has been varied, while the three remaining ones have been kept constant.

The measured development of permanent strain $\varepsilon^{\rm acc}$ with increasing number of cycles N is presented in the diagrams on the left-hand side of Fig. 2. Evidently, the residual strains $\varepsilon^{\rm acc}$ increase with increasing stress amplitude (Fig. 2a), decreasing initial density (Fig. 2c), and increasing average stress ratio (Fig. 2g). In case of a constant amplitude-pressure-ratio $\zeta = q^{\rm ampl}/p^{\rm av}$ similar accumulation curves are usually measured in tests with different average mean pressures (Fig. 2e).

The right-hand side of Fig. 2 contains diagrams in which the residual strain after certain numbers of cycles is plotted versus the parameter varied in the test series, i.e. against the strain amplitude $\bar{\varepsilon}^{\rm ampl}$, void ratio $\bar{e}$, average mean pressure $p^{\rm av}$ and average stress ratio $\eta^{\rm av}$. $\bar{\varepsilon}^{\rm ampl}$ and $\bar{e}$ represent mean values of strain amplitude and void ratio up to the number of cycles under consideration. On the ordinate the residual strain has been divided by the functions $f_{\rm ampl}$ and/or f_e of the HCA model in order to remove any influence of slightly different strain amplitudes or void ratios. These functions are also evaluated with $\bar{\varepsilon}^{\rm ampl}$ and $\bar{e}$, which is indicated by the superposed bar $\bar{\Box}$. Figure 2f shows that the intensity of accumulation decreases with increasing average mean pressure if a constant strain amplitude is considered. The other tendencies in Fig. 2b, d, h agree well with those in Fig. 2a, c, g.

The dependencies of the intensity of strain accumulation $\dot{\varepsilon}^{\rm acc}$ on strain amplitude, void ratio, average mean pressure and average stress ratio are described by the functions $f_{\rm ampl}$, f_e, f_p and f_Y of the HCA model. The influence of the cyclic preloading history (i.e. of the number of cycles N in the simplest case of a cyclic loading with $\varepsilon^{\rm ampl}$ = constant) is captured by the function $\dot{f}_N = \partial f_N/\partial N$. For the detailed equations of the HCA model the interested reader is referred to [10] or [19]. The parameters $C_{\rm ampl}$, C_e, C_p, C_Y, C_{N1}, C_{N2} and C_{N3} used within the functions describing the intensity of accumulation can be calibrated "by hand" following the procedure described e.g. in [19,23]. Afterwards an optimization can be done by means of an element test program. The results of such simulations with the optimum set of parameters have been added as thick solid curves in Fig. 2. A good agreement between the measured data and the prediction by the HCA model can be concluded from Fig. 2.

Figure 3 presents the paths of accumulated deviatoric versus accumulated volumetric strain measured in the test series with a variaton of $\eta^{\rm av}$. Evidently, the deviatoric portion of strain accumulation increases with growing $\eta^{\rm av}$. Numerous similar test series have shown that a purely volumetric accumulation of strain occurs for isotropic average stresses ($\eta^{\rm av} = 0$), while a purely deviatoric one can be expected at the critical state line ($\eta^{\rm av} = M_{cc} = 6 \sin\varphi_{cc}/(3 - \sin\varphi_{cc})$). This dependence is described by the flow rule of the HCA model with the parameter φ_{cc} ($\varphi_{cc} = 33.1°$ for KFS, see the predictions added as solid curves in Fig. 3). In contrast, the direction of accumulation is only marginally influenced by variations in stress amplitude, soil density and average mean pressure.

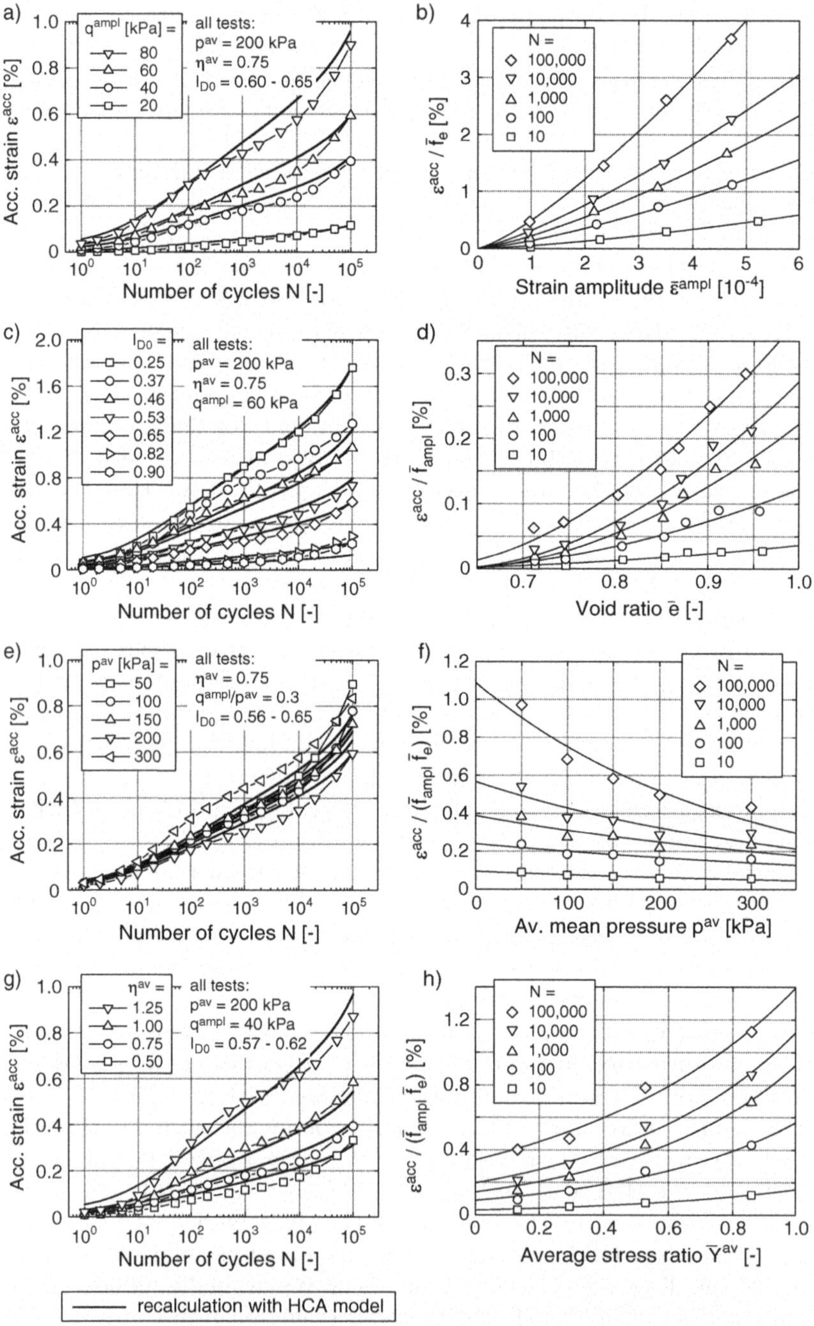

Fig. 2. Results of drained cyclic tests with different (a, b) stress amplitudes q^{ampl}, (c, d) initial relative densities I_{D0}, (e, f) average mean pressures p^{av} and (g, h) average stress ratios η^{av}

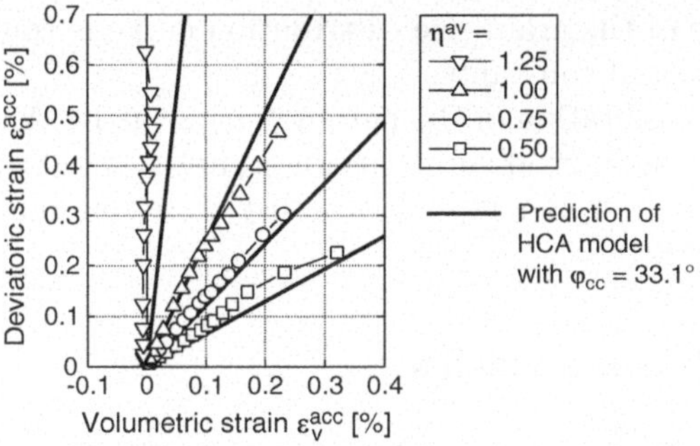

Fig. 3. $\varepsilon_q^{acc} - \varepsilon_v^{acc}$ strain paths measured in drained cyclic triaxial tests with different average stress ratios η^{av}; Description by the direction of accumulation (flow rule) in the HCA model

3 Grain Size Distribution Curve

An extensive laboratory testing program has been performed on special mixtures of a quartz sand with subangular grains, having linear grain size distribution curves with mean grain sizes in the range 0.1 mm $\leq d_{50} \leq$ 6 mm and uniformity coefficients in the range $1.5 \leq C_u \leq 8$ [21–23]. These tests have been performed in the laboratory at RUB, using samples of 100 mm diameter and 200 mm height and a loading frequency of 1 Hz for the regular cycles. In Fig. 4 the residual strains after 10,000 cycles are plotted versus d_{50} or C_u, respectively. All specimens were medium dense and subjected to the same average and cyclic stresses. Figure 4 reveals that the residual strains are larger for sands with smaller mean grain sizes and higher uniformity coefficients. Also the shape of the strain accumulation curves $\varepsilon^{acc}(N)$ depends on the granulometric properties [21,23], while the direction of accumulation is rarely affected [22]. In simulations with the HCA

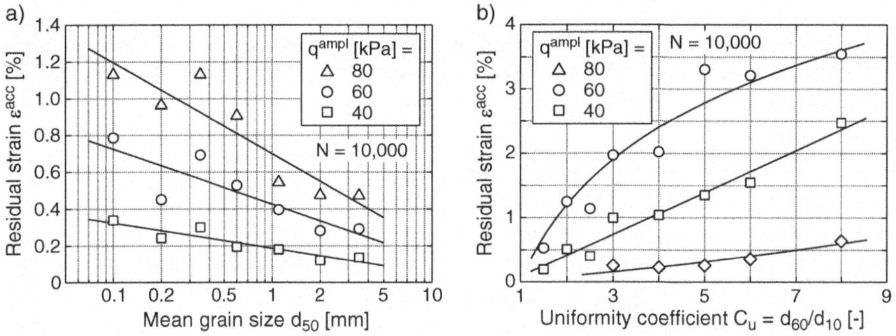

Fig. 4. Accumulated strain ε^{acc} after 10,000 cycles as a function of (a) mean grain size d_{50} and (b) uniformity coefficient C_u (all tests: p^{av} = 200 kPa, η^{av} = 0.75, medium dense samples)

model the influence of the grain size distribution curve is considered by chosing different sets of material constants.

For a simplified calibration of the parameters of the HCA model, correlations with the grain size distribution curve (mean grain size d_{50}, uniformity coefficient C_u) or index properties from simple laboratory tests (minimum void ratio $e_{\min}$) have been proposed [21–23].

4 Particle Characteristics

The influence of grain shape and surface roughness has been studied in drained cyclic triaxial tests on glass beads, natural sand and crushed sand. All three materials were tested with the same grain size distribution curve ($d_{50} = 0.6$ mm, $C_u = 1.5$). Each of the three materials has been tested with different initial relative densities I_{D0}, deviatoric stress amplitudes q^{ampl} and average stresses (p^{av}, η^{av}) [26]. The diagrams in Fig. 5 present the accumulated strain $\varepsilon^{\mathrm{acc}}$ after $N = 10^5$ cycles as a function of the stress amplitude q^{ampl} or average mean pressure p^{av}, respectively. While for glass beads and natural sand the residual strain grows faster than proportional with increasing stress amplitude, the opposite is true in the case of the crushed sand.

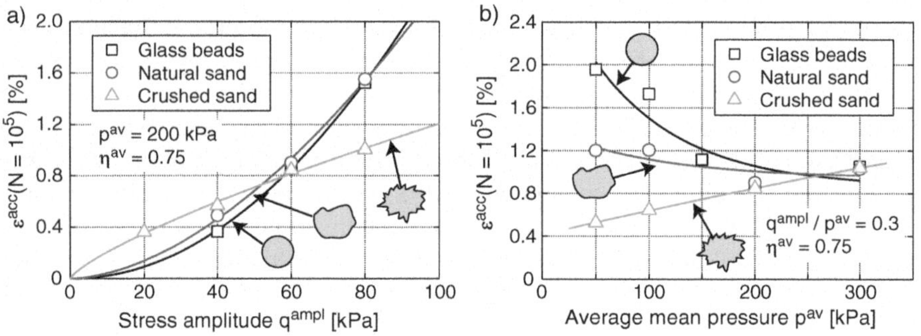

Fig. 5. Results from drained cyclic tests on medium dense samples of three materials with different grain shape: Accumulated strain $\varepsilon^{\mathrm{acc}}$ after $N = 10^5$ cycles as a function of (a) stress amplitude q^{ampl} and (b) average mean pressure p^{av}

The largest differences are evident in the series with different average mean pressures. The tests on glass beads showed a significant decrease of the residual strain with increasing values of p^{av}. Also the natural sand exhibited a moderate reduction. The opposite, i.e. a considerable increase of $\varepsilon^{\mathrm{acc}}$ with p^{av} was observed for the crushed sand. At low pressures ($p^{\mathrm{av}} = 50$ kPa) the rates of strain accumulation were much higher in the tests on the glass beads compared to the natural sand and in particular to the crushed sand. At higher pressures ($p^{\mathrm{av}} = 300$ kPa) no significant differences between the residual strains measured for the three different materials could be detected. This is probably due to the fact, that larger contact forces lead to a flattening of the contact zones. Therefore, with increasing stresses, the shape of the contacts developing in the different materials converges. The original shape of the particles involved in the contact then becomes less important. At lower contact stresses, reorientations of the particles due to slipping and rotation are easier in the assemblies of round glass beads with their smooth surface than in the angular crushed sand with a distinct interlocking between adjacent grains.

5 Content of Non-cohesive Fines or Shell Fragments

The influence of a non-cohesive fines content on the cumulative deformations has been investigated in a study with drained cyclic tests on fine sands possessing different amounts of fines (grains with sizes $d < 0.063$ mm), namely $FC = 0$, 11.3 and 19.6%. At grain sizes $d > 0.063$ mm the grain size distribution curves of all mixtures were parallel to each other ($C_u = 1.5$). The residual strains after $N = 10^5$ cycles measured for different initial relative densities and fines contents are compared in Fig. 6a. Evidently, for a given I_{D0} value larger residual strains are obtained for the mixtures with a higher content of non-cohesive fines.

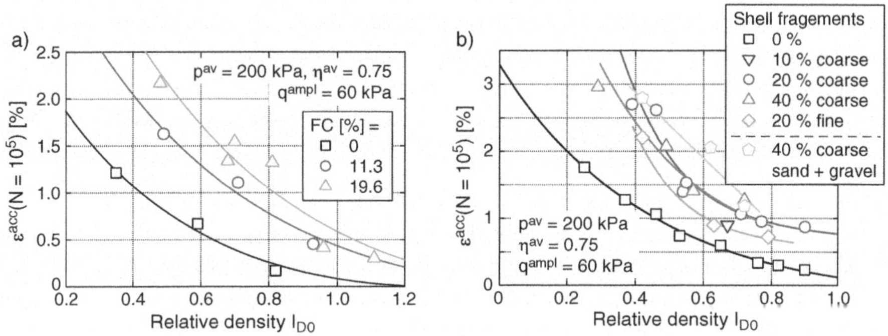

Fig. 6. Residual strain ε^{acc} after $N = 10^5$ cycles as a function of initial relative density I_{D0} in tests on (a) three fine sands with different amounts of non-cohesive fines and (b) mixtures of Karlsruhe fine sand with different percentage of shell fragments

In order to study the influence of a content of shell fragments, mixtures of Karlsruhe fine sand with different portions and sizes of North sea shell fragments (10, 20 or 40% coarse shell fragments with particle sizes $0.063 \leq d \leq 8$ mm or 20% fine shell fragments with $0.063 \leq d \leq 2$ mm) have been subjected to a drained high-cyclic loading [27]. Figure 6b presents the measured residual strain after 10^5 cycles as a function of I_{D0}. For similar densities and identical average and cyclic stresses, the residual strain grows with increasing content of shell fragments. This is probably mainly attributed to the fact that the mixtures with higher contents of shell fragments are more well-graded, cf. [23]. This hypothesis is supported by the fact that another mixture of KFS with 40% coarse sand and gravel particles delivered similar cumulative rates as the mixture containing 40% shell fragments (Fig. 6b). Both mixtures had identical grain size distribution curves. Furthermore, the mixture with 20% fine shell fragments, possessing a more uniform gradation, exhibited lower strain accumulation rates than the mixture with 20% coarse shell fragments. Therefore, the crushability and the platy particle shape of the shell fragments seem to be of secondary importance in this case.

In order to examine a possible breakage of the shell fragments during cyclic loading, a sieve analysis has been performed on a sample of the mixture with 40% shell fragments before and after a cyclic test. The grain size distribution curves determined before and after the test were practically identical, giving no hints for particle breakage. The shell fragments involved in the present study, primarily originating from cockles, seem to be too hard to allow for a significant amount of breakage under the applied average and cyclic loads, in particular when they are embedded into a fine sand matrix.

6 Sample Preparation Method

It is well known that the initial fabric, generated in the laboratory by the chosen sample preparation method, has a large influence on the behaviour of sand, in particular under undrained monotonic and cyclic loading [3,6,7,14,15,17]. Therefore, the influence has been investigated for a drained high-cyclic loading. The samples of KFS were prepared by the following eight methods, which are depicted by the schemes in Fig. 7:

- Air pluviation, where dry sand falls out of the outlet of a funnel
- Homogeneous dry sand rain over the whole cross section of the sample using a multiple sieve apparatus similar to the one proposed in [6]
- Air pluviation by means of a large sand spreader (diameter $d \approx 1$ m), usually applied for model testing at the IBF [28].
- Water pluviation, leading to rather low initial densities not exceeding $I_{D0} = 0.3$, with subsequent compaction by vibration applied with a shaking table (for densities $I_{D0} > 0.3$)
- Water pluviation with subsequent compaction by tapping of the mold using a rubber hammer
- Dry deposition in the loosest state with subsequent compaction by vibration applied with a shaking table
- Dry deposition in the loosest state with subsequent compaction by tapping of the mold using a rubber hammer
- Moist tamping in eight layers using the undercompaction method proposed in [4] with a water content of 10% and a degree of undercompaction of 5%. For samples with $I_{D0} > 0.8$ a constitution procedure analogous to the Proctor test has been alternatively used.

For each sample preparation method several drained cyclic triaxial tests with different relative densities but same average and cyclic stresses ($p^{av} = 200$ kPa, $\eta^{av} = 0.75$, $q^{ampl} = 60$ kPa) have been performed. The residual strains at the end of 10^5 cycles are given in Fig. 7 as a function of I_{D0}.

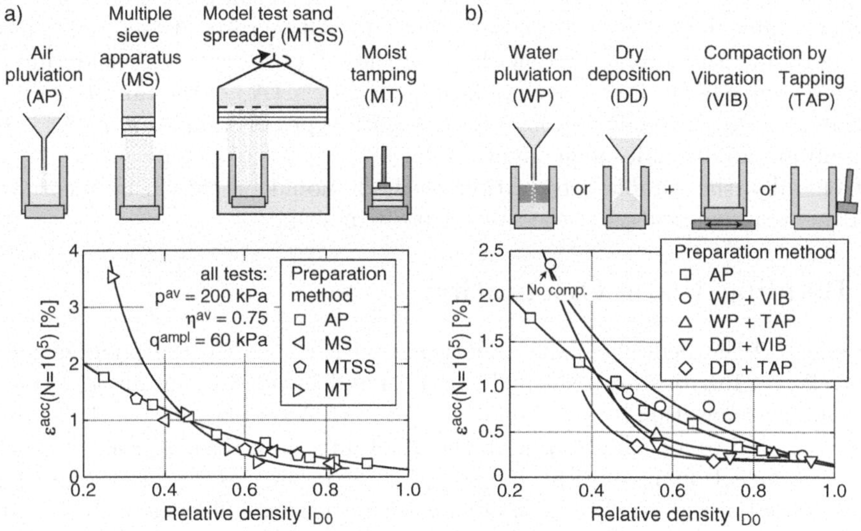

Fig. 7. Accumulated strain after 10^5 cycles versus initial relative density for various sample preparation methods

The samples prepared by the different dry pluviation techniques (AP, MS, MTSS) gave almost identical results (Fig. 7a). The conventional air pluviation technique (AP) and the procedure involving a multiple-sieve apparatus (MS), leading to a homogeneous sand rain over the whole sample cross section, seem thus to generate a similar initial fabric. The good agreement between the results for the conventional air-pluviated samples (AP) and those constituted by means of the model test sand spreader (MTSS) confirms that parameters (e.g. material constants of a constitutive model) obtained from conventional laboratory samples are representative for the sand in the model tests, i.e. may be regarded as appropriate for simulations of the model tests (see e.g. the FE re-calculations of small-scale model tests on OWPP foundations presented in [28] and [29]).

The diagram in Fig. 7b confirms that for I_{D0} = constant, water-pluviated samples compacted by vibration (WP + VIB) show similar cumulative strains than those constituted by air pluviation (AP). A compaction by tapping (WP + TAP) seems to lead to a fabric being more resistant to cumulative deformations than that of the vibrated samples, i.e. leading to lower accumulated strains. These differences are most striking for medium dense sand. In that range of densities, even lower residual strains are generated in samples prepared by dry deposition in the loosest state followed by compaction (DD + VIB and DD + TAP, Fig. 7b).

Analyzing the data for moist tamping (MT) in Fig. 7a, it is a matter of density whether these samples deliver higher or lower intensities of strain accumulation compared to the air-pluviated specimens. The tamped samples show higher permanent strains in the loose state, but lower cumulative rates at medium density, leading to a considerably different shape of the $\varepsilon^{\text{acc}}(I_{D0})$ curve. Considering the different course of the strain accumulation curves $\varepsilon^{\text{acc}}(N)$ of the tamped and the pluviated samples, the deviations between both types of samples in a $\varepsilon^{\text{acc}} - I_{D0}$ diagram are additionally dependent on N.

Generally, the sample preparation method should be chosen in accordance with the assumed genesis of the sand deposit in situ.

7 Bedding Plane Orientation

The effect of the bedding plane orientation with respect to the polarization of the cyclic loading has been studied using a mold inclined by an angle α during the sample preparation by air pluviation (Fig. 8a). After preparation the samples were subjected to a cyclic loading in the axial direction. Results from tests with $\alpha = 0°$ (conventional samples) and $\alpha = 45°$ performed on medium dense KFS are presented in Fig. 8b. The data demonstrate that the influence of the bedding plane inclination α on the cumulative behaviour of KFS is rather small, probably due to the predominantly compact particles of this sand.

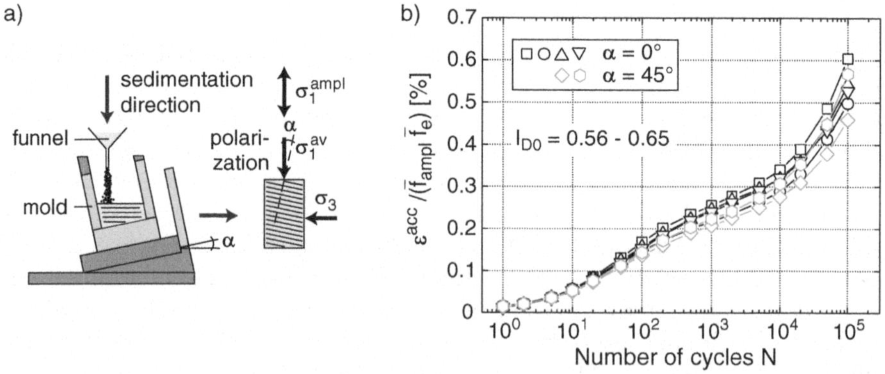

Fig. 8. Comparison of samples with bedding plane inclinations $\alpha = 0°$ and $\alpha = 45°$ subjected to a drained high-cyclic loading in the vertical direction

8 Sample Geometry and Dimensions

The influence of sample geometry and dimensions has been investigated in a series of tests on KFS. Five types of samples have been tested, four of them being cylindrical and one cuboid-shaped (see the schemes in Fig. 9). In case of the cylindrical specimens, beside the IBF standard ($d = h = 100\,\text{mm}$) further 1:1 samples with smaller ($d = h = 50\,\text{mm}$) and larger size ($d = h = 150\,\text{mm}$) have been tested. Furthermore, also 2:1 samples ($d = 100\,\text{mm}$, $h = 200\,\text{mm}$) have been included in the testing program. The cuboid-shaped specimens measured $a \times b = 90 \times 90\,\text{mm}$ in cross section and $h = 180\,\text{mm}$ in height. They are further used for local strain measurements in Sect. 12.

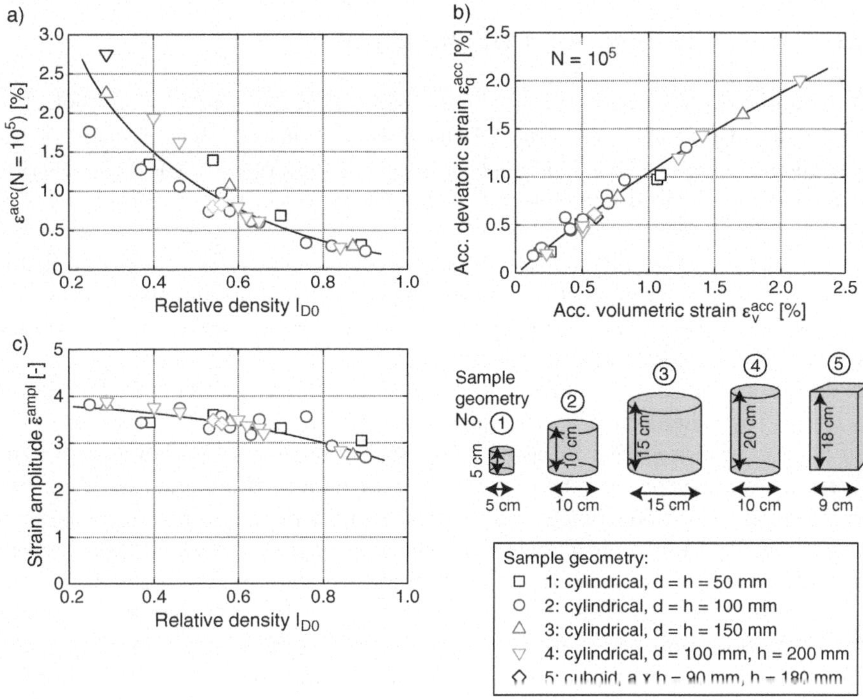

Fig. 9. Results from a test series with different sample geometries: (a) Accumulated strain ε^{acc} after 10^5 cycles versus initial relative density I_{D0}, (b) Accumulated deviatoric strain $\varepsilon_q^{\text{acc}}$ versus accumulated volumetric strain $\varepsilon_v^{\text{acc}}$ both measured at $N = 10^5$, (c) mean value of strain amplitudes $\bar{\varepsilon}^{\text{ampl}}$ over 10^5 cycles

Figure 9 presents the data from tests with the same average and cyclic stresses ($p^{\text{av}} = 200\,\text{kPa}$, $\eta^{\text{av}} = 0.75$, $q^{\text{ampl}} = 60\,\text{kPa}$) but different initial relative densities. For the cylindrical 2:1 and the cuboid-shaped samples only medium densities were tested, while a larger I_{D0} range was investigated for the three other types. The accumulated strain ε^{acc} after 10^5 cycles measured for the different types of samples were quite similar for medium to high densities, but showed stronger variations at lower densities, with higher residual strains for the larger samples ($d = h = 150\,\text{mm}$ and $d = 100\,\text{mm}$, $h = 200\,\text{mm}$), Fig. 9a. The reasons are not clear so far. Since several years lay between the tests on the standard samples ($d = h = 100\,\text{mm}$) and those on all other types of samples, the differences could be caused by changes of the re-used test material (e.g. by abrasion) or by the different technicians preparing the samples (i.e. slightly different fabric). In contrast, the influence of the sample geometry and dimensions on the direction of strain accumulation (Fig. 9b) and on the strain amplitudes (Fig. 9c) is rather small in this case. The differences in $\bar{\varepsilon}^{\text{ampl}}$ may be larger for coarser granular materials, as demonstrated by another series of tests in [19].

9 Multiple Changes of Polarization

Based on multi-dimensional simple shear tests an earlier study documented in [18, 20] demonstrated that a change of the polarization, i.e. the direction of the cycles in the strain space by 90° leads to a temporary increase of the rate of strain accumulation. These findings were later confirmed in [5], where also changes of the polarization ≠90° were tested. However, all tests in [5, 18, 20] were restricted to a single change of the polarization only. In order to study the effect of multiple polarization changes, a new test series on Karlsruhe fine sand has been conducted in the triaxial apparatus [25]. In all tests the cycles were superposed to an average stress with $p^{\text{av}} = 200\,\text{kPa}$ and $\eta^{\text{av}} = 0.5$.

Six different polarizations have been tested. The stress paths in the P-Q plane are shown in the scheme in Fig. 10a. The isometric variables $P = \sqrt{3}p$ and $Q = \sqrt{3/2}q$ [9] are advantageous in connection with studies on the influence of the polarization because the lengths of the stress paths and the angles between two polarizations are preserved when transferred from a principal stress coordinate system to the P-Q plane, in contrast to the p-q representation. In the present test series, two neighbored polarizations differed by an angle $\Delta\alpha_{PQ} = 30°$ in the P-Q plane. Polarizations 1 and 4 were parallel to the P- or Q-axis, respectively, i.e. they represent purely isotropic or purely deviatoric stress cycles. In order to achieve the stress paths depicted in Fig. 10a a simultaneous oscillation of the axial and the lateral effective stress was necessary. Due to the simultaneous σ_1' and σ_3' variation a relative low frequency of 0.01 Hz has been chosen for the regular cycles in all tests within this study. This implicates a lower maximum number of cycles compared to other test series presented herein.

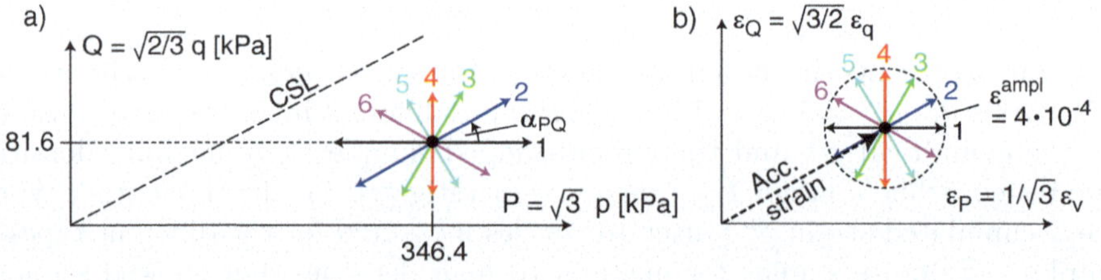

Fig. 10. Stress and strain paths in the tests with multiple polarization changes (CSL = critical state line)

The strain paths are shown schematically in a $\varepsilon_P - \varepsilon_Q$ diagram in Fig. 10b, using the isometric strain variables $\varepsilon_P = \varepsilon_v/\sqrt{3}$ and $\varepsilon_Q = \sqrt{3/2}\varepsilon_q$. The tests were performed stress-controlled. Based on preliminary tests the stress amplitudes have been chosen in order to achieve a strain amplitude $\varepsilon^{\text{ampl}} = \sqrt{(\varepsilon_P^{\text{ampl}})^2 + (\varepsilon_Q^{\text{ampl}})^2} \approx 4 \cdot 10^{-4}$ for all six polarizations at a medium density $I_{D0} \approx 0.6$.

52 tests with initial relative densities in the range $0.42 \leq I_{D0} \leq 0.84$ have been performed, among them 41 with multiple polarization changes and 11 reference tests with a constant polarization throughout the test. The 20 different sequences applied in the tests with polarization changes are illustrated in Fig. 11. In case of sequences Nos. 1 to 5 the polarization was changed every 1000 cycles. All six polarizations were involved in the first four sequences while only purely isotropic and purely deviatoric stress cycles (polarizations 1 and 4) were alternatingly applied in the tests with the fifth sequence. The polarization was altered every 2500 cycles in case of sequences Nos. 6 to 16, while the polarization was kept constant for 5000 cycles in case of sequences Nos. 17 to 20. In all these tests a total number of at least 10,000 cycles has been applied. The reference tests were performed with 11,000 cycles along polarizations 1, 4 or 5.

The influence of the polarization with respect to the sedimentation direction can be studied based on the first 1,000 cycles of each test, i.e. the data recorded prior to the first change of the polarization. Figure 12a demonstrates that the residual strain after 1,000 cycles is only marginally affected by the polarization. Based on this knowledge, next the influence of *changes* of the polarization can be studied.

Figure 12b presents the curves of accumulated strain $\varepsilon^{\text{acc}}(N)$ during the tests with 11 or 12 packages of cycles, i.e. 10 or 11 polarization changes. For comparison purpose, the data from tests with a constant polarization (1D tests) performed on samples with similar initial density have been added in Fig. 12b. The differences between the curves $\varepsilon^{\text{acc}}(N)$ obtained in the individual tests are primarily due to the variations in the initial densities. A temporary increase of the rate of strain accumulation usually occurred due to the first change of the polarization only. All subsequent alterations of the cyclic loading direction had only a moderate impact on the further course of the curves $\varepsilon^{\text{acc}}(N)$.

In Fig. 12c the residual strain after 10,000 cycles is given as a function of initial relative density. The different symbols distinguish between different numbers of polarization changes during the cyclic loading. The data do hardly show any clear differences between the tests with 9 polarization changes and those with constant polarization. The same applies to the tests with a single change. Surprisingly, the effect is more pronounced in case of the tests with three such alterations of polarization. Almost independent of initial density, the tests with three polarization changes delivered an about $\Delta\varepsilon^{\text{acc}} = 0.1$ to 0.2% larger residual strain than those with a lower or higher number of α_{PQ} alterations.

The $\varepsilon_q^{\text{acc}} - \varepsilon_v^{\text{acc}}$ data after $N = 10{,}000$ cycles shown in Fig. 12d do not show any noticeable effect of the number of polarization changes conducted during these cycles. Also the $\varepsilon_q^{\text{acc}} - \varepsilon_v^{\text{acc}}$ strain paths measured in the individual tests

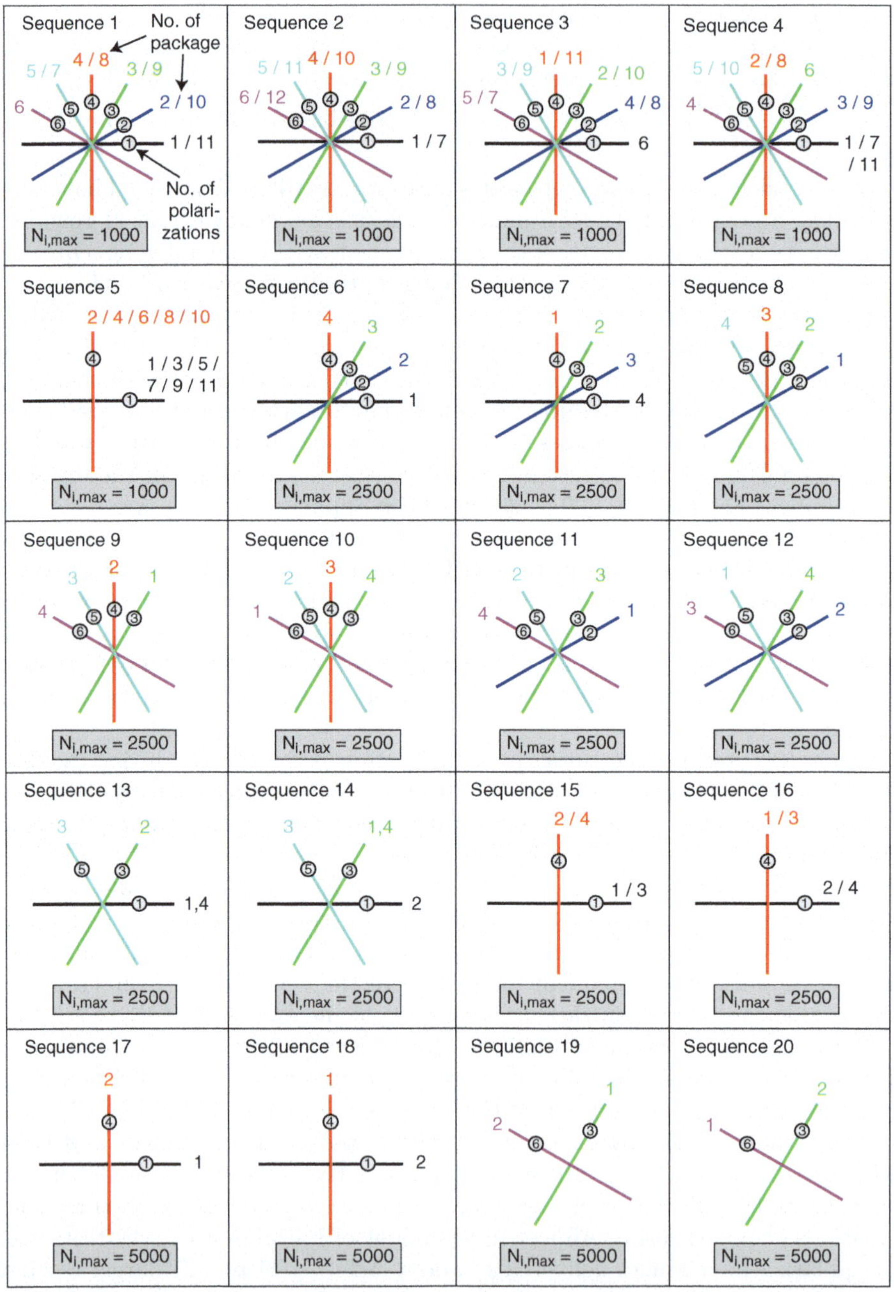

Fig. 11. Sequences of polarizations applied in the tests ($N_{i,\max}$ = number of cycles in each package with a certain polarization)

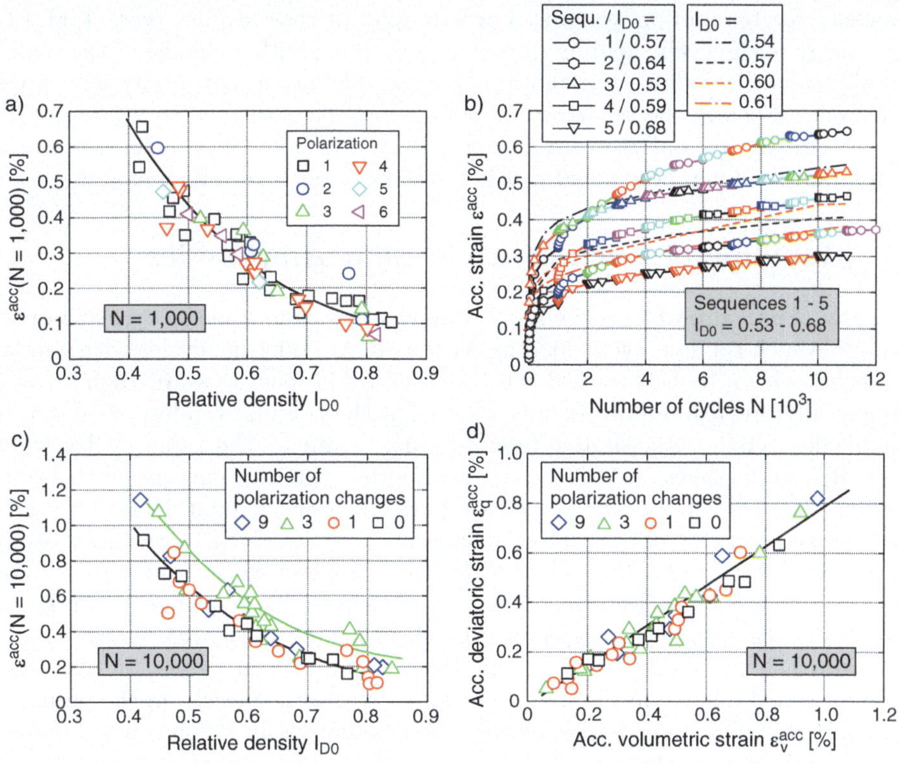

Fig. 12. (a) Accumulated strain ε^{acc} after 1000 cycles as a function of initial relative density I_{D0}, (b) Development of accumulated strain $\varepsilon^{\text{acc}}(N)$ in five tests with different sequences of polarizations and in the corresponding reference tests with constant polarization, (c) Accumulated strain ε^{acc} after 10,000 cycles as a function of initial relative density I_{D0}, (d) Accumulated deviatoric strain $\varepsilon_q^{\text{acc}}$ as a function of accumulated volumetric strain $\varepsilon_v^{\text{acc}}$ after 10,000 cycles

confirm that the direction of strain accumulation is not altered by the multiple changes of the direction of the cycles.

Based on Fig. 12c it can be concluded that the influence of multiple polarization changes on the final residual strain may be regarded as less important than previously thought, based on the experiments presented in [20] or [5]. It seems relevant for the first change of the polarization only and has only a moderate effect on the residual strain after a larger number of cycles. Thus the factor f_π of the HCA model, describing the effect of polarization changes, could possibly be omitted. A moderate increase of the rate of strain accumulation (e.g. via a 20% increase of parameter C_{N1}) for problems with multiple polarization changes could be sufficient for practical purposes.

It should be kept in mind, however, that although the effect of polarization changes may be of minor importance on the element test level, it can be of relevance for foundation systems. For example, in case of piles (e.g. [1,11,12]) the change of the cyclic loading direction goes along with a change of the region of soil involved in strain accumulation or stress relaxation. Zones with low cumulative effects before the polarization change may be subjected to a large cyclic impact afterwards, and vice versa. Such influence of polarization changes on the foundation level can be studied in FE simulations with the HCA model.

10 Bundling of Cycles with Similar Amplitudes

In many practical problems the amplitude changes from cycle to cycle. In order to handle such random cyclic loading with a HCA model the cycles with similar amplitudes have to be grouped into packages. The packages are then treated sequentially. Such bundling is only allowed if the original random sequence of amplitudes generates a similar final residual strain as the same cycles when ordered into packages. A respective examination was the purpose of the tests presented in this section. All tests were performed on medium dense KFS with an average stress of $p^{\mathrm{av}} = 200\,\mathrm{kPa}$ and $\eta^{\mathrm{av}} = 0.75$. Each of the four stress amplitudes $q^{\mathrm{ampl}} = 20$, 40, 60 and 80 kPa was applied with 12,500 cycles. The total number of cycles was thus 50,000 in all tests. The sequence of the stress amplitudes has been varied, as shown by the schemes in Fig. 13a.

Figure 13b presents the measured strain accumulation curves for the eight different signals shown in Fig. 13a. In order to purify the data from the influence of the slightly different initial densities, the residual strain $\varepsilon^{\mathrm{acc}}$ has been divided by the void ratio function f_e of the HCA model, which was evaluated with the initial void ratio e_0 of a sample. Obviously, all tests with signals of types 1 to 4, i.e. with a frequently changing amplitude, deliver a similar residual strain $\varepsilon^{\mathrm{acc}}/f_e(e_0)$ after 50,000 cycles, irrespective of the order of the amplitudes. The tests with cycles grouped into four or eight packages (types 5 to 8) lead to even somewhat larger residual strains at the end of the test (Fig. 13b). This is also evident in Fig. 13c where the final residual strain after 50,000 cycles has been plotted versus I_{D0}. It can be concluded that the bundling of cycles with frequently changing amplitudes into a limited number of packages with constant amplitudes leads to a slight overestimation of the residual deformations, i.e. the procedure is conservative.

11 Monotonic Loading Between Bundles of Cycles (Loss of Cyclic Preloading Memory)

This test series [24] was originally dedicated to the question if the HCA model parameters can be calibrated from multi-stage tests, i.e. tests with a subsequent application of packages of cycles at various average stresses and with different stress amplitudes. Such multi-stage tests were thought to have the potential to significantly reduce the experimental effort necessary for the calibration of a full set of HCA model parameters. However, as demonstrated in the following, the test results revealed an effect not captured by the HCA model yet.

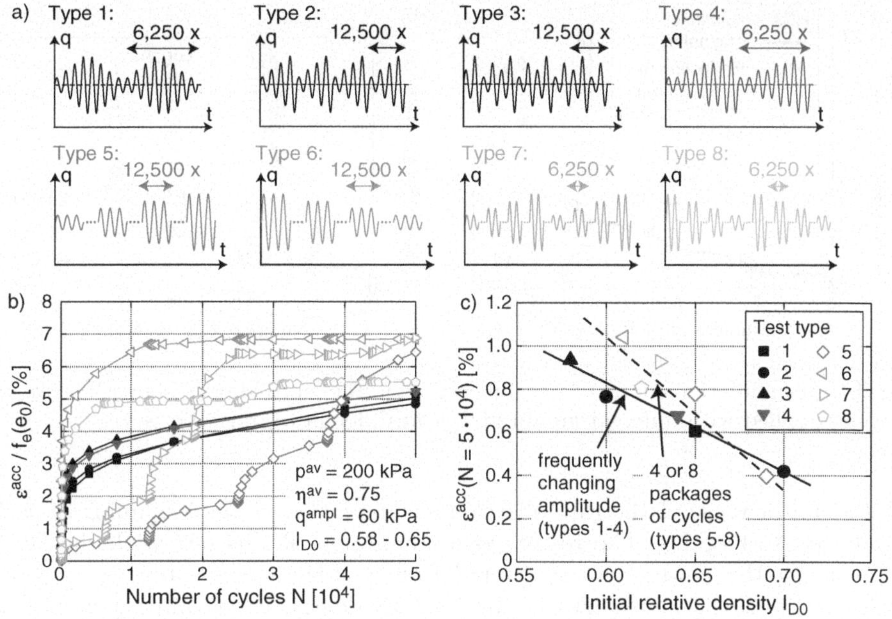

Fig. 13. Comparison of tests with frequently changing amplitude and tests with packages of cycles: (a) tested cyclic load signals, (b) accumulated strain $\varepsilon^{\mathrm{acc}}$ as a function of the number of cycles N, (c) residual strain after 50,000 cycles versus initial relative density I_{D0}

Medium dense samples of a natural fine sand ($d_{50} = 0.10$ mm, $C_u = 1.6$) taken near-shore in Cuxhaven, Germany have been tested in this series. The HCA model parameters of this sand had been already calibrated from 17 single-stage drained cyclic triaxial tests with different amplitudes, densities and average stresses [28].

The results from two of the new multi-stage tests are presented in Fig. 14. Figure 14a shows the development of accumulated strain $\varepsilon^{\mathrm{acc}}$ with increasing number of cycles N in a test with cycles applied at a constant average stress ($p^{\mathrm{av}} = 200$ kPa, $\eta^{\mathrm{av}} = 0.75$). The four packages with 25,000 cycles each and stress amplitudes $q^{\mathrm{ampl}} = 20$, 40, 60 and 80 kPa were applied in ascending order. The measured development of residual strain with N agrees well with similar test series in the literature [2,13,18]. $\varepsilon^{\mathrm{acc}}(N)$ data stemming from a recalculation of this test using the HCA model with the parameters calibrated from the 17 single-stage tests are shown as the thick solid curve in Fig. 14a. Due to its preloading variable g^A the HCA model is able to reproduce well the experimental data.

In the multi-stage test presented in Fig. 14b the average stress ratio has been increased between the successive packages of cycles, keeping the average mean pressure constant ($p^{\mathrm{av}} = 200$ kPa). Each of the five packages comprised 25,000 cycles with a stress amplitude $q^{\mathrm{ampl}} = 60$ kPa. The $\varepsilon^{\mathrm{acc}}(N)$ curve in Fig. 14b

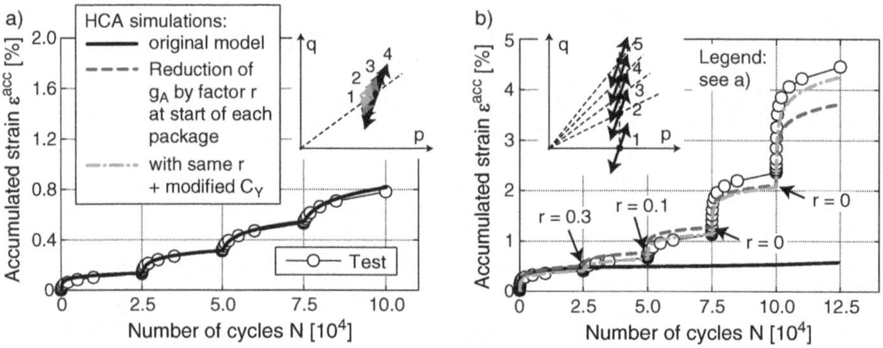

Fig. 14. Accumulation curves $\varepsilon^{\mathrm{acc}}(N)$ measured in the multi-stage tests with packages of cycles applied with different amplitudes or at different average stresses

exhibits an increase of the rate of strain accumulation at the beginning of each new package of cycles. The results of a simulation of this test with the HCA model using the parameters calibrated from the single-stage tests have been added as the thick solid curve in Fig. 14b. Evidently, the strain accumulation rates measured in the later packages of this test are significantly underestimated by the HCA model. This is in contrast to simulations of the single-stage tests, where the stress ratio-dependence experimentally observed could be adequately reproduced. Therefore, the differences between the experimental results and the model prediction apparent in Fig. 14b must be due to an effect not captured by the HCA model yet.

A drained cyclic loading leads to subtile changes in the orientations of the grains or grain contact normals, usually rendering the sand fabric more resistant to the subsequent cycles, i.e. leading to an adaption of the fabric to the actual cyclic loading and thus to a reduction of the strain accumulation rate $\dot{\varepsilon}^{\mathrm{acc}}$ with increasing number of cycles. In the HCA model this is phenomenologically captured by the preloading variable g^A. In the multi-stage test presented in Fig. 14b the change of the average stress between two succeeding packages of cycles represents a monotonic loading. It is likely that re-orientations of the grains caused by this monotonic loading erase some parts or the whole memory of the sand regarding the preceding cyclic loading. After sufficiently large monotonic strains the cumulative behaviour of the sand sample is probably similar to that of a freshly pluviated one, because the cyclic preloading history has been completely forgotten. In the context of the HCA model this means that the preloading variable g^A is reduced or even completely erased (to $g^A = 0$) by a monotonic loading. This effect has not been considered in calculations with the HCA model so far, i.e. the preloading variable g^A has been assumed to increase continuously only.

A better prediction of the experimental data in Fig. 14b (thick dashed curve) can be indeed achieved if g^A is reduced by a certain factor r at the start of each subsequent package. $r = 1$ means that the preloading variable remains unchanged and $r = 0$ corresponds to a total loss of the cyclic preloading memory. The r

values necessary for a satisfying reproduction of the test data are provided in Fig. 14b. A complete loss of memory can be concluded from the parameter $r = 0$ obtained for the packages applied at higher stress ratios. An almost perfect agreement with the test data can be achieved if additionally the parameter C_Y of the function f_Y, originally calibrated from the 17 single-stage tests, is slightly increased (thick dot-dashed curve in Fig. 14b). Based on the data from all tests within this series, it has been demonstrated in [24] that r correlates with the strain occurring during the monotonic loading phases. A more detailed experimental investigation of this effect followed by an incorporation in simulations with the HCA model is planned for the future.

The (partial) loss of the cyclic preloading memory due to monotonic loading may be of great practical relevance for foundations subjected to a cyclic loading with varying amplitudes and average stresses. Furthermore, a foundation is usually subjected to a monotonic loading during the construction phase caused by the own weight of the structure. This monotonic loading could partly or fully erase the memory of a cyclic loading history of the subsoil, probably leading to larger cumulative deformations during a subsequent cyclic loading of the foundations. Therefore, the conservative assumption $g_0^A = 0$ for the initial state in predictions with the HCA model, usually made because the cyclic preloading of the in situ soil is unknown and no suitable determination method is available yet, may be not as far from reality as previously thought. Beside that, the observed effect could be utilized for a reduction of the number of cyclic tests necessary for a calibration of the HCA model parameters. If a sample already tested under a certain cyclic loading condition can be reset to a state with $g_0^A = 0$ (similar to a freshly pluviated sample), simply by applying a monotonic loading, several average stresses or amplitudes could be tested in succession on a single sample.

12 Multi-dimensional Cyclic Loading

Earlier test series [18,20] had shown that two-dimensional (2D) stress or strain paths applied in a cyclic DSS or triaxial device deliver higher cumulative strains than 1D paths of similar span [20]. In the HCA model this is considered by a special tensorial definition of a multidimensional strain amplitude for (up to) 6D convex strain paths [9]. A scalar value of the strain amplitude $\varepsilon^{\text{ampl}}$ is obtained from a procedure involving multiple projections of the original strain loop. As an example, projections from 3D to 1D are shown in Fig. 15. The projection is done in the direction of a line connecting the two points of the strain path with the largest distance in that direction. The (up to 6) values of maximum distance and the corresponding vectors describing the projection direction are finally used for calculating $\varepsilon^{\text{ampl}}$. For details the interested reader is referred to [9]. The amplitude definition of the HCA model was only confirmed for 1D and 2D paths so far [18,20]. The experimental program described in this section has been performed with the aim of a validation for 1D to 4D paths. 4D paths mean an independent variation of four of the six components of the stress or strain tensor.

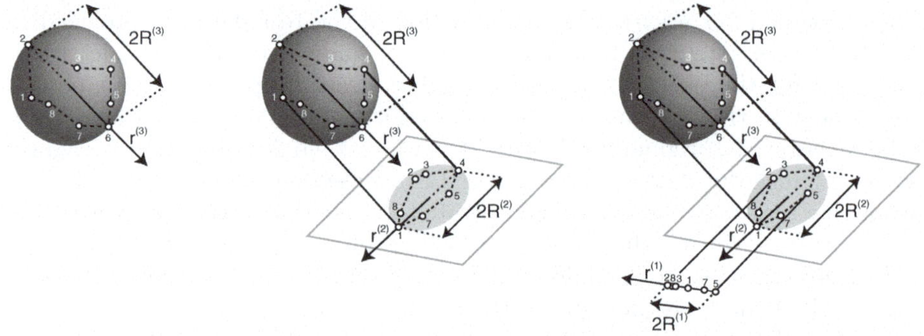

Fig. 15. Schemes showing the projections of a strain loop from 3D to 2D and from 2D to 1D for the determination of strain amplitude $\varepsilon^{\mathrm{ampl}}$ according to the procedure described in [9]

The drained cyclic tests have been performed in three different types of test devices, using three different sample geometries and two different saturation conditions. 64 tests were conducted in a cyclic triaxial device as it has been used for most of the studies reported earlier in this paper (see the scheme in Fig. 1). Full cylinder samples (diameter d = 100 mm, height h = 200 mm, see a photo in Fig. 16a) have been used in these tests. 30 additional tests were performed in a hollow cylinder triaxial device (Fig. 16c). The samples had an outer diameter d_a = 100 mm, an inner diameter d_i = 60 mm and a height h = 200 mm. The volume changes of both the full and the hollow cylinder samples were measured via the pore water. Thus, the samples were tested fully water-saturated.

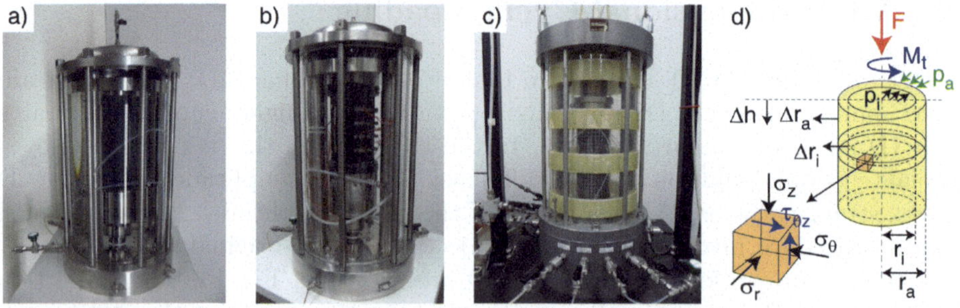

Fig. 16. Three types of devices used in the present study: (a) cyclic triaxial device with water-saturated full cylinder samples, (b) cyclic triaxial device with local strain measurements using LDTs on dry cuboid-shaped samples, (c) hollow cylinder device, (d) external loads and stresses acting on an element of soil in the hollow cylinder test

The application of 2D to 4D stress paths necessitates the cyclic variation of cell pressure, which can lead to membrane penetration effects [8, 16]. These and other sources of errors of global deformation measurements (e.g. system compliance, bedding error) can be avoided by applying local strain measurements. Therefore, for comparison purpose 26 further tests have been performed in a cyclic triaxial device similar to that shown in Fig. 1, but with local measurements of deformation on cuboid-shaped samples ($a \times b = 90$ mm, $h = 180$ mm). LDTs, i.e. strips of stainless steel equipped with strain gauges (Fig. 16b) were applied for that purpose. The cuboid-shaped samples were tested in the dry condition and without water in the pressure cell, in order to guarantee the long-term stability of the LDT measurements. The cell air pressure was kept constant to maintain a constant temperature in the cell, which is necessary for reliable LDT measurements. The cyclic variation of lateral effective stress was realized by an oscillation of pore air pressure.

In the triaxial devices (Fig. 16a, b) 1D stress paths with different inclinations in the P-Q plane and 2D stress paths can be applied by a simultaneous cyclic variation of effective axial stress (applied via the load piston) and effective lateral stress (applied as a cyclic variation of cell or pore pressure). 3D and 4D paths can be tested in the hollow cylinder device only. The independent cyclic variation of vertical force F, torsional moment M_T, outer cell pressure p_a and inner cell pressure p_i (Fig. 16d) allows independent oscillations of three normal stresses (σ_z in vertical, σ_r in radial and σ_θ in circumferential direction) and one shear stress component $\tau_{\theta z}$. These four stresses can be converted to three principal stresses σ_1, σ_2 and σ_3 by means of a rotation of the coordinate system by the angle α.

Air-pluviated samples of Karlsruhe fine sand have been used in all tests of this series. The samples have been prepared with two different initial relative densities ($I_{D0} \approx 0.40$ and 0.70). In all tests $N = 10.000$ regular cycles were applied at an average stress with $p^{\text{av}} = 200$ kPa and $\eta^{\text{av}} = 0.50$, using a loading frequency of $f = 0.02$ Hz.

In the cyclic triaxial tests on water-saturated cylindrical and on dry cuboid-shaped samples 1D, 2D elliptical and 2D circular stress paths in the P-Q plane have been applied. All stress paths had a sufficient distance to the failure lines. For the 1D stress paths the left-hand side of Table 1 specifies the tested spans (lengths, double amplitudes) l_{PQ} and the inclinations (polarizations) α_{PQ} in the P-Q plane. A polarization $\alpha_{PQ} = 54.75°$ means a conventional cyclic axial loading at constant lateral effective stress. The dots in the table indicate that spans between $l_{PQ} = 60$ kPa and $l_{PQ} = 200$ kPa were tested in steps of $\Delta l_{PQ} = 10$ kPa. For 2D stress paths the spans a_{PQ} and b_{PQ} in two orthogonal axes and the orientation α_{PQ} of the longer axis are given on the right-hand side of Table 1. For circular 2D stress paths $a_{PQ} = b_{PQ} = r_{PQ}$ holds. A polarization α_{PQ} cannot be specified for circular paths.

The program of the tests in the hollow cylinder device is shown in Table 2. The first test represents a reference test. It has been performed in analogy to the circular 2D stress paths applied in the cyclic triaxial tests with a radius $r_{PQ} = 40$ kPa in the P-Q plane. Such stress loop is achieved by a cyclic variation of the vertical and the radial stress with the amplitudes $\sigma_z^{\text{ampl}} = 40$ kPa and $\sigma_r^{\text{ampl}} = 28.28$ kPa. These amplitudes have been kept constant within the whole

Table 1. Program of the cyclic triaxial tests with 1D or 2D stress paths

Dimension	Polarization α_{PQ} [°]	Span l_{PQ} [kPa]
1D	54.75	60.00
1D	54.75	70.00
1D	...	...
1D	54.75	200.00
1D	0	80.00
1D	45	80.00
1D	90	80.00
1D	135	80.00

Dimension	Polarization α_{PQ} [°]	Span a_{PQ} [kPa]	Span b_{PQ} [kPa]
2D	-	28.28	28.28
2D	-	40.00	40.00
2D	-	60.00	60.00
2D	-	80.00	80.00
2D	0	35.78	17.78
2D	0	80.00	40.00
2D	45	80.00	40.00
2D	90	80.00	40.00
2D	135	80.00	40.00

Table 2. Program of the cyclic hollow cylinder tests with 2D, 3D and 4D stress paths

No.	Dim.	Stress amplitudes				Phase shift			
		σ_z^{ampl} [kPa]	σ_r^{ampl} [kPa]	$\sigma_\theta^{\text{ampl}}$ [kPa]	$\tau_{\theta z}^{\text{ampl}}$ [kPa]	ϕ_{σ_z} [°]	ϕ_{σ_r} [°]	ϕ_{σ_θ} [°]	$\phi_{\tau_{\theta z}}$ [°]
1	2D	40.00	28.28	28.28	0.00	35.26	125.26	125.26	0.00
2	3D	40.00	28.28	56.56	0.00	35.26	125.26	125.26	0.00
3	3D	40.00	28.28	28.28	28.28	35.26	125.26	125.26	0.00
4	4D	40.00	28.28	56.56	28.28	35.26	125.26	62.63	62.63
5	4D	40.00	28.28	56.56	28.28	35.26	125.26	125.26	125.26

test series. By a cyclic variation of the two remaining stress components σ_θ and $\tau_{\theta z}$ with the amplitudes $\sigma_\theta^{\text{ampl}}$ and $\tau_{\theta z}^{\text{ampl}}$ and by chosing different values of phase shift ϕ between the different stress components up to 4D stress paths could be investigated. The loading frequency was the same for all four stress components.

Figure 17 compares data from the triaxial tests performed on either the fully water-saturated full cylinder samples with global strain measurements (left column of diagrams) or the dry cuboid-shaped samples with local strain measurements (right column). In both cases results for two different 1D stress paths with inclinations α_{PQ} of either $-45°$ or $45°$, two elliptical 2D paths with different inclinations of the larger axis ($\alpha_{PQ} = 0°$ and $90°$) and a circular 2D path are shown. A comparison of the diagrams in the first and second row of Fig. 17 reveals that circular stress paths in the P-Q plane cause elliptical strain paths in the $\varepsilon_P - \varepsilon_Q$ diagram, with a longer axis in the deviatoric direction. The strain amplitudes given as a function of the number of cycles in Fig. 17c, g have been derived following the procedure described in [9] and illustrated in

Fig. 15. According to that definition, the strain amplitude increases with increasing area encompassed by the strain path in the $\varepsilon_P - \varepsilon_Q$ diagram. The smallest strain amplitudes are obtained for the 1D stress path with $\alpha_{PQ} = 45°$, while the circular 2D stress path delivers the highest values. Evidently, for the same stress paths, the strain amplitudes obtained from the local strain measurements on dry cuboid-shaped samples (Fig. 17g) are slightly smaller than those derived from the global deformation records in case of fully water-saturated cylindrical samples (Fig. 17c). The differences are probably mainly caused by the different states of saturation rather than by the differences in the method of strain measurement or sample geometry. The higher stiffness of KFS in the dry condition compared to the water-saturated state has been already observed in previous experimental studies, using cylindrical samples and global measurements in all tests [19]. A comparison of the diagrams in Fig. 17c, g and d, h reveals that for both types of samples and saturation conditions, higher strain amplitudes lead to larger cumulative strains.

A comparison of the results from the tests with the same circular stress paths applied on the three types of samples is shown on the left-hand side of Fig. 18. The data confirms the smaller strain amplitudes for the dry cuboid-shaped samples compared to the water-saturated full cylinder samples (Fig. 18c). The highest strain amplitudes were recorded for the hollow cylinder samples, however. Again, larger strain amplitudes (Fig. 18c) lead to larger strain accumulation rates (Fig. 18d), at least at $N > 100$.

The results of the five different types of hollow cylinder tests listed in Table 2, with 2D, 3D or 4D stress paths are provided on the right-hand side of Fig. 18. The smallest strain amplitudes and cumulative strains were obtained for the 2D stress path, while the largest values resulted from the 4D one. Thus, with increasing dimensionality of the stress path, both the strain amplitudes $\varepsilon^{\mathrm{ampl}}$ and the cumulative strains $\varepsilon^{\mathrm{acc}}$ grow.

Finally, in Fig. 19 the cumulative strains $\varepsilon^{\mathrm{acc}}$ after 10,000 cycles from all 120 tests are given as a function of the strain amplitude $\bar{\varepsilon}^{\mathrm{ampl}}$ of that test. The strain amplitude is calculated as a mean value over 10,000 cycles. The diagram contains data for the two different relative densities $I_{D0} \approx 0.40$ and $I_{D0} \approx 0.70$. For a certain density, the $\varepsilon^{\mathrm{acc}} - \bar{\varepsilon}^{\mathrm{ampl}}$ data for the different types of samples, saturation conditions and dimensionality of the stress or strain path (1D up to 4D) lie on a unique curve. Thus, the definition of the strain amplitude according to Niemunis [9] allows the description of the $\varepsilon^{\mathrm{acc}} - \bar{\varepsilon}^{\mathrm{ampl}}$ relationship by a single function. In the present case, the function $\varepsilon^{\mathrm{acc}} \sim (\bar{\varepsilon}^{\mathrm{ampl}})^n$ has been fitted to the data, leading to exponents $n = 1.40$ for $I_{D0} = 0.40$ and $n = 1.48$ for $I_{D0} = 0.70$. Based on Fig. 19 the amplitude definition used in the HCA model can be seen as validated for strain paths from 1D to 4D.

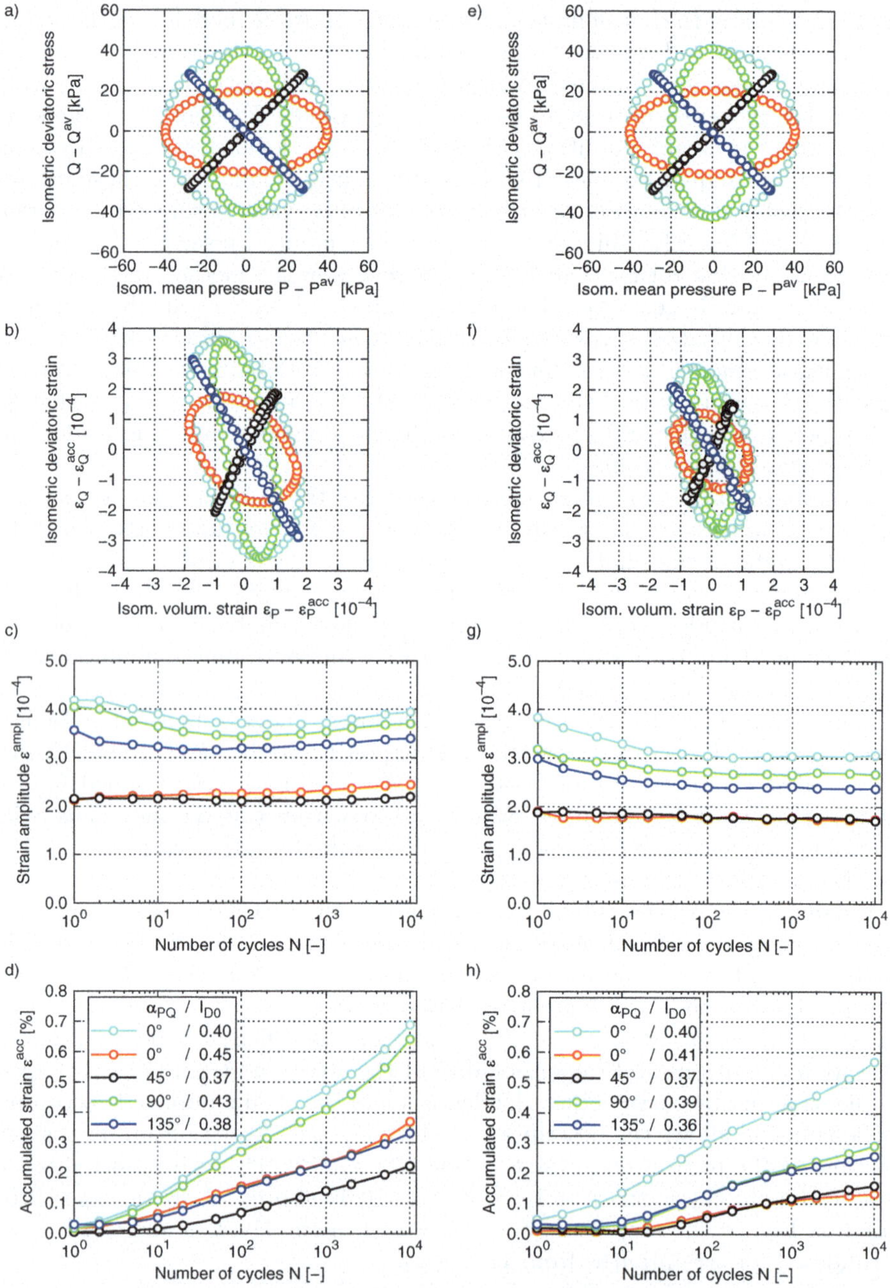

Fig. 17. Selected cyclic triaxial tests on fully water-saturated full cylinder samples (left-hand side) and dry cuboid-shaped samples (right-hand side): (a, e) P-Q stress paths, (b, f) $\varepsilon_P - \varepsilon_Q$ strain paths at $N = 10{,}000$, (c, g) development of strain amplitude $\varepsilon^{\text{ampl}}$ and (d, h) accumulated strain ε^{acc} with increasing number of cycles N in tests on medium dense samples ($I_{D0} \approx 0.40$)

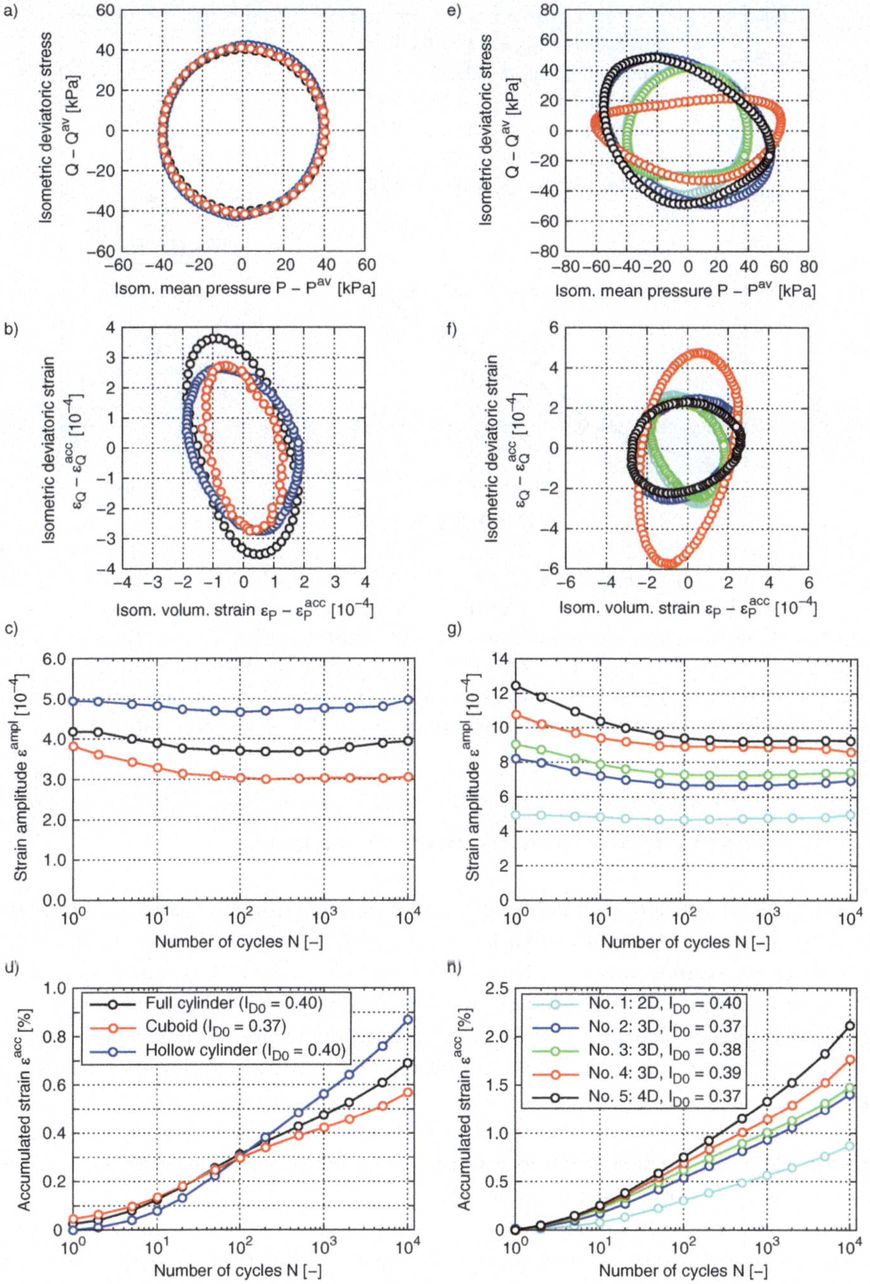

Fig. 18. Left-hand side: Comparison of results from the three types of tests for a circular 2D stress path (hollow cylinder test No. 1 in Table 2); Right-hand side: Results from hollow cylinder tests Nos. 1 to 5 with 2D to 4D stress paths; (a, e) P-Q stress paths, (b, f) ε_P-ε_Q strain paths at $N = 10{,}000$, (c, g) development of strain amplitude $\varepsilon^{\text{ampl}}$ and (d, h) accumulated strain ε^{acc} with increasing number of cycles N in tests on medium dense samples ($I_{D0} \approx 0.40$)

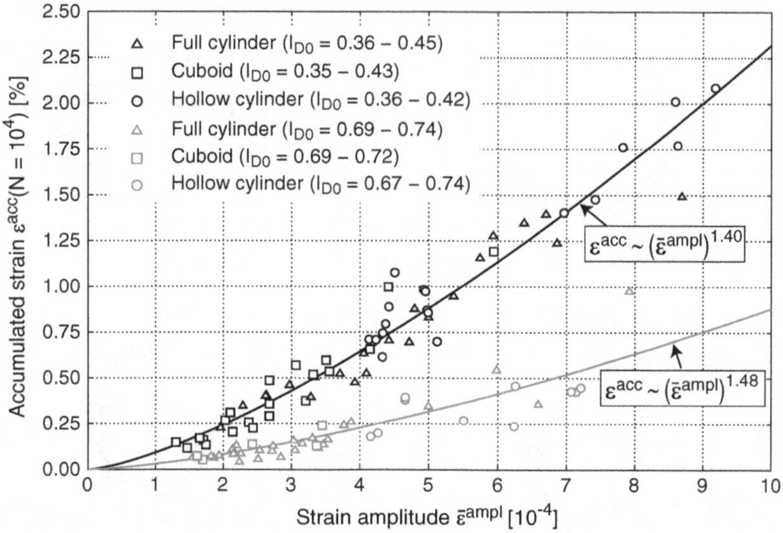

Fig. 19. Compilation of the data of all 120 cyclic tests with 1D to 4D stress paths, performed on either fully water-saturated full cylinder samples, dry cuboid-shaped samples with local strain measurements or water-saturated hollow cylinder samples: Accumulated strain ε^{acc} after 10,000 cycles as a function of the strain amplitude $\bar{\varepsilon}^{ampl}$ for relative densities $I_{D0} \approx 0.40$ and $I_{D0} \approx 0.70$. The strain amplitude has been evaluated by applying the procedure described in [9]. $\bar{\varepsilon}^{ampl}$ is a mean value over 10,000 cycles.

13 Summary, Conclusions and Outlook

The rates of strain accumulation in sand under high-cyclic loading grow with increasing stress or strain amplitude, decreasing density and increasing average stress ratio. At similar values of relative density, average stress and stress amplitude, finer and more well-graded clean sands show higher cumulative strains. A content of non-cohesive fines or shell fragments up to 20% further increases the rates of strain accumulation. Tests on glass beads, natural sand and crushed sand demonstrate that the influence of grain shape depends on stress level. At low pressures, materials with round particles show considerable larger strain accumulation rates than those with angular grains. At higher pressures, the particle shape seems to become rather irrelevant.

A considerable influence of the sample preparation method, in particular at medium density, has been observed in a series where the samples have been constituted either by air pluviation, water pluviation or moist tamping and compacted by vibration or tapping. At medium density air pluviated samples and water-pluviated samples compacted by vibration showed the largest cumulative rates, while the samples compacted by tapping or prepared by moist tamping turned out to be the most resistant ones. In tests on air-pluviated samples of a fine sand the effect of the bedding plane orientation was found to be of minor importance. The influence of the sample geometry was rather small for medium dense to dense fine sand, but some differences in the cumulative strains for loose samples need further investigations.

The effect of multiple polarization changes has been found less significant as assumed based on previous studies with only a single change of the direction of the cycles. The results of a new test series demonstrated that only the first alteration of the direction leads to an increase of the strain accumulation rate. A comparison of tests with frequently changing amplitudes and tests with the same cycles grouped into packages with constant amplitude confirms that such bundling is conservative, i.e. leads to a slight overestimation of the permanent deformations. Based on another test series it has been demonstrated that monotonic loading paths between bundles of cycles, as they are caused by a change of the average stress, can erase the cyclic preloading history of the sand, i.e. lead to larger cumulative rates in the subsequent bundle of cycles.

Finally, based on axisymmetric triaxial or hollow cylinder triaxial tests, performed on either water-saturated full cylinder samples, dry cuboid-shaped samples with local strain measurements, or water-saturated hollow cylinder samples, the amplitude definition of Niemunis [9] used in the HCA model could be confirmed for 1D to 4D strain paths.

In future, the loss of cyclic preloading history caused by monotonic loading paths will be further investigated. The influence of plastic fines on the sand behaviour under high-cyclic loading has to be clarified as well. The high-cyclic behaviour of partially saturated sands and lightly cemented granular materials will be a matter of future research, too.

Acknowledgements. Parts of the research presented in this paper was performed within the framework of the project "Improvement of an accumulation model for high-cyclic loading" funded by German Research Council (DFG, project Nos. TR218/18-1/2, WI 3180/3-1/2). The authors are grateful to DFG for the financial support. The majority of the cyclic tests have been performed by H. Borowski, P. Gölz and N. Demiral in the IBF soil mechanics laboratory. The tests presented in Section 3 have been conducted by M. Skubisch in the soil mechanics laboratory at RUB.

References

1. Dührkop, J., Grabe, J.: Monopilegründungen von Offshore-Windenergieanlagen - Zum Einfluss einer veränderlichen zyklischen Lastangriffsrichtung. Bautechnik **85**(5), 317–321 (2008)
2. Kaggwa, W.S., Booker, J.R., Carter, J.P.: Residual strains in calcareous sand due to irregular cyclic loading. J. Geotech. Eng. ASCE **117**(2), 201–218 (1991)
3. Ladd, R.S.: Specimen preparation and liquefaction of sands. J. Geotech. Eng. Div. ASCE **100**(GT10), 1180–1184 (1974)
4. Ladd, R.S.: Preparing test specimens using undercompaction. Geotech. Test. J. ASTM **1**(1), 16–23 (1978)
5. Le, V.H.: Zum Verhalten von Sand unter zyklischer Beanspruchung mit Polarisationswechsel im Einfachscherversuch. Dissertation, Fachgebiet Grundbau und Bodenmechanik, Technische Universität Berlin, Issue No. 66 (2015)

6. Miura, S., Toki, S.: A sample preparation method and its effect on static and cyclic deformation-strength properties of sand. Soils Found. **22**(1), 61–77 (1982)
7. Mulilis, J.P., Seed, H.B., Chan, C.K., Mitchell, J.K., Arulanandan, K.: Effects of sample preparation on sand liquefaction. J. Geotech. Eng. Div. ASCE **103**(GT2), 91–108 (1977)
8. Nicholson, P.G., Seed, R.B., Anwar, H.A.: Elimination of membrane compliance in undrained triaxial testing. I. Measurement and evaluation. Can. Geotech. J. **30**(5), 727–738 (1993)
9. Niemunis, A.: Extended hypoplastic models for soils. Habilitation thesis, Publications of the Institute of Soil Mechanics and Foundation Engineering, Ruhr-University Bochum, Issue No. 34, 2003.
10. Niemunis, A., Wichtmann, T., Triantafyllidis, T.: A high-cycle accumulation model for sand. Comput. Geotech. **32**(4), 245–263 (2005)
11. Rudolph, C., Bienen, B., Grabe, J.: Effect of variation of the loading direction on the displacement accumulation of large-diameter piles under cyclic lateral loading in sand. Can. Geotech. J. **51**(10), 1196–1206 (2014)
12. Rudolph, C., Grabe, J.: Untersuchungen zu zyklisch horizontal belasteten Pfählen bei veränderlicher Lastrichtung. Bautechnik **36**(2), 90–95 (2013)
13. Stewart, H.E.: Permanent strains from cyclic variable-amplitude loadings. J. Geotech. Eng. ASCE **112**(6), 646–660 (1986)
14. Sze, H.Y., Yang, J.: Failure modes of sand in undrained cyclic loading: Impact of sample preparation. J. Geotech. Geoenviron. Eng. ASCE **140**(1), 152–169 (2014)
15. Tatsuoka, F., Ochi, K., Fujii, S., Okamoto, M.: Cyclic undrained triaxial and torsional shear strength of sands for different sample preparation methods. Soils Found. **26**(3), 23–41 (1986)
16. Tokimatsu, K.: System compliance correction from pore pressure response in undrained triaxial tests. Soils Found. **30**(2), 14–22 (1990)
17. Towhata, I.: Geotechnical Earthquake Engineering. Springer, Heidelberg (2008)
18. Wichtmann, T.: Explicit accumulation model for non-cohesive soils under cyclic loading. Ph.D. thesis, Publications of the Institute of Soil Mechanics and Foundation Engineering, Ruhr-University Bochum, Issue No. 38 (2005)
19. Wichtmann, T.: Soil behaviour under cyclic loading - experimental observations, constitutive description and applications. Habilitation thesis, Publications of the Institute of Soil Mechanics and Rock Mechanics, Karlsruhe Institute of Technology, Issue No. 181 (2016)
20. Wichtmann, T., Niemunis, A., Triantafyllidis, T.: On the influence of the polarization and the shape of the strain loop on strain accumulation in sand under high-cyclic loading. Soil Dyn. Earthq. Eng. **27**(1), 14–28 (2007)
21. Wichtmann, T., Niemunis, A., Triantafyllidis, T.: Validation and calibration of a high-cycle accumulation model based on cyclic triaxial tests on eight sands. Soils Found. **49**(5), 711–728 (2009)
22. Wichtmann, T., Niemunis, A., Triantafyllidis, T.: Flow rule in a high-cycle accumulation model backed by cyclic test data of 22 sands. Acta Geotechnica **9**(4), 695–709 (2014)
23. Wichtmann, T., Niemunis, A., Triantafyllidis, T.: Improved simplified calibration procedure for a high-cycle accumulation model. Soil Dyn. Earthq. Eng. **70**(3), 118–132 (2015)
24. Wichtmann, T., Triantafyllidis, T.: Strain accumulation due to packages of cycles with varying amplitude and/or average stress - on the bundling of cycles and the loss of the cyclic preloading memory. Soil Dyn. Earthq. Eng. **101**, 250–263 (2017)

25. Wichtmann, T., Triantafyllidis, T.: On the influence of multiple polarization changes on the cumulative deformations in sand under drained high-cyclic loading. Geotech. Test. J. ASTM **43**, 1–18 (2019)
26. Wichtmann, T., Triantafyllidis, T., Späth, L.: On the influence of grain shape on the cumulative deformations in sand under drained high-cyclic loading. Soils Found. **59**, 208–227 (2019)
27. Wichtmann, T., Triantafyllidis, T., Ziesmann, L.: On the influence of platy shell particles on the cumulative deformations in sand under drained high-cyclic loading. Soil Dyn. Earthq. Eng. **117**, 1–15 (2019)
28. Zachert, H.: Zur Gebrauchstauglichkeit von Gründungen für Offshore-Windenergieanlagen. Dissertation, Veröffentlichungen des Institutes für Bodenmechanik und Felsmechanik am Karlsruher Institut für Technologie, Heft Nr. 180 (2015)
29. Zachert, H., Wichtmann, T., Triantafyllidis, T.: Soil structure interaction of foundations for offshore wind turbines. In: 26th International Ocean and Polar Engineering Conference, ISOPE-2016, Rhodos (2016)

Soil Dynamics and Structural Dynamics

Shear Wave Based Screening Method for Liquefaction Evaluation

J. Yang(✉)

Department of Civil Engineering, The University of Hong Kong,
Pok Fu Lam, Hong Kong
junyang@hku.hk

Abstract. This paper describes a new screening method for evaluating the liquefaction potential of sand deposits with varying percentages of fines. The method is based on measurements of shear wave velocity (V_s), and is developed from a comprehensive experimental program comprising small-strain shear wave testing and large-strain undrained shear tests for sand samples with different quantities of non-plastic fines. A novel point of the method is the unified characterization of shear wave velocity for both clean sand and silty sand through a state parameter that properly combines the effects of void ratio and confining stress in a sound theoretical context. As modern technology has made it more convenient and reliable to measure the shear wave velocity both in the laboratory and in the field, and since the state parameter is a rational index for characterizing various aspects of soil behavior, the proposed method is promising in a wide range of geotechnical applications.

Keywords: Liquefaction · Sands · Fines · Shear wave velocity · In situ state

1 Introduction

Soil liquefaction has been a subject of long standing interest in soil mechanics and geotechnical engineering. In the past decades, soil liquefaction has been investigated mainly in the context of earthquake loading [8, 9, 16, 17, 30], due to the significant liquefaction-related damage observed during the Niigata and Alaska earthquakes in 1964. Some aspects of soil liquefaction have been well understood, many others, however, remain confusing, controversial or mysterious. The liquefaction phenomenon widely observed in laboratory undrained cyclic triaxial tests [8, 14] is characterized by repeated loss and regain of stiffness along with the development of excessive deformation. This behavior, arguably, should not be referred to as liquefaction but, more appropriately, be called cyclic mobility [4, 5]. Based on a comprehensive experimental program, Yang and his co-workers have found that undrained cyclic behavior of sand is much more complicated than previously thought and can be categorized into five major types: flow-type failure, cyclic mobility, plastic strain accumulation, limited flow followed by cyclic mobility, and limited flow followed by strain accumulation [15, 19, 26, 27]. The question as to which pattern will occur depends on a number of inter-related factors, including packing density, effective confining stress, soil fabric, initial static shear stress level and cyclic load level.

T. Triantafyllidis (Ed.): *Recent Developments of Soil Mechanics and Geotechnics in Theory and Practice*, LNACM 91, pp. 169–182, 2020.
https://doi.org/10.1007/978-3-030-28516-6_9

Among the five deformation patterns, the flow-type failure is most critical and is characterised by abrupt, run-away deformation without any precautionary signal. This failure pattern is pertinent to sufficiently loose sand and in many ways is similar to the static or flow liquefaction phenomenon observed in monotonic triaxial tests [21, 27]. Actually, a reasonable correspondence exists between the two, as shown in Fig. 1 where the results from a monotonic test and a cyclic triaxial test on Toyoura sand at a similar post-consolidation state are plotted together. Flow liquefaction, triggered by either cyclic or monotonic loading, can produce the most catastrophic effects of all liquefaction-related phenomena and therefore is a major concern in the deign and construction of large sand structures such as earth or tailings dams and artificial islands [4, 9, 10].

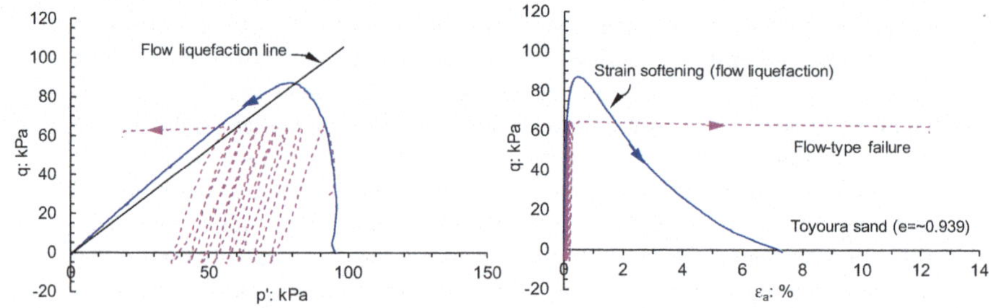

Fig. 1. Correspondence of liquefaction behavior under cyclic and monotonic loading

Whether a sand is in a loose or dense state depends not only on its density (or void ratio) but also on the effective confining stress applied. There is now a general agreement that the behavior of a sand is more closely related to the proximity of its initial state to the critical state or steady state locus [2, 21, 23], which can be described by a state parameter (ψ). If the initial state of a saturated sand lies above the critical state locus with a positive ψ value (state A in Fig. 2), it tends to contract when sheared undrained, accompanied by strain softening and a buildup of high pore water pressure. If the initial state lies below the critical state locus with a negative ψ value (state D or B), it tends to undergo a net dilation, accompanied by strain hardening to a much higher strength. The initial state defined by the void ratio and mean effective stress with reference to the critical state locus is therefore a meaningful index that can be used to identify the potential for liquefaction.

Along this line, we have developed a new method which allows a unified evaluation of *in situ* state of both clean sand and silty sand and thereafter their potential for liquefaction through shear wave velocity measurements. Theoretically, shear wave velocity is a fundamental property with clear physical meaning, directly related to the small-strain shear stiffness of soil. Practically, modern technology has advanced such that shear wave velocity can be measured more conveniently and reliably both in the laboratory and in the field [6]. This paper describes the main aspects of this promising method along with validation.

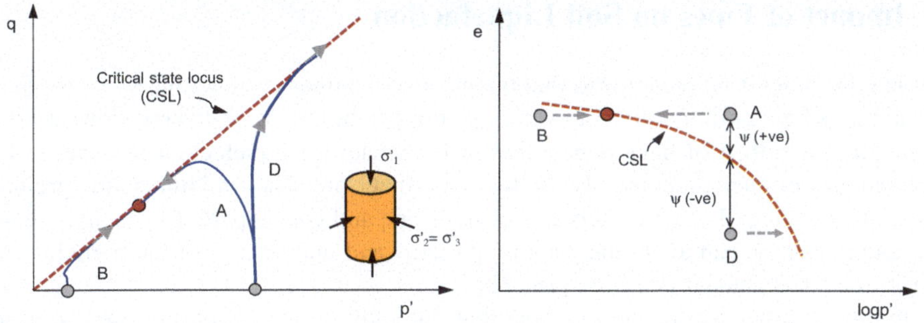

Fig. 2. Schematic illustration of state-dependent behavior of sand under undrained monotonic loading

2 Problems with Current Methods

It remains a big challenge to obtain undisturbed sand samples using conventional methods for the determination of their *in situ* state. As a result, efforts have been made to use field tests, particularly the cone penetration test (CPT), to evaluate the state of sand [1, 3]. Central to the CPT-based method is an empirical correlation between relative density (or void ratio), effective stress level and cone tip resistance, established mainly from laboratory chamber tests on clean uniform sands. Attempts have also been made to use shear wave velocity (V_s) to estimate the state of sand [13]. Similarly, the key to the V_s-based method is an empirical correlation linking void ratio, effective stress level and shear wave velocity, derived from laboratory measurements of shear wave velocity in clean sand samples. Often natural sand deposits or fills are not clean, but contain a certain amount of fines (referred to as silty sand in practice). Even within a single deposit of sand, the percentage of fines may vary appreciably. For example, the fill materials used in construction of the artificial islands in the Beaufort Sea contained non-plastic fines at percentages varying from 0 to 12%. Application of the existing CPT- or V_s-based methods implicitly requires the assumption that the empirical correlations are not affected by the presence of fines.

The assumption is, however, questionable. There is increasing evidence that for a given void ratio and confining stress, the shear wave velocity or associated shear modulus of clean sand will change with the addition of fines [20, 24]. The cone tip resistance is also known to be sensitive to the presence of fines [11, 12], which can be a cause for the considerable uncertainty with the correction factors for interpreting chamber tests. Within this context, caution should be exercised when applying the existing empirical correlations to sand deposits with fines, since they could in some circumstances cause potentially catastrophic consequences.

3 Impact of Fines on Soil Liquefaction

While it has long been recognized that the presence of fines can alter the shear behavior of sand under either monotonic or cyclic loading conditions, very diverse views exist as to whether the effect of fines is negative or beneficial for liquefaction resistance. The problem is complex due mainly to the fact that sand-fines mixtures are granular materials in nature. The characteristics of both fine and coarse particles, such as shape, size and plasticity, can affect the packing patterns and interactions of the particles and, hence, their mechanical behavior [18, 28].

Figure 3 shows a concrete example that the addition of non-plastic fines (crushed silica) to Toyoura sand can lead to a notable increase in strain softening compared with the base sand on its own at a similar void ratio. Under otherwise similar conditions, the specimen at fines content (FC) of 15% (denoted as TSS15) underwent complete liquefaction when subjected to undrained shear, whereas the clean sand specimen (TS) exhibited a strong dilative response with no liquefaction. Furthermore, Fig. 4 compares the response of a loose Toyoura sand specimen under cyclic loading with that of a loose specimen of the same sand mixed with 10% fines (TSS10) under similar testing conditions. While both specimens exhibited the flow-type behavior, the specimen TSS10 quickly failed in less than 10 cycles of loading but liquefaction of the TS specimen was not initiated until ~78 loading cycles.

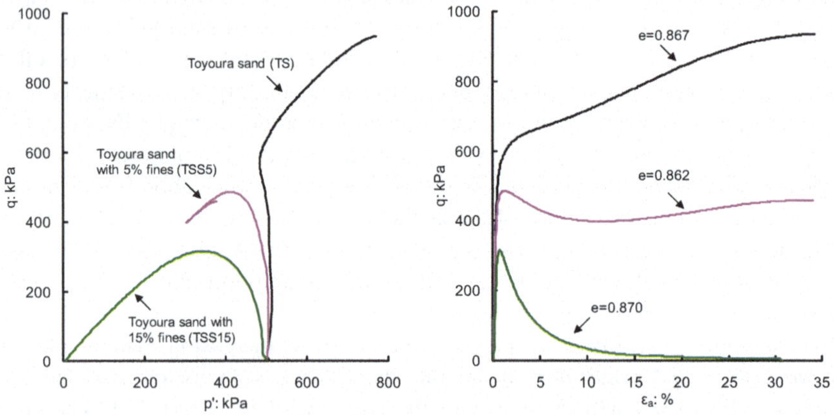

Fig. 3. Effect of fines on undrained behavior of sand under monotonic loading

It is worth noting that for both monotonic and cyclic loading tests, the effect of non-plastic fines is always an increase in liquefaction potential or a decrease in liquefaction resistance if post-consolidation void ratio is used consistently as the comparison basis. Yang et al. [29] have elaborated that the conventional global void ratio is a more rational state variable compared with such others as relative density and skeleton void ratio. Also, the fines contents concerned here are all less than a threshold value (typically ~30%) such that the sand-fines mixtures are sand dominant.

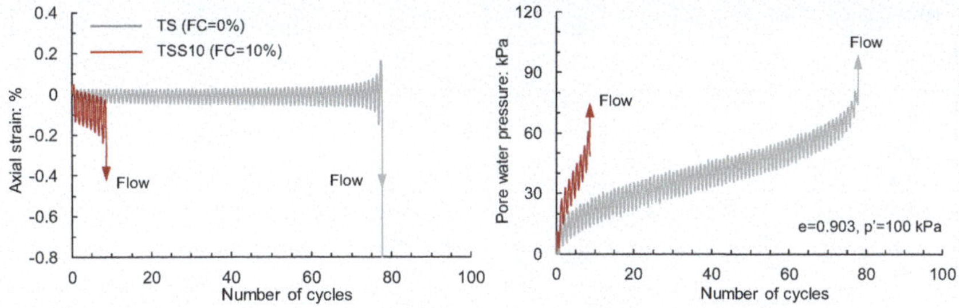

Fig. 4. Effect of fines on undrained behavior of sand under cyclic loading

4 Shear Wave Velocity Affected by Fines

A systematic investigation of state-dependent shear wave velocity in saturated sand samples with different percentages of fines has been carried out using piezoelectric bender elements. A schematic illustration of the setup for shear wave measurements is shown in Fig. 5, and more details about the apparatus and the method for data interpretation can be found in [22]. As an example, Fig. 6(a) presents a set of received signals in a TS specimen consolidated to 50 kPa, at excitation frequencies varying from 1 to 80 kHz, while Fig. 6(b) shows received signals in a TSS10 specimen at a similar state. For a given frequency, the wave form in the TS specimen appears to be similar to that in the TSS10 specimen; a careful inspection of the wave forms indicates, however, that the travel times of shear waves differ. To better view this difference, Fig. 7 shows shear wave signals recorded in four specimens (TS, TSS10, TSS20, TSS30) at the excitation frequency of 10 kHz; all the four specimens were at a similar consolidation state. It is immediately evident that the presence of a small amount of fines can lead to a notable delay in the arrival time of shear wave and, thereafter, a reduction of shear wave velocity.

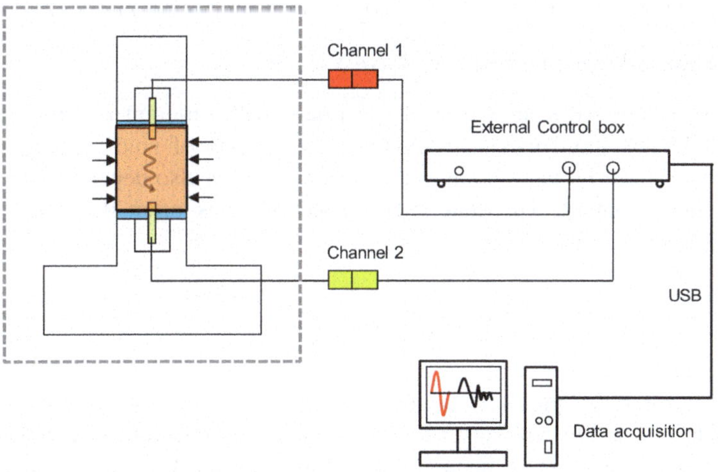

Fig. 5. Schematic illustration of laboratory setup for shear wave measurement

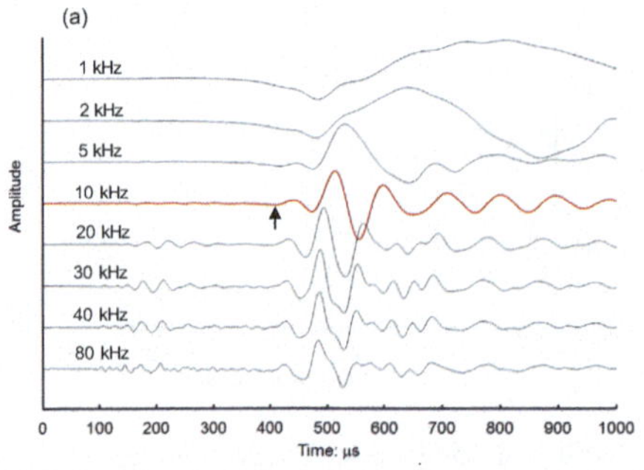

Fig. 6. Shear wave signals at various excitation frequencies for a given state (p′ = 50 kPa, e = ~0.802): (a) TS specimen; (b) TSS10 specimen

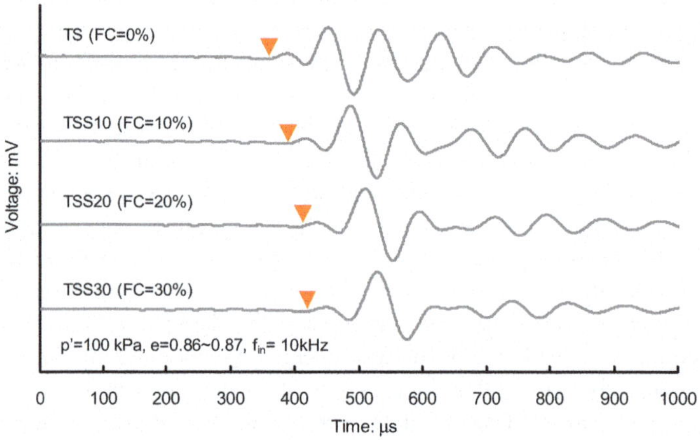

Fig. 7. Shear wave signals in saturated sand specimens with different percentages of fines

4.1 Void Ratio-Dependent Shear Wave Velocity

Measured V_s values for saturated sand specimens with different percentages of fines (TS, TSS5, TSS10 and TSS20) are plotted as a function of void ratio for two stress levels in Fig. 8. Clearly, in addition to the void ratio, V_s is also dependent on confining stress and fines content. The state dependence of V_s is often described using the following relationship [7, 13]:

$$V_s = (b_1 - b_2 e)\left(\frac{p'}{p_a}\right)^{1/4} \tag{1}$$

where p_a is a reference pressure, typically taken as the atmospheric pressure. Notice that parameters b_1 and b_2 will vary with fines content. By introducing a stress-corrected shear wave velocity (V_{s1}), a linear relationship between V_{s1} and e can readily be established as:

$$V_{s1} = V_s \left(\frac{p_a}{p'} \right)^{1/4} = (b_1 - b_2 e) \tag{2}$$

The above relationship is commonly used in geotechnical engineering practice. When all measured V_s values are corrected for stress and then plotted against void ratio, as shown in Fig. 9, one can see that this linear relationship fails to work in an acceptable way although a general trend exists that V_{s1} decreases with increasing void ratio.

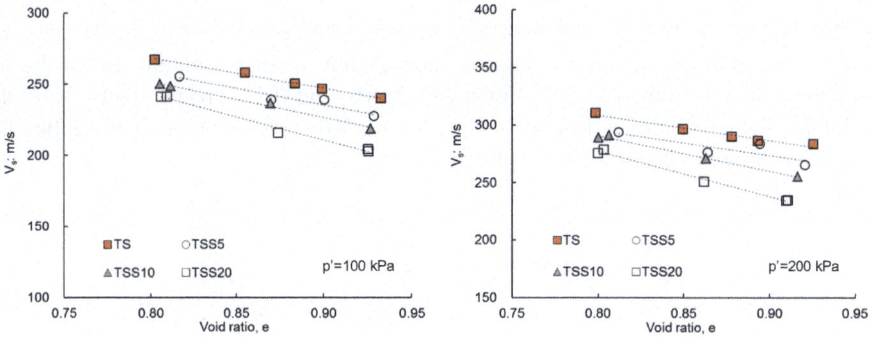

Fig. 8. Shear wave velocity versus void ratio at two stress levels

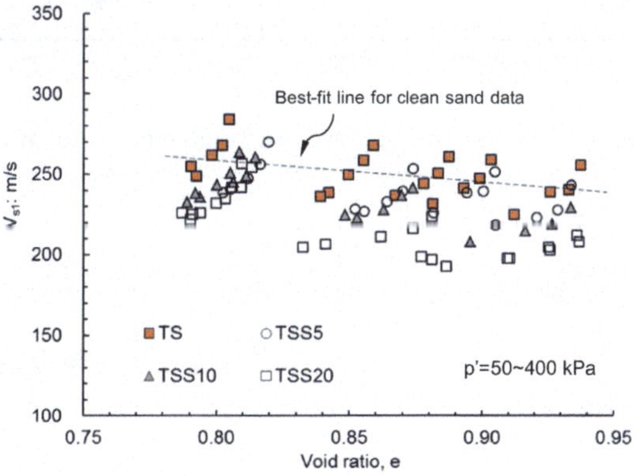

Fig. 9. Stress-corrected shear wave velocity versus void ratio

4.2 State Parameter-Dependent Shear Wave Velocity

All of the shear wave tests were conducted on saturated specimens and covered a wide range of states in terms of void ratio and effective confining stress [24, 25]. As an example, the two plots of Fig. 10 show the states of TS and TSS10 specimens with reference to their critical state loci in the e-lop' space, respectively. Note that the critical state locus moves downward with increasing fines content [28]. By calculating state parameters, measured V_s values for clean and mixed sand specimens (TS, TSS5, TSS10, TSS20) are shown as a function of ψ in Fig. 11 at two stress levels. A striking finding is that for either stress level, all data points fall on a single straight line, regardless of fines content. This finding is significantly different from the observation in Fig. 8, obtained from the traditional method of analysis. Furthermore, using the concept of stress-corrected shear wave velocity, a unique relationship can be established between V_{s1} and ψ, which works well for all data, as shown in Fig. 12. The relationship is a simple, linear function:

$$V_{s1} = V_s \left(\frac{p_a}{p'}\right)^n = A - B\psi \tag{3}$$

where A, B and n are material constants independent of fines content. From the test results they are determined as 237.2 m/s, 281.2 m/s, and 0.222, respectively. Note that if a default value of 0.25 is assigned to n, the performance of the above relationship remains reasonably good (see the right plot of Fig. 12).

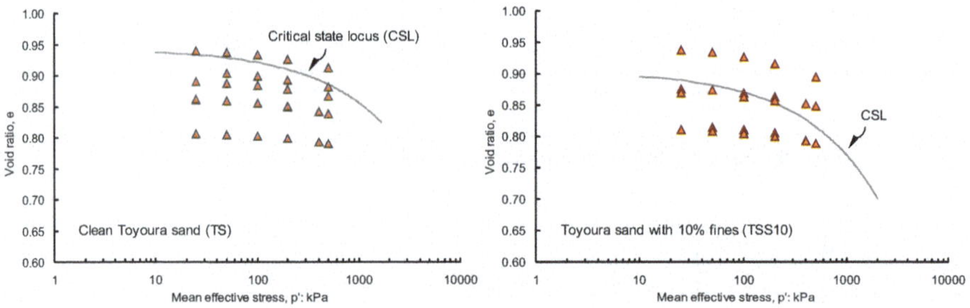

Fig. 10. States of TS and TSS10 specimens with reference to their critical state loci

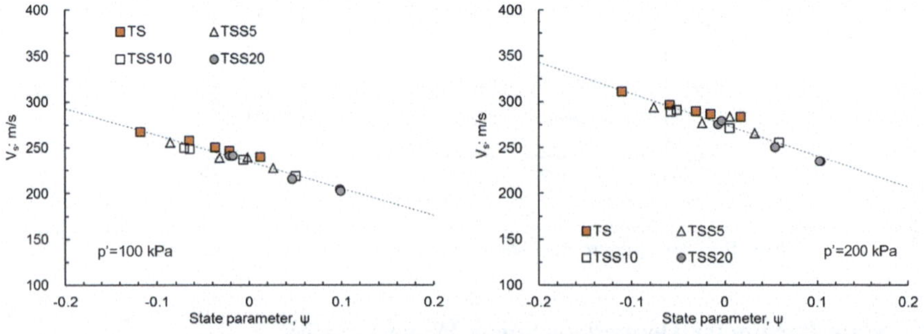

Fig. 11. Shear wave velocity versus state parameter at two stress levels

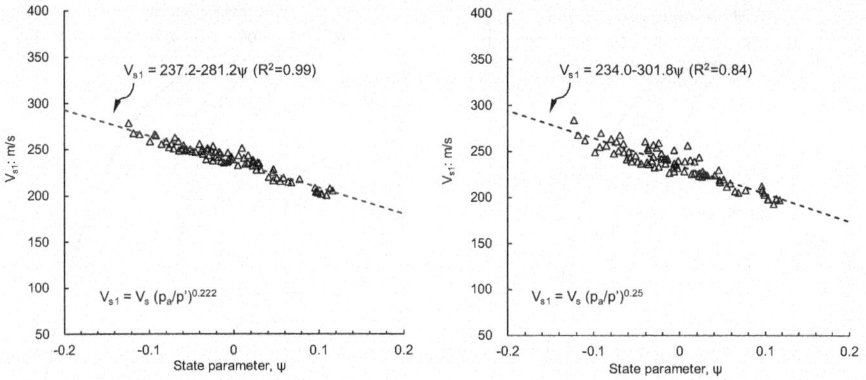

Fig. 12. Stress-corrected shear wave velocity versus state parameter for all test data

5 Evaluation of In Situ State

Freshly deposited sands are generally under normally consolidated, anisotropic conditions. The mean effective stress p' at a given depth can be estimated as

$$p' = \left(\frac{1+2K_0}{3}\right)\sigma'_v \tag{4}$$

where σ'_v is effective vertical stress at the depth and K_0 is the coefficient of earth pressure at rest. Combining Eq. (4) with Eq. (3) gives rise to an alternative form for state-dependent V_s as follows

$$V_s = (A - B\psi)\left(\frac{\sigma'_v}{p_a}\right)^n\left(\frac{1+2K_0}{3}\right)^n \tag{5}$$

Note that values of K_0 for normally consolidated, loose and medium dense sands typically vary between 0.4 and 0.6.

Using the above relationship, a set of state profiles for the sand-fines mixtures tested can be constructed in the plane of V_s and σ'_v, as shown in Fig. 13. The two plots here correspond respectively to two different K_0 values (0.4 and 1), and in each plot the set of curves represent sand states varying from $\psi = -0.2$ (highly dilative) to $\psi = +0.2$ (highly contractive). Particularly, the curve for $\psi = 0$ can be used to approximately define the boundary between a dilative-response zone and a contractive-response zone. If sand state lies in the contractive zone, caution should be exercised that the sand has a potential for liquefaction when subjected to undrained shear.

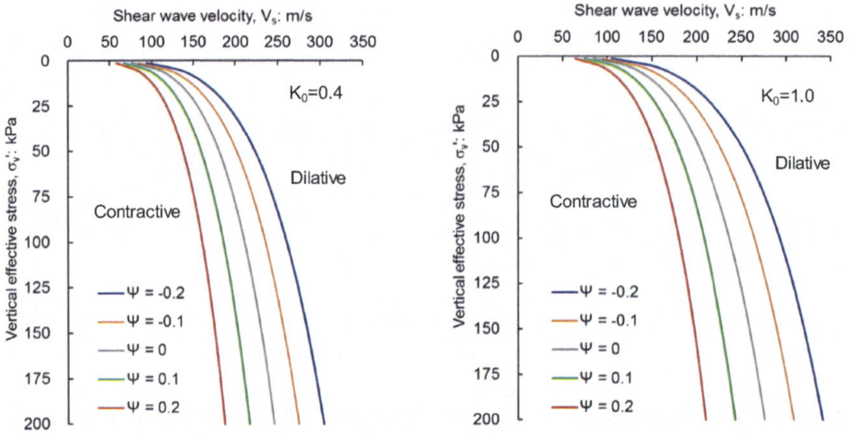

Fig. 13. State profiles in V_s–σ_v' plane at two different K_0 values

6 Validation Using Independent Tests

The significance of Eq. (3) or (5) is that it provides a unified relationship linking V_s and ψ for both clean sand and silty sand. To validate this relationship, two independent specimens were prepared: one composed of clean Toyoura sand only (TS) and the other composed of 90% Toyoura sand and 10% fines (TSS10). The TS specimen was saturated and then isotropically consolidated to the state of $e = 0.846$ and $p' = 100$ kPa, whereas the TSS10 specimen was consolidated to the state of $e = 0.904$, $p' = 300$ kPa. For both specimens, their shear wave velocities were measured using the piezoelectric bender elements. Figure 14 shows received signals at the excitation frequency of 10 kHz. The measured V_s values for the TS and TSS10 specimens are 257.5 m/s and 281.2 m/s, respectively. Provided similar mineral composition, one often makes a direct comparison of V_s measurements and concludes that the TSS specimen with higher V_s is at a denser state and hence is less susceptible to liquefaction. However, the conclusion derived from this practice is misleading.

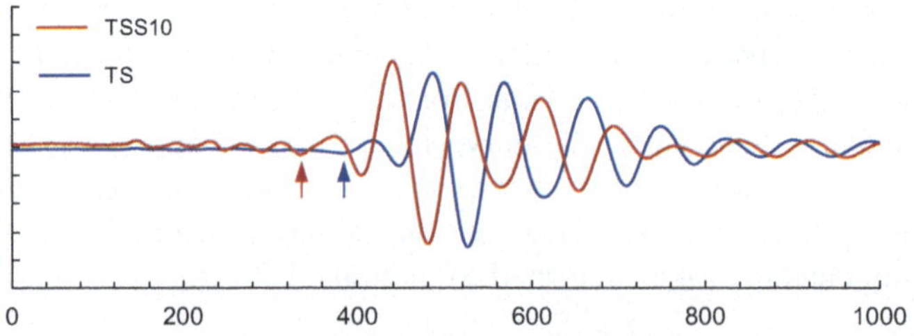

Fig. 14. Shear wave signals in two independent specimens at excitation frequency of 10 kHz

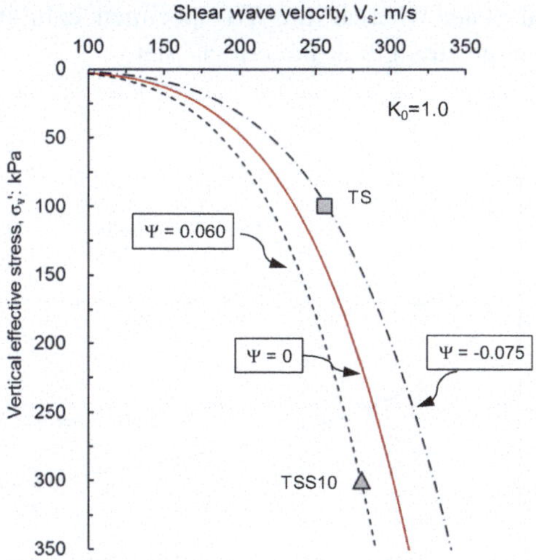

Fig. 15. Predicted state profiles in $V_s{-}\sigma_v'$ plane along with measured data

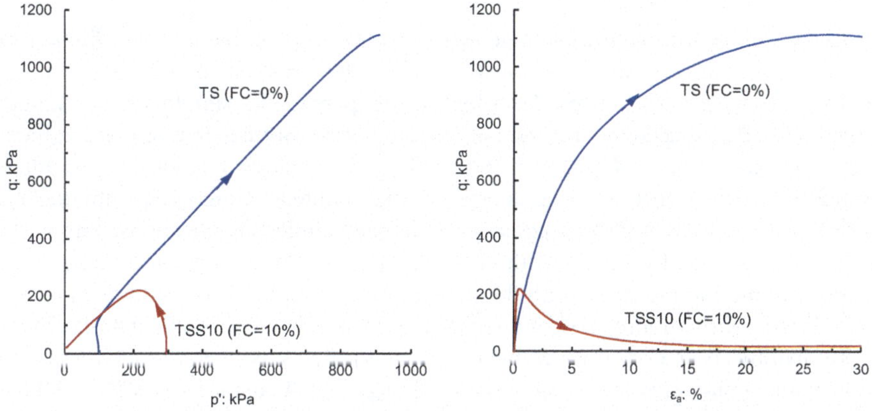

Fig. 16. Undrained shear responses of two independent specimens

The measured V_s values are plotted on the plane of V_s versus σ'_v in Fig. 15, where a set of state profiles are also given using the relationship in Eq. (5). It is noted that the TS specimen lies on the state profile for $\psi = -0.075$ while the TSS10 specimen lies on the profile of $\psi = +0.060$. As the TSS specimen lies far left of the state boundary ($\psi = 0$), one can predict that it is susceptible to liquefaction when subjected to undrained shear. On the other hand, the TS specimen lying to the right of the boundary is expected to dilate when loaded undrained.

To verify the prediction, undrained triaxial tests were conducted on the two specimens. Figure 16 shows the stress-strain curves and stress paths, and Fig. 17 compares the responses of pore water pressures. It is highly encouraging that the TSS specimen, as predicted, underwent almost complete liquefaction, with an extremely low strength at the critical state, whereas the TS specimen exhibited a highly dilative response, achieving a large strength at the critical state.

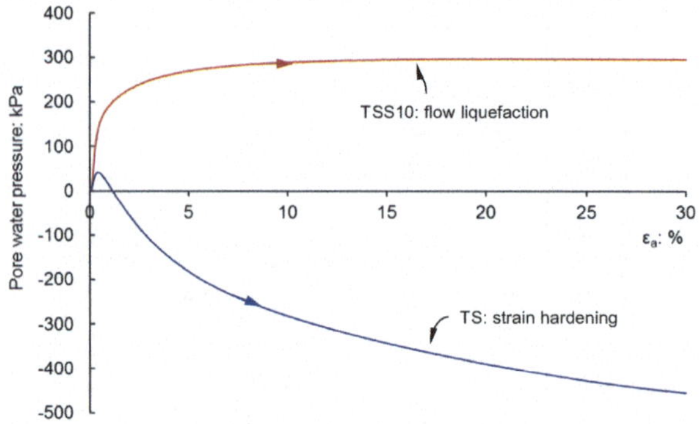

Fig. 17. Responses of pore water pressures recorded in two independent specimens

7 Conclusions

Evaluating the *in situ* state of a sand deposit and its potential for liquefaction is a challenging problem. A new method to address this problem based on shear wave velocity measurements has been described in this paper. The significant advantage of the method is that it accounts for the complicated effect of fines in a unified manner in the critical state context. Central to the method is a unique relationship established between V_{s1} and ψ that is independent of fines content. Compared with the cone penetration resistance widely used in current geotechnical applications, shear wave velocity is a basic soil property with clear physical meaning, and can be measured in soils that are hard to penetrate with the penetrometer or at sites where borings may not be permitted. Furthermore, modern technology has advanced such that shear wave velocity can be measured more conveniently and reliably both in the laboratory (e.g. piezoelectric bender elements) and in the field (e.g. SASW and seismic CPT). All these advantages make the new method appealing to a variety of geotechnical applications.

Further studies to refine the method based on laboratory and field tests on different sands with different grain characteristics are worthwhile. Also, it is necessary to mention that the method is mainly applicable to fresh, unaged sand deposits or fills. To what extent it is applicable to aged or cemented sand deposits is yet unclear; future research along the proposed line to explore this issue would be of interest. Nevertheless, it is important to bear in mind that it is the fresh, unaged sand deposits that are susceptible to liquefaction and excessive deformation.

Acknowledgments. This work is part of long-term research on soil liquefaction and related phenomena from multi-scale perspectives at The University of Hong Kong (HKU). The author is grateful for the financial support provided by the Research Grants Council of Hong Kong (Nos. 17206418; 17250316; 719105), the National Science Foundation of China (No. 51428901), the Seed Fund for Basic Research of HKU, and the Strategic Research Development Fund of the Department of Civil Engineering over the years.

References

1. Baldi, G., Bellotti, R., Chionna, V., Jamiolkowski, M., Pasqualini, E.: Design parameters for sands from CPT. In: Proceedings of 2nd European Symposium on Penetration Testing, Amsterdam, Holland, pp. 425–432 (1982)
2. Been, K., Jefferies, M.G.: A state parameter for sands. Géotechnique **35**(2), 99–102 (1985)
3. Been, K., Crooks, J.H.A., Becker, D.E., Jefferies, M.G.: The cone penetration test in sands. Part 1: state parameter interpretation. Géotechnique **36**(2), 239–249 (1986)
4. Casagrande, A.: Liquefaction and cyclic deformation of sands, a critical review. In: Proceedings of 5th Pan-American Conference on Soil Mechanics Foundation Engineering, Buenos Aires, vol. 5, pp. 79–133 (1975)
5. Castro, G.: Liquefaction and cyclic mobility of saturated sands. J. Geotech. Eng. Div. ASCE **101**(GT6), 551–569 (1975)
6. Clayton, C.R.I.: Stiffness at small-strain: research and practice. Géotechnique **61**(1), 5–38 (2011)
7. Hardin, B.O., Richart, F.E.: Elastic wave velocities in granular soils. J. Soil Mech. Found. Div. ASCE **89**(SM1), 39–56 (1963)
8. Idriss, I.M., Boulanger, R.W.: Soil liquefaction during earthquakes. EERI Monograph (2006)
9. Ishihara, K.: Liquefaction and flow failure during earthquakes. Géotechnique **43**(3), 351–451 (1993)
10. Jamiolkowski, M.: Soil mechanics and the observational method: challenges at the Zelazny Most copper tailings disposal facility. Géotechnique **64**(8), 590–619 (2014)
11. Lunne, T., Robertson, P.K., Powell, J.J.M.: Cone Penetration Testing in Geotechnical Practice. Blackie Academic & Professional (Chapman & Hall), Glasgow (1997)
12. Mayne, P.W.: Cone penetration testing. NCHRP Synthesis 368, Transportation Research Board, Washington, D.C. (2007)
13. Robertson, P.K., Sasitharan, S., Cunning, J.C., Sego, D.C.: Shear wave velocity to evaluate flow liquefaction. J. Geotech. Eng. ASCE **121**(3), 262–273 (1995)
14. Seed, H.B., Lee, K.L.: Liquefaction of saturated sands during cyclic loading. J. Soil Mech. Found. Div. ASCE **92**(SM6), 105–134 (1996)
15. Sze, H.Y., Yang, J.: Failure modes of sand in undrained cyclic loading: Impact of sample preparation. J. Geotech. Geoenviron. Eng. ASCE **140**(1), 152–169 (2014)
16. Triantafyllidis, Th. (ed.): Cyclic Behavior of Soils and Liquefaction Phenomena: Proceedings of the International Conference, Bochum, Germany. CRC Press (2004)
17. Vaid, Y.P., Stedman, J.D., Sivathayalan, S.: Confining stress and static shear effects in cyclic liquefaction. Can. Geotech. J. **38**(3), 580–591 (2001)
18. Wei, L.M., Yang, J.: On the role of grain shape in static liquefaction of sand-fines mixtures. Géotechnique **64**(9), 740–745 (2014)
19. Wei, X., Yang, J.: Cyclic behavior and liquefaction resistance of silty sands with presence of initial static shear stress. Soil Dyn. Earthq. Eng. (2018). https://doi.org/10.1016/j.soildyn.2018.11.029
20. Wichtmann, T., Hernandez, M., Triantafylidis, T.: On the influence of a non-cohesive fines content on small strain stiffness, modulus degradation and damping of quartz sand. Soil Dyn. Earthq. Eng. **69**, 103–114 (2015)
21. Yang, J.: Non-uniqueness of flow liquefaction line for loose sand. Géotechnique **52**(10), 757–760 (2002)
22. Yang, J., Gu, X.Q.: Shear stiffness of granular material at small strains: does it depend on grain size? Géotechnique **63**(2), 165–179 (2013)

23. Yang, J., Li, X.S.: State-dependent strength of sands from the perspective of unified modeling. J. Geotech. Geoenviron. Eng. ASCE **130**(2), 186–198 (2004)
24. Yang, J., Liu, X.: Shear wave velocity and stiffness of sand: the role of non-plastic fines. Géotechnique **66**(6), 500–514 (2016)
25. Yang, J., Liu, X., Guo, Y., Liang, L.B.: A unified framework for evaluating in situ state of sand. Géotechnique **68**(2), 177–183 (2018)
26. Yang, J., Sze, H.Y.: Cyclic behaviour and resistance of saturated sand under non-symmetrical loading conditions. Géotechnique **61**(1), 59–73 (2011)
27. Yang, J., Sze, H.Y.: Cyclic strength of sand under sustained shear stress. J. Gotech. Geoenviron. Eng. ASCE **137**(12), 1275–1285 (2011)
28. Yang, J., Wei, L.M.: Collapse of loose sand with the addition of fines: the role of particle shape. Géotechnique **62**(12), 1111–1125 (2012)
29. Yang, J., Wei, L.M., Dai, B.B.: State variables for silty sands: global void ratio or skeleton void ratio? Soils Founds **55**(1), 99–111 (2015)
30. Youd, T.L., Idriss, I.M., Andrus, R.D., Castro, G., Christian, J.T., Dobry, R., Finn, W.D.L., Harder Jr., L.F., Hynes, M.E., Ishihara, K., Koester, J.P., Liao, S.S.C., Marcuson III, W.F., Martin, G.R., Mitchell, J.K., Moriwaki, Y., Power, M.S., Robertson, P.K., Seed, R.B., Stokoe II, K.H.: Liquefaction resistance of soils: summary report from the 1996 NCEER and 1998 NCEER/NSF workshops on evaluation of liquefaction resistance of soils. J. Geotech. Geoenviron. Eng. ASCE **127**(10), 817–833 (2001)

A Study on the Wave Dispersion in Concrete Due to Damage; Numerical Observations and Theoretical Predictions

Georgios Livitsanos[1], Abdul Bashit Ahmadi[1], Dimitrios G. Aggelis[1], Theodore V. Gortsas[2], and Demosthenes Polyzos[2](✉)

[1] Department of Mechanics of Materials and Constructions, Vrije Universiteit Brussel, Brussels, Belgium

[2] Department of Mechanical Engineering and Aeronautics, University of Patras, Patras, Greece

polyzos@upatras.gr

Abstract. Experimental results have shown that waves propagating in non-homogeneous and composite materials exhibit dispersive behavior even in cases where they are characterized by isotropic effective properties. Concrete is such a material and at both its fresh and hardened state wave dispersion is observed. The dispersion becomes more pronounced when microdefects like microcracks and air voids appear in the main body of the concrete. The goal of the present work is threefold: first to examine numerically the contribution of various defects to the wave dispersion in concrete, second to show through numerical experiments that circular voids embedded in a concrete matrix is a good approximation for simulating wave dispersion in damaged concrete and third to capture the dispersive behavior of a longitudinal plane wave propagating in damaged concrete with the aid of two theories, namely the multiple wave scattering model of Waterman and Truell and the generalized elastic theory of a strain gradient elastic material. Numerical, experimental and theoretical results are compared and discussed throughout the paper.

1 Introduction

Ultrasonic pulse velocity is a widely spread non-destructive testing parameter that has proven very reliable and relatively easy to use for decades. The empirical correlations with strength of materials (especially concrete), although not theoretically justified, seriously contribute in the evaluation of structural condition [1]. In the domain of cementitious media the wave velocity is used as the first demonstrator of the structural health and allows strength prediction either in hardening or hardened concrete [2]. The majority of the scientific articles and even more of the in-situ cases, employ ultrasonic pulse velocity as the main parameter for characterization. The reason is the relative easiness to apply and the reliability on the strength estimation. Commercial, high-voltage ultrasonic systems employ transducers of a fixed frequency (usually 54 kHz) and detect the onset of the waveform at the receiving transducer through its first crossing above a fixed threshold. Many correlations have seen publicity showing that pulse velocities above 4000 m/s are measured in good quality concrete while velocities

T. Triantafyllidis (Ed.): *Recent Developments of Soil Mechanics and Geotechnics in Theory and Practice*, LNACM 91, pp. 183–198, 2020.
https://doi.org/10.1007/978-3-030-28516-6_10

below 3000 m/s indicate problematic state. Relatively recently, the influence of frequency has been studied. It is shown that as the heterogeneity of the medium increases, the dependence of velocity on frequency (dispersion) becomes stronger. The heterogeneity in concrete is related to the impedance mismatch between its constituents. This has been theoretically treated using scattering models and more recently with gradient elastic theories [3–5]. To a limited extent, a mismatch is noted between the cement matrix and the stiffer aggregates although in general, their density discrepancies are not strong. More intensive scattering or microstructural effects are expected by air bubbles (entrapped or entrained air) and cracks, as in these cases the impedance mismatch is huge and the inclusions are in essence voids. The easier way to model these heterogeneities is the spherical shape. This is supposed to result in a wave velocity and overall behavior, in general, close to a case with random inclusions and of random distribution, provided of course that the volume content and the mechanical properties are constant. Though this perception is common in the field of wave scattering, there is no concrete study in the international literature to prove this for a variety of shapes of inclusions. In this paper, numerical simulations are conducted using several shapes of voids (inclusions with properties of air), distribution patterns and volume contents. The results allow for comparison between the shapes, showing that even with the same content, the inclusion shapes are responsible even for 20% difference in UPV. Thus, one aim of the study concerns the basic level of wave propagation and is conducted through numerical simulations, a second aim is the comparison with experimental data from cementitious mortar with different volume content of polystyrene inclusions, while a third aim is to verify the observed wave dispersion via the multiple wave scattering model of Waterman and Truell [6] and the use of the simple strain gradient elastic theory of Mindlin [7].

2 Numerical Simulation Model

In this section, a numerical experiment dealing with the propagation of a plane longitudinal elastic wave in a concrete specimen containing air voids in different shapes and orientation is carried out. For the shake of simplicity plane strain conditions are considered. The cases examined were solved numerically in order to produce the waveform of the receiver and extract the transit time of the onset of the wave. The fundamental equation governing the two-dimensional propagation of elastic waves in an elastic medium, ignoring viscous losses, is as follows:

$$\mu\nabla^2\mathbf{u} + (\lambda + \mu)\nabla\nabla\cdot\mathbf{u} = \rho\frac{\partial^2\mathbf{u}}{\partial t^2} \tag{1}$$

where $\mathbf{u} = \mathbf{u}(x, y, t)$ is the time-varying displacement vector, ρ is the material density, λ and μ are the first and second Lamé constants, respectively, and t is time. The simulations were conducted with commercially available software wave2000 [8], which solves the above equation using the finite difference method in the plane strain case. Equation (1) is solved at discrete points with respect to the boundary conditions of the model, which include the input source that has predefined time-dependent displacements at a given location and a set of initial conditions [9]. For heterogeneous geometries like the ones studied herein, wave propagation in each distinct homogeneous phase is solved according to Eq. (1), while the continuity conditions for stresses and strains must be satisfied on the interfaces [9].

The mechanical properties of the different materials are shown in Table 1. In total three different types of materials were considered: cementitious matrix, voids (inclusions of air), and aggregates.

Table 1. Material properties used for the numerical model.

Material	Density ρ (kg/m^3)	First Lame λ (GPa)	Second Lame μ (GPa)	Ultrasonic velocity, C (m/s)
Cement matrix	2400	11.1	16.7	4216
Air	1.2	0.0001	0.00001	316

The basic geometry was a square of 100 mm side on which the different inhomogeneities were incorporated, as shown in Fig. 1. Infinite boundary conditions were applied to all sides of the geometry. The "pulser" was triggered by a displacement excitation of 10 cycles of constant amplitude and the simulation was repeated for several frequencies, namely 50 kHz, 150 kHz, 300 kHz and 500 kHz in order to estimate the effect of the different wavelengths (roughly from 80 mm down to 8 mm). The "receiver" was placed on opposite side of the geometry allowing for a propagation distance of 100 mm between pulser and receiver. One example of a geometric model of a matrix with few circular air inclusions is shown in Fig. 1.

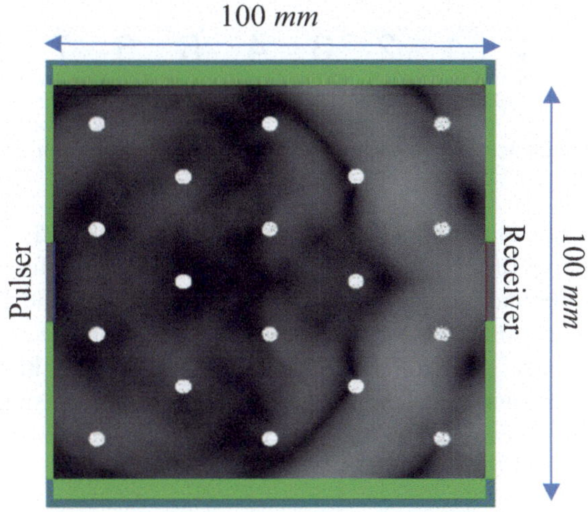

Fig. 1. Geometrical model for the numerical simulations.

The “receivers” provide the average vertical displacement over their length, meaning that the receiver signal represents the average response over a number of nodes. The simulation accuracy depends on the density of the points on which the equation of motion is solved. Thus, preliminary simulations were conducted on a specific geometry (with circular inclusions) to estimate the effect of mesh size on accuracy as well as computation time. Mesh sizes from 8 mm down to 0.8 mm were selected. It is seen that the resulted velocity is certainly influenced by the selected resolution showing however convergence at 1 mm (Fig. 2). Therefore, the analysis was done with the resolution of 1 mm, which was just 0.0066% different than the finest resolution tried, but considerably faster. This resolution is much lower than the shortest applied wavelength of 8 mm corresponding to the highest frequency applied of 500 kHz. The time resolution is 0.01537 μs, corresponding to sampling rate of 65 MHz, well above the applied frequencies.

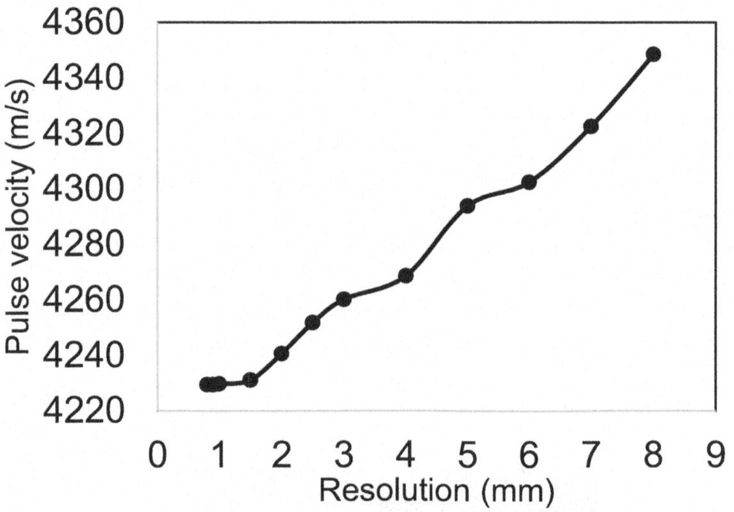

Fig. 2. Convergence of pulse velocity results for finer resolution (mesh size).

Three different contents of inclusions were used, namely 2%, 5% and 10%. In addition, several shapes of inclusions were applied. In total ten different cases of single shaped voids (and orientations when applicable) were applied, and Fig. 3 shows five indicative cases, concerning circles, squares, ellipses, rectangular at different orientation. For the two latter, four different orientations of the longitudinal axis of the inclusion were applied (horizontal, vertical and ±45°). The dimensions of the different inclusion shapes are shown in Table 2. The common characteristic for all shapes is their area which is fixed to 11.11 mm^2.

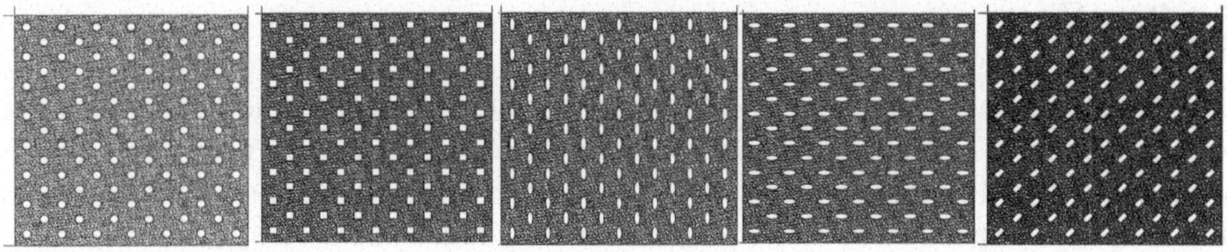

Fig. 3. Indicative patterns of inclusions in the model (from left to right: circular, square, vertical ellipses, horizontal ellipses, diagonal rectangles).

Table 2. Basic shapes and dimensions of inclusions

Circle		radius 1.9 mm
Rectangle		4.54x2.45 mm
Oval		2.9x1.2 mm
Square		side 3.33 mm

Apart from the single shape of inclusions, five different patterns were created by mixing the different shapes. Three of them are shown in Fig. 4a. Finally, two cases with random patterns of circular voids were created as shown in Fig. 4b, making a total of 16 different cases.

(a)

(b)

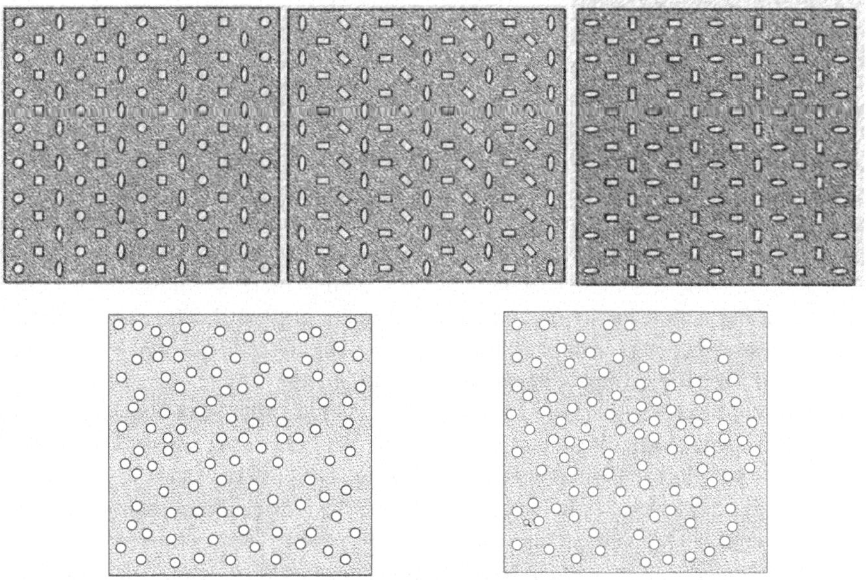

Fig. 4. (a) Indicative patterns of mixed inclusions, (b) random pattern of spherical inclusions.

3 Results

The transit time used for velocity calculations was determined by the first disturbance on the receiver's waveform (no threshold needed as all previous waveforms points were zero). Two indicative waveforms are shown in Fig. 5 concerning the response to 150 kHz for the pure matrix and the matrix with circular inclusions to the content of 10%. The effect of heterogeneity is seen both in the delayed onset and the amplitude of the waveform of 10% compared to 0%.

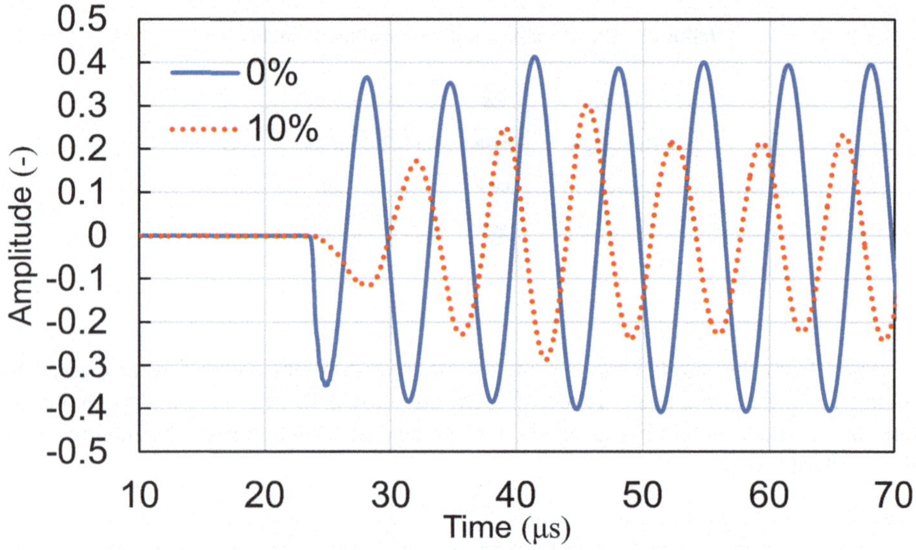

Fig. 5. Simulated waveforms for 150 kHz excited frequency for matrix with 0% and 10% of circular inclusions.

Figure 6 shows the pulse velocity vs. frequency curves for some indicative heterogeneity patterns with 10% content. In all cases the velocity increases towards the 500 kHz, something that as will be seen later is also in agreement with the solution of the scattering model of Waterman and Truell [6]. Nevertheless, similar results have been reported for the field of concrete, suggesting increase of wave velocity with increase of frequency for sound and damaged material [3, 10]. The cases with the least dispersion are the ones with the horizontal inclusions (ellipses and rectangles). The strongest dispersion along with the lower velocity curves are shown by the vertical orientation of the same shapes. The diagonal orientation shows an intermediate velocity curve. The circle shape of inclusions comes very close to the square shape starting at 3986 m/s and reaching 4210 m/s for 500 kHz. It is evident that as the frequency increases, the discrepancies between the different shapes becomes negligible except for the extreme cases of vertical inclusions which exhibit the strongest reflection fronts against the propagating wave. Excluding these two extreme cases, the circular shape seems to be a good approximation yielding velocities in the average of the other cases.

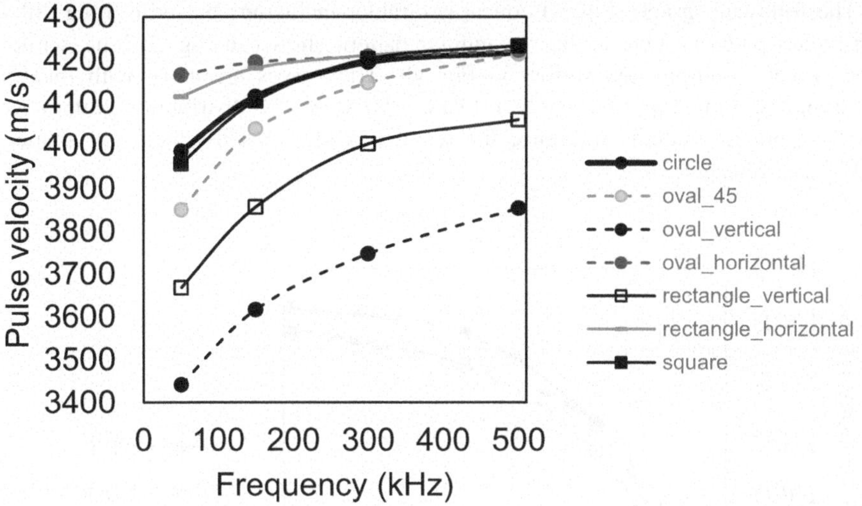

Fig. 6. Pulse velocity vs. frequency curves for different inclusion shapes and patterns of content 10%.

Figure 7 shows the corresponding results for the mixed shapes cases, including again the circular inclusion shape for comparison. The basic trend of the increasing velocity with frequency holds for all cases and while there are patterns exhibiting higher or lower velocity curve than the circular shape, the latter's curve is placed closer to the top rather than the bottom curve, something also noted earlier in Fig. 6.

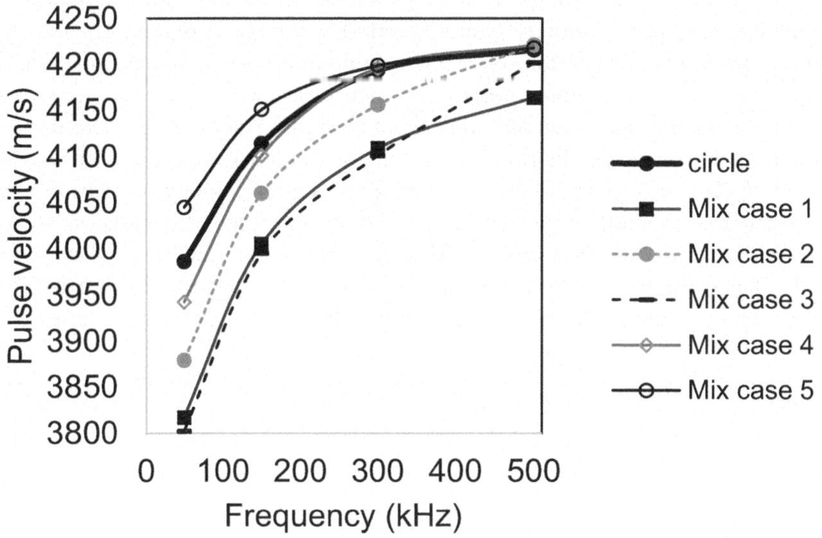

Fig. 7. Pulse velocity vs. frequency curves for different mixed shape patterns of content 10%.

The following graph (Fig. 8) concerns circular inclusions but with three different distribution patterns. One is the rectangular pattern shown in Fig. 3 (left), while the other two are random and shown in Fig. 4b. The curves are close with maximum deviation of 50 m/s (approximately 1.25%) at 50 kHz. The distribution pattern seems therefore, not to crucially influence the result provided that the inclusion shape and content are constant.

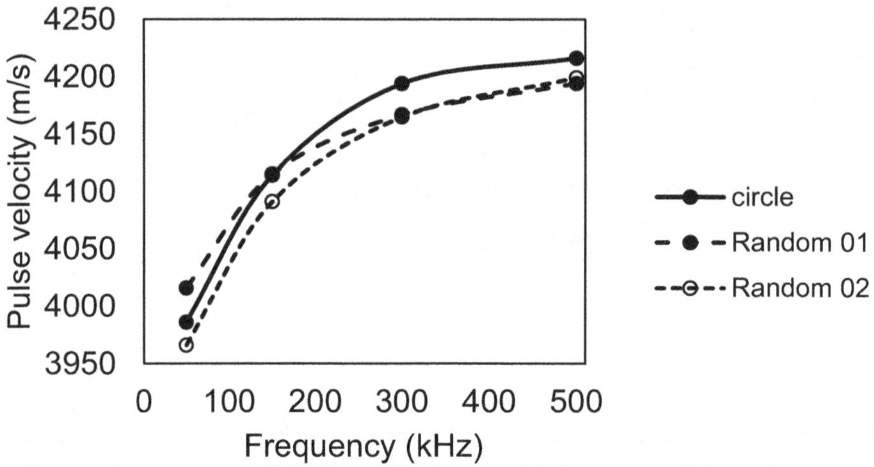

Fig. 8. Pulse velocity vs. frequency curves for circular inclusions and different patterns of content 10%.

In order to have a clear comparison for different shapes and combinations of them, Fig. 9 includes the pulse velocity values at 50 kHz for 10% content. All the applied patterns are shown for completeness. What is again evident is that apart from the cases of ellipsoidal and rectangular horizontal inclusions which do not pose strong obstruction to wave propagation and thus result in higher velocities, the circular case of scatterer (either in specific distribution pattern or random) presents velocities at the high level of 4000 m/s, only 5% less than the pure matrix velocity of 4216 m/s. Other inclusion shapes/orientations present lower velocities like the ellipsoidal at 45° to the propagation direction, which is close to 3850 m/s, while long inclusions with their axis perpendicular to the propagation direction drop to values close to 3400 m/s. The circular case presents the 5th highest pulse velocity from the presented 16 cases.

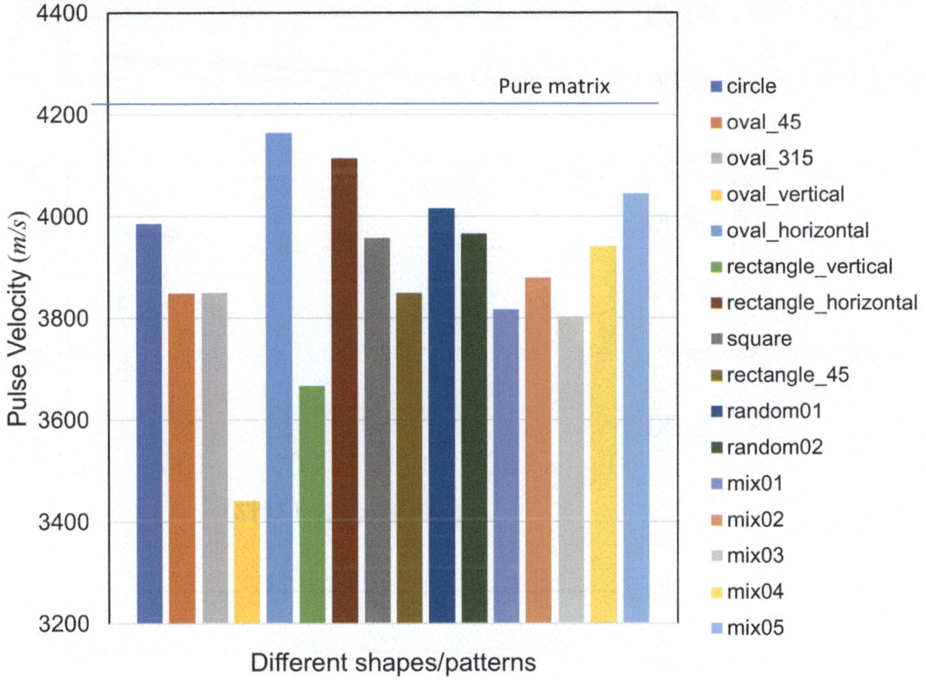

Fig. 9. Pulse velocities for 50 kHz for all cases of inclusion shapes and geometrical distribution patterns for content of 10%.

4 Comparison with Experimental Data

The above numerical results are compared with experimental results on the effect of frequency on pulse velocity. The experimental data are taken from [11], where film-shape light inclusions of 15 × 15 × 0.5 mm were added during mixing of cementitious mortar in volume contents from 0% (reference), 1%, 5% to 10%. Figure 10a shows the experimental pulse velocity vs. frequency curves. All curves (even plain mortar) exhibit a dispersive trend, which is expected due to the inherent heterogeneity of mortar. This dispersion is more pronounced for higher number of voids leading to quite strong discrepancies between the velocity at 30 kHz (3390 m/s) and the velocity of 500 kHz (3900 m/s) for 10% vol. content. The corresponding numerical results are shown in Fig. 10b. There, the curves corresponding to the ellipsoidal vertical shape are presented revealing many similarities to the experimental results. The frequency dependence increases for high amount of voids, while even the values are quite close. Obviously the wave dispersion observed in Fig. 10a is due not only to the presence of the randomly oriented film-shape light inclusions but also to the heterogeneity of the cementitious mortar matrix. Since all numerical experiments consider non-dispersive matrix, it is expected that the mixed models of Fig. 4a provide wave behavior being less dispersive than that observed experimentally. This explains why in Fig. 10(b) the numerical results are close to experimental ones, when vertical ellipsoidal voids are

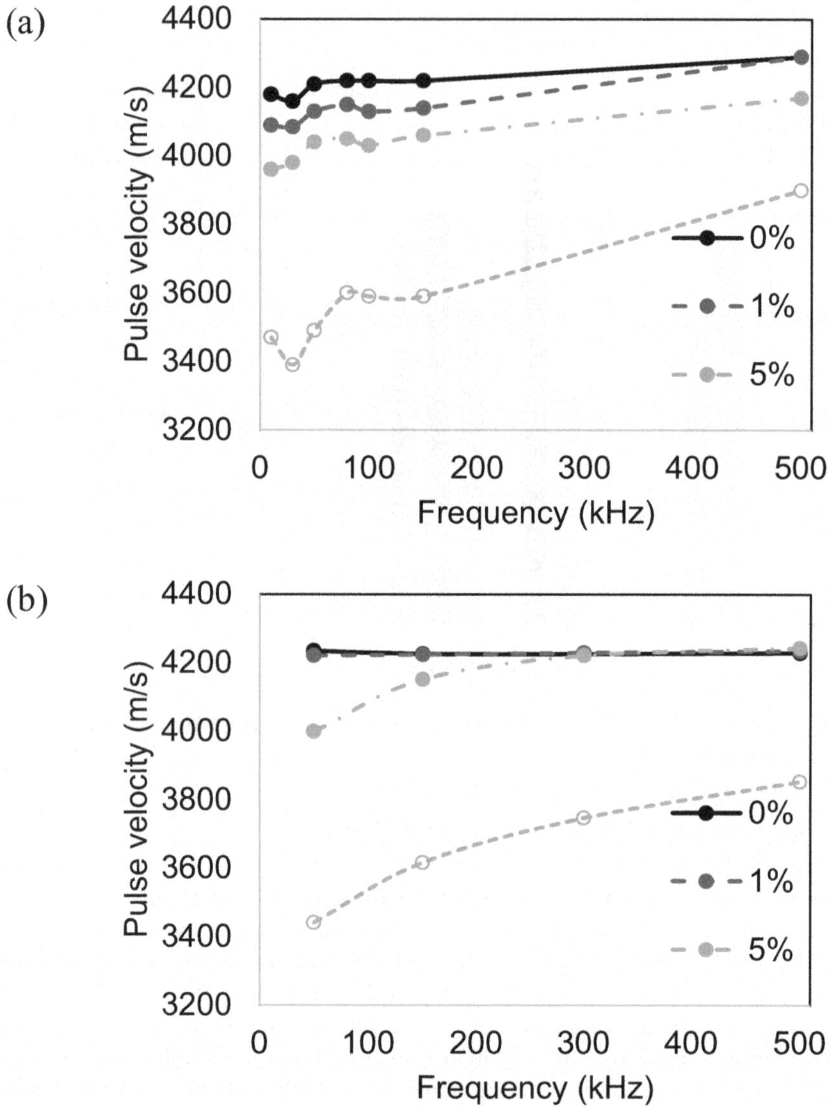

Fig. 10. (a) Experimental pulse velocity vs. frequency curve for concrete with different percentage of light plate inclusions [10]. (b) Numerical pulse velocity vs. frequency curve for ellipsoidal voids vertical to propagation.

considered. Indicatively for the 10% voids case, the velocity at low frequencies for both numerical and experimental approach is at the level of 3400 m/s, while for the highest frequency it reaches 3800–3900 m/s. In addition, it is interesting to note that the discrepancy between 10% and 5% is much more than between 5% and 0% both in numerical and experimental, showing the accumulated effect of damage.

5 Theoretical Predictions

The main purpose of this section is to develop an understanding of the mechanism responsible for the dispersive behavior of the longitudinal elastic waves observed numerically and experimentally in Sects. 3 and 4, respectively. Depending on the frequency, all the results obtained so far converge to the conclusion that voids of various shapes affect drastically, even in very small volume concentrations, the velocity of propagating waves. Many researchers have proposed the multiple scattering of the incident wave by the distributed voids as the main mechanism of the observed wave dispersion [3, 12–14]. Others believe that the microstructure of the nonhomogeneous material is responsible for the phenomenon. However, due to the lack of internal length scale parameters that correlate the microstructure with the macrostructure, the classical theory of elasticity is not able to capture wave dispersion in those materials. For this reason enhanced elastic theories like the micropolar and stainn gradient elasticity are preferred [5, 15–17]. In this section the scattering theory of Waterman and Truell [6] and the strain gradient elastic theory of Mindlin [7] are employed in order to explain the numerical results presented in Sect. 3. For demonstration purpose, only circular voids are considered.

5.1 Wave Dispersion Predictions via the Waterman and Truell Scattering Model

According to Waterman and Truell [6], the wave dispersion of an elastic wave propagating in a nonhomogeneous elastic medium is undertaken through the multiple scattering theory which, solves problems dealing with the scattering of incident plane waves by a cloud of inclusions embedded in an elastic matrix. According to this theory, the wave dispersion and attenuation in a nonhomogeneous medium can be obtained in terms of the particle/void concentration and the forward as well as the backward far field scattering amplitudes. More precisely, the wave dispersion and attenuation is represented via the frequency-dependent complex wavenumber, $k(\omega)$, which is expressed in terms of the void concentration and the forward as well as the backward far-field scattering amplitudes, i.e.

$$\left(\frac{k(\omega)}{k_c}\right)^2 = 1 + \frac{3\varphi}{k_c^2 R^3} f(0) + \frac{9\varphi^2}{4k_c^4 R^6} \left[f^2(0) - f^2(\pi)\right] \tag{2}$$

where k_c is the real wavenumber of the matrix material, φ the volume fraction of the inclusion, R the particle radius when spherical particles are considered and $f(0)$ and $f(\pi)$ are the complex single scattering forward and backward scattering amplitudes, respectively. The latter are evaluated analytically for circular voids [18] and numerically for arbitrarily shaped cavities.

It should be noticed here that the Waterman and Truell theory ignores the interaction between the inclusions and for this reason its applicability is confined to concentrations of particles or voids being below to 12%. For higher concentrations can be replaced by the iterative effective medium approximation technique illustrated in [19].

For the concrete matrix perforated by 10% of circular voids, described in Sect. 2, the phase velocity of a longitudinal plane wave is evaluated according to (2) and depicted as a function of frequency in Fig. 11. The results are compared to corresponding numerical ones obtained by the wave2000 commercial code [8] as it is explained in Sect. 3. As it is apparent, relation (2) provides a reasonable explanation for the wave dispersion occurred in the concrete due to the presence of 10% circular cavities.

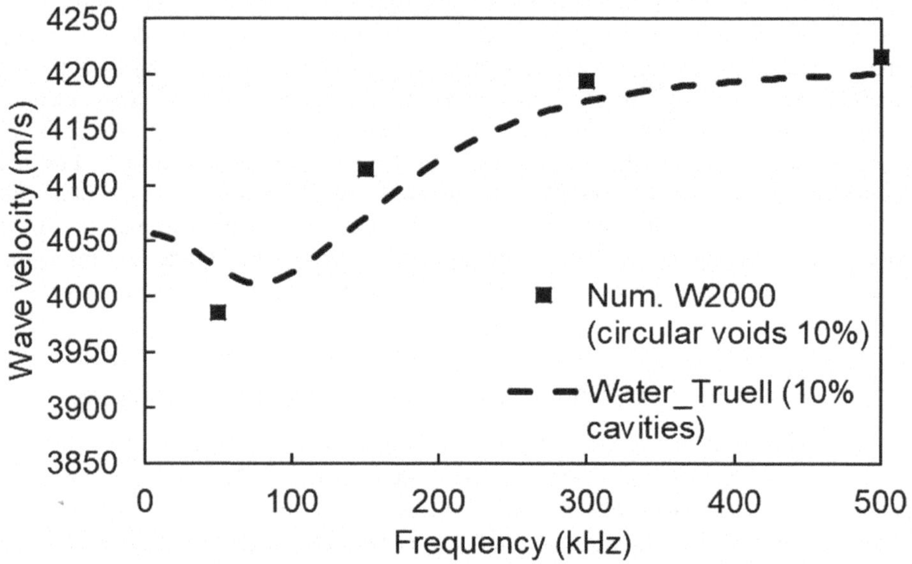

Fig. 11. Phase velocity vs. frequency curve for 10% vol. assuming circular shape of inclusions, based on the scattering model of Waterman and Truell [6].

5.2 The Concrete as a Strain Gradient Elastic Material

Mindlin [7], in the Form-II version of his strain gradient elastic theory, considered that the potential energy density is a quadratic form not only of the strains but also of the gradient of strains. Similarly, the kinetic energy appears to be a quadratic function of both velocity and gradient of velocity. Under those consideration, his theory concludes to an equation of motion, which in its simplest form, can be written as

$$(1 - g^2\nabla^2)[\mu\nabla^2\mathbf{u} + (\lambda + \mu)\nabla\nabla \cdot \mathbf{u}] = \rho(\ddot{\mathbf{u}} - h^2\nabla^2\ddot{\mathbf{u}}) \tag{3}$$

where $\mathbf{u}$ stands for the displacement vector, λ, μ represent the classical elastic Lame constants, ∇ and ∇^2 indicate the gradient and Laplace operator, respectively, dots indicate differentiation with respect to time and g, h are the stiffness and inertia, respectively, internal length scale parameters that correlate the microstructure with the dynamic macrostructural behavior of the strain gradient elastic material.

Considering displacements as harmonic plane waves of frequency ω, Eq. (3) leads to the following dispersion relation

$$V_{p,s} = \frac{\omega}{k_{p,s}} = C_{p,s}\sqrt{\frac{1+g^2k_{p,s}^2}{1+h^2k_{p,s}^2}} \tag{4}$$

where $V_{p,s}$ represents the phase velocity of a longitudinal (p) or transverse (s) plane wave propagating in the strain gradient elastic material, while $C_{p,s}$ and $k_{p,s}$ stand for the phase velocity and the wave number, respectively, of a plane wave traveling in a classical linear elastic medium characterized by the Lame constants λ, μ.

Equation (4) reveals that, unlike the classical elastic case characterized by constant velocities of longitudinal and shear waves and hence nondispersive wave propagation, the gradient elastic is characterized by phase velocities for longitudinal and shear waves, which are functions of the wave number, indicating wave dispersion. This dispersion is imposed with the presence of the two intrinsic parameters g^2 and h^2. It is apparent that for $g = h = 0$, Eq. (4) reads $V_{p,s} = C_{p,s}$, but what is interesting here is that dispersion disappears when the micro-stiffness coefficient g is equal to micro-inertia one h. Dispersion relation (4) can be written in terms of frequency as follows:

$$V_{p,s} = C_{p,s}\frac{1-\left(\frac{\omega h}{C_{p,s}}\right)^2+\sqrt{\left(1-\left(\frac{\omega h}{C_{p,s}}\right)^2\right)^2+\left(2\frac{\omega g}{C_{p,s}}\right)^2}}{2} \tag{5}$$

Considering the material of the previous subsection as a homogenized strain gradient elastic medium and utilizing the dispersion relation (5), the frequency dependent phase velocity of a longitudinal plane wave propagating in the considered strain gradient elastic material is evaluated and depicted in Fig. 12.

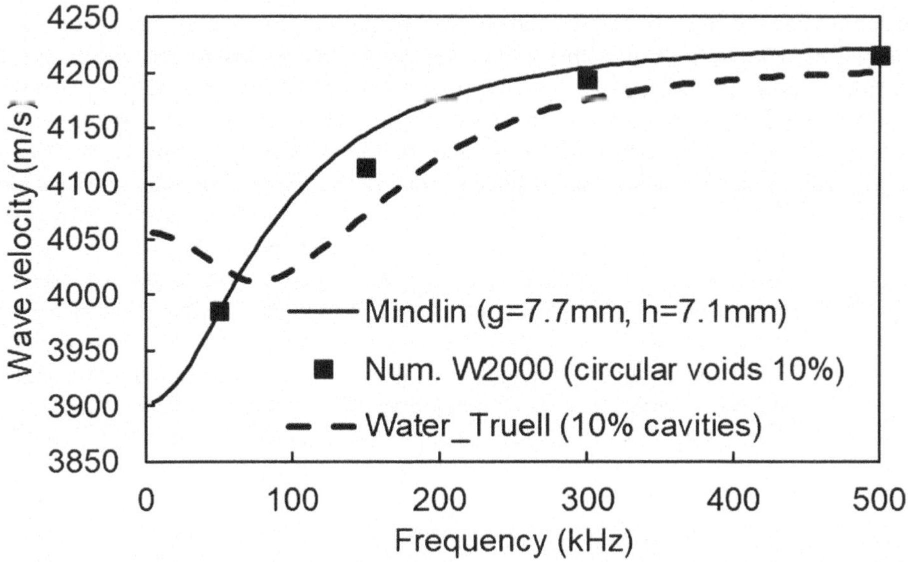

Fig. 12. Wave velocity vs. frequency according to Mindlin's gradient elastic theory [7]

It is obvious that both Waterman and Truell and strain gradient elastic approaches exhibit a similar trend with increasing velocity for higher frequency up to 500 kHz while some differences only exist at the low band of frequency below 50 kHz. The values of the Mindlin's microstructural coefficients leading to a good agreement with numerical results are g (micro-stiffness) = 7.7 mm and h (micro-inertia) = 7.1 mm. It is noted that the circular void (the case of which is included in the comparison of Fig. 12) has a diameter of 3.8, being approximately the half of the value of microstiffness.

Due to the interaction of a propagating elastic wave with the voids of the concrete material, the choice of the Waterman and Truell multiple scattering model as a candidate theory to explain the observed wave dispersion was quite reasonable. The spontaneous question here is why to consider one the concrete with the circular voids as a strain gradient elastic material. It is well known that the strain gradient elasticity is a theory, which by definition can be applied only in materials where significant strain gradients occur. Thus, in order to give an answer to that question, the traction field that follows the direction of a propagating harmonic longitudinal wave is evaluated for the cases of 50 kHz, 150 kHz and 300 kHz. To this end, 175 circular cavities in a hexagonal arrangement, embedded in an infinitely extended cementitious matrix with the properties of Table 1, are considered. The representative volume element of this periodic structure corresponds to a void concentration of 10%. A longitudinal plane harmonic wave at the frequencies of 50 kHz, 150 kHz, 300 kHz and wavelengths 0.085 m, 0.0283 m, 0.0141 m, respectively, impinges upon the rows of cavities. The tractions fields created between the voids for the aforementioned frequencies are evaluated with the aid of the Boundary Element Method (BEM), which undoubtedly is the most accurate method for calculating stress fields in linear elastic structures. An advanced BEM, explained in [20, 21], is employed to this purpose. The traction fields, with respect to the propagation direction, are evaluated and shown in Fig. 13. Normalizing with the highest traction the results are presented, for each frequency, in a dimensionless scale from −1(compressive stress) to 1(tensile stress). Since the matrix is considered as a classical elastic material, it is apparent that high gradients of stresses correspond to high gradients of strains. As it can be seen from Fig. 13(a), qualitatively, there are certain areas where high stress and thus high strain gradients occur. These strain gradients appear as sudden changes of the contour colors from blue to cyan and red to yellow in the compressive and tensile regions, respectively. Mainly, they appear in the region between the voids and in a direction being perpendicular to the wave propagation. Similar results obtained in the recent work [22] where strain gradients appear in the body of bended perforated plates. Although the figure seems to be different in Fig. 13(b) and (c), a careful observation reveals that almost all the regions between the holes appear sudden changes from tensile to compressive tractions, which are associated with significant strain gradients. Such a behavior justifies the use of strain gradient elasticity as a theory that can capture dispersion phenomena in damaged concrete.

(a) (b) (c)

Fig. 13. Distribution of tractions along the direction of a longitudinal elastic wave propagating in a concrete matrix containing 10% hexagonally distributed circular voids. The propagating wave is harmonic at the frequencies of (a) 50 kHz, (b) 150 kHz and (c) 300 kHz.

6 Conclusions

The present numerical investigation sheds light into the effect of shape of inclusions (voids) in the wave propagation velocity in an elastic matrix. Results highlight the dispersive nature of the heterogeneous material. The pulse velocity increases with frequency between 50 and 500 kHz by as much as 12%, making frequency a very crucial parameter of any ultrasonic test in concrete. Furthermore, the differences between inclusion shapes are more evident at low frequencies where the wavelength is much longer than the heterogeneity size and tend to diminish for higher frequency where the wavelength is of the same order with the inclusion size. Results show that the circular shape of inclusion is a generally good approximation resulting however, in velocities a few percent higher than the average of the different shapes. Care should be taken when there is a specific orientation preference of heterogeneity. Then the wave direction relatively to the voids orientation becomes relevant and the result may differ by almost 30% between favorable and unfavorable direction. In addition, experimental results obtained by plate-like inclusions are presented showing strong similarity with the numerical cases concerning ellipsoidal vertical inclusions. Finally, the numerical results are compared with theoretical from scattering models and Mindin's gradient elastic theory, showing a convergence and implying that all different approaches are different prisms to study the same phenomenon.

References

1. Naik, T.R., Malhotra, V.M., Popovics, J.S.: The ultrasonic pulse velocity method. In: Malhotra, V.M., Carino, N.J. (eds.) Handbook on Nondestructive Testing of Concrete. CRC Press, Boca Raton (2004)

2. Anderson, D.A., Seals, R.K.: Pulse velocity as a predictor of 28-and 90-day strength. ACI J. **78–9**, 116–122 (1981)
3. Chaix, J.F., Garnier, V., Corneloup, G.: Ultrasonic wave propagation in heterogeneous solid media: theoretical analysis and experimental validation. Ultrasonics **44**(2), 200–210 (2006)
4. Punurai, W., Jarzynski, J., Qu, J., Kurtis, K.E., Jacobs, L.J.: Characterization of entrained air voids in cement paste with scattered ultrasound. NDT E Int. **39**(6), 514–524 (2006)
5. Iliopoulos, S.N., Aggelis, D.G., Polyzos, D.: Wave dispersion in fresh and hardened concrete through the prism of gradient elasticity. Int. J. Solid Struct. **78**, 149–159 (2016)
6. Waterman, P.C., Truell, R.: Multiple scattering of waves. J. Math. Phys. **2**, 512–537 (1964)
7. Mindlin, R.D.: Microstructure in linear elasticity. Arch. Rat. Mech. Anal. **16**, 51–78 (1964)
8. http://www.cyberlogic.org/wave2000.html. Accessed April 2019
9. Kaufman, J.J., Luo, G., Siffert, R.S.: Ultrasound simulation in bone. IEEE Trans. Ultrason. Ferroelectr. Freq. Control **55**(6), 1205–1218 (2008)
10. Philippidis, T.P., Aggelis, D.G.: Experimental study of wave dispersion and attenuation in concrete. Ultrasonics **43**(7), 584–595 (2005)
11. Aggelis, D.G., Shiotani, T.: Effect of inhomogeneity parameters on wave propagation in cementitious material. ACI Mater. J. **105**(2), 187–193 (2008)
12. Aggelis, D.G., Polyzos, D., Philippidis, T.P.: Wave dispersion and attenuation in fresh mortar: comparisons of theory with experimental measurements. J. Mech. Phys. Solids **53**, 857–883 (2005)
13. Villarreal, A., Solis-Najera, S., Medina-Gomez, L.: Application of Waterman-Truell and the dynamic generalized self-consistent models on concrete. Phys. Proc. **70**, 437–441 (2015)
14. Yu, T., Chaix, J.-F., Komatitsch, D., Garnier, V., Audibert, L., Henault, J.-M.: 2D numerical modeling of ultrasonic wave propagation in concrete: a parameterization study in a multiple-scattering medium. In: AIP Conference Proceedings, vol. 1806, p. 080011 (2017). https://doi.org/10.1063/1.4974636
15. Iliopoulos, S.N., Malm, F., Grosse, C.U., Aggelis, D.G., Polyzos, D.: Interpretation of concrete's microstructural coefficients through Mindlin's strain gradient elastic theory and controlled experiments. J. Acoust. Soc. Am. **142**(1), 89–94 (2016)
16. Dona, M., Lombardo, M., Barone, G.: Experimental study of wave propagation in heterogeneous materials. In: Kruis, J., Tsompanakis, J., Topping, B.H.V. (eds.) Proceedings of the Fifteenth International Conference on Civil, Structural and Environmental Engineering Computing, Prague, 1–4 September, Paper 208 (2015)
17. Hassanpour, S., Heppler, G.R.: Micropolar elasticity theory: a survey of linear isotropic equations, representative notations, and experimental investigations. Math. Mech. Solids 1–19 (2015). https://doi.org/10.1177/1081286515581183
18. Lewis, T.S., Kraft, D.W., Hom, N.: Scattering of elastic waves by cylindrical cavity in a solid. J. Appl. Phys. **47**(5), 1795–1798 (1976)
19. Aggelis, D.G., Tsinopoulos, S.V., Polyzos, D.: An iterative effective medium approximation (IEMA) for wave dispersion and attenuation predictions in particulate composites suspensions and emulsions. J. Acoust. Soc. Am. **116**(6), 3443–3452 (2004)
20. Polyzos, D., Tsinopoulos, S.V., Beskos, D.E.: Static and dynamic boundary element analysis in incompressible linear elasticity. Eur. J. Mech. A/Solids **17**(3), 515–536 (1998)
21. Gortsas, T.V., Tsinopoulos, S.V., Polyzos, D.: An advanced ACA/BEM for solving 2D large-scale problems with multi-connected domains. CMES: Comput. Model. Eng. Sci. **107** (4), 321–343 (2015)
22. Gortsas, T.V., Tsinopoulos, S.V., Rodopoulos, D., Polyzos, D.: Strain gradient elasticity and size effects in the bending of fiber composite plates. Int. J. Solids Struct. **143**, 103–112 (2018)

Longitudinal Shock Waves in Soil

Vladimir A. Osinov(✉)

Institute of Soil Mechanics and Rock Mechanics, Karlsruhe Institute of Technology, 76128 Karlsruhe, Germany
vladimir.osinov@kit.edu

Abstract. The stiffness of soils in compression increases with increasing pressure. This property makes the strain-stress relation nonlinear and strongly influences the propagation of compression waves: a smooth wave front steepens and turns into a shock front. This paper discusses some theoretical questions related to the formation and propagation of longitudinal shock waves in soil within the pressure range typical of geotechnical engineering problems (up to a few megapascals). In particular, the topics discussed in the paper are the critical distance for dry and saturated soils (the distance covered by a smooth front before it becomes discontinuous), the jump conditions on a discontinuity, and smooth viscous shocks.

1 Introduction

A shock wave in a continuum in a narrow mathematical sense is a discontinuity in the stress and velocity which moves with respect to the medium. In a broader sense, a shock wave is the part of a continuous wave where the stress and velocity change sharply on an interval much smaller than the characteristic length of the problem under study. Shock waves as moving discontinuities can exist in both linearly elastic and nonlinear media. In a linearly elastic medium, a shock wave can be induced by a discontinuous step-like boundary condition, e.g. through an impact or a collision with another body. In a nonlinear medium, a shock front as a moving discontinuity can form even if the boundary condition is continuous. The property of the medium responsible for the shock formation in this case is the increase in the compressional stiffness with increasing pressure. This property is characteristic of the constitutive response of soils.

This paper discusses some theoretical aspects of the formation and propagation of shock waves in soil. Shock waves are often associated not only with high strain rates but also with high pressures produced, for instance, in situ by explosions. Actually, neither the emergence nor the propagation of a shock front require the pressure to be high. The pressure amplitude does play a role for the choice of the constitutive model for the mathematical description of the dynamic deformation and, in particular, a shock wave in the soil. We assume that the pressure does not exceed a few megapascals and thus lies within the range typical of geotechnical engineering problems. We will consider plane longitudinal waves with the simplest deformation mode – uniaxial compression (except

T. Triantafyllidis (Ed.): *Recent Developments of Soil Mechanics and Geotechnics in Theory and Practice*, LNACM 91, pp. 199–217, 2020.
https://doi.org/10.1007/978-3-030-28516-6_11

for some reference made to spherical and cylindrical longitudinal waves as well). The main two issues addressed in the paper are

- the formation of a shock front from a continuous wave and
- the propagation of a shock front.

2 Shock Formation. Plane Waves

We will study plane longitudinal waves propagating along the x_1-axis of a Cartesian coordinate system (x_1, x_2, x_3). The waves have one nonzero velocity component v_1 and three nonzero stress components $\sigma_{11}, \sigma_{22}, \sigma_{33}$. In what follows, we will write for brevity x for x_1, v for v_1 and σ for σ_{11}. The stress components σ_{22}, σ_{33} will not be considered. The governing equations will be written in the small-strain approximation with the partial time derivatives instead of the material ones. The density of the medium will be treated as a constant. Compressive stresses and deformations are taken to be negative according to the sign convention of continuum mechanics.

The equation of motion for the waves reads

$$\frac{\partial \sigma}{\partial x} = \varrho \frac{\partial v}{\partial t}, \tag{1}$$

where ϱ is the density and t is the time variable. For rate-independent behaviour, the constitutive equation can be written in rate form as

$$\frac{\partial \sigma}{\partial t} = k \frac{\partial \varepsilon}{\partial t}, \tag{2}$$

where ε is the axial deformation and k is the stiffness. Using the kinematic relation

$$\frac{\partial \varepsilon}{\partial t} = \frac{\partial v}{\partial x}, \tag{3}$$

the constitutive equation can be written as

$$\frac{\partial \sigma}{\partial t} = k \frac{\partial v}{\partial x}. \tag{4}$$

The wave propagation is described by the hyperbolic system of two Eqs. (1), (4) for two unknown functions $v(x,t), \sigma(x,t)$. The system has two families of characteristics with the slopes $dt/dx = \pm\sqrt{\varrho/k}$.

Consider a half-space $x \geq 0$ with a homogeneous initial stress σ^+ and zero initial velocity. Let a stress $\sigma_b(t)$ with $\sigma_b(0) = \sigma^+$ be applied at the boundary $x = 0$. This stress induces a wave propagating in the half space. If the medium is linearly elastic with $k = \text{const}$, the system (1), (4) is linear, and we obtain a travelling-wave solution: the wave propagates with the characteristic wave speed $c = \sqrt{k/\varrho}$ without changing its shape. Let σ^+ be equal to $-100\,\text{kPa}$ and $\sigma_b(t)$ represent an increasing compressive stress as shown in Fig. 1. The travelling wave induced by this boundary condition is shown in Fig. 2 for $c = 400\,\text{m/s}$.

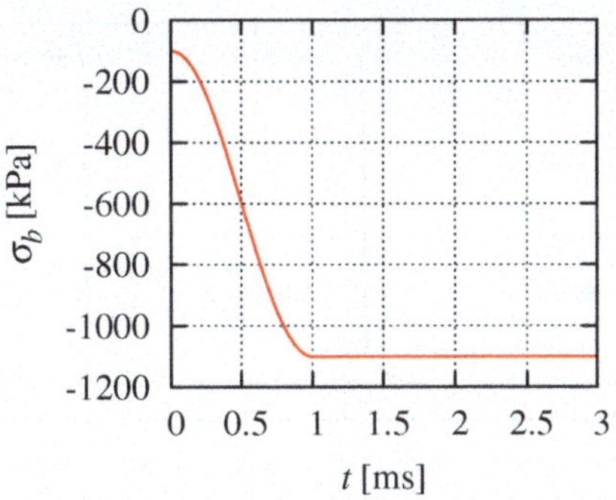

Fig. 1. Boundary condition at $x = 0$ for the solutions in Figs. 2 and 3

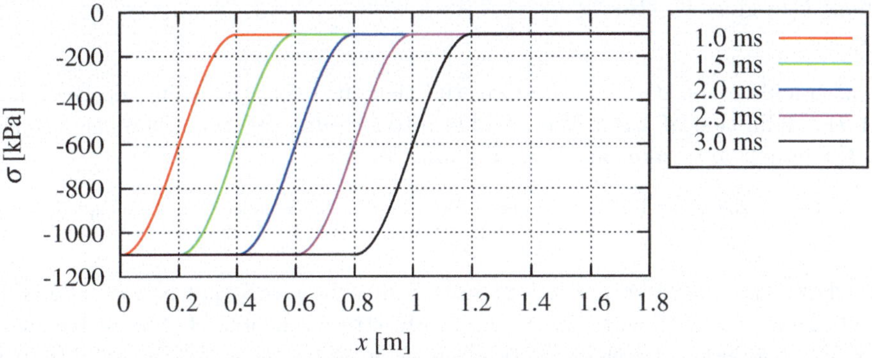

Fig. 2. Wave profiles in the half-space $x \geq 0$ at different times for constant stiffness k with the boundary condition in Fig. 1

The situation becomes qualitatively different if the wave propagates, for instance, in a dry granular soil and the stress increase is sufficiently large, like in the present example. The uniaxial (oedometric) compression of the soil up to a few megapascals yields a nonlinear strain-stress relation $\sigma(\varepsilon)$ with the increasing stiffness $k = d\sigma/d\varepsilon$ (higher pressures will not be considered here, as they may change the shape of the strain-stress curve because of grain crushing). For a given curve $\sigma(\varepsilon)$ obtained for the compression, k may be thought of as a function of either ε or σ. Writing $k(\sigma)$ allows us to consider again the system (1), (4) for the same unknown functions v, σ. However, the system becomes quasilinear: the coefficient k depends on the solution σ.

An important question is whether the quasilinear system with the stress-dependent stiffness $k(\sigma)$ possesses travelling-wave solutions like the linear system. Let $\sigma(\xi), v(\xi)$ be such a solution, where $\xi = t - x/c$, and c is the speed of propagation. The partial derivatives of the functions $\sigma(\xi), v(\xi)$ satisfy the equalities

$$\frac{\partial \sigma}{\partial t} + c\frac{\partial \sigma}{\partial x} = 0, \quad \frac{\partial v}{\partial t} + c\frac{\partial v}{\partial x} = 0. \tag{5}$$

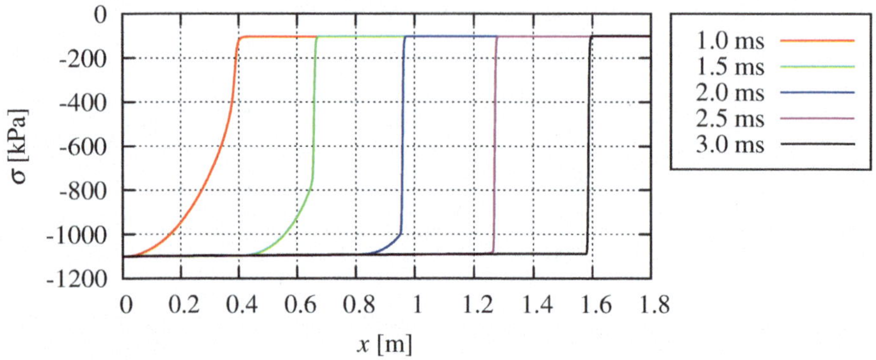

Fig. 3. Wave profiles in the half-space $x \geq 0$ at different times for the stress-dependent stiffness $k(\sigma)$ given by (16)

Let us calculate the stiffness of a material element using this solution. Beginning with the definition of k and then taking into account (5), the kinematic relation (3), the equation of motion (1) and again (5), we obtain

$$k = \frac{\partial \sigma}{\partial t}\left(\frac{\partial \varepsilon}{\partial t}\right)^{-1} = -c\,\frac{\partial \sigma}{\partial x}\left(\frac{\partial v}{\partial x}\right)^{-1} = \varrho c^2\,\frac{\partial v}{\partial t}\left(\frac{\partial v}{\partial t}\right)^{-1} = \varrho c^2. \tag{6}$$

This shows that, because c is a constant, k must be a constant as well. Thus, for travelling-wave solutions to exist, the strain-stress relation $\sigma(\varepsilon)$ must be linear.

For a nonlinear strain-stress relation such as in dry soil, two neighbouring points on a moving wave profile propagate with different wave speeds: the point with a higher stress moves faster than the point with a lower stress. For the boundary condition with the increasing compressive stress shown in Fig. 1, the wave profile steepens during the propagation until the gradients $\partial\sigma/\partial x$, $\partial v/\partial x$ become infinite at some point of the profile. At that instant, a discontinuity in the solution (a shock) arises, and the solution differentiable everywhere in the domain does not exist any longer. Figure 3 shows the numerical solution for the same boundary loading as in Fig. 1 with a nonlinear constitutive function $\sigma(\varepsilon)$ for dry soil. The constitutive function and the method of solution will be discussed below in Sects. 4 and 5. Here the focus is on the fact that the continuous front turns into a shock at a distance of about 0.5 m from the boundary.

The mechanism of the transition from a continuous compression front to a shock is well known and is in essence the same for fluids and solids [1–5]. It is also a general property of quasi-linear systems that the solution may become discontinuous in a finite time even if the initial and boundary data are continuous [6]. The time at which the solution loses continuity is called the critical time. In the case of the boundary value problem for the half-space formulated above, the coordinate x at which the gradients $\partial\sigma/\partial x$, $\partial v/\partial x$ become infinite at the critical time is called the critical distance.

The solution to the boundary value problem for the half-space with stress-dependent stiffness is continuous up to the critical time. This solution is a

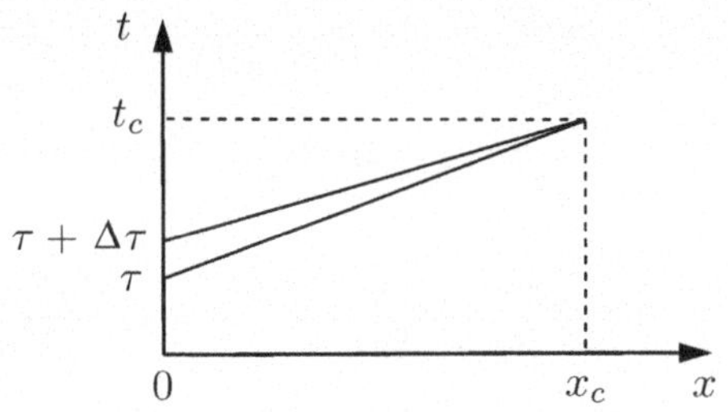

Fig. 4. Two straight characteristics on the (x, t)-plane

so-called simple wave [1]: the characteristics with the positive slopes $dt/dx = \sqrt{\varrho/k(\sigma_b)}$ emanating from the t-axis on the (x, t)-plane are straight lines, and the functions σ, v are constant on these characteristics, Fig. 4. The values of v on the t-axis, $v_b(t) = v(0, t)$, are found from the solution of the ordinary differential equation

$$\frac{dv_b}{dt} = -\frac{1}{\sqrt{\varrho k(\sigma_b)}} \frac{d\sigma_b}{dt} \tag{7}$$

with $v_b(0) = 0$ (the initial condition for v). It can be verified directly that this simple-wave solution satisfies the system (1), (4). For details and for a more general system of two quasilinear equations, see [7].

The simple-wave solution allows us to derive an ordinary differential equation for a certain linear combination of the gradients $\partial\sigma/\partial x$, $\partial v/\partial x$ as functions of x along a given characteristic with a positive slope [7]. The differential equation obtained is of the Bernoulli type. The equation has the property that the solution may become infinite at a finite value of the independent variable. The equation can be integrated to find the coordinate x at which the gradients tend to infinity. This coordinate may be interpreted as the critical distance for a particular characteristic and will be denoted by $x_c(\tau)$, where τ is the time at which the characteristic emanates from the t-axis, see Fig. 4.

If the knowledge of the gradients along the characteristic is not needed, the value of $x_c(\tau)$ alone can be found in a simpler way. Consider two neighbouring characteristics with positive slopes emanating from the t-axis at times τ and $\tau + \Delta\tau$ as shown in Fig. 4. For the increasing stress at the boundary shown in Fig. 1, the slope dt/dx of the characteristic for τ with $\sigma_b(\tau)$ will be greater than the slope of the characteristic for $\tau + \Delta\tau$ with $\sigma_b(\tau + \Delta\tau)$, because the latter corresponds to a higher stiffness and a higher wave speed. The two characteristics cross at some x giving two different values of σ at this x, and the same is true for v (recall that σ and v are constant on these characteristics in the simple-wave solution). The solution must therefore be discontinuous at this point. If $\Delta\tau \to 0$, the intersection point tends to a point which gives the critical distance $x_c(\tau)$ for this characteristic. The gradients $\partial\sigma/\partial x$, $\partial v/\partial x$ become infinite at $x_c(\tau)$. The two characteristics in Fig. 4 are described on the (x, t)-plane by the equations

$$t = \frac{x}{c(\tau)} + \tau, \quad t = \frac{x}{c(\tau + \Delta\tau)} + \tau + \Delta\tau, \tag{8}$$

where $c(\tau) = \sqrt{k(\sigma_b(\tau))/\varrho}$ is the wave speed as a function of τ. If $\Delta\tau \to 0$, the two lines (8) cross at

$$x_c(\tau) = \lim_{\Delta\tau\to 0} \frac{c(\tau)\, c(\tau + \Delta\tau)\, \Delta\tau}{c(\tau + \Delta\tau) - c(\tau)} = c^2 \left(\frac{dc}{d\tau}\right)^{-1}. \tag{9}$$

This is the same $x_c(\tau)$ as the one that would be obtained from the solution to the Bernoulli equation mentioned above.

The time $t_c(\tau)$ at which the gradients become infinite at $x_c(\tau)$ is

$$t_c(\tau) = \tau + \frac{x_c(\tau)}{c(\tau)}. \tag{10}$$

The critical time for the boundary value problem, denoted by $\hat{t}_c$, is therefore

$$\hat{t}_c = \min_{\tau} t_c(\tau). \tag{11}$$

The critical distance $\hat{x}_c$, defined above as the coordinate at which the gradients become infinite at the critical time, is given by

$$\hat{x}_c = x_c(\hat{\tau}), \tag{12}$$

where $\hat{\tau}$ satisfies the equation $\hat{t}_c = t_c(\hat{\tau})$.

Finding the critical distance $\hat{x}_c$ with (9)–(12) for a particular problem may be a rather intricate procedure. Theoretically, $\hat{x}_c$ may even be nonunique: the gradients may become infinite simultaneously at two different points of the wave profile. In applications, where the strain-stress relation $\sigma(\varepsilon)$ and the boundary condition $\sigma_b(t)$ are known only approximately, it suffices to obtain an estimation of the critical distance from (9) rather than the exact value determined by (12).

Using the definition of c, (9) can be written as

$$x_c(\tau) = \frac{2\,k^{3/2}}{\sqrt{\varrho}} \left(\frac{dk}{d\sigma}\right)^{-1} \left(\frac{d\sigma_b}{d\tau}\right)^{-1} = \frac{2\,k^{5/2}}{\sqrt{\varrho}} \left(\frac{dk}{d\varepsilon}\right)^{-1} \left(\frac{d\sigma_b}{d\tau}\right)^{-1}, \tag{13}$$

where $k, dk/d\sigma, dk/d\varepsilon$ represent the values at the boundary and are functions of τ due to their dependence on $\sigma_b(\tau)$. Let us take a closer look at these expressions for x_c. Since compressive stresses and deformations are taken to be negative, $dk/d\sigma$ and $dk/d\varepsilon$ are negative. For x_c to be positive, the stress rate $d\sigma_b/d\tau$ at the boundary must be negative, that is, the boundary condition must induce a compression front. In the case of an expansion front, formula (13) gives a negative value of x_c. Such a front flattens out as it propagates and does not turn into a shock. For a given strain-stress relation $\sigma(\varepsilon)$, the critical distance is inversely proportional to the stress rate $d\sigma_b/d\tau$: the higher the rate, the shorter x_c. The critical distance is also inversely proportional to the second derivative of the function $\sigma(\varepsilon)$. If $k = \text{const}$, x_c is infinite in accordance with the fact that the wave profile in a linearly elastic medium does not change during the propagation. The critical distance increases with increasing k if all other terms are fixed.

3 Shock Formation. Cylindrical and Spherical Waves

A different way to study the formation of shocks from continuous solutions is to consider acceleration waves, i.e. discontinuities in the first partial derivatives of the functions $\sigma(x,t), v(x,t)$ (called weak discontinuities, as distinct from strong discontinuities, or shocks, when the functions σ, v themselves are discontinuous). This approach is used in [8,9].

A weak discontinuity turns into a shock when the gradients $\partial\sigma/\partial x$, $\partial v/\partial x$ become infinite on one side of the discontinuity. The curve on the (x,t)-plane where the partial derivatives are discontinuous is a characteristic curve of the system. The vector of the jumps, $([\![\partial\sigma/\partial x]\!], [\![\partial v/\partial x]\!])$, is proportional to an eigenvector of the matrix of the system, with the factor being defined as the amplitude of the discontinuity. The amplitude as a function of the distance covered by the front satisfies an ordinary differential equation of the Bernoulli type. The equation can be integrated if the front propagates into a quiescent region with a known stress σ. The coordinate at which the amplitude becomes infinite is the critical distance. As is readily seen from this description, the acceleration-wave method requires the boundary condition $\sigma_b(t)$ to have a nonzero rate $d\sigma_b/dt$ at $t = 0$ for which the critical distance is calculated. The point where infinite gradients arise is the leading point of the wave profile. Although the formula obtained for the critical distance coincides with (13), the method cannot be directly applied to other points of the wave profile.

An advantage of the acceleration-wave method over the previous one is that it is applicable to cylindrical and spherical waves. The equations for such waves include the circumferential stress as an additional unknown function. The equation of motion contains additional terms without derivatives. Continuous cylindrical and spherical waves can turn into shocks as well as plane waves if the stiffness of the medium is stress-dependent. The critical distance is understood as the distance from the cavity wall where the wave is generated by a radial stress $\sigma_b(t)$ applied to the wall. If the constitutive response of the medium satisfies certain isotropy conditions, both plane and cylindrical, or both plane and spherical, waves can exist. In this case, given a loading rate $d\sigma_b/dt$ at the boundary, a question arises as to how the critical distances for the three types of waves differ from each other. As shown in [8], unique relations exist between the critical distances independently of the particular choice of the constitutive function. The relations are also valid for a saturated porous solid in which the pore fluid bulk modulus is a function of the pore pressure.

Let l_P, l_C, l_S be the critical distances calculated with the acceleration-wave method for plane, cylindrical and spherical fronts, respectively. The values of l_C and l_S are connected with l_P by the relations

$$l_C = l_P \left(1 + \frac{l_P}{4r_0}\right), \tag{14}$$

$$l_S = r_0 \left[\exp\left(\frac{l_P}{r_0}\right) - 1\right], \tag{15}$$

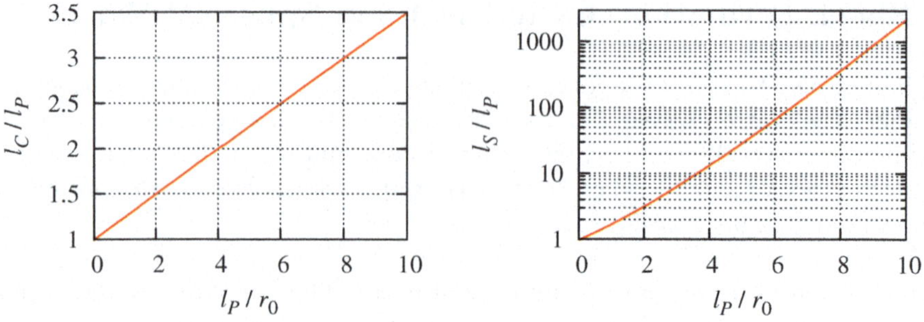

Fig. 5. Relations between the critical distances for plane, cylindrical and spherical fronts [8]

where r_0 is the cavity radius in the cylindrical and spherical problems [8]. It is convenient to plot the ratios l_C/l_P and l_S/l_P as functions of l_P/r_0. These functions are shown in Fig. 5. The distances l_C and l_S are always greater than l_P. If l_P is much smaller than r_0, both l_C and l_S are close to l_P. The growth of l_S/l_P with increasing ratio l_P/r_0 is much faster than that of l_C/l_P.

4 Critical Distances in Soil

In dry granular soils such as sand, the stiffness as a function of the confining pressure is known to obey a power law with an exponent of about 0.5–0.6 [10]. Based on this fact, we consider a simple constitutive model in which the stiffness k for uniaxial deformation is a function of σ of the form

$$k(\sigma) = k_0 \left(\frac{\sigma}{\sigma_0} \right)^m , \tag{16}$$

where k_0 is the value of k at $\sigma = \sigma_0$, and m is a constant. It may be convenient to use, instead of k_0, the wave speed $c_0 = \sqrt{k_0/\varrho}$ in the dry soil at $\sigma = \sigma_0$, where $\varrho = (1-n)\varrho_s$ is the soil density, ϱ_s is the density of the solid phase (grains), and n is the porosity. The strain-stress relation $\sigma(\varepsilon)$ obtained with (16) is shown in Fig. 6 (the curve for $S_r = 0$) with the model parameters given in Table 1.

Table 1. Model parameters

c_0 [m/s]	σ_0 [kPa]	m	n	ϱ_s [kg/m^3]	ϱ_f [kg/m^3]
400	−100	0.6	0.375	2650	1000

Figure 7 shows the critical distance in dry soil as a function of σ calculated with (13), (16) for the boundary rate $d\sigma_b/dt$ equal to -10^3 MPa/s. The boundary loading shown in Fig. 1 would have this rate if the stress increase were linear in

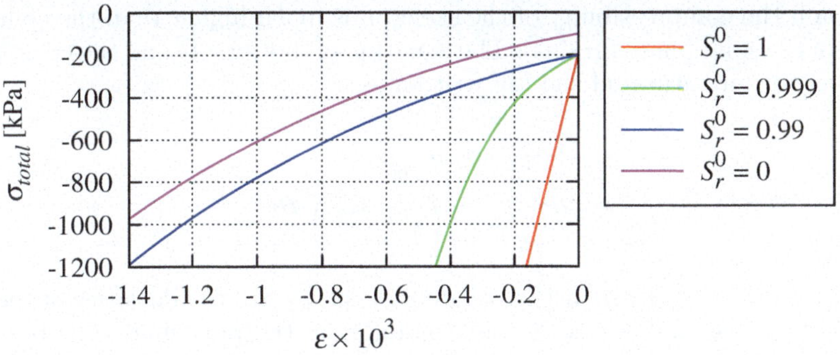

Fig. 6. Strain-stress relations for dry and saturated soils obtained with (16) for an initial effective stress of -100 kPa and, if $S_r \neq 0$, for an initial pore pressure of 100 kPa

time. Since the critical distance is inversely proportional to the stress rate, the values for other rates can easily be found by multiplication. The shock formation observed in Fig. 3 is in good agreement with Fig. 7.

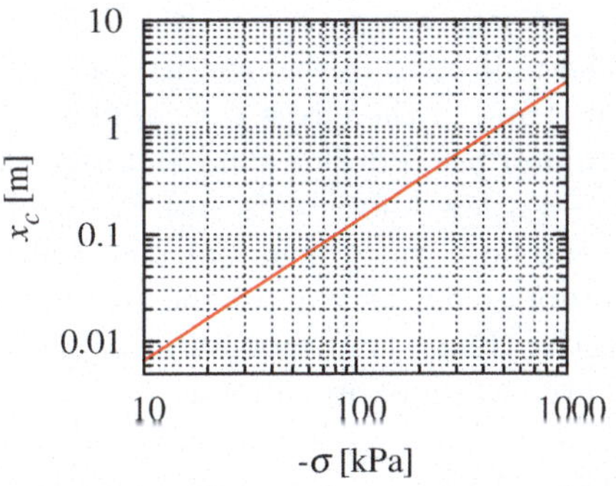

Fig. 7. Critical distance in dry soil as a function of the stress for the boundary rate $d\sigma_b/dt = -10^3$ MPa/s

Formula (13) for the critical distance is applicable to saturated soil if the soil permeability is low enough to assume locally undrained conditions (for nonzero permeability, see [9]). For saturated soil, the stress σ with the stiffness (16) is the effective stress defined as

$$\sigma = \sigma_{total} + p, \tag{17}$$

where σ_{total} is the total stress, and p is the pore pressure (positive for compression). The effective stress σ in (17) is determined by the deformation of the skeleton if the compressibility of the skeleton is much higher than the compressibility of the solid phase (grains). The equation of motion (1) and the constitutive equation (4) for saturated soil are written as

$$\frac{\partial \sigma_{total}}{\partial x} = \varrho \frac{\partial v}{\partial t}, \tag{18}$$

$$\frac{\partial \sigma_{total}}{\partial t} = \left(k + \frac{K_f}{n}\right) \frac{\partial v}{\partial x}, \tag{19}$$

where $\varrho = (1-n)\varrho_s + n\varrho_f$ is the soil density, ϱ_s, ϱ_f are the densities of the solid and fluid phases, and K_f is the bulk modulus of the pore fluid. The boundary loading $\sigma_b(t)$ is understood as the total stress. The stiffness k in Eq. (13) for the critical distance should be replaced with the total stiffness $k + K_f/n$. Since the compression of saturated soil from a given initial state yields a definite strain-stress function $\sigma_{total}(\varepsilon)$, the total stiffness can be viewed as a function of either the total stress or the skeleton deformation.

Suppose that the pore water contains a small amount (a few volume percent) of free gas. The bulk modulus of such a water-gas mixture, considered as a single fluid, is [11–13]

$$K_f = \left(\frac{S_r}{K_w} + \frac{1 - S_r}{K_g}\right)^{-1}, \tag{20}$$

where S_r is the degree of saturation, K_w is the bulk modulus of pure water (2.2 GPa), and K_g is the bulk modulus of the free gas. The difference between the pressures in the water and in the gas is the capillary pressure due to surface tension at the water-gas interface. For fine sand and high saturation, the pressure difference does not exceed a few kilopascals and is neglected. Furthermore, the mass of the free gas in the pore water is assumed to remain constant (this may not be true for slow compression, as part of the free gas would be dissolved in the water according to the Henry law). For an ideal gas, the bulk modulus K_g is proportional to the absolute gas pressure: $K_g = \gamma(p + p_a)$, where p_a is the atmospheric pressure (100 kPa), $\gamma = 1$ for isothermal processes and $\gamma = 1.4$ for adiabatic processes for air. We take $\gamma = 1$ assuming that the pore water hinders the temperature rise in the gas. For a given degree of saturation S_r^0 at an initial pore pressure p^0, the modulus K_f is a function of the current pressure p [13].

The strain-stress relation for saturated soil is shown in Fig. 6 for three values of S_r^0. If the soil is fully saturated, the pore fluid bulk modulus, K_f, is equal to the modulus of pure water, K_w. Since this modulus is much higher than the skeleton stiffness, the strain-stress relation for the total stress is practically linear. The presence of a small amount of free gas in the pore water substantially reduces the bulk modulus of the pore fluid. As follows from (20), one volume percent of free gas at $p^0 = 100$ kPa reduces K_f from 2.2 GPa to 20 MPa. This makes the soil stiffness nearly the same as in the dry soil, see the curves for $S_r^0 = 0.99$ and $S_r^0 = 0$ in Fig. 6. Even the case with $S_r^0 = 0.999$ is markedly different from full saturation in that the strain-stress relation is nonlinear.

The nonlinearity of the strain-stress relation leads to the formation of shocks for compression fronts. Figure 8 shows the critical distance in saturated soil as a function of the gas content $1 - S_r$ for the boundary rate $d\sigma_b/dt = -10^3$ MPa/s. Comparison of Figs. 8 and 7 shows that the critical distance in saturated soil is close to that in dry soil for the degrees of saturation up to 0.999. A similar plot with slightly different constitutive parameters can be found in [8].

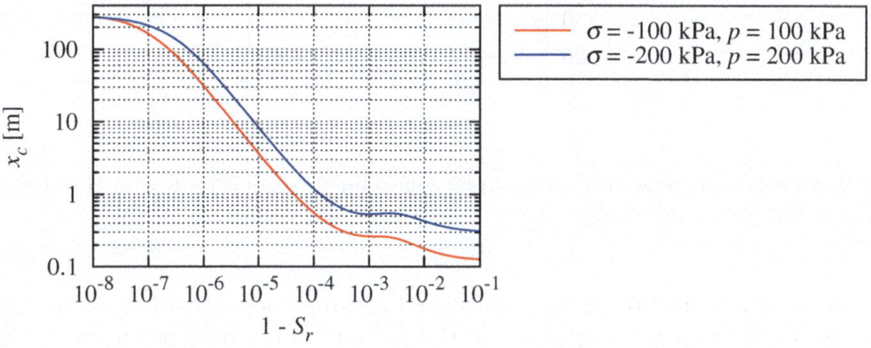

Fig. 8. Critical distance in saturated soil as a function of the gas content for the boundary rate $d\sigma_b/dt = -10^3$ MPa/s

So far we have considered a specific boundary value problem for the half-space with a compressive boundary condition. Of course, if the boundary condition is not compressive but the dynamic process develops in such a way that the soil undergoes compression, shock fronts may also arise. We have seen that the critical distance for fully saturated soil is much larger than in the other cases considered because of the dominant contribution of the bulk modulus of water to the total stiffness of the soil. The constant modulus of water makes the soil stiffness nearly constant as well. However, under certain conditions shock fronts can also arise very quickly in problems with fully saturated soil. If the soil is subjected to tensile deformation, the pore pressure may become sufficiently low to give rise to pore water cavitation. At that instant, the pore fluid bulk modulus abruptly falls from 2.2 GPa to a nearly zero value determined by the compressibility of the vapour in the bubbles. Assuming that the cavitation begins at zero absolute pore pressure and neglecting the vapour pressure, the tensile deformation of the soil produces a strain-stress relation like that shown in Fig. 9. The kink on the curves corresponds to the beginning of cavitation, after which the pore fluid bulk modulus is zero and the soil stiffness is determined solely by the stiffness of the skeleton. As the soil is compressed after the expansion, the strain-stress path will be slightly different from that shown in Fig. 9 due to the plastic behaviour of the skeleton, but the curve will also have a kink at the point where all cavitation bubbles disappear and the fluid bulk modulus is restored to 2.2 GPa. Because of the sharp increase in the stiffness, the compression will produce a shock front. The jump in the stiffness in this simple cavitation model means an infinite second derivative $d^2\sigma_{total}/d\varepsilon^2$ and, as seen from (13), a zero critical distance.

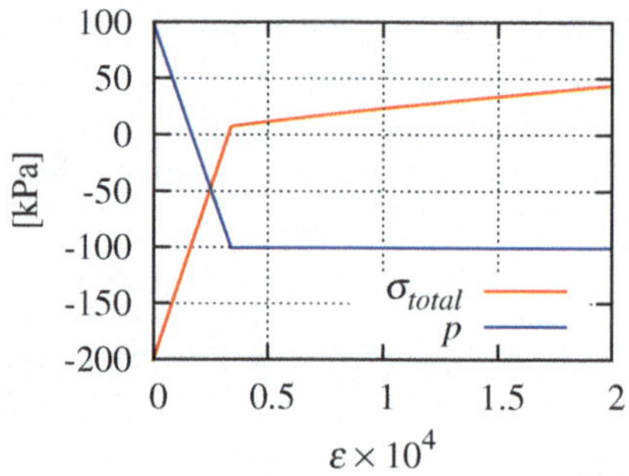

Fig. 9. Strain-stress relation for the expansion of fully saturated soil with the stiffness (16) and the pore water cavitation

Pore water cavitation in fully saturated soil may occur if a large-amplitude pressure wave propagates upwards in the ground and is reflected from the free surface. Examples of such waves induced by an explosion in a tunnel can be found in [14,15]. If the ground were elastic, the reflected wave would have a tensile stress of the same magnitude as the incident wave. In fully saturated soil, large tensile stresses are impossible because neither the skeleton nor the pore water can withstand such stresses. As a result of the interaction of the incident wave with the free surface, a large cavitation zone forms in the soil [14,15]. The subsequent shrinkage of the cavitation zone is accompanied by the formation of shock fronts at the boundary of the cavitation zone according to the mechanism described above. Although the amplitudes of these shock fronts are much smaller than the amplitude of the primary pressure wave, these shocks, like any others, entail numerical problems because of spurious numerical oscillations which are difficult to eliminate.

5 Shock Propagation

After a shock front has formed from the continuous wave, the dynamic deformation can no longer be described by differential equations in the whole domain as before the shock formation. The differential equations cannot be applied directly to a point where the solution is discontinuous. We can still use the differential equations for the smooth part of the wave, but need new relations on the shock to be able to solve the problem further. In this Section we discuss the question of how plane longitudinal shock fronts can be described in the context of soil mechanics and to what extent the description is correct.

Relations on discontinuities can be derived if the dynamic process is governed by differential equations in the form of conservation laws

$$\frac{\partial u_i}{\partial t} + \frac{\partial}{\partial x} f_i(u_1, \ldots, u_N) = 0, \quad i = 1, \ldots, N, \tag{21}$$

where u_i are the unknown functions, and f_i are sufficiently smooth functions of $u_1, \ldots, u_N$. System (21) can be integrated with respect to x and written as

$$\frac{d}{dt}\int_{x_a}^{x_b} u_i dx + f_i(u_1, \ldots, u_N)\Big|_{x_a}^{x_b} = 0, \quad i = 1, \ldots, N, \tag{22}$$

where x_a, x_b are two fixed points. As distinct from (21), the integral form (22) is applicable to discontinuous solutions. Let $[\![u_i]\!] = u_i^+ - u_i^-$ denote the jump in the function u_i across the shock, where u_i^+ and u_i^- are the values ahead of and behind the shock, respectively, as shown in Fig. 10. If the solution has a discontinuity moving with a speed s, the values of the functions on the shock satisfy the Rankine-Hugoniot jump conditions [2–5, 16–18]

$$[\![u_i]\!]s = [\![f_i]\!], \quad i = 1, ..., N. \tag{23}$$

Conditions (23) are obtained by considering a discontinuity between x_a and x_b in (22) and taking the limit $x_b \to x_a$.

We restrict ourselves to plane longitudinal waves in the half-space described by Eqs. (1) and (4) for two unknown functions σ, v, with a monotonic boundary condition like that shown in Fig. 1. To see how many jump conditions are necessary to solve the problem with a shock, consider large times at which the wave consists of the shock and two spatially homogeneous states ahead of and behind the shock, see Fig. 3, $t = 3$ ms and Fig. 10 left. The known quantities are σ^+, v^+ (the initial conditions in the half-space) and σ^- (the boundary condition). The unknown quantities are the velocity v^- behind the shock and the shock speed s. Hence, we need two equations to determine the two unknowns. It can be shown that, if the smooth part of the wave is not homogeneous (Fig. 10 right), we also need two jump conditions to solve the problem in the whole domain.

Whereas Eq. (1) is in the conservation-law form (21) (we assumed $\varrho = \text{const}$), Eq. (4) with variable k is not in this form and does not give any jump condition, unless the medium is linearly elastic with $k = \text{const}$. A second jump condition can be found if we take ε and v as unknown functions and use Eq. (3) instead of (4). The stress σ in (1) is then viewed as a known function of ε obtained from the strain-stress relation for the uniaxial compression. The two Eqs. (1) and (3) give two required jump conditions

$$[\![v]\!]\varrho s = -[\![\sigma]\!], \quad [\![\varepsilon]\!]s = -[\![v]\!]. \tag{24}$$

As follows from (24), the shock speed s is determined by the jumps in the strain and stress as

$$s^2 = \frac{[\![\sigma]\!]}{[\![\varepsilon]\!]\varrho}. \tag{25}$$

A question that arises in connection with the shock description is whether σ^- as a function of ε immediately behind the shock can be obtained from the same strain-stress curve $\sigma(\varepsilon)$ as for the smooth part of the solution. These two functions need not be the same. Moreover, certain considerations suggest that they are different. The difference is due to the way that the soil deforms in time

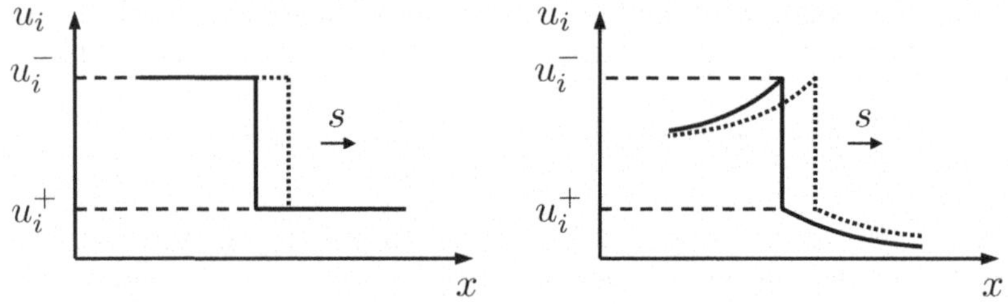

Fig. 10. A propagating shock with homogeneous (left) and inhomogeneous (right) states in the neighbourhood of the shock

and space. Constitutive models of soil mechanics and their calibration are based on the results of element tests (in the present case – the oedometric test). Shock front as a discontinuity in the solution is a mathematical notion pertinent to hyperbolic conservation laws. A real shock front in the soil is continuous with a nonzero thickness which may be assumed not to exceed the size of the soil sample in an element test. The strain rate in an element test is much lower than the strain rate on a shock front. Since the strain rate is equal to the velocity gradient, it tends to infinity as the shock thickness tends to zero. Another essential difference lies in the strain rate gradient. The deformation in an element test is more or less homogeneous, at least for small strains. Although the residual deformation behind a shock can also be spatially homogeneous, the strain rate during the shock passage is inherently inhomogeneous. The smaller the shock thickness, the higher the degree of inhomogeneity in the strain rate. This is illustrated in Fig. 11 for a smooth shock front of thickness Δx. The strain rate in a soil layer of the same thickness Δx is always inhomogeneous, unless it is zero before and after the shock passage. The deformation like that shown in Fig. 11 cannot be reproduced in element tests. More sophisticated dynamic experiments with propagating shocks and inhomogeneous strain rates are required. A high strain rate alone does not suffice. Taking the function $\sigma(\varepsilon)$ for the shock description is an approximation with as yet unknown inaccuracy.

If we accept the necessity of two different functions $\sigma(\varepsilon)$ and $\sigma^-(\varepsilon)$, the overall mathematical description of the dynamic deformation will become more complicated. This concerns, in particular, the analysis of the shock existence and the task of incorporating the two functions into a single boundary value problem.

If the shocks and the continuous deformation are described by the same function $\sigma(\varepsilon)$, one can establish an inequality relation between the shock speed and the characteristic wave speeds ahead of and behind the shock. Assume the function $\sigma(\varepsilon)$ to be such that $d^2\sigma/d\varepsilon^2 < 0$. Since the shocks are compressive, we have $\sigma^+ > \sigma^-$, and it is easy to see that the shock speed given by (25) is greater than the characteristic wave speed $\sqrt{k(\sigma^+)/\varrho}$ ahead of the shock, and less than the characteristic wave speed $\sqrt{k(\sigma^-)/\varrho}$ behind the shock. This result constitutes a necessary condition for the existence of the shock [2–5, 16–18] and is

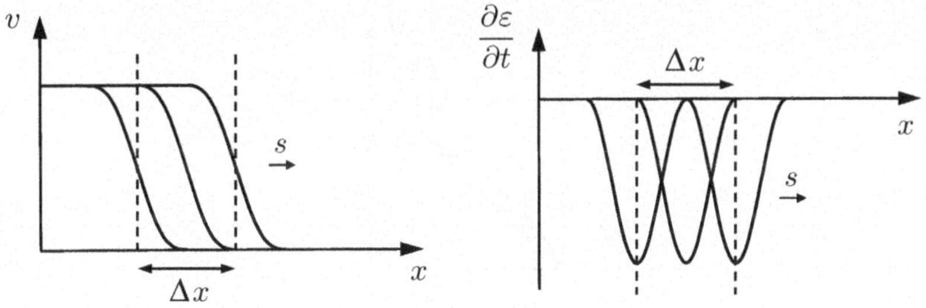

Fig. 11. Velocity and strain rate profiles at different times

also well known from gas dynamics: a shock propagates with a supersonic speed with respect to the medium ahead of the shock, and with a subsonic speed with respect to the medium behind the shock [1].

A better insight into the shock propagation phenomenon can be gained if we try to find the strain-stress path that a soil element follows between ε^+ and ε^- as the shock crosses the element. Note that we could say that it has no sense to speak of a strain-stress path as such, but only of the initial and final states on the shock, $(\varepsilon^+, \sigma^+)$ and $(\varepsilon^-, \sigma^-)$. However, the question of the strain-stress path arises naturally if we take into account that a real shock in the soil is continuous, has a nonzero thickness and, if modelled within the framework of continuum mechanics, must have a definite strain-stress path. This path can be identified even for an idealised zero-thickness shock with the help of the energy balance.

For the analysis of shock propagation we have used two conservation laws (1), (3) for two unknown functions σ, ε, which yields two jump conditions (24). Given σ^+, σ^-, v^+ and a function $\sigma^-(\varepsilon)$, the two jump conditions enable us to determine the shock, i.e. to find two unknowns v^-, s. The energy balance is another conservation law which has not been used so far. For the present one-dimensional case, the energy balance can be written either in the integral form

$$\frac{d}{dt} \int_{x_a}^{x_b} \left(E + \frac{1}{2} \varrho v^2 \right) dx - (\sigma v) \Big|_{x_a}^{x_b} = 0 \tag{26}$$

or in the differential form

$$\frac{\partial}{\partial t} \left(E + \frac{1}{2} \varrho v^2 \right) - \frac{\partial}{\partial x} (\sigma v) = 0, \tag{27}$$

where E is the internal energy of the medium per unit volume. For continuous solutions, the energy balance (27) together with the equation of motion (1) and the kinematic relation (3) leads to the familiar expression for the energy rate:

$$\frac{\partial E}{\partial t} = \sigma \frac{\partial \varepsilon}{\partial t}. \tag{28}$$

For discontinuous solutions, the conservation law (27) yields the jump condition

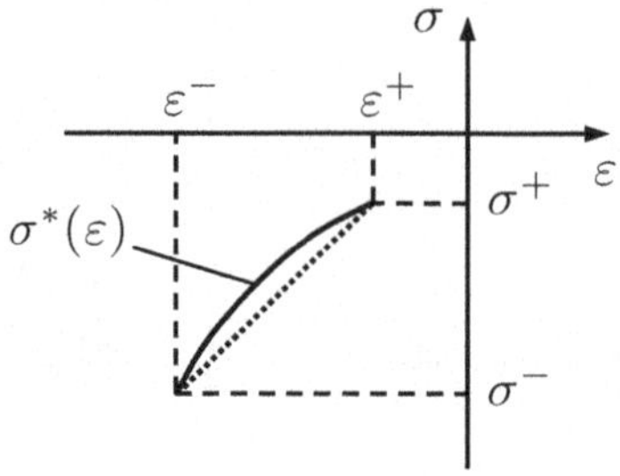

Fig. 12. Strain-stress path on the shock

$$\left[\!\left[E + \frac{1}{2}\varrho v^2 \right]\!\right] s = -[\![\sigma v]\!]. \tag{29}$$

Using (29), (24) and the formulae for the jump of a product,

$$[\![ab]\!] = [\![a]\!]b^+ + [\![b]\!]a^- = \frac{1}{2}(b^+ + b^-)[\![a]\!] + \frac{1}{2}(a^+ + a^-)[\![b]\!], \tag{30}$$

we can derive an expression for the jump in the internal energy:

$$[\![E]\!] = \frac{1}{2}(\sigma^+ + \sigma^-)[\![\varepsilon]\!]. \tag{31}$$

If the deformation path on the shock is described by some function $\sigma^*(\varepsilon)$, then, according to (28), the increase in the internal energy across the shock is

$$\Delta E^{(1)} = \int_{\varepsilon^+}^{\varepsilon^-} \sigma^*(\varepsilon) d\varepsilon > 0. \tag{32}$$

On the other hand, the increase in the internal energy across the shock is $\Delta E^{(2)} = -[\![E]\!] > 0$. As follows from the comparison of (31) and (32), for any curve $\sigma^*(\varepsilon)$ that lies above the linear path (Fig. 12),

$$\Delta E^{(1)} < \Delta E^{(2)}, \tag{33}$$

and the energy balance is violated. Hence, the strain-stress path across the shock must be linear independently of the form of the function $\sigma^-(\varepsilon)$ between ε^+ and ε^-. This seeming discrepancy is resolved in the next Section where we consider smooth shocks of nonzero thickness.

6 Viscous Shocks

Smooth shocks can be obtained by introducing an additional viscous stress $\sigma_{vis} = \mu\, \partial v/\partial x$, where μ is a viscosity coefficient. The equation of motion (1) then becomes

$$\frac{\partial \sigma}{\partial x} + \frac{\partial \sigma_{vis}}{\partial x} = \varrho \frac{\partial v}{\partial t}, \tag{34}$$

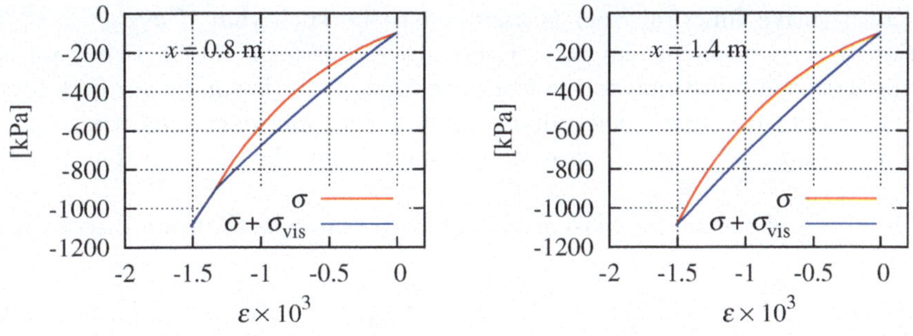

Fig. 13. Stresses as functions of the strain at two points in the numerical solution shown in Fig. 3

where σ is the rate-independent constitutive stress determined by a function $\sigma(\varepsilon)$. This approach has been used for the numerical solution of the boundary value problem for the half-space formulated in Sect. 2. The stress σ is given by the constitutive model (16) with the parameters from Table 1. The solution for dry soil shown in Fig. 3 is obtained by the finite-difference method with the viscosity coefficient $\mu = 10^{-3}\,\mathrm{MPa \cdot s}$ and 1300 discretization points between $x = 0$ and $x = 1.8\,\mathrm{m}$.

As can be inferred from Fig. 3, the wave front that passes, for instance, through the point $x = 0.8\,\mathrm{m}$ has not yet fully turned into a shock and has both a (theoretically) discontinuous part from $\sigma^+ = -100\,\mathrm{kPa}$ to $\sigma^- = -900\,\mathrm{kPa}$ and a continuous part from -900 to $-1100\,\mathrm{kPa}$. The front that passes through the point $x = 1.4\,\mathrm{m}$ is entirely discontinuous from $\sigma^+ = -100\,\mathrm{kPa}$ to $\sigma^- = -1100\,\mathrm{kPa}$. Figure 13 shows the stresses as functions of the strain at these two points. While the stress σ follows the curve determined by the constitutive equation, the total stress $\sigma + \sigma_{vis}$ follows a linear strain-stress path on the discontinuous part of the front. The viscous stress on the shock develops to such an extent that the total stress becomes a linear function of the strain, and the energy balance (31), (32) is satisfied. Given σ^+ and σ^-, the viscous stress on the shock is the difference between the linear function and the constitutive curve $\sigma(\varepsilon)$ and, therefore, does not depend on the viscosity coefficient μ. The value of μ determines the shock thickness: the greater μ, the thicker the shock. The discontinuous shock is obtained as a limit for $\mu \to 0$ with the same viscous stress and the linear strain-stress relation on the shock.

As the numerical viscous shock shown in Fig. 3 propagates, it tends to a travelling-wave solution (a steady shock). The problem of finding this asymptotic solution is formulated as follows: given σ^+, σ^-, v^+, find the wave speed c and the functions $\varepsilon(y), v(y)$, where $y = x - ct$, $y \in (-\infty, +\infty)$, which satisfy (3), (34) and the boundary conditions

$$\lim_{y \to +\infty} \sigma(y) = \sigma^+, \qquad \lim_{y \to -\infty} \sigma(y) = \sigma^-, \qquad \lim_{y \to +\infty} v(y) = v^+, \tag{35}$$

$$\lim_{y \to +\infty} \sigma_{vis}(y) = 0, \qquad \lim_{y \to -\infty} \sigma_{vis}(y) = 0. \tag{36}$$

The constitutive function $\sigma(\varepsilon)$ is assumed to be such that $d^2\sigma/d\varepsilon^2 < 0$. The solution to this problem can be derived in explicit form. For brevity, we omit intermediate computations and present the final result. It can be verified directly that the solution satisfies the equations and the boundary conditions. In the formulae below, ε^+ and ε^- are determined from the constitutive function: $\sigma(\varepsilon^+) = \sigma^+, \sigma(\varepsilon^-) = \sigma^-$.

The speed of the smooth viscous shock is the same as that of a discontinuous shock:

$$c^2 = \frac{\sigma^+ - \sigma^-}{\varrho(\varepsilon^+ - \varepsilon^-)}. \tag{37}$$

The inverse to the function $\varepsilon(y)$ is given by the integral

$$y(\varepsilon) = \mu c \int_{\varepsilon_0}^{\varepsilon} \frac{d\eta}{\sigma(\eta) - \sigma^- - \varrho c^2(\eta - \varepsilon^-)}, \tag{38}$$

where $\varepsilon \in (\varepsilon^-, \varepsilon^+)$, and ε_0 is an arbitrary fixed strain between ε^- and ε^+. The velocity profile is obtained from the function $\varepsilon(y)$:

$$v(y) = -c\varepsilon(y) + c\varepsilon^+ + v^+. \tag{39}$$

This gives the unknown quantity $v^- = \lim_{y\to-\infty} v(y)$:

$$v^- = -c\varepsilon^- + c\varepsilon^+ + v^+. \tag{40}$$

Equation (40) coincides with the second jump condition (24). The first jump condition (24) also holds true and can readily be derived from (37) and (40). As we have seen in Sect. 2, the strain-stress relation in a travelling wave is necessarily linear. In the present case, this relation is

$$\sigma(y) + \sigma_{vis}(y) = \frac{\sigma^+ - \sigma^-}{\varepsilon^+ - \varepsilon^-} \left(\varepsilon(y) - \varepsilon^+\right) + \sigma^+. \tag{41}$$

7 Concluding Remarks

The strain-stress relation for soil in compression is nonlinear with increasing stiffness. This nonlinearity strongly influences the propagation of compression waves: a smooth wave front steepens and turns into a shock front. If the constitutive behaviour is rate-independent, the solution becomes discontinuous. The distance covered by a smooth front before it becomes discontinuous (the critical distance) is determined by (13). The critical distance depends on the stress rate applied at the boundary and on the curvature of the strain-stress function. For given medium and boundary loading, the critical distances for the cylindrical and spherical waves are related to the critical distance for the plane wave by (14), (15). The relations are independent of the constitutive function. The propagation of waves with discontinuities is governed, besides the differential equations for the smooth part of the wave, by the jump conditions (24) on the shock fronts. The same jump conditions hold true for steady viscous shocks.

The soil deformation on the shock is essentially different from the deformation on the smooth part of the wave. The strain rate on the shock is not only high but also highly inhomogeneous in space, Fig. 11. Using the same strain-stress curve for the shock and for the smooth part of the wave should be regarded as approximation.

References

1. Courant, R., Friedrichs, K.O.: Supersonic Flow and Shock Waves. Springer, New York (1976)
2. LeVeque, R.J.: Numerical Methods for Conservation Laws. Birkhäuser, Basel (1992)
3. Chorin, A.J., Marsden, J.E.: A Mathematical Introduction to Fluid Mechanics. Springer, New York (1993)
4. Smoller, J.: Shock Waves and Reaction-Diffusion Equations. Springer, New York (1994)
5. Toro, E.F.: Riemann Solvers and Numerical Methods for Fluid Dynamics: A Practical Introduction. Springer, Berlin (2009)
6. Courant, R., Hilbert, D.: Methods of Mathematical Physics: Partial Differential Equations, vol. 2. Interscience Publishers, New York (1965)
7. Osinov, V.A.: On the formation of discontinuities of wave fronts in a saturated granular body. Continuum Mech. Thermodyn. **10**(5), 253–268 (1998)
8. Osinov, V.A.: Critical distances for the formation of strong discontinuities in fluid-saturated solids. Arch. Appl. Mech. **80**, 581–592 (2010)
9. Osinov, V.A.: Transition from acceleration waves to strong discontinuities in fluid-saturated solids: drained versus undrained behaviour. Acta Mech. **211**, 181–193 (2010)
10. Lambe, T.W., Whitman, R.V.: Soil Mechanics. Wiley, New York (1969)
11. Richart, F.E., Hall, J.R., Woods, R.D.: Vibrations of Soils and Foundations. Prentice-Hall, Englewood Cliffs (1970)
12. Santamarina, J.C., Klein, K.A., Fam, M.A.: Soils and Waves. Wiley, Chichester (2001)
13. Osinov, V.A.: Blast-induced waves in soil around a tunnel. Arch. Appl. Mech. **81**, 543–559 (2011)
14. Osinov, V.A., Chrisopoulos, S., Triantafyllidis, Th.: Numerical analysis of the tunnel-soil interaction caused by an explosion in the tunnel. Soil Dyn. Earthq. Eng. **122**, 318–326 (2019)
15. Osinov, V.A., Chrisopoulos, S.: Two neighbouring tunnels in saturated soil under blast loading. This volume
16. Lax, P.D.: Hyperbolic Systems of Conservation Laws and the Mathematical Theory of Shock Waves. SIAM, Philadelphia (1973)
17. Lax, P.D.: Hyperbolic Partial Differential Equations. Courant Institute of Mathematical Sciences, New York (2006)
18. Trangenstein, J.A.: Numerical Solution of Hyperbolic Partial Differential Equations. Cambridge University Press, Cambridge (2007)

Renaturation of Open Cast Mining Areas

Geotechnical Aspects in Designing the Permanent Slopes of Future Lakes in Opencast Mines in the Rhenish Lignite Mining Area

Dieter Dahmen(✉)

RWE Power AG, Cologne, Germany
dieter.dahmen@rwe.com

Abstract. The use of former opencast mines as lakes has a long tradition in the Rhenish lignite mining area, as does agricultural and forest rehabilitation. There are 57 lakes, that have been created from the final mine voids of a multitude of small-scale opencast mines, varying in size between 0.2 ha and 100 ha. Added to this are lakes that were planned specifically for reasons of landscape preservation in the context of a recultivation scheme (known as landscape lakes). All lakes are used today for a variety of different purposes, i.e. nature conservation, recreation and aquatic sports.

With the operation concentrated on a small number of large-scale opencast mines on the one hand and complete backfilling of opencast mines in the central part of the mining area with boxcut-masses from newer opencast mines (mainly Hambach) on the other hand, a concept was developed for the entire mining area which provides for the creation of three large opencast mine lakes in the active opencast mines of Inden, Garzweiler and Hambach. The planned lake sizes vary between 12 km^2 and 40 km^2 at depths of between 180 m and 330 m.

In the creation scheme for the opencast mine lake slopes with a general inclination of 1:5 (below the future wave braking zone) by using main mine equipment are planned. The area where the final water levels will later connected with the groundwater level is to be turned into a wave zone with a minimum width of 100 m and a compensation inclination of between 1:20 and 1:25. Since the lakes do not have a reservoir function and are not situated higher than the surrounding terrain either, the requirements of DIN 19700 ('Dam plants') are not relevant. Instead, geotechnical dimensioning of the slopes is carried out in compliance with the Guideline for Stability Analyses (RfS) issued by the Arnsberg regional government as NRW's competent mining authority.

The filling scheme for the planned opencast mine lakes provides for one-time filling with external water from local streams and rivers. It is flanked by a purposeful continuation of opencast mine dewatering, preventing seepage pressures from the slopes in the direction to the lake.

The stability-related boundary conditions were already taken into account when the slope design was decided on, with seismic impacts also being considered in accordance with the requirements of the RfS. This was done by means

Paper for presentation at the 2019 Colloquium on Geotechnical Engineering. "Recent Contributions of Soil Mechanics and Geotechnics in theory and practice – celebratory colloquium held on the occasion of the 65th birthday of Prof. Dr.-Ing. habil. Theodoros Triantafyllidis".

T. Triantafyllidis (Ed.): *Recent Developments of Soil Mechanics and Geotechnics in Theory and Practice*, LNACM 91, pp. 221–236, 2020.
https://doi.org/10.1007/978-3-030-28516-6_12

of a quasi-static approach according to GOLDSCHEIDER that has been further developed and takes account of the acceleration effects of seismic impacts on the mass of the pore water as well.

A current field of investigation in line with an RfS requirement is the verification of the safety of lake slopes constructed by backfilling against possible liquefaction effects resulting from seismic impacts. A corresponding verification procedure is at present being developed in conjunction with Prof. Triantafyllidis at the Institute of Soil and Rock Mechanics (IBF) of the Karlsruhe Institute of Technology (KIT). It consists of a combination of field investigations, technical-centre and laboratory tests as well as numerical calculations. By comparing the dependency, derived from these tests, of the cyclic stress ratio with shear wave velocity and cone-penetration-test tip pressure with cyclic impacts determined by means of numerical calculations, we will in future be able to establish safety against liquefaction effects with sufficient reliability.

1 Recultivation in the Rhenish Mining Area

For more than 100 years, lignite has been extracted in the Rhenish mining area (Fig. 1) and used as a source of energy and as a chemical substance. It has played a pivotal role in the industrial and economic development of the region. Today, with the Garzweiler, Hambach and Inden opencast mines, there are three highly productive operations that supply the lignite-fired power stations and refining plants with between 80 and 90 million tonnes of raw lignite per year. This is converted into about 65 terawatt-hours of electricity and roughly 5 million tonnes of refined products that are sold on the market. According to the recommendations of the Commission on Growth, Structural Change and Employment (KWSB) appointed by the German federal government, lignite plant capacity is planned to be reduced successively in the years ahead on the basis of statutory provisions, and generation of electricity from lignite is currently expected to run in Germany until 2038.

Lignite extraction in the Rhenish mining area has so far affected approximately 32,500 ha of land. With regards to the agricultural characteristics of the Lower Rhine Basin, more than 65% of the land temporarily claimed by mining operations is accounted for by farmland, 25% by forests. Settlement areas and industrial sites, traffic routes and water surfaces make up most of the remaining 10%. At present, the opencast mines in the Rhenish mining area, whose active operations affect an area of about 9,750 ha, use around 250 ha for mining purposes annually. Recultivation is roughly on the same scale.

The necessity of constructing opencast mine lakes arises from a mass deficiency due to coal extraction. In the early days of the Rhenish lignite industry, which was marked mainly by a large number of small opencast mines, the emphasis was largely on forestry and water-body recultivation, as evidenced, for instance, by the multitude of small lakes in the southern part of the mining area.

On account of the concentration on a small number of large-scale opencast mines, the focus in recent decades – in accordance with the land used – has been on agricultural rehabilitation. The techniques for the recultivation of high-quality arable land

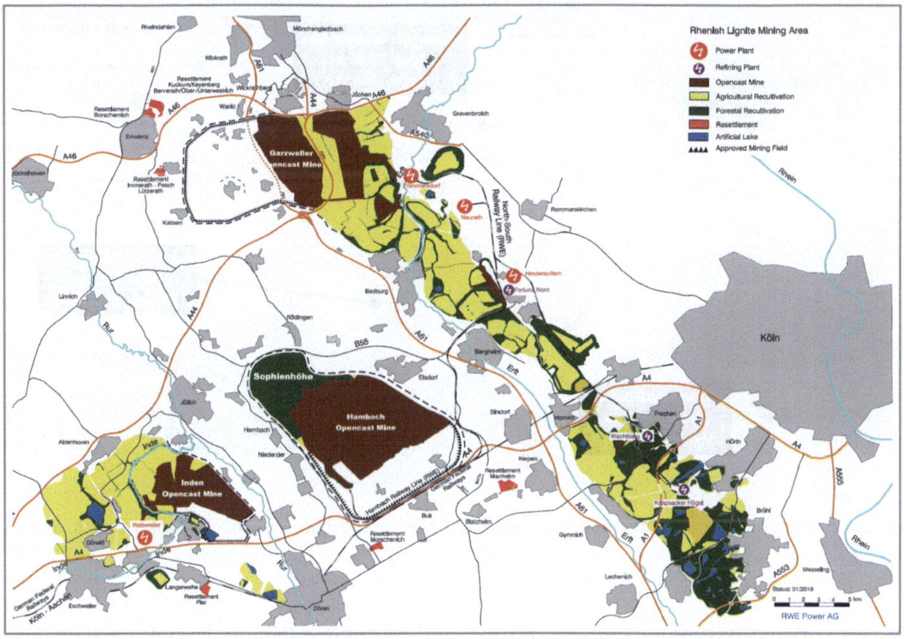

Fig. 1. Recultivation in the Rhenish lignite mining area [1]

were steadily further developed in the Rhenish mining area and today guarantee farmland of very high quality that can be utilised without restriction.

The focus on agricultural rehabilitation also includes an interchange scheme for recultivating material between the opencast mines with the aim of producing fertile arable land on the most comprehensive basis possible. Owing to this clear setting of priorities, approximately 55% of a total of 22,800 ha or so of recultivated surface (cf. coloured illustration in Fig. 1) was rehabilitated for agricultural purposes.

The interchange of mass between the various mines that this requires have various consequences for the mining area. On the one hand, it has been possible for numerous opencast mines to be completely backfilled, so that the creation of opencast mine lakes was not necessary there. On the other, the mass deficiencies in the current mines associated with long-time coal extraction add up. In line with the objectives defined in the existing lignite mining plans, large-scale residual lakes will therefore be created in the Inden, Hambach and Garzweiler opencast mines after the end of operations, with the planned lake sizes varying between 11 km^2 and 40 km^2 at lake depths of between 180 m and 330 m (Fig. 2).

For reasons of ecology and local recreation, water surfaces of different sizes, besides agricultural and forest areas, were also created in the backfilled opencast mines. To distinguish them from opencast mine lakes, they are usually referred to as landscape lakes because they were not created due to mass deficiencies but for landscaping reasons.

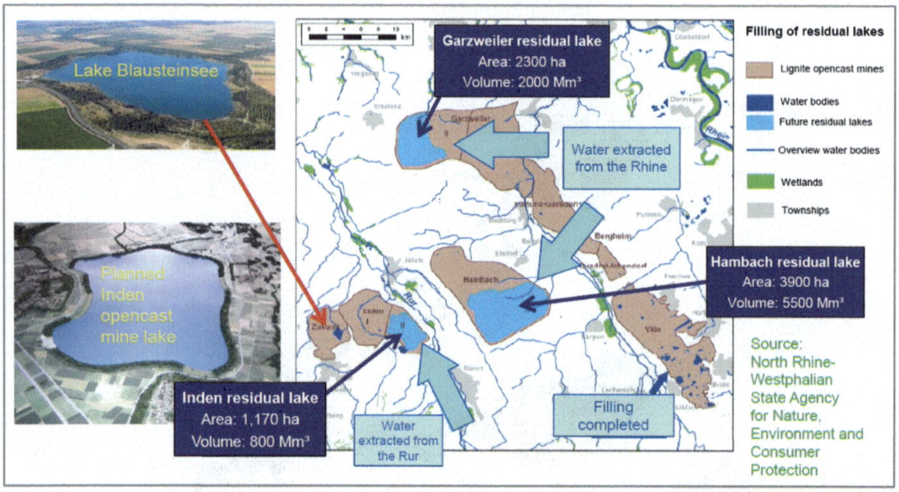

Fig. 2. Existing and planned opencast mine and landscape lakes in the Rhenish lignite mining area [2].

Both opencast mine lakes and landscape lakes today enhance the natural scenery of the Lower Rhine Basin formerly marked by forestry and agriculture, offering a wealth of benefits for both ecology and recreational use, and, hence, for the region's economic life as a whole.

This similarly applies to permanent high dumps (located above the terrain surface) in the Rhenish mining area, which were necessary in the case of some opencast mines because not all of the overburden material produced in the start-up phase could be used to backfill depleted mines. These permanent dumps were integrated into the recultivated areas (in the form of so-called outside dumps or superelevated inside dumps) as permanent landscape features and enhance the formerly flat fertile plain with their topography. As a rule, the slopes of such dumps have been and will be recultivated for forestry purposes, while agricultural recultivation is most common on the plateaus.

On account of the slope areas, which remain as permanent landscape features, and the resulting high safety interests of follow-up users, opencast mine lakes, just like surface dumps, call for an especially careful approach in their planning, approval and creation.

This ensures that all recultivated areas in the Rhenish lignite mining area can be utilised without restriction. The fact that the dump areas are used for a wide variety of purposes, e.g. for traffic routes, residential buildings and trade and industry, is a case in point.

2 Opencast Mine Lakes in the Rhenish Mining Area

At present, a total of 57 opencast mine and landscape lakes exist in the area of former opencast mines [3]. The various water bodies vary greatly in size. A total of 17 of them occupy a surface of over 20 ha. With a water surface of about 100 ha, lake Blausteinsee, which falls into the category of landscape lake, is the largest water body of this type to date in the Rhenish mining area. All existing lakes have already reached their target water level. The water quality of the various lakes is good enough to enable a broad use spectrum which has, in fact, been implemented. Reflecting the various local objectives, it extends from ecological and landscape emphases to leisure and bathing uses [3].

In stability terms, the examined lakes show no irregularities. Only in two lakes did minor slope deformations occur in the past. The lakes concerned include the Neurather See, 13 ha in size, on whose shores a deformation surface measuring some 900 m^2 occurred in the wake of what is known as the 'Roermond earthquake' in 1992. No remediation was necessary. Likewise, there was local deformation during the filling phase of the lake Zülpicher See, encompassing 85 ha, which was dealt with by placing suitable drainage materials. At no point in time were material goods or people in danger from these two events [3].

Overall, the shores of the examined lakes were thus shown to be as stable as planned. In order to find out how the lake basins profiled in dumping operations behave below the water surface, two exemplary echo-sounding measurements were carried out in the western part of the mining area. Lakes Lucherberger See and Blausteinsee were selected for the measurements. With an area measuring some 100 ha and a maximum depth of 41 m, Blausteinsee is the biggest and one of the youngest landscape lakes in the Rhenish mining area. A true opencast mine lake, Lucherberger See, by contrast – being some 70 years old, 62 ha in size and having a maximum depth of 25 m – is among the oldest water surfaces in the Rhenish mining area created for lignite-extraction reasons. In addition, its case is special in that, presumably as of 2021, it will be emptied for the Inden lignite opencast mine and partly consumed by mining operations. A comparison of the echo-sounding measurements with the underlying planning and the mapping system after completion of the lake basins shows no abnormalities. According to specialist interpretation of the measurement results, minor changes in the range of a few decimetres to the lake contour of Lucherberger See are mainly due to organic sedimentation at deep points. These phenomena are above all due to the lake's age and not unusual for standing waters [3].

3 Principles of Opencast Mine-Lake Creation

The planned creation of three large lakes in the Inden, Garzweiler and Hambach opencast mines requires lake slopes with a general inclination of 1:5 to be constructed by dumping overburden with spreaders in front of the opencast mines' rim slopes situated at the edge of the mining fields (Fig. 3). In theory, it would also be possible to dig lake slopes using bucket-wheel excavators, but this would substantially reduce the lignite quantity that can be extracted from the approved mining fields.

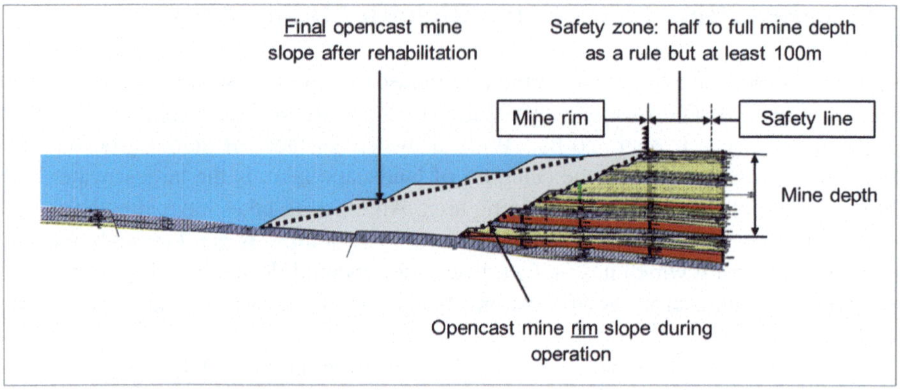

Fig. 3. Construction of a lake slope by backfilling in front of the mine's excavated rim slope [4].

In their final stage, the opencast mine lakes will be directly connected with the groundwater, respectively streams and rivers, allowing them to take part in groundwater flow processes. This requires a wave breaking zone with a width of at least 100 m and a compensation inclination of between 1:20 and 1:25 to be pre-profiled in the area of the lake's planned final water line. Apart from enabling waves to peter out, this zone ensures flows between the lake water and groundwater without erosion effects.

A key determinant in the stability of slopes during the filling process is the position of the water table in the lake relative to the water table in the surrounding strata or the overburden dump (Fig. 4). In the case of slow, uncontrolled rising of the groundwater, the lake table would be below the groundwater table, resulting in adverse flow forces from the lake slopes into the lake, which would necessitate very gentle slopes with suitable drainage fills.

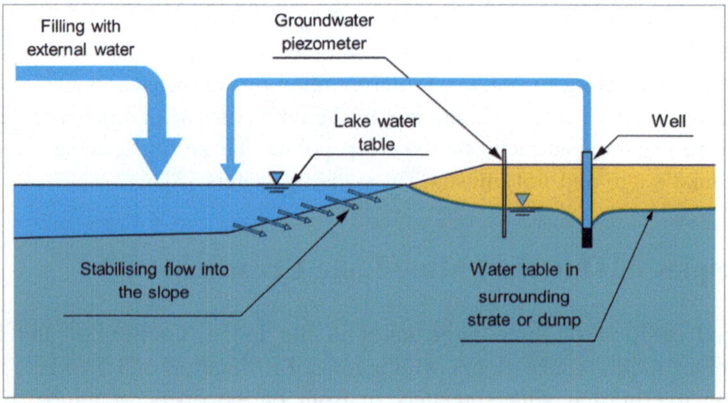

Fig. 4. Control of groundwater during the filling phase [4].

To prevent such adverse flow forces and very gentle slopes on the one hand and allow for quick filling of the lakes on the other, the lake basins are to be filled with 'external water' supplied from the outside. In this way, the groundwater body in the area of the lake slopes is refilled in a controlled manner by filling the lake, with the continued operation of pumps ensuring that the groundwater table in the dump and the adjacent strata is always kept below the water table of the lake concerned. Flow forces are directed from the lake into the slope, which has a stabilising effect on the slopes.

The external water for lake filling is supplied from neighbouring or remote flowing waters by pipes. For the Inden opencast mine, water will be drawn from the Rur river and for the Garzweiler and Hambach opencast mines from the Rhine.

Planning calls for the lakes to be filled once until the target water levels and final groundwater levels have been reached. Owing to the direct groundwater connection, fluctuations in the lake table after completion of the filling process will only occur as part of the natural, seasonal changes and be limited to the area within the planned wave breaking zones.

Using the lakes for water or energy storage would require technical facilities (e.g. slope revetments) that would be at odds with the aimed-at near-natural design; such uses are thus explicitly ruled out.

Since the lakes, which in their final stage will be directly integrated into the groundwater regime, do not have a reservoir function and are not situated higher than the surrounding terrain either, the requirements of DIN 19700 ('Dam plants') are not relevant. Eurocode 7 (EC 7) 'Geotechnical design' does not apply to lignite opencast mine slopes either because DIN 1054 ('Subsoil - Verification of the safety of earthworks and foundations'), being part of the national annex and, as such, supplementary regulation to Eurocode 7, explicitly rules them out when defining its scope:

> "Lignite opencast mines are not covered by the scope of this standard since other safety conditions exist here owing to different geological and geotechnical explorations, preliminary studies and monitoring."

4 Guideline for Stability Investigations

Thus, the slopes are dimensioned in compliance with the Guideline for Stability Investigations (RfS) issued by the Arnsberg regional government as NRW's competent mining authority [5].

This guideline has set the framework for proving and examining the stability of rim and final slopes in Rhenish lignite mining, and for slope monitoring, for more than three decades.

The minimum safety levels to be verified with regard to the stability of opencast mine slopes are specified as global safety factors in the RfS. They correspond to the partial safety concept of EC 7 (derived from global safety). For mine-rim slopes with properties to be protected and for permanent slopes, the RfS minimum safety levels exceed the requirements of EC 7 [5].

On account of the three parties (mining company, Geological Survey of North Rine Westphalia - NRW - and mining authority) involved in the review and approval process, the Guideline provides for a so-called 'six-eyes principle', which ensures a safety level in accordance to the importance of such slopes. In this way, RWE Power as mining company provides evidence of stability in accordance with RfS requirements for the slopes to be created with an adequate time run-up, and the documents are examined by the mining authority and – by order of the authority – by the NRW Geological Survey.

The requirements for stability analyses and slope monitoring set out in the Guideline are regularly reviewed with regard to technical and scientific progress in geomechanics and adapted accordingly. Since 2010, the working group 'Geomechanics in Rhenish lignite mining' has been responsible for this task. It comprises specialists from the mining authority, the NRW Geological Survey and RWE Power as mining company, and there is a regular exchange of views with scientific bodies and experts from Germany's other lignite-mining areas. It is also in charge of reviewing and revising the RfS in terms of up-to-dateness and consideration of new scientific findings [5]. Collaborating with Prof. Triantafyllidis from the KIT, the working group laid down the boundary conditions for dimensioning permanent opencast-mine slopes with respect to earthquakes, which formed the basis for the 2013 revised version of the RfS including a supplement regarding earthquakes.

5 Seismic Impacts in Accordance with RfS

As the Rhenish lignite mining area is situated in a zone belonging to Germany's highest earthquake category, the RfS, as of its 2003 version already, has expressly required seismic impacts on permanent slopes to be taken into account. This demand was specified in the first earthquake-related addendum to the Guideline in the version dated 8 August 2013, which is based on an expert opinion by Prof. Triantafyllidis [6] and a recommendation of the NRW Geological Survey.

In respect of the seismic impacts to be considered it states that opencast mine lakes, in their final, fully filled stage, must be dimensioned by analogy with the safety level of the retaining structure of a large dam plant above ground according to DIN 19700 or technical leaflet 58 'Consideration of seismic loading in accordance with DIN 19700' (State Environmental Agency) [5].

This requirement must be met although permanent slopes in the Rhenish lignite area situated lower than the surrounding terrain can be expected to pose a lower secondary hazard in the event of a failure than retaining dams. Since the requirements of DIN 19700 in principle apply only to the actual retaining structure but not to the (man-made) embankments or (natural) hillsides of dam lakes that are situated below the surrounding terrain, stronger earthquakes are assumed for slope systems of opencast mine lakes than for embankments or hillsides of dam lakes that are subject to Eurocode 8 (EC 8) 'Design of structures for earthquake resistance'.

Thus, all slope systems around opencast mine lakes need to satisfy safety standards similar to those applying only to retaining structures (dams) above ground when dimensioning dam plants.

For the limited duration of the filling phase for a opencast mine lake, which is more or less identical to the life of a building structure, a recurrance period of 475 or 500 years has been assumed as safety level for the slopes of opencast mine lakes in line with Eurocode 8. Opencast mine lake slopes are thus dimensioned with a view to a possible future use for building purposes, to which the requirements of DIN EN 1998 (Eurocode 8) apply as well [5].

6 Slope Stability Analyses for Mine Lakes

For stability analyses of mine lake slopes and permanent surface-dump slopes, a quasi-static approach developed by GOLDSCHEIDER is used that takes account of the impact of seismic acceleration not only on the mass of unconsolidated rock but also on pore water and external water [7]. The assumptions upon which BISHOP's method and the method of composite failure mechanisms (rigid body method) are based on the assumption that they do not change when seismic loading is taken into consideration. Thus, both the discretisation assumptions (stress distribution along the shear zone and slice width) and the static assumptions (horizontal earth pressures) are retained in BISHOP's method. In this way, the approach according to GOLDSCHEIDER, which includes earthquakes and external water, takes account of the additional acceleration acting on all inertia forces (granular skeleton, pore water and external water) and also the impact of short-term excess pore-water pressures caused by a change in these inertia forces [7], ensuring static correctness.

Compared with the assumption of undrained shear strength c_u for cohesive soils common for such issues, this approach is an improvement as it allows the change in water pressures to be realistically reproduced and taken into account [7].

Horizontal and vertical acceleration as a result of earthquakes, which, for DIN 19700, must be requested for each site in accordance with the RfS via the web application of the GFZ Research Centre for Geosciences in Potsdam, are used as input parameters for stability calculations [8]. Seismic forces are assumed additionally and as a function of weight forces by applying relative seismic acceleration (relative to gravity acceleration). When assuming total soil unit weights and pore water pressures, this procedure can be implemented easily and intuitively. As the vertical force can have either a favourable or an unfavourable effect depending on the position of the slice or rigid body, both directions of action (upwards and downwards) are considered; the horizontal force, on the other hand, is only assumed to act unfavourably, i.e. in the direction of the slope gradient. For stability calculations, the decisive factor is the division of the additional total soil unit weight caused by earthquakes into effective stresses and pore water pressures because only the additional effective stress influences frictional resistance. The calculation method is described in detail in [7].

The use of the quasi-static calculation method requires the granular skeleton to remain stable (conservative skeleton) in the event of an earthquake, ruling out liquefaction [6]. Consequently, the first addendum to the RfS points out that permanent slopes are to be designed and constructed in a way that soil liquefaction is not to be a concern.

7 Verification Procedure for Establishing Safety Against Soil Liquefaction

The issue of liquefaction as a result of earthquakes is dealt with in an expert opinion by Prof. Triantafyllidis prepared in 2015. It concludes that although there is no risk of such liquefaction effects occurring in the opencast mine lakes located in the Rhenish mining area, the exclusion of such effects could not be verified with certainty at the time the expert opinion was produced. For this reason, the expert opinion recommends that a verification procedure adapted to the conditions prevailing in the Rhenish lignite mining area and to the planned lake slopes should be developed. In 2016, the working group 'Geomechanics in Rhenish lignite mining' decided that Prof. Triantafyllidis, as authorised expert appointed by the Arnsberg regional government, was to implement this recommendation at the KIT.

The current state of development of the verification procedure is described in the following.

The procedure consists of a combination of field investigations, technical-centre and laboratory tests as well as numerical calculations. It is described in detail in [9]. Owing to its comprehensive approach, it takes both local and global material properties into account and can therefore be used for future verification in operational practice, even for opencast mine-lake-scale slopes.

7.1 Opencast-Mine Field Investigations

Using different methods, field samples were taken from the dump areas of the future lake slopes, including undisturbed cylindrical soil samples and frozen soil samples that were subsequently subjected to both static and cyclical loads until failure in the laboratory.

Fig. 5. Dumping of test cones in the Inden opencast mine [10, 11].

Since the structure of the samples plays a crucial role, large-scale test cones (Fig. 5) that were thoroughly examined as to their properties were built in the opencast mine.

In addition, surface seismic tests, seismic cone-penetration tests, shear wave analyses and cross-hole measurements were carried out.

7.2 Laboratory Tests

In the course of developing the verification procedure, cyclical undrained stress- or displacement-controlled triaxial tests were conducted to determine the cyclic stress ratio. In addition, static triaxial tests were performed where shear wave velocities within the pressure cell were measured by means of bender elements.

Samples included both very high-quality undisturbed samples obtained by means of the freezing method and reconstituted samples produced from disturbed sample material. To ensure that the latter have similar properties to the frozen samples, comprehensive test series were used to check different sample preparation methods for suitability. In the end, the decision was made, to prepare the samples by free fall into a sample former. This procedure is similar to the dumping process in opencast mines and, at the same density, supplies cyclic stress ratios that are the same or slightly lower than those of the undisturbed samples. Alternatively in order to produce homogeneous samples the moist tamping method may be used.

Figure 6 shows a typical test result for a frozen sample. The tests were carried out for different relative densities, with the results being used to derive the dependence of the cyclic stress ratio on the number of cycles until failure (ε = 10%) for different relative densities.

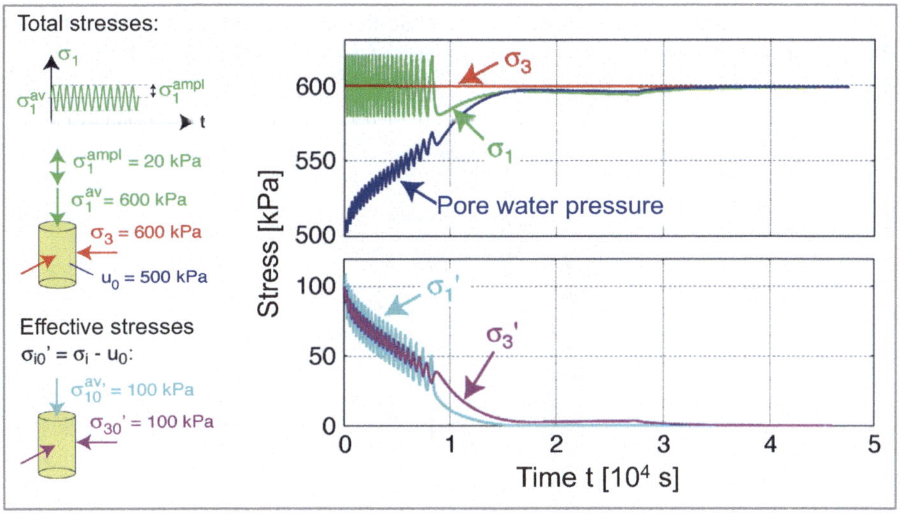

Fig. 6. Cyclic test on a frozen sample [2].

This data was used to derive the correlation between the cyclic stress ratio (CSR) and the relative density, see Fig. 7. The decisive number of cycles for the Rhenish mining area is N = 10. The cyclic resistance ratio (CRR) therefore is equivalent to the CSR-value for N = 10.

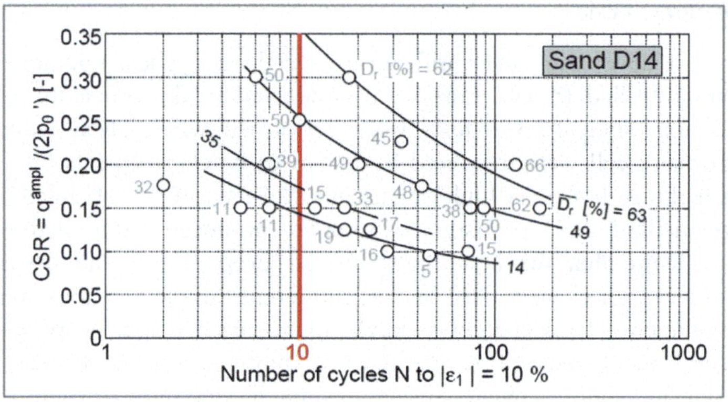

Fig. 7. Correlation between CSR and relative density [12].

7.3 Technical-Centre Tests

To correlate the results from the cone penetration tests (CPT) carried out in the opencast mine with the cyclic laboratory tests, cone penetration tests were conducted in a calibration chamber at the IBF under precisely defined boundary conditions (Fig. 8). Using the preparation method described (moist tamping) for homogeneous distribution of porosity, the materials were placed in the chamber at different relative densities and CPTs performed at varying pressures by means of an original-scale CPT tip.

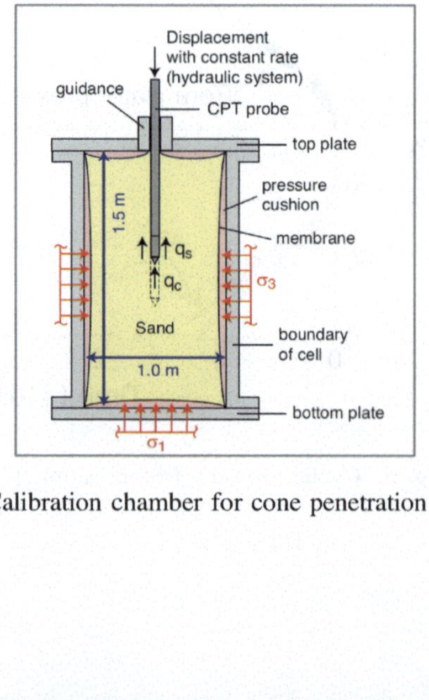

Fig. 8. Calibration chamber for cone penetration tests [12]

After the cone penetration tests, samples taken from the calibration chamber have been used to determine the CRR with the aid of undrained triaxial tests for verification purposes.

The following correlations were derived from the field investigations and laboratory and technical-centre tests:

- Cyclic laboratory tests: CSR as a function of relative density (for the preparation method found)
- Static laboratory tests: shear wave velocity as a function of relative density and pressure levels
- CPTs in calibration chamber: correlation between relative density and tip resistance

By combining accordingly, these correlations permit the following extended correlations to be derived for the various materials tested (see combined diagrams below):

- Dependence of CRR on tip resistance (Fig. 9)
- Dependence of CRR on shear wave velocity (Fig. 10)

These extended correlations have enabled the dump materials' CSR to be determined with the aid of commonly used field tests. Overall, the results appear quite plausible and are consistent with the data/diagrams found in the relevant literature.

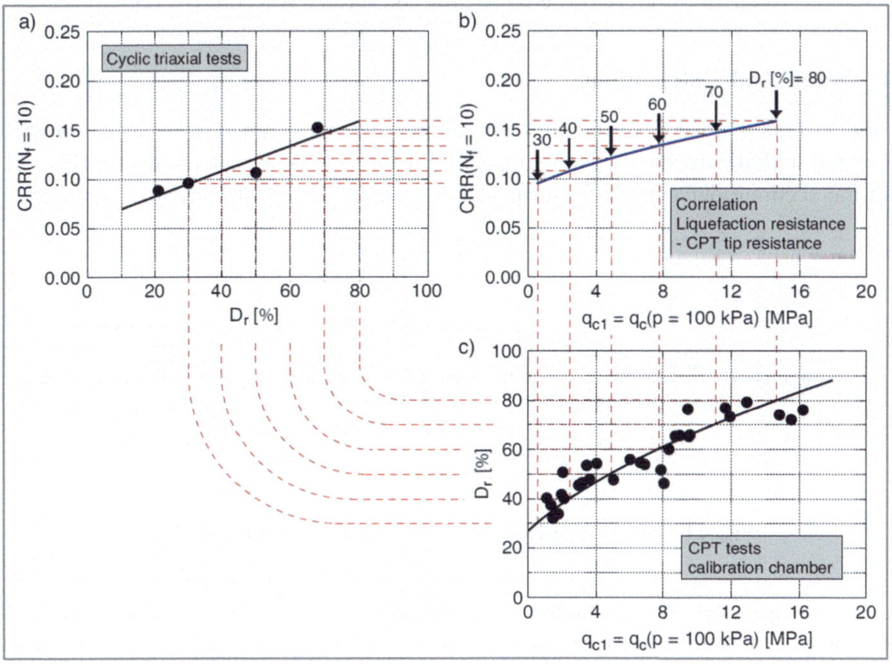

Fig. 9. Correlation between CRR and CPT tip resistance [12].

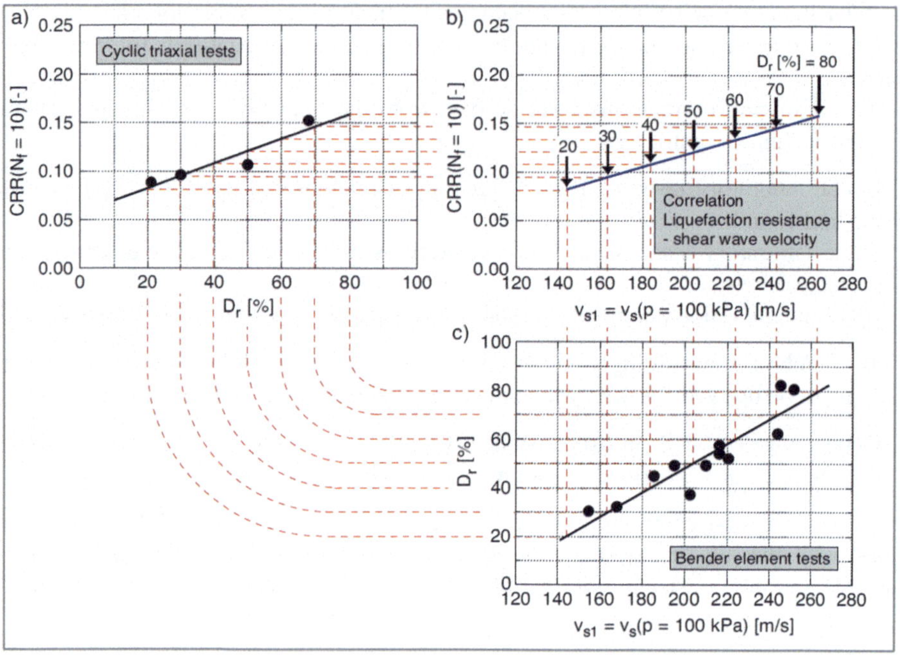

Fig. 10. Correlation between CRR and shear wave velocity [12].

7.4 Numerical Calculations

In the last step of the verification procedure, the ascertained resistances must be compared with the decisive seismic impact by means of numerical simulation (Fig. 11), with the accurate determination of the seismic impact being of particular significance.

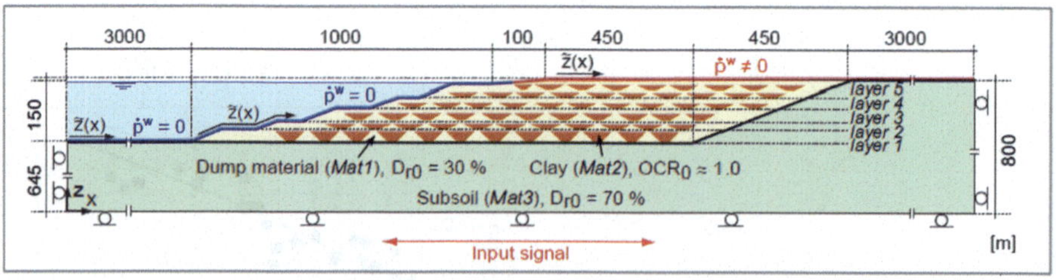

Fig. 11. Model of a lake slope for numerical simulation [13].

The comparison of the impacts determined by numerical calculations with the resistances that were previously ascertained from the laboratory tests permits safety against soil liquefaction to be evaluated in accordance with the state of knowledge in relevant literature.

For the shear wave velocities, a numerical calculation is unnecessary. The measuring results obtained by field trials can be directly compared with the liquefaction thresholds (e.g. using the procedure according to Stokoe [14] considered suitable pursuant to Eurocode 8), which allows safety against liquefaction to be assessed.

The correlations derived on the basis of the investigations will in future permit the CRR of the dumped soil/unconsolidated rock to be determined by means of field trials such as standard cone penetration tests or field measurements of shear wave velocities. The described verification procedure thus has the advantage that not only individual outcrops are utilised (CPTs) but also wide areas covered (shear wave analyses).

Next Steps and Acknowledgements. The three planned 'large' opencast mine lakes to be created in the Inden, Garzweiler and Hambach mining fields will in future, with their wide variety of possible uses, substantially enhance the fertile plain in the Rhenish mining area.

One essential prerequisite for the implementation and viability of such major projects is the appropriate assessment and handling of geotechnical boundary conditions. In this connection, our colleague Professor Dr.-Ing. habil. Dr. h. c. Theodoros Triantafyllidis deserves our thanks for a long-standing and fruitful collaboration and his valuable assistance in helping us to tackle the most challenging tasks in geotechnical terms on an opencast mine scale.

References

1. DEBRIV – Deutscher Braunkohlen-Industrieverein, Köln (2016)
2. Wichtmann, T.: Prognosen der Standsicherheit für Böschungen von Tagebau-Restseen unter Erdbebenbeanspruchung. Habilitationsvortrag am KIT, Karlsruhe (2016)
3. Petri, R., Stein, W., Dahmen, D.: Sustainable follow-up use of recultivated surfaces Evaluation of residual lakes and high dumps in the Rhenish lignite-mining area after the end of mining supervision. World Min. Surf. Undergr. **65**(2), 92–101 (2013)
4. Eyll-Vetter, M.: Significance of geotechnical boundary conditions in planning and designing residual lakes in the Rhenish lignite mining area illustrated by the example of the Inden opencast mine. World Min. Surf. Undergr. **67**(6), 371–378 (2015)
5. Mittmann, A., Petri, R., Buschhüter, K.: Examination of geotechnical aspects in the approval of residual lakes. World Min. **67**(6), 379–388 (2015)
6. Triantafyllidis, Th.: Gutachterliche Stellungnahme zu Standsicherheitsberechnungen mit Ansatz von "Erdbebenbeschleunigungen für Böschungen im Rheinischen Braunkohlenbergbau" Mai, 2012 mit Ergänzungen vom Juni 2013
7. Goldscheider, M., Dahmen, D., Karcher, C.: Berücksichtigung von Erdbeben bei Standsicherheitsberechnungen für tiefe Endböschungen unter Wasser. World Min. Surf. Undergr. **62**(5), 252–261 (2010)
8. GFZ: Interaktive Abfrage von Beschleunigungs-Antwortspektren für die Gefährdungsniveaus gemäß DIN19700 für beliebige Orte in Deutschland und von nationalen Karten der Erdbebengefährdung. http://dx.doi.org/10.5880/GFZ.2.6.2012.001
9. Triantafyllidis, Th.: Dynamic effects from earthquakes acting on residual lakes. World Min. Surf. Undergr. **67**(6), 389–397 (2015)
10. Wichtmann, T., Back, M.: Geomechanisches Verhalten von Böschungen geplanter Tagebauseen im Rheinischen Braunkohlerevier unter Erdbebenbeanspruchung, Veröffentlichung im Tagungsband der 3. Deutschen Bodenmechanik Tagung 2018 (in memorian Professor Tom Schanz)

11. Wichtmann, T., Steller, K., Triantafyllidis, Th., Back, M., Dahmen, D.: An experimental parametric study on the liquefaction resistance of sands in spreader dumps of lignite opencast mines. Soil Dyn. Earthq. Eng. **122**, 290–309 (2019)
12. Wichtmann, T., Kimmig, I., Steller, K., Triantafyllidis, Th., Back, M., Dahmen, D.: Correlations of the liquefaction resistance of sands in spreader dumps of lignite opencast mines with CPT tip resistance and shear wave velocity. Soil Dyn. Earthq. Eng. **124**, 184–196 (2019)
13. Machaček, J., Triantafyllidis, Th.: Numerical investigation of a partially saturated opencast mine subjected to earthquake loading. In: Veröffentlichung im Tagungsband der 3. Deutschen Bodenmechanik Tagung 2018 (in memorian Professor Tom Schanz). Springer (2018)
14. Stokoe, K.H., Roesset, J.M., Bierschwale, J.G., Aouad, M.: Liquefaction potential of sands from shear wave velocity. In: Proceedings Ninth World Conference on Earthquake Engineering, vol. 3, pp. 213–218 (1988)

Coupled Numerical Simulations for Seismic Hazard Assessment of Large Opencast Slopes

Jan Machaček(✉), Carlos Eduardo Grandas Tavera, and Theodoros Triantafyllidis

Institute of Soil Mechanics and Rock Mechanics (IBF), Karlsruhe Institute of Technology (KIT), Karlsruhe, Germany
jan.machacek@kit.edu

Abstract. Earthquake stability assessments of large opencast mine slopes (dimensions exceeding several hundred meters) are complex and non-linear problems, often addressed using pseudo-static approaches that neglect material-induced failures and the role of pore-fluids. In this study, a numerical approach is used to understand the dynamic response of saturated and unsaturated soils. Since the requirements for such simulations are often not yet met by commercial software packages, user-defined finite elements and user-defined material models have been implemented in Abaqus/Standard. To account for the large depth of the finite element model, a scaling procedure of the system of equations is proposed to purge the influence of initial stress. The constitutive model parameters are calibrated based on laboratory tests and by back calculation of downhole measurements. Large-scale fully coupled finite element simulations are performed to study the response of a flooded opencast mine under earthquake loading. The present work illustrates the importance of the pore-fluids treated as independent phases in the context of seismic analysis of slopes. The simulations show strong wave diffraction effects for inhomogeneous dump structures, resulting in smaller displacements in near-surface areas of the slope. Furthermore, it was found that large areas of the dump show a (significant) temporary decrease of effective mean pressure.

1 Introduction

When opencast mines exploitation ceases, the residual holes are often refilled with soil or water, in the latter case creating so-called residual lakes. The enclosing slopes of such lakes are made of uncompacted material obtained from the excavation pit. Due to the deposition method, the inner structure strongly differs from the one of cut-off or natural slopes. Large scale failures of such slopes, some of them leading to fatalities, have been documented and studied over the last decades. There is a strong debate about the circumstances under which seismic analyses are necessary for ensuring the stability of slopes in opencast mines [1–3]. However, in the long term stability assessment, major seismic events must be consid-

T. Triantafyllidis (Ed.): *Recent Developments of Soil Mechanics and Geotechnics in Theory and Practice*, LNACM 91, pp. 237–264, 2020.
https://doi.org/10.1007/978-3-030-28516-6_13

ered as the period of exploitation is exceeded. Studying the behavior of opencast mine slopes under earthquake loading becomes even more crucial as the depth of opencast mines increases.

The most widespread methods for accounting the seismic impact on the stability of opencast edges and opencast slopes are the pseudo static approach (PSA) and the Newmark sliding block method [4,5]. In the PSA, the dynamic effects of an earthquake are replaced by a static force acting in a potential sliding mass in a limit-equilibrium analysis [6]. Different earthquake accelerations are applied until the factor of safety is reduced to 1.0, which leads to the so-called yield acceleration. No information on potential slope deformations are provided within the PSA. However, these analysis only represent one instant in time (in general the time of max. strain amplitude). Newmark's method treats the potential slip masses of a landslide as a rigid-plastic body, i.e. the mass does not deform internally, but deforms plastically along a discrete shear surface when a critical acceleration is exceeded. No permanent displacements at accelerations below the critical or yield level are considered. However, opposite to the PSA, the analysis calculates the cumulative permanent displacement of the block as it is subjected to the effects of an earthquake acceleration-time history. In both methods, the reduction of shear strength during an earthquake [7] and the influence of relative small but longstanding oscillations on the slopes stability is ignored.

When water fills the pores of the deposited soils during the see filling, a second group of failure mechanisms becomes possible. Depending on the drainage conditions, cyclic shearing of the grain skeleton induced by earthquakes may lead to a reduction of effective stress and, consequently, to a decrease of shearing resistance of the soil. In case of a total reduction of effective stress, a disintegration of grain skeleton and a separation of soil phases may occur. Additionally, such slopes may experience increasing displacements due to cyclic mobility. Contrarily to the system induced failure covered by the PSA and Newmark's method, material induced failure is a gradual process. To study the stability of the opencast mines being subject to earthquake loading, dynamic and fully coupled pore fluid flow and stress analyses are necessary. In general, they are based on the finite-element method. However, the reliability of these calculations depends on the accuracy of the physical model upon which the finite element (FE) model is based on and the accuracy of the constitutive laws describing the mechanical (and hydraulic) material behavior. Furthermore, the numerical integration of such FE models requires fine spatial discretization and small time increments, to account for the complexity of the internal dump geometry (layers, structure) and material composition and to be able to accurately capture the dynamic behavior. For saturated and unsaturated soils, a variety of numerical formulations for the finite element method has been presented [8,9], from which the *u-p* formulation has been adopted in numerous commercial finite element codes such as e.g. *Abaqus*[1] [10] and *Plaxis* [11]. However, these FE codes are only capable of treating the quasi-static case and consider only a constant value of pore-gas pressure in unsaturated conditions. To overcome these restrictions, user-defined

[1] The versions used in this paper is *Abaqus/Standard* 6.14 update 5.

elements based on the *Theory of Porous Media* as well as user-defined material models have been implemented in *Abaqus* enabling for static, quasi-static and dynamic analysis of unsaturated soils. In relation to to slopes subjected to earthquake loading, this theory has been successfully applied to back-calculations of centrifuge tests (e.g. [12–15]) and for seismic analysis of dams (e.g. [16]). To the authors best knowledge, no coupled dynamic simulations of large-scale models (dimensions exceeding several hundred meters) of residual lakes (flooded open-cast mines) have been reported in the literature.

This paper presents first the physical model and the governing equations upon which the user-defined finite elements are based. Then, numerical instabilities during large-scale geotechnical FE calculations due to the influence of initial geostatic stress are addressed proposing a special scaling procedure of the system of equations to increase the accuracy and stability of the numerical solution. The calibration of the constitutive model parameters for the deposited excavation soil based on laboratory experiments is described. For the subsoil beneath the dump, which is composed of up to 84 layers, the related parameter are calibrated by means of back calculation of downhole measurements. Then, the paper describes the performed FE calculation, which not only covers the excitation due to the earthquake, but also includes the deposition of the excavation material and post-seismic consolidation. Finally, a thorough discussion of the results is given and some key findings are highlighted.

2 Governing Equations and Finite-Element Implementation

In order to derive the governing equations of the multiphase porous medium, the balance of linear momentum for the whole mixture together with the balance of mass for the two fluid phases are used. In addition, the constitutive equation for the soil skeleton, for the pore-fluids and the constitutive relationships between the degree of saturation and the suction as well as the equations relating the relative velocities to pressure gradients via the generalized Darcy law are also needed.

2.1 Theory of Porous Media

The unsaturated porous medium behaves as a multiphase system where wetting (pore-water w) and non-wetting (pore-air a) porous fluids simultaneously exist in the pores of the solid skeleton (s). To derive the governing equations describing the multiphase system, the *Theory of Porous Media* [17–22] (*TPM*) serves as a starting point. A known shortcoming of the three-phase formulation is that the contribution of the suction to the effective stress becomes infinity for increasing suction. To overcome this issue, we follow the approach of [23] and extend three-phase formulation by the residual water phase (wr). The residual water is considered to be attached to the particles or trapped in micro-pores (e.g. [24, 25]), forming a discontinuous phase. Consequently, pressure dependent flow models are not valid for the residual water. Neglecting evaporation and diffusion processes, the residual water has the same intrinsic velocity as the solid phase. For the sake of simplicity air-water interfaces are ignored and isothermal conditions are assumed. Furthermore, no phase transformations and mass transfers between the pore-fluids are accounted for.

Based on the fundamental concepts of the *TPM*, all constituents are assumed to be in a state of ideal disarrangement and thus statistically distributed over the control space. The local composition of the mixture is described by the volume fractions $\varphi^\alpha = \mathrm{d}V^\alpha/\mathrm{d}V$ relating the (intrinsic) volume $\mathrm{d}V^\alpha$ occupied by the phase $\alpha = \{s, w, a, wr\}$ to the volume of the overall mixture $\mathrm{d}V$. The volume of the mixture is given by the sum of all partial volumes $\mathrm{d}V = \sum_\alpha \mathrm{d}V^\alpha$. Within the *TPM*, an unsaturated soil is treated as a saturated three-phase material with the partial volumes of the pore-fluids related to the pore-space. Therefore the total porosity is defined as $n = 1 - \varphi^s$. The effective porosity is $n^e = 1 - \varphi^s - \varphi^{wr}$. The effective degrees of saturation for the free pore-fluids $\beta = \{w, a\}$ are defined as $S_e^\beta = \varphi^\beta / n^e$ with the corresponding constraints $\sum_\beta S_e^\beta = 1$ and $\sum_\beta \mathrm{d}S_e^\beta/\mathrm{d}t = 0$. The degree of saturation is $S = (\varphi^w + \varphi^{wr})/n$ and the residual degree of saturation $S^{wr} = \varphi^{wr}/n$, respectively. The density of the mixture is

$$\rho^{tot} = \sum_\alpha \rho^\alpha = \rho^s + \rho^{wr} + \rho^w + \rho^a, \tag{1}$$

with the partial densities of the solid $\rho^s = (1-n)\bar{\rho}^s$, residual water $\rho^{wr} = \varphi^{wr}\bar{\rho}^{wr}$, free water $\rho^w = n^e S_e^w \bar{\rho}^w$ and pore-air $\rho^a = n^e(1 - S_e^w)\bar{\rho}^a$ respectively. $\bar{\rho}^\alpha$ is the intrinsic density of each phase α.

The *TPM* assumes a superimposed continuum in which any spatial point in the current configuration x_i, is simultaneously occupied by material points from different phases $x_i = x_i^\alpha(\boldsymbol{X}^\alpha, t)$. Each phase possesses an independent and unique function of movement x_i^α, thus, their original (reference) position $\boldsymbol{X}^\alpha$ may be different. Using a Lagrangian description, the intrinsic velocity $v_i^\alpha = \dot{u}_i^\alpha$ and the intrinsic acceleration $a_i^\alpha = \ddot{u}_i^\alpha$ of phase α in the direction i read:

$$v_i^\alpha = \dot{u}_i^\alpha = \frac{\mathrm{d}x_i^\alpha(\boldsymbol{X}^\alpha, t)}{\mathrm{d}t}; \quad a_i^\alpha = \ddot{u}_i^\alpha = \frac{\mathrm{d}^2 x_i^\alpha(\boldsymbol{X}^\alpha, t)}{\mathrm{d}t^2}. \tag{2}$$

For the sake of convenience, the solid displacement $\bar{u}_i^s = \bar{x}_i^s - \bar{X}_i^s$ is chosen as the primary kinematic variable. The pore-fluids β are favorably expressed in a modified Eulerian setting relative to the solid motion. The velocities for both flowing pore-fluids $\beta = \{w, a\}$ can be written as follows:

$$\dot{u}_i^\beta = \dot{u}_i^s + \bar{w}_i{}^\beta = \dot{u}_i^s + \frac{w_i^\beta}{\varphi^\beta}. \tag{3}$$

Therein $\bar{w}_i^\beta$ is the intrinsic and w_i^β is the *Darcy* velocity of the pore-fluid β. In other words, the Darcy velocity represents the relative volumetric rate of discharge per unit area of the mixture.

2.2 Balance Equations and Effective Stress Principle

Effective Stress Principle

The stress controlling the deformation and the soil strength for unsaturated porous media can be expressed through the concept of the modified effective stress [26] which introduces the suction s to the well known effective stress for saturated porous media [27–29]:

$$\sigma_{ij} = \sigma_{ij}^{tot} + \alpha\,(p^a - \chi s)\,\delta_{ij}, \tag{4}$$

where σ_{ij}^{tot} is the total stress, p^a is the pore air pressure, p^w is the pore water pressure, $s = p^a - p^w$ is the suction and $\chi(S_e^w)$ is the *Bishop* parameter. Assuming that the bulk modulus of the solid grains (K^s) is much higher than the bulk modulus of the skeleton ($K^s \gg K$), the *Biot* constant α is $\alpha = 1 - K/K^s \approx 1$.

Balance of Linear Momentum

The balance of linear momentum for the overall mixture of the multiphase porous medium can be written as:

$$\rho^{tot} b_i + \sigma_{ij,j}^{tot} - (\rho^s + \rho^{wr})\,\ddot{u}_i^s - \rho^w \ddot{u}_i^w - \rho^a \ddot{u}_i^a = 0_i. \tag{5}$$

Therein, b_i is the body force (e.g. due to gravity). Under isothermal conditions, the linear momentum balance equation of each individual (free) pore-fluid phase $\beta = \{w, a\}$ results in the generalized Darcy equation for multiphase flow [20,30] (a detailed derivation is given in [31]):

$$w_i^\beta = \frac{k^\beta K_{ij}}{\mu^\beta}\left(-p_{,j}^\beta + \bar{\rho}^\beta\left(b_j - \ddot{u}_j^\beta\right)\right) \qquad \text{with:}\quad \beta = \{w, a\}, \tag{6}$$

with the intrinsic permeability matrix of the porous medium K_{ij}, the relative permeability k^β and the dynamic viscosity μ^β of the pore-fluid β. The relative permeability $k^\beta(S_e^w)$ is related to the effective degree of saturation through an experimental function with the corresponding constraint $0 < k^\beta \leq 1$.

Balance of Mass

Based on the assumption of a homogeneous distribution of the solid density ($\bar{\rho}_{,i}^s = 0_i$) and the solid grains to be incompressible $K^s \to \infty$, the balance of mass of the solid phase results in the evolution law of the effective porosity [23]:

$$\dot{n}^e = (1 - n^e)\dot{u}_{i,i}^s. \tag{7}$$

Adopting a pressure dependent function for the intrinsic compressibility of the pore-fluid phases [17,32], i.e. $\bar{\rho}^\alpha = \bar{\rho}^\alpha(p^\alpha)$, the mass balance equations for the pore-water and pore-air read:

$$n^e \dot{S}^e + n^e S^e \frac{1}{\bar{K}^w}\dot{p}^w + \frac{1}{\bar{K}^w} p_{,i}^w w_i^w + S^e \dot{u}_{i,i} + w_{i,i}^w = 0, \text{ and} \tag{8}$$

$$-\,n^e \dot{S}^e + n^e(1 - S^e)\frac{1}{\bar{K}^a}\dot{p}^a + \frac{1}{\bar{K}^a} p_{,i}^a w_i^a + (1 - S^e)\dot{u}_{i,i} + w_{i,i}^a = 0, \tag{9}$$

where $\bar{K}^w$ and $\bar{K}^a$ are intrinsic bulk moduli of the pore-water and pore-air, respectively. The bulk modulus of pore-water $\bar{K}^w$ is considered as a parameter, with a typical value of $\bar{K}^w = 2.2\,\text{GPa}$. The pore-air is modelled as an ideal gas, therefore $\bar{K}^a$ is governed by a linear relation (Boyle-Mariotte's law) $\bar{K}^a = p^{atm} + p^a$. Therein p^{atm} is the (absolute) atmospheric pressure and p^a is the (relative) pore-air pressure. No additional parameters are needed for this model.

2.3 Finite Element Implementation

A proper choice of primary variables (degrees of freedom) in terms of which the governing equations are solved, is a crucial step in efficiently modelling of non-linear problems of multi-phase flow in porous media. The choice of primary variables affects the conditioning of the Jacobian matrix and, hence, the number of iterations required to solve the implicit system of equations leading to poor performance of the computer code. It may also determine the feasibility of numerical modelling. In the present work, we choose the widely used u-p-p formulation[2] (e.g. [16,17,22]) where the problem is discretised in terms of the solid skeleton displacements u_i^s and the pore-fluid pressures p^β.

The *u-p-p* formulation is derived by neglecting the relative accelerations between the pore-fluids and the solid: $\ddot{u}_i^\beta = \ddot{u}_i^s$. The system of equations consists of the momentum balance equation for the mixture (Eq. (5)) and the mass balance equations for the free pore-fluid phases β (Eqs. (8)–(9)). Due to the Ladyshenskaya-Babuška-Brezzi (LBB) condition [35] related to the solvability and the convergence, a different order of interpolation for both displacement and pressure fields has been used[3]. Therefore, the solid displacements $u_i^s = \hat{N}_I u_{Ii}^s$ are interpolated from their nodal values u_{Ii}^s using quadratic shape functions $\hat{N}_I$ for each node I while the pore-fluid pressures $p^\beta = N_J p_J^\beta$ are interpolated using linear shape functions N_J. The residua of the weak forms of the linear momentum balance equation for the overall mixture r_{Ii}^s and the mass balances of the pore-fluid phases r_K^β are written as:

$$\begin{aligned} r_{Ii}^s = &\int_\Omega \hat{N}_{I,j}\sigma_{ij}\,\mathrm{d}\Omega + \int_\Omega \hat{N}_I \rho^{tot} b_i\,\mathrm{d}\Omega - \int_\Omega \hat{N}_I \rho^{tot} \ddot{u}_{iJ}^s \hat{N}_J\,\mathrm{d}\Omega + \int_{\Gamma^s} \hat{N}_I t_{iJ} \hat{N}_J\,\mathrm{d}\Gamma^s \\ &- \int_\Omega \hat{N}_{I,j} B\left(N_K p_K^a - S_e^w\left(N_K p_K^a - N_K p_K^w\right)\right) 1_{ij}\,\mathrm{d}\Omega, \end{aligned} \tag{10}$$

[2] A second set of elements has been implemented based on the so-called u-U-U approximation, discretising the displacements of the solid and the pore-fluids and accounting for the relative accelerations. Comparison of simulation results have proven the applicability of the u-p-p approximation. These findings are reported by the first author in [15]. Additional examinations regarding the influence of the chosen approximation are reported in [12,13,33,34].

[3] It could be noted that if the incompressible or undrained limit state is never approached, equal-order interpolations can be used for both solid displacements and pore-fluids pressures [16].

$$r_K^\beta = \int_\Omega N_K n^e \frac{\mathrm{d}S_e^\beta}{\mathrm{d}t}\,\mathrm{d}\Omega + \int_\Omega N_K \frac{n^e S_e^\beta}{K^\beta} \dot{p}_L^\beta N_L\,\mathrm{d}\Omega + \int_\Omega N_K S_e^\beta \dot{u}_{Ii}^s \hat{N}_{I,i}\,\mathrm{d}\Omega$$
$$+ \int_\Omega N_{K,i} \frac{k^\beta K_{ij}}{\mu^\beta} \left(N_{L,j} p_L^\beta - \bar{\rho}^\beta \left(b_j - \hat{N}_I \ddot{u}_{Ij}^s \right) \right) \mathrm{d}\Omega \tag{11}$$
$$+ \int_\Omega N_K \frac{1}{K^\beta} N_{L,i} p_L^\beta w_i^\beta\,\mathrm{d}\Omega + \int_{\Gamma^\beta} N_K q^\beta\,\mathrm{d}\Gamma^\beta.$$

Therein t_{iJ} corresponds to the total force vector acting on element boundary Γ^s and q^β is the flow of pore-water and pore-air perpendicular to the element boundary Γ^β. $B = 1 - \bar{K}/K^s$ is similar to the Biot parameter, with $\bar{K}$ being the bulk modulus of the solid matrix.

2.4 Solution Procedure and SOE-Scaling

The linearisation of the weak form of the system of equations (*SOE*) can be summarized as follows (see [15] for more details):

$$\boldsymbol{r}\left(\boldsymbol{d}, \dot{\boldsymbol{d}}, \ddot{\boldsymbol{d}}\right) = s_D \boldsymbol{M}\ddot{\boldsymbol{d}} - \boldsymbol{F}\left(\boldsymbol{d}, \dot{\boldsymbol{d}}\right). \tag{12}$$

Therein $\boldsymbol{r} = [\boldsymbol{r}^s, \boldsymbol{r}^w, \boldsymbol{r}^a]^T$ is the residuum, $\boldsymbol{d}$ are the unknown nodal variables (displacement of solid skeleton, pore-water pressure and pore-air pressure), $\boldsymbol{M}\ddot{\boldsymbol{d}}$ corresponds to the inertia forces and $\boldsymbol{F}\left(\boldsymbol{d}, \dot{\boldsymbol{d}}\right)$ gathers the internal and external nodal forces. For dynamic simulations $s_D = 1$ and for quasi-static simulations $s_D = 0$ holds. The *Hilber-Hughes-Taylor* (*HHT*) time integration scheme [36] is used to integrate the spatially discretised governing equations. The non-linear *SOE* is solved using the iterative *Newton-Raphson*-Method, where the estimated solution $\boldsymbol{d}^{(i)}$ at the beginning of the iteration (i) is enhanced after each iteration $(i+1)$ by adding a correction $\boldsymbol{c}^{(i+1)}$. The non-linear system at time $t + \Delta t$ is linearised using the first two terms of a *Taylor* series:

$$\frac{\mathrm{d}\boldsymbol{r}}{\mathrm{d}\boldsymbol{d}} \boldsymbol{c}^{(i+1)} = -\boldsymbol{r}\left(\boldsymbol{d}^{(i)}\right). \tag{13}$$

Convergence is reached when Eq. (13) and $c_{max}^\alpha < R_c^\alpha d_{max}^{\alpha,(i)}$ is satisfied within a given tolerance, where $d_{max}^{\alpha,(i)}$ is the maximum increment of the nodal solution at the ith iteration. Moreover it is required, that the largest residual $r_{\max}^\alpha$ in the balance equation for the unknown α is smaller than some percentage R^α of an overall reference value of out-of-balance forces $\bar{q}^\alpha(t)$ space-averaged over the entire model: $r_{max}^\alpha < R^\alpha \bar{q}^\alpha(t)$. For the calculation of $\bar{q}^\alpha(t)$, Abaqus sums up the absolute values of the element contributions $|q_{\text{elem}}^\alpha|$ to the residual vector $\boldsymbol{r}$. In large-scale geotechnical simulations the acting average forces are dominated by the initial geostatic stress state which increases with the depth. In the case of the opencast mine the stress can be larger than 8 MPa at a depth of 800 m. Since the convergence criteria are defined as a function of the mean average force $\bar{q}^\alpha(t)$, the correctness of the solution in the region of interest (being located near the surface and thus exhibiting large changes in stresses compared to the geostatic stress state) may be misjudged. Therefore it is proposed to scale the element contributions $(\sqcup)^{\text{elem}}$ to the *SOE* according to the depth of the their nodes. Using the depth h_I a mean pressure is calculated at each node. The modified element contributions read:

$$\boldsymbol{r}(\boldsymbol{d}^{(i)})^{\text{elem,new}} = -\boldsymbol{Q}^{\text{elem}} \cdot \boldsymbol{r}(\boldsymbol{d}^{(i)})^{\text{elem}}, \tag{14}$$

$$\left(\frac{\mathrm{d}\boldsymbol{r}(\boldsymbol{d}^{(i)})^{\text{elem}}}{\mathrm{d}\boldsymbol{d}}\right)^{\text{new}} = \boldsymbol{Q}^{\text{elem}} \cdot \frac{\mathrm{d}\boldsymbol{r}(\boldsymbol{d}^{(i)})^{\text{elem}}}{\mathrm{d}\boldsymbol{d}}, \tag{15}$$

where $(\cdot)$ denotes a single contraction and $\boldsymbol{Q}^{\text{elem}}$ the scaling matrix, which is defined as:

$$Q_{IJ}^{\text{elem}} = \left((1 + 2K_0)\tilde{\gamma} h_I^{\text{over}}\right)^{-1} \delta_{IJ} \tag{16}$$

where K_0 is the earth pressure coefficient, $\tilde{\gamma}$ is the equivalent unit weight and h_I^{over} is the overburden pressure at the nodal location of *dof* I. The solution $\boldsymbol{d}$ as well as the correction to the solution $\boldsymbol{c}$ remain unchanged after this scaling. For a more detailed explanation see [15].

2.5 Compliant Base

Setting the appropriate boundary condition at the base of a soil sediment for wave propagation problems is not a trivial task. Application of the earthquake motion using a 'rigid base' (Dirichlet) boundary condition (usually as a time-history of acceleration[4]) leads to reflections of downward propagating waves back into the model as depicted in Fig. 1. These reflections are often not readily apparent in complex non-linear analyses, as they can be masked by the damping at larger strains in non-linear soil models or by 'natural' reflections caused by dynamic impedance contrasts at layer boundaries (dashed lines in Fig. 1). Therefore, a rigid base is only appropriate for cases with a large impedance contrast at the base of the model [40,41].

Lysmer and Kuhlemeyer [42] proposed a viscous traction as boundary condition to absorb incident waves. It can be used without further modification when the source of excitation is within the model, e.g. simulations of vibratory pile driving or vibrating machinery. To simulate excitations originating from the outside of the model, such as incoming seismic waves, this boundary condition needs to be extended. To derive the boundary condition, the one-dimensional wave

[4] Notice that in case of the 'rigid base' boundary condition, the so-called 'within' motion has to be prescribed. The 'within' motion is the actual motion at that location, the superposition of the upward and downward propagating waves and can be obtained using 1D wave propagation codes such as SHAKE [37], EERA [38] or SUA [39].

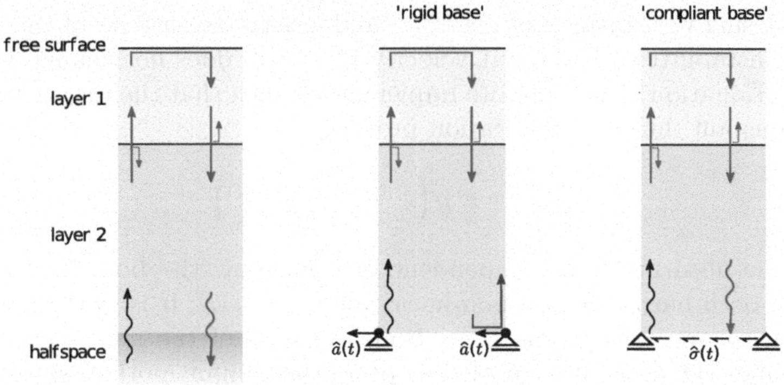

Fig. 1. Schematic representation of the half space subjected to a seismic input motion (left) and its corresponding numerical model using a rigid base (middle) or a compliant base (right) boundary condition.

propagation of an harmonic wave through an elastic homogeneous solid is considered. Starting from the d'Alembert's general solution of the one-dimensional wave equation

$$u(x,t) = f(x+ct) + g(x-ct) \tag{17}$$

where f and g are two waves with constant velocity c moving in opposite directions (i.e. f moves to the bottom and g to the top of the column depicted in Fig. 1). Differentiation of Eq. (17) leads to

$$u_{,t} = v = c\,(f' - g') \quad \text{and} \quad u_{,x} = \varepsilon = f' + g' \tag{18}$$

The traction needed to suppress any upwards reflections and, at the same time, to prescribe the input motion is found by substituting Eq. $(18)_1$ in Eq. $(18)_2$

$$\varepsilon = v/c + 2g' \tag{19}$$

Limiting the discussion to the shear direction only ($v = v^s$, $c = c^s$ and $\varepsilon = \gamma$), using $\gamma = \tau/G$ and $G = (c^s)^2\rho$, and imposing the input motion as a prescribed velocity such that $\hat{v}^s = \hat{g} = c^s\hat{g}'$, the equivalent traction $\hat{\tau}$ is obtained as

$$\hat{\tau} = c^s\rho v^s + 2c^s\rho\hat{v}^s = c^s\rho\,(v^s + 2\hat{v}^s) \tag{20}$$

Therein ρ is the density of the material, v^s is the particle velocity at the boundary and c^s is the shear wave velocity. The prescribed velocity $\hat{v}$ corresponds to the upward propagating motion of the seismic input signal. Although the particle velocity $v^{s,(t+\Delta t),(i)}$ at each iteration i of the current time step $t+\Delta t$ is known[5], we have found (by trial and error) that it is better to use the solution of the previous

[5] This holds true for the use of Abaqus with user defined elements or the FE program `numgeo`. When incorporating the compliant base boundary condition to Abaqus by means of the subroutine `utracload`, $v^{s,(t+\Delta t),(i)}$ is unknown.

step $v^{s,(t)}$. Using $v^{s,(t)}$ instead of $\bar{v}^{s,(t+\Delta t),(i)}$ reduces the sensitivity of the model to numerical instabilities. The input velocity $\bar{v}^{s,(t+\Delta t),(i)}$ does not change within an increment. Equation 21 is therefore implemented such that the magnitude $\hat{\tau}^{(t+\Delta t)}$ remains constant during the iteration process

$$\hat{\tau}^{(t+\Delta t)} = c^s \rho \left(v^{s,(t)} + 2\hat{v}^{s,(t+\Delta t)} \right) \tag{21}$$

Since the absorption is independent of frequency, the boundary condition can absorb both harmonic and non-harmonic waves [42]. It is worth noting that in addition to preventing reflections from the base of the model, a compliant base simplifies the computation of the appropriate input motion, since only the upward propagating motion is required. The performance of the compliant base boundary condition is shown in Appendix A.1.

3 Finite Element Model

A schematic illustration of an investigated opencast mine is given in Fig. 2. The idealized cross section of the dumps (fill) consists of loosely deposited sand (*D1*) interrupted by isolated areas of clay (*D2*) (brown areas in Fig. 2). The subsoil consists of 84 layers in total, which are grouped into six main materials *S1*–*S6*: gravel, sand, till, clay, coal and marl (the latter one is only located at the bottom of the model).

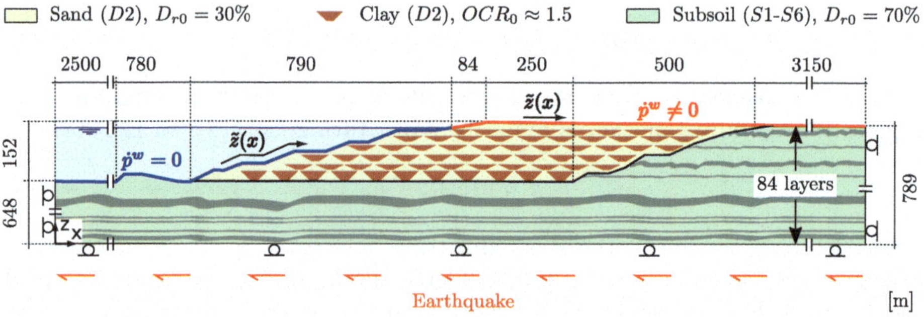

Fig. 2. Schematic illustration of an opencast mine. The illustrated state corresponds to the state prior to the dynamic calculation. The ground water level is located at $z = 785$ m.

3.1 Spatial Discretization

For the numerical analysis large areas outside the mine are included in the simulations to reduce reflections from the vertical model boundaries into the dump area of interest during the dynamic steps. The area of the dump was discretised using ~10,000 triangular shaped elements resulting in a spatial discretization of

$\Delta x = \Delta z \approx 3\,\text{m}$. The subsoil area beneath the dump is discretised using quadratic shaped elements ($\Delta x = \Delta z = 3\,\text{m}$), with increasing distance from the dump, the element size in horizontal direction increases as well (up to $\Delta x^{\max} = 25$ and $\Delta z^{\max} = 3\,\text{m}$ for the outer most elements). The FE model consists of about 65,000 user defined elements in total (for the visualisation in Abaqus another 65,000 so-called overlay-elements are required).

3.2 Material

To describe the stress-strain behavior of granular soil the ISA-Sand constitutive model has been chosen [43, 44]. The constitutive model for clay is based on the *ISA-Clay* model proposed by *Fuentes et al.* [45]. As member of the ISA-plasticity family, both models are able to describe the monotonic and cyclic behavior of saturated soils, as shown by [46, 47]. The material properties of the pore-fluids are provided in Table 1.

Table 1. Material properties of the pore-water and pore-air (p^{atm} = atmospheric pressure).

	$\bar{\rho}^{\beta}$ [g/cm^3]	η^{β} [kPas]	K^{β} [kPa]
Pore-water	1.0	10^{-6}	$2.2 \cdot 10^{6}$
Pore-air	$1.18 \cdot 10^{-3}$	$1.8 \cdot 10^{-8}$	$p^{\text{atm}} + p^{a}$

Dump

The deposited material consist of two materials: *D1* is a silty sand with a mean grain diameter of $d_{50} = 0.55\,\text{mm}$, a coefficient of uniformity of $C_u = 11.2$ and approximately 10% of fines. The second material (*D2*) is a clay characterized by a flow limit of $w_L = 45.6\%$, a plastic limit $w_P = 19.5\%$ and a plasticity index of $I_P = 26.1\%$. The permeability tensor K_{ij} is assumed to be isotropic $K_{ii} = k$ and takes the values $k^{D1} = 10^{-12}\,\text{m}^2$ and $k^{D2} = 5 \cdot 10^{-16}\,\text{m}^2$. The parameters of the ISA-Sand model have been calibrated in [48] on pluviated cones of sand, index tests, oedometric tests and drained monotonic tests following the procedure proposed by *Fuentes* [43] and then optimized by using the element test program *Incremental Driver* by *Niemunis* [49]. The parameters of the *ISA-Clay* constitutive model are calibrated based on oedometric tests, monotonic triaxial tests and monotonic undrained triaxial tests with strain rate jumps. For more details, the interested reader is referred to [45]. In addition to the static tests, the constitutive material laws are calibrated by means of undrained cyclic triaxial tests. The material parameters of the constitutive models are summarized in Table 2.

Subsoil

Contrary to the deposited material, soil samples from the subsoil were not available. According to soil investigation reports (including several drilling logs), the subsoil beneath the dump is composed of up to 84 soil layers varying also in horizontal direction. For the sake of simplicity, however, a horizontal layering based on only one drilling log is assumed. Due to the lack of soil samples, the parameters for the constitutive models were calibrated based on elastic soil properties of a representative soil profile located near the opencast mine. These soil properties include depth profiles of both the shear wave velocity as well as the density and have been gained by means of back calculation of downhole measurements performed by Weber [50]. The parameters of the ISA models have been iteratively adjusted in order to achieve the best possible fit with regard to the shear wave velocity c_s. A comparison between simulation results and the depth-profile reported in [50] is shown in Fig. 3.

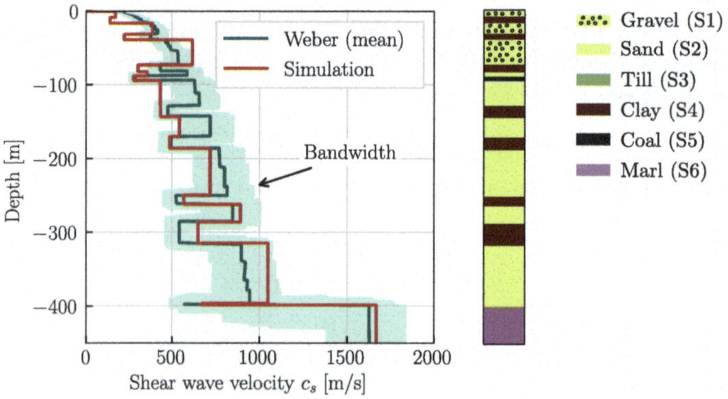

Fig. 3. Comparison of shear wave depth profiles obtained by simulations using the ISA constitutive models and the depth-profile by Weber [50] obtained by back calculation of downhole measurements. The light green area shows the range of different (bandwidth) of these back calculations, whereas the solid dark green line represents the mean value.

Table 2. Constitutive model parameters for both the deposited materials (Di) and the subsoils (Si): $D1$ = silty sand, $D2$ = clay, $S1$ = gravel, $S2$ = sand, $S3$ = till, $S4$ = clay, $S5$ = coal and $S6$ = marl.

ISA-Sand																
	λ [10^{-3}]	n^p [-]	n^e [-]	e_{i0s} [-]	λ_c [10^{-3}]	n_{pc} [-]	e_{c0s} [-]	ν [-]	Mc [-]	n_d [-]	f_{b0} [-]	R [10^{-4}]	m_R [-]	β_R [-]	χ_0 [-]	c_z [10^3]
D1	3	0.5	3	0.878	7	0.5	0.63	0.31	1.458	1	1.35	1	2	0.25	9.5	5
S1 S2	1 5	0.5	3	0.878	7	0.5	0.63	0.31	1.458	1	1.35	1	2	0.25	9.5	5

ISA-Clay													
	λ [-]	κ [$^{-3}$]	e_{i0s} [-]	ν [-]	Mc [-]	f_{b0} [-]	I_v [-]	R [10^{-3}]	m_R [-]	β_h [-]	χ [-]	χ_{max} [-]	C_a [-]
D2	0.13	50	1.76	0.33	1.0	1.5	0.025	2	5	0.076	1	40	0.005
S3	0.012	2.4	0.536				0.015						
S4	0.012	2.4	1.429	0.412	1.5	1.25	0.025	1	2	1	7	40	0.005
S5	0.024	4.8	1.388				0.025						

4 Initial State and Loading History

4.1 Initial State

The initial state of the FE model corresponds to the state before the final deposition is constructed by successive depositions of sand and clay material. From a numerical point of view this is achieved by applying a reduced gravitational force $g^* = 0.2g$ to the overburden area, whereas the gravity acting on the subsoils is $g = 9.81\,\text{m/s}^2$. The ground water table is located at $z = 588\,\text{m}$. The initial relative density of the non-cohesive overburden material (*D1*) after deposition is $D_{r0} = 30\%$[6] and of the subsoils (*S1*–S6) $D_{r0} = 70\%$. The initial void ratio follows the Bauer's equation [52], therein, the depth dependent initial void ratio $e_0(\tilde{z})$ is expressed as a function of the initial void ratio e_{00} under a mean effective pressure p_0, the actual mean effective stress $\sigma_{ii}(x, \tilde{z})$ in depth $\tilde{z}$ under consideration of the hypoplastic parameters $h_s = 8 \cdot 10^4\,\text{kPa}$ and $n = 0.5$:

$$e_0(\tilde{z}) = e_{00}\, \chi \cdot \exp\left[-\left(\frac{-\sigma_{ii}}{h_s}\right)^n\right], \tag{22}$$

where $\chi = (e_{max} - (e_{max} - e_{min})D_{r0})\,/e_{c0}$ is introduced to account for the initial relative density, e_{c0} is a material parameter and denotes the critical state void ratio. Note that the height of overburden $\tilde{z}(x) - z$ varies severely. The initial void ratio of the cohesive materials is calculated following Eq. 23[7]:

$$e_0(\tilde{z}) = e_{i0} - \lambda \cdot \log\left(\frac{-\sigma_{ii}/3}{p_{ref}} \cdot OCR_0\right), \tag{23}$$

where e_{i0} and λ are material parameters and $p_{ref} = 1\,\text{kPa}$. Notice that in arriving in Eq. (23), an isotropic initial stress state ($K_0 = 1$) was assumed for the cohesive materials.

4.2 Loading History

The mechanical behaviour of soils strongly depends on its density (void ratio) and the recent loading history. To reproduce the field of stress and strain of the opencast mine prior to the dynamic loading, the stepwise deposition as well as the groundwater raising is simulated in 17 subsequent static steps. A brief description of each step of applied loading history is given in the following and a schematic overview is given in Fig. 4.

---

[6] Wichtmann et al. determined the (initial) relative densities in the deposition by means of field testing campaigns, as reported in [51]. Therefore, samples were taken at the base of small pits excavated to a depth of about 2 m. Furthermore, the relative density of freshly deposited cones were investigated. The relative density was determined to range from $D_r \approx 24\%$ to $D_r \approx 42\%$ for the samples, and $D_r \approx 30\%$ for the deposited cones.

[7] Equation (23) is derived from the constitutive relations of the ISA-Clay model for the special case of an isotropic initial stress states ($K_0 = 1$).

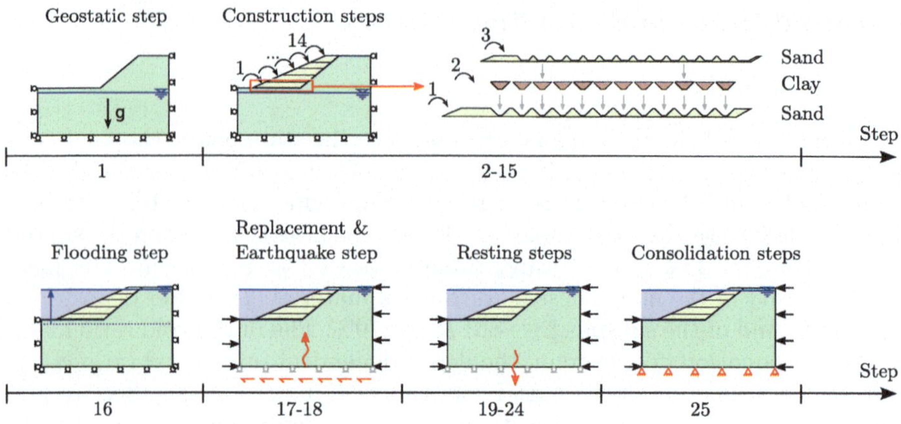

Fig. 4. Schematic illustration of the loading sequence.

`Step 1:`
First, the initial stress state due to the self-weight of the soil is applied in a so-called *Geostatic Step*, without generating any deformations. The displacements of the bottom boundary are constrained in vertical direction, whereas the lateral boundaries are constrained in horizontal direction. An artificial traction $\hat{t}_{stab} =$ 5 kPa is applied perpendicular to the top surface for numerical purposes (notice that the magnitude remains constant during the following steps).

`Steps 2–15:`
In the next stage, the construction of the dump by deposition of the excavated soils by using 'spreaders' is considered. This process is simulated in 14 subsequent steps. In successive layers (starting with layer 1, see Fig. 4, and one layer per step) the initially reduced gravity g^* in the dump area is increased linearly with step time until $g^* = g$ is reached. The step is solved under quasi-static conditions.

`Step 16:`
After the dump is constructed the raise of the groundwater level is simulated by gradually increasing the pore-pressure at the bottom and the side boundaries of the model. Simultaneously a pressure $\hat{t}^w$ corresponding to the dead-weight of the overlying water acting perpendicular to the top surface is applied. The distribution of initial stress, void ratio and over-consolidation ratio for the area of the slope are illustrated in Fig. 14.

`Step 17:`
Prior to the dynamic analysis, the horizontal boundary conditions at both sides of the model are replaced by static forces equivalent to the reaction forces of the horizontal boundary conditions. These traction boundaries may alter the shape

of the propagation front and introduce unrealistic surface waves. These effects can be minimized by placing the boundaries far enough from the slope[8].

`Step 18:`
During the *Earthquake* step, a compliant base boundary condition $\hat{\tau}^{(t+\Delta t)}$ is prescribed along the bottom boundary of the model. This von Neumann boundary condition prescribes the earthquake input motion and at the same time absorbs downward travelling reflections. Drainage along the top surface is assumed, meaning the water table inside the residual lake is assumed not to be affected by the earthquake. During the *Earthquake* step a fixed time increment of $\Delta t = 2.0 \cdot 10^{-4}$ s. The total time step for the earthquake input signal is $t^{\text{step}} = 8.46$ s.

`Steps 19–24:`
After the *Earthquake* step a series of *Resting* steps is calculated, where further horizontal displacements at the bottom boundary are prevented. During the first *Resting* step, the time increment is allowed to increase from $\Delta t = 2.5 \cdot 10^{-4}$ to $\Delta t = 5 \cdot 10^{-4}$ s, the total step time is $t^{\text{step}} = 10$ s. For the following *Resting* step the time increment is allowed to increase from $\Delta t = 5 \cdot 10^{-4}$ to $\Delta t = 10^{-3}$ s.

`Step 25:`
After the last *Resting* step the acceleration magnitudes in the dump area are of magnitude $|a| = 10^{-2}$ m/s^2 but the excess pore water pressures are still non-zero. The consolidation process is simulated using a transient analysis procedure (neglecting inertia effects).

5 Results

Figure 5 shows the acceleration magnitude at different times of the dynamic analysis ($t = \{1.86, 2.42, 5.1, 45\}$ s). Until the wave front reaches the dump at $t \approx 1.86$ s (Fig. 5a)), a pattern similar to a vertical wave-propagation problem is observed. However, due to the variable overburden height, the propagation speed varies as well.

The acceleration magnitude increases as the wave front travels to the top. As the wave front impinges on the division level between the subsoil and the deposition, the calculation shows reflection and refraction effects. Additionally, as an effect of the geometric layout of the deposition, a strong wave diffraction is observed as a result of the different stiffnesses of sand (*D1*) and clay (*D2*) as shown in Fig. 5b) (this strong diffraction is absent in calculations with a homogeneous dump geometry, as shown in [15,31]). The internal structure of

[8] Alternatively, so called absorbing boundary conditions could be applied. Application of these have shown an increase or decrease of the amplitude of incident waves (non-perpendicular to the element face), which may lead to results "on the unsafe side", if the boundary is located too close to the dump. Due to the inhomogeneity of the model in horizontal direction, a kinematic coupling of the left and the right lateral boundaries is not possible.

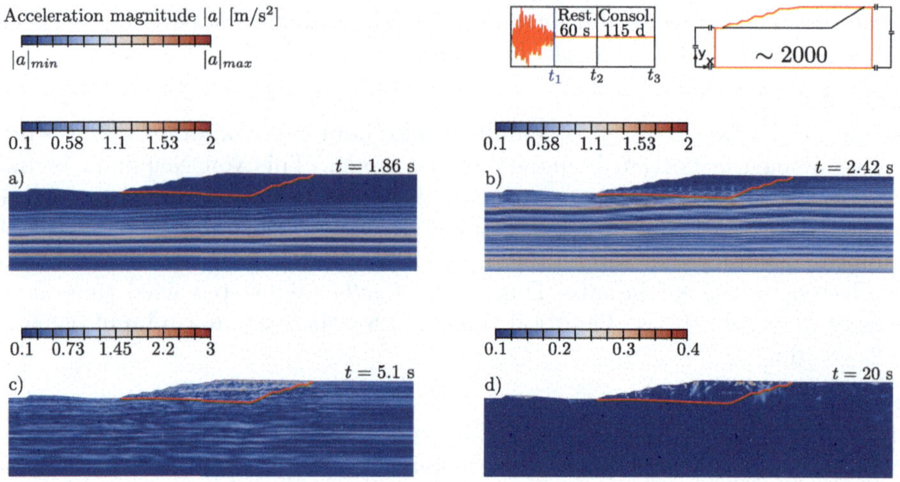

Fig. 5. Acceleration magnitude at different times of the dynamic analysis. The red line in figures highlights the abrupt junction between the deposited material and the subsoil.

the inhomogeneous dump has a shielding effect, leading to reduced accelerations at the top surface of the slope compared to a homogeneous dump. However, as a result of the abrupt junction between the deposited material and the subsoil (and thus change in stiffness) reflections lead to a slower decay of the dynamic loading with time in the area of the dump compared to the subsoil as depicted in Fig. 5c).

Figure 6 presents some plots of the horizontal acceleration-time history during the earthquake loading and the subsequent resting steps.

The record point *P*1 is located at the bottom of the model, thus the acceleration-time history corresponds to the input motion. The acceleration at point *P*2 is recorded at the transition between the subsoil and the dump, thus represents the alteration of the earthquake signal due to the inhomogeneities of the subsoil. Compared to *P*1 the recorded acceleration signal shows similar amplitudes but a larger content of high frequencies. These changes in frequency are a product of the layered subsoil with varying stiffnesses and densities. The free surface motion is given for two representative points *P*3 and *P*4, of which the first is located above *P*2 on the surface of the deposition. Point *P*4 is located above the subsoil at a distance of about 2 km of the deposition. As a result of the (comparable) small stiffness of the deposition material, large amplifications of the input signal (showing peak accelerations with $|a_x| > 3\,\text{m/s}^2$) are observed for *P*3. Due to the higher stiffness of the subsoils, these amplifications are only of magnitude 2 for the free surface motion recorded at *P*4. The mean velocity of the wave front is estimated based on the time history recorded at point *P*4 and is approximately $\bar{c}_s \approx 789\,\text{m}/2.25\,\text{s} = 350\,\text{m/s}$.

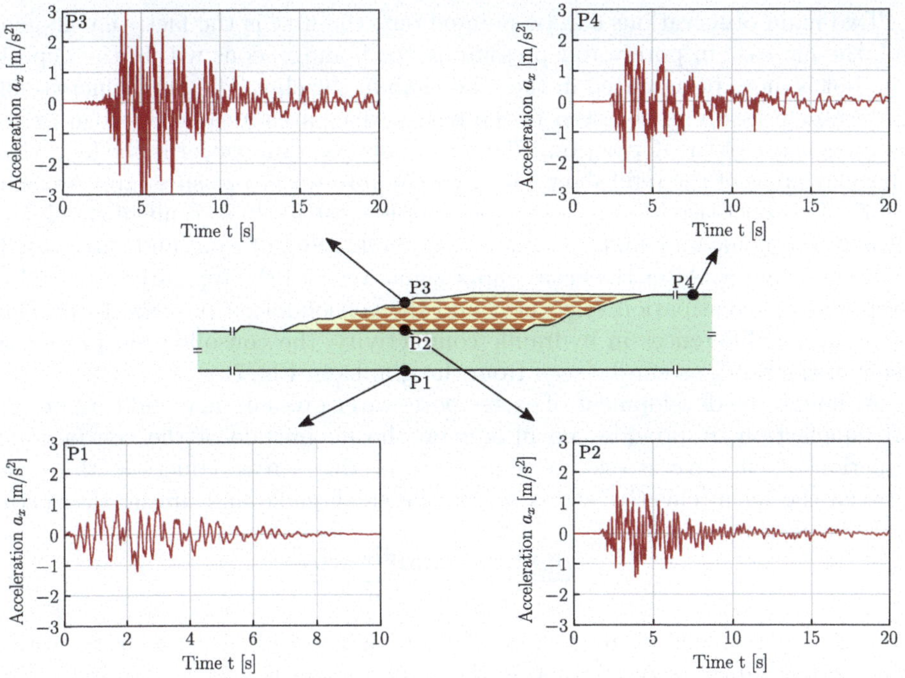

Fig. 6. Horizontal acceleration over time recorded at four points of the FE model.

During the dynamic analysis a build-up of excess pore-water pressure takes place. The field of excess pore-water pressure after the dynamic excitation (t = 8.46 s) and a graph showing the development of pore-water pressure with time are depicted in Fig. 7.

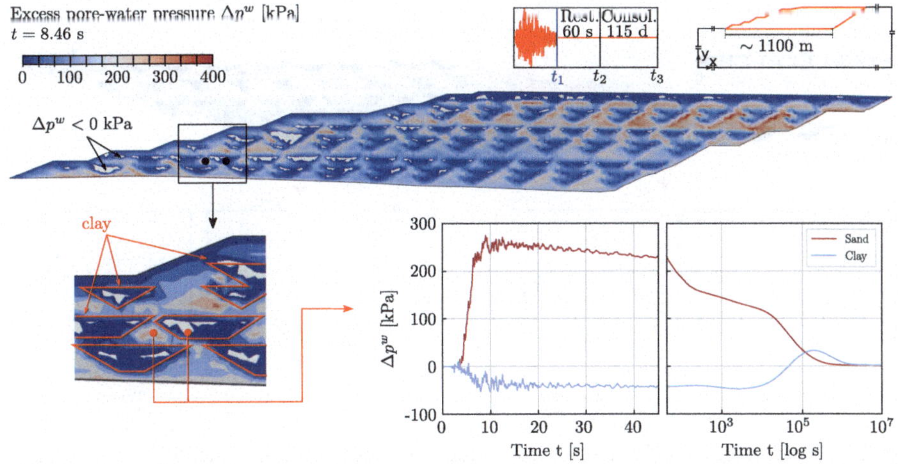

Fig. 7. Change of pore-water pressure $\Delta p^w = p^w(t) - p^w(t = 0)$ at the end of the dynamic excitation $t = 8.46$ s.

Two main observations can be pointed out: the first is the fact, that in general, the increase in pore-water pressure is not homogeneous within the deposition, but is more pronounced in the sand than in the clay. The major increase in pore-water pressure is observed in the near-surface area located 30–60 m below the top surface of the deposition. The second observation concerns the fact, that some clay areas of the refill show a decrease in pore-water pressure (grey areas in Fig. 7). This decrease in pore-water pressure is most likely a result of the different material behaviour under (dynamic) cyclic loading of sand and clay, as will be discussed later. After the entire input signal reaches the top surface ($t \approx 11$ s) the pore-water dissipation takes over and the consolidation process starts. Due to the major differences in hydraulic conductivity, the consolidation process is faster in the sand, as can be seen from the graphs in Fig. 7.

Although the development of excess pore-water pressure may indicate potential liquefaction, it provides no insight on the magnitude of the concomitant reduction of effective stress. As a measure of this stress-reduction, the ratio between the mean effective stress before the earthquake and during the earthquake is chosen:

$$\zeta = \frac{\sigma_{ii}(t^{dyn}) - \sigma_{ii}(t^{dyn} = 0)}{\sigma_{ii}(t^{dyn} = 0)}, \tag{24}$$

where $\zeta = -1$ corresponds to the case of a complete loss of effective stress. Notice that the mechanical sign convention (i.e. compression is negative) is used. The ζ-field after the dynamic excitation ($t = 11$ s) as well as the time history for two regions $A1$ and $A2$ of the deposition are shown in Fig. 8. Although the spreading of areas affected by a stress-reduction is strongly influenced by the clay (due to the smaller hydraulic conductivity), the effect of stress-reduction is restricted to the sandy material, as shown in Fig. 8. Despite nearly undrained conditions, the clayey material does not suffer from a strong decrease of effective stress as shown in Fig. 8. This behaviour has also been observed in experimental studies (see e.g.

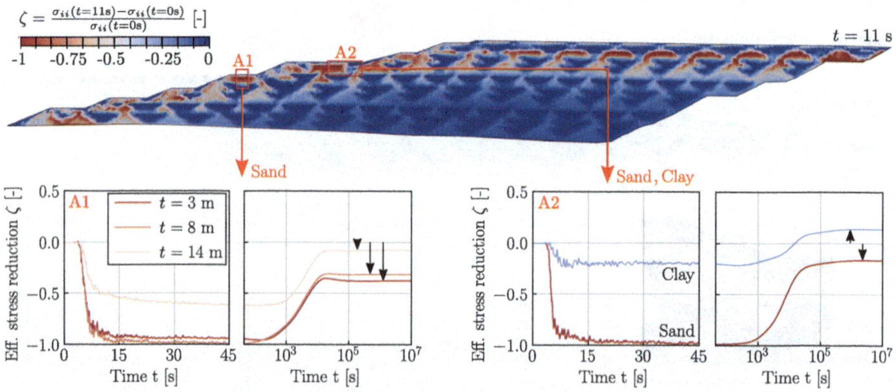

Fig. 8. Illustration of effective stress reduction after 11 s of resting (top) and time history of the stress ratio ζ for two near-surface regions of the deposition (bottom).

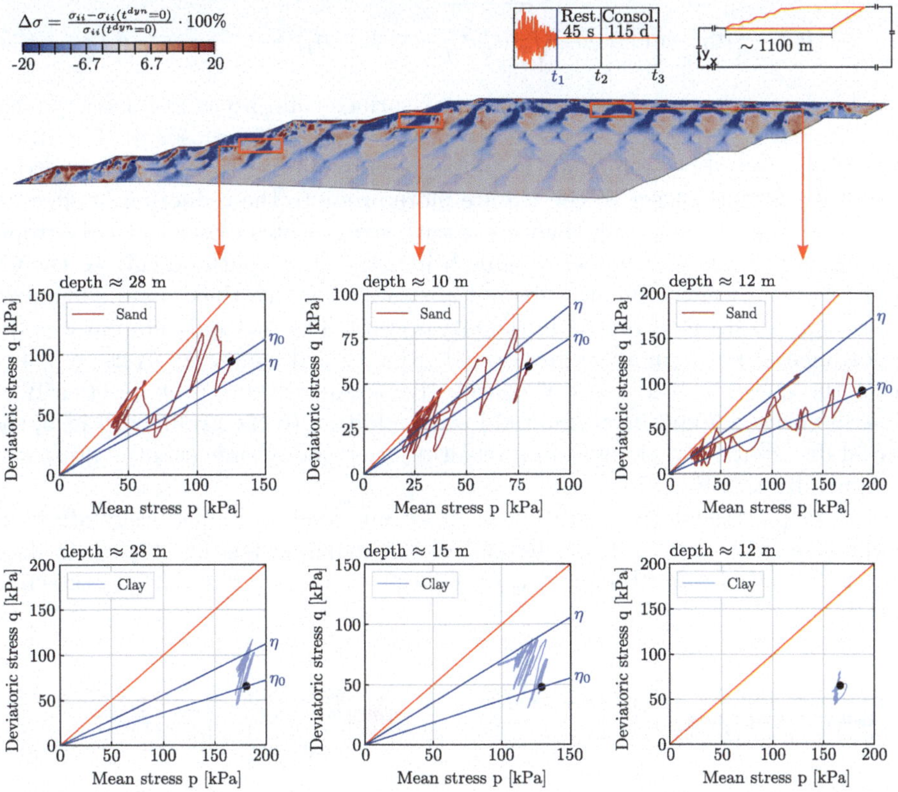

Fig. 9. Change of mean effective stress at the end of the consolidation process in relation to the initial mean effective stress (top) and stress path in $p-q$-plane for near-surface areas. The circle labels the start of the dynamic analysis.

[53]). In contrast to sands, clayey materials are more prone to failure due to the development of rapidly increasing strains, also known as *cyclic softening*. It is important to note, that although the stress ratio ζ shown for the areas $A1$ and $A2$ in Fig. 8 reaches its minimum at $t \approx 15\,\mathrm{s}$ the areas affected by the stress reduction spread with ongoing calculation time ($t > 11\,\mathrm{s}$) as reported in [15].

A stress state, a stress redistribution in the area of the dump is observed at the end of the consolidation phase. The magnitude of mean effective pressure in formerly disintegrated sand areas ($\zeta \approx -1$, red areas in Fig. 8) is reduced in comparison to the initial state. On the other hand, the red areas in Fig. 9 on the other hand experience a permanent increase of effective pressure. Changes in mean effective stress affect the barotropic stiffness E due to $\mathsf{E} \sim (-\sigma_{\mathrm{ii}}/3)^{\mathrm{n}}$, which should be accounted for, if subsequent loading (such as a second earthquake) are of interest. In the present case, the observed growing inhomogeneity of stresses leads to an increasing homogeneity of the barotropic stiffness. Furthermore, changes in stress ratio $\eta = q/p$ are observed. Therein p and q are the Roscoe stress invariants:

$$p = -\frac{1}{3}\sigma_{ii} \quad \text{and} \quad q = \sqrt{\frac{3}{2}\sigma^*_{ij}\sigma^*_{ij}} \quad \text{with:} \quad \sigma^*_{ij} = \sigma_{ij} - \frac{4}{3}\sigma_{mm}\delta_{ij}. \tag{25}$$

The decrease of mean stress p in near-surface sand areas are accompanied by an increase in deviator stress q, leading to stress states closer to the critical state (CS). As can be seen from the plots presented in Fig. 9, sand areas starting from a stress state closer to the CS are more prone to the reduction of effective stress due to cyclic shearing than areas with stress states closer to the isotropic state. In other words, a higher η value brings the stress path quickly to the CS line and begins to accumulate substantial strains. Clearly, the dynamic behavior is (strongly) controlled by the initial stress state. This not only emphasizes the importance of static analysis performed prior to a dynamic analysis, but also raises the question if a series of consecutive seismic events (not necessarily of same intensity) should be accounted for. In addition to the graphs in Fig. 9, the spatial distribution of changes in η resulting from the seismic event is presented in Appendix A.2, Fig. 15.

The displacements Δu resulting from seismic loading immediately after the earthquake ($t^{dyn} = 8.46$ s), at the end of the dynamic resting steps ($t = 45$ s) and at the end of the subsequent consolidation analysis ($t = 115$ d) are shown in Fig. 10.

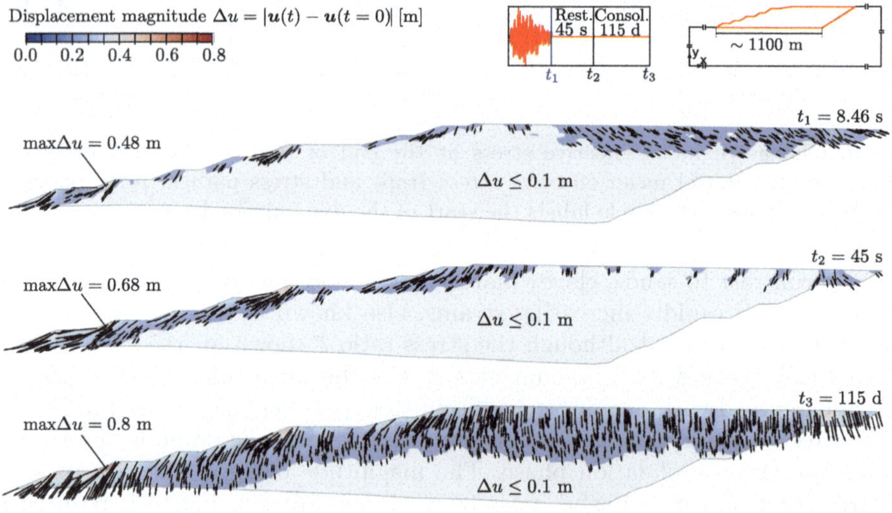

Fig. 10. Vector field of displacements resulting from dynamic loading at the end of the excitation time ($t_1 = 8.46$ s), at the end of the resting step ($t_s = 45$ s) and at the end of the consolidation process ($t_3 = 115$ d).

The resulting displacements reach a maximum value of about 0.8 m near the toe area of the deposition. The vector field of displacements evaluated at the end of the dynamic analysis ($t^{dyn} = 45$ s) shows a movement of surface near areas directed towards the residual lake. Areas which experienced a strong reduction of effective stress (blue areas in Fig. 8) show increasing vertical displacements (settlements) during the consolidation analysis, driven by the expulsion of water until the excess pore-water pressure vanishes and emphasizing the importance of the observation time exceeding the excitation time. Although some superficial regions (especially in the toe region of the deposition) undergo large displacements and deformations due to the earthquake loading and the subsequent consolidation, the deeper regions of the deposition experience large changes in stress but small deformations.

6 Summary and Future Perspective

User-defined elements have been developed within the software package Abaqus/Standard to perform dynamic analysis of saturated and unsaturated soils. To understand aspects of the dynamic slope behavior, large-scale finite element calculations of a flooded lignite opencast mine were conducted. Based on these calculations, the following key conclusions can be drawn:

- Some numerical problems have been observed during simulations, which could be traced back to the influence of initial (geostatic) stress on the average force of the model used to judge the convergence of the estimated solution. A scaling of the element contributions to the overall system of equations, without changing its solution, is therefore proposed and applied in all simulations.
- As a result of the inhomogeneous internal dump structure, strong wave diffraction effects are observed as the wave front impinges the dump. Additional damping effects occur due to the viscosity of the clayey material. Both effects result in smaller displacements in near-surface areas of the slope (compared to a homogeneous build-up of the dump).
- Although some superficial regions (especially in the toe region of the dump) undergo large displacements and deformations due to the earthquake loading and the subsequent consolidation, the deeper regions of the slope experience large changes in stress but small deformations.
- These large changes in stress is followed by a permanent stress redistribution during the post-seismic consolidation, leading to a homogenisation of the stiffness in the dump. This effect should be accounted for if the impact of severe seismic events with high recurrence periods are studied (and smaller seismic events before the main earthquake are likely to occur).
- Large areas of the dump show a temporary decrease of effective stress. Since no experimental data are available for low stress states ($0 < p < 10\,\text{kPa}$), the response of the constitutive models has to be judged as uncertain. Laboratory tests investigating the behavior of (silty) sands at low stress states (targeting the transition from an intact grain fabric to a kind of suspension) are therefore necessary.
- Furthermore, the execution of dynamic in-situ tests in an opencast mine or in a reduced scale should be concerned as well as prior simulations of these tests in form of a Class-A-Prediction. Recent work devoted to the comparison of

class-A and class-C predictions of centrifuge tests indicate that the parameter calibration based on small element tests does not necessarily lead to a successful simulation of seismic site response [54–56]. However, the overall trends of pore-water pressure and displacements were captured reasonable well.
- Also, emerging large deformations and the liquefaction-induced soil failure should be addressed in future numerical simulations. Therefore the coupled Material-Point-Method (e.g. [57,58]) or the multi-material arbitrary Lagrangian-Eulerian method (e.g. [59]) could be a promising approach.

Appendix

A.1 Compliant Base Boundary Condition

The performance of the presented compliant base boundary condition (BC) is now examined with two additional simulation examples of one-dimensional soil columns on top of bedrock subjected to seismic excitation at the column base. All simulations have been carried out using a rigid base (Dirichlet) BC as well as the compliant BC presented in this work. The first simulation considers a homogeneous linear elastic soil column as depicted in Fig. 11. A sinusoidal input motion is prescribed at the bottom of the column. The velocity-time histories at the bottom and the top of the soil column are shown in Fig. 11. It is obvious that using the rigid base BC to prescribe the input motion, downward-propagating waves are artificially reflected back to the sediment. These reflections are absent when using the compliant base BC instead, as $\hat{\tau}(t)$ absorbs the downward-propagating waves, see Eq. (21).

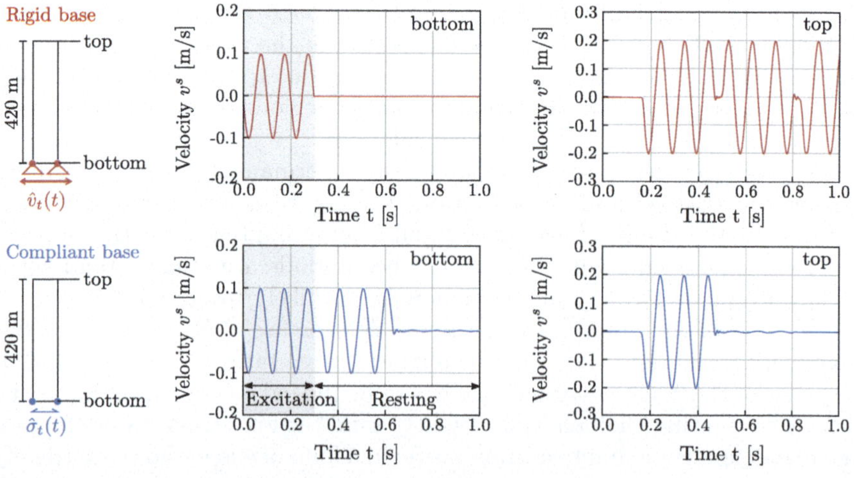

Fig. 11. Recorded velocity-time history at the bottom and the top of a one-dimensional linear elastic soil column. The input motion is either prescribed using a Dirichlet boundary condition (top) or the compliant base boundary condition (bottom).

For the second example, the subsoil is considered to be composed of in total 54 different soil layers, classified as gravel, sand, silt, coal and clay. The non-cohesive soils (sand and gravel) are simulated using the ISA-Sand model, whereas the ISA-Clay constitutive model is chosen for the cohesive soils (silt, coal and clay). For details on the material parameter and their calibration see Sect. 3.2. For these simulations, the bedrock is considered by adding an additional layer of variable length ($d_1^{br} = 10$ m and $d_1^{br} = 100$ m) at the bottom of the soil column. A linear elastic constitutive model is chosen for the bedrock. The excitation time of the input motion is $t^{inp} = 8.5$ s. Figure 12 presents the acceleration-time histories and the corresponding response spectra recorded at the top of the soil column.

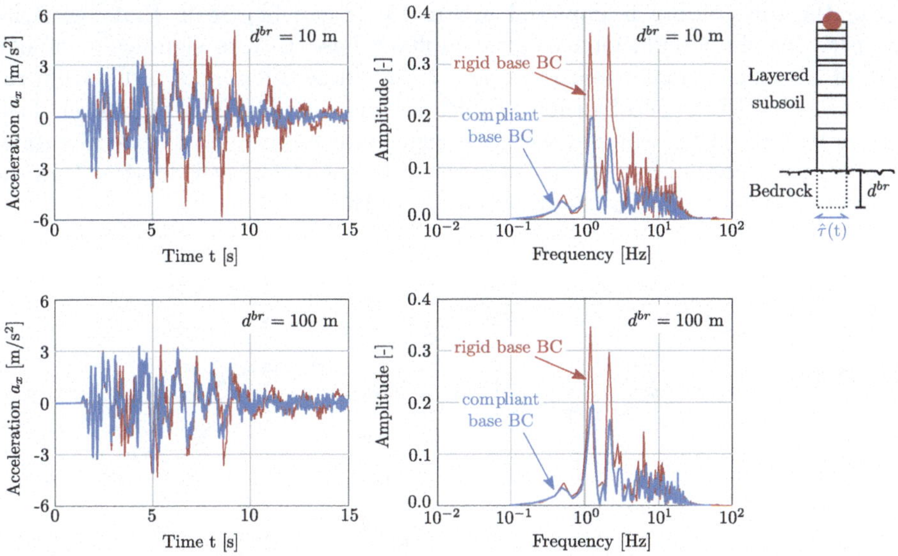

Fig. 12. Acceleration-time history (left) and corresponding response spectrum (right) for simulations with variable bedrock depth and boundary condition.

As shown in Fig. 12, a strong dependence of the depth of the included bedrock layer d^{br} on the simulation results is observed for the simulations using the rigid base BC. This dependence is absent in the simulations with compliant base BCs. Furthermore, the calculated accelerations at the top of the column obtained from simulations with rigid base BCs exceed those from the simulations using compliant base BCs. This is particularly evident for frequencies of $f \approx 1$ Hz and $f \approx 2$ Hz. However, these differences become smaller with increasing depth d^{br} of the bedrock layer.

To quantify the influence of the BC on the actual loading intensity in the simulations (for the same input motion), the Arias Intensity is evaluated. The Arias Intensity [60] is defined as the cumulative energy per unit weight absorbed by an infinite set of undamped single-degree-of-freedom oscillators at the end of an earthquake. The Arias Intensity for a ground motion in the horizontal (x-) direction is calculated as follows:

$$I_A = \frac{\pi}{2g} \int_{t_0}^{t} a_x^2(t) \mathrm{d}t \tag{26}$$

where $a_x(t)$ is the corresponding acceleration time history, g is the coefficient of acceleration due to gravity and t_0 is the total duration of the ground motion. The Arias Intensity for the calculated horizontal acceleration recorded at the top of the soil column is depicted in Fig. 13. It can bee seen, that the Arias Intensity for the simulations using compliant base BCs is significant smaller than in the calculations using rigid base BCs. These differences result from the overestimation of the acceleration amplitudes and the reduced attenuation of the seismic loading in the simulations with rigid base BCs. As in the acceleration-time history shown in Fig. 12, no influence of the bedrock layer depth d^{br} on the results is observed when using compliant base BCs.

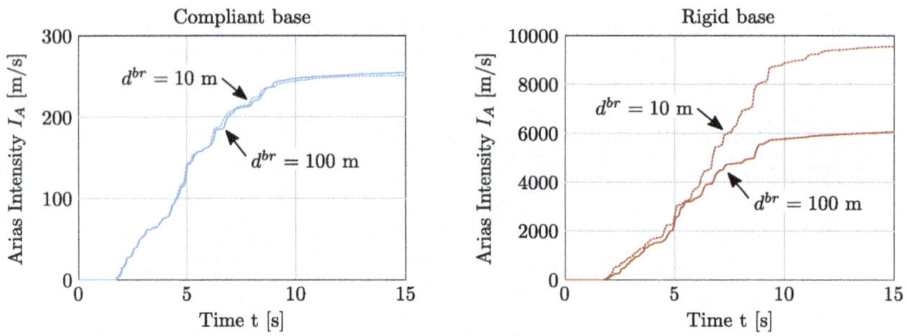

Fig. 13. Arias Intensity I_A of the calculated signal recorded at the top of the column for simulations with different bedrock layer depths and boundary conditions.

A.2 Additional Simulation Results

Figure 14 represents the distribution of initial vertical stress σ_v, void ratio e_0 and over-consolidation ratio OCR_0 for the area of the slope after the flooding step (`Step 16`). The initial void ratio is only provided for the sandy areas (Fig. 14 middle), whereas spatial distribution of OCR_0 is only depicted for the clayey areas of the dump. This state is referred to as the 'initial state' with regard to the dynamic (earthquake) step.

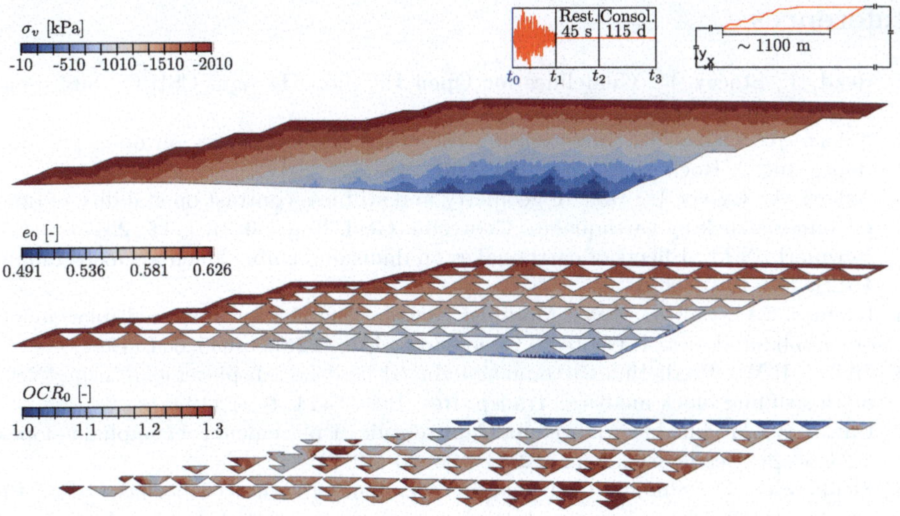

Fig. 14. Initial effective vertical stress σ_{v0} in the area of the dump (top). Distribution of the initial void ratio e_0 for the sandy material of the dump (middle) and the initial over-consolidation ratio OCR_0 for the clayey material (bottom). The initial state refers to the state prior to the dynamic analysis but after the flooding process.

In addition to the reduction of effective stress during the excitation and the post-seismic consolidation analysis, changes in stress ratio $\eta = q/p$ are observed. The spatial distribution of the change in stress ratio $\Delta\eta$ (compared to the initial state) evaluated at the end of the consolidation analysis is provided in Fig. 15.

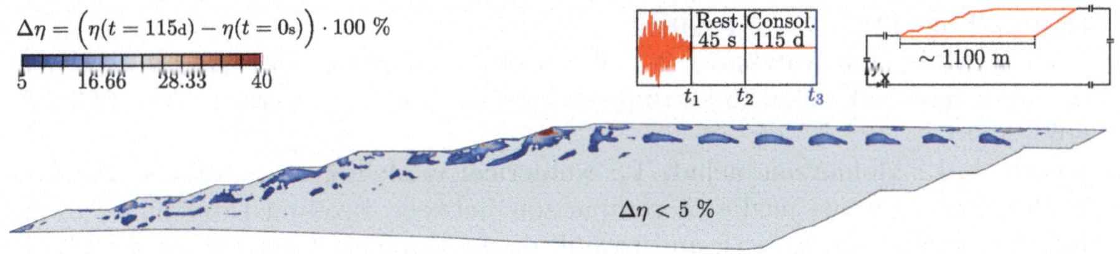

Fig. 15. Change of stress ratio $\Delta\eta = \Delta q/\Delta p$ (referred to the initial stress ratio before the earthquake) at the end of the consolidation process.

References

1. Read, J., Stacey, P.: Guidelines for Open Pit Slope Design. CSIRO Publishing, Clayton (2009)
2. Azhari, A., Ozbay, U.: Investigating the effect of earthquakes on open pit mine slopes. Int. J. Rock Mech. Min. Sci. **100**, 218–228 (2017)
3. Azhari, A., Ozbay, U.: Role of geometry and stiffness contrast on stability of open pit mines struck by earthquakes. Geotechn. Geol. Eng. **36**(2), 1–18 (2017)
4. Newmark, N.M.: Effects of earthquakes on dams and embankments. Geotechnique **15**(2), 139–160 (1965)
5. Kramer, S.L., Smith, M.W.: Modified Newmark model for seismic displacements of compliant slopes. J. Geotech. Geoenviron. Eng. **123**(7), 635–644 (1997)
6. Jibson, R.W.: Predicting earthquake-induced landslide displacements using Newmark's sliding block analysis. Transp. Res. Rec. **1411**, 9–17 (1993)
7. Day, R.W.: Modified Newmark model for seismic displacements of compliant slopes. J. Geotech. Geoenviron. Eng. **125**(1), 86–90 (1999)
8. Zienkiewicz, O., Shiomi, T.: Dynamic behaviour of saturated porous media; the generalized Biot formulation and its numerical solution. Int. J. Numer. Anal. Methods Geomech. **8**(1), 71–96 (1984)
9. Ravichandran, N., Muraleetharan, K.K.: Dynamics of unsaturated soils using various finite element formulations. Int. J. Numer. Anal. Methods Geomech. **33**(5), 611–631 (2009)
10. Dassault Systemes Simulia Corp.: Abaqus 6.14-5 (2014)
11. Plaxis BV: PLAXIS-2D (2017)
12. Muraleetharan, K.K., Deshpande, S., Adalier, K.: Dynamic deformations in sand embankments: centrifuge modeling and blind, fully coupled analyses. Can. Geotech. J. **41**(1), 48–69 (2004)
13. Ravichandran, N.: Fully coupled finite element model for dynamics of partially saturated soils. Soil Dyn. Earthq. Eng. **29**(9), 1294–1304 (2009)
14. Madabhushi, S., Haigh, S.K., Madabhushi, G.: LEAP-GWU-2015: centrifuge and numerical modelling of slope liquefaction at the University of Cambridge. Soil Dyn. Earthq. Eng. **113**, 671–681 (2016)
15. Machaček, J., Triantafyllidis, T., Staubach, P.: Fully coupled simulation of an opencast mine subjected to earthquake loading. Soil Dyn. Earthq. Eng. **115**, 853–867 (2018)
16. Khoei, A.R., Mohammadnejad, T.: Numerical modeling of multiphase fluid flow in deforming porous media: a comparison between two-and three-phase models for seismic analysis of earth and rockfill dams. Comput. Geotech. **38**(2), 142–166 (2011)
17. Schrefler, B.A., Simoni, L., Xikui, L., Zienkiewicz, O.C.: Mechanics of partially saturated porous media. In: Numerical Methods and Constitutive Modelling in Geomechanics, pp. 169–209. Springer (1990)
18. De Boer, R.: Theory of porous media - past and present. ZAMM-J. Appl. Math. Mech./Zeitschrift für Angewandte Mathematik und Mechanik **78**(7), 441–466 (1998)
19. Ehlers, W., Diebels, S., Blome, P.: Modelling of partially saturated soils by the theory of porous media. In: Teilgesättigte Böden Weimar 2001, vol. Heft 6, pp. 91–112. Bauhaus-Universität Weimar (2001)
20. Ehlers, W., Blome, P.: A triphasic model for unsaturated soil based on the theory of porous media. Math. Comput. Modell. **37**(5–6), 507–513 (2003)

21. Borja, R.I.: On the mechanical energy and effective stress in saturated and unsaturated porous continua. Int. J. Solids Struct. **43**(6), 1764–1786 (2006)
22. Ehlers, W., Bluhm, J.: Porous Media: Theory, Experiments and Numerical Applications. Springer, Heidelberg (2013)
23. Fuentes, W., Triantafyllidis, T.: On the effective stress for unsaturated soils with residual water. Géotechnique **63**(16), 1451 (2013)
24. Bear, J.: Dynamics of Fluids in Porous Media. Courier Corporation, Chelmsford (2013)
25. Pande, G.N., Pietruszczak, S.: On unsaturated soil mechanics - personal views on current research. Studia Geotechnica et Mechanica **37**(3), 73–84 (2015)
26. Bishop, A.W., Blight, G.E.: Some aspects of effective stress in saturated and partly saturated soils. Geotechnique **13**(3), 177–197 (1963)
27. Terzaghi, K.: The shearing resistance of saturated soils. In: Proceedings of the 1st International Conference on Soil Mechanics and Foundation Engineering, Cambridge, Massachusetts, vol. 1, pp. 54–56 (1936)
28. Terzaghi, K., Frölich, O.K.: Theory of settlement of clay layers. Franz Deuticke (1936)
29. Fillunger, P.: Erdbaumechanik? Selbstverl. d. Verf. (1936)
30. Lewis, R.W., Schrefler, B.A.: The Finite Element Method in the Static and Dynamic Deformation and Consolidation of Porous Media. Wiley, Hoboken (1998)
31. Machaček, J.: Contribution to the numerical modeling of saturated and partially saturated soils. Veröffentlichung des Instituts für Bodenmechanik und Felsmechanik am Karlsruher Institut für Technologie (KIT) (2019)
32. Malvern, L.E.: Introduction to the Mechanics of a Continuous Medium. Prentice-Hall, Upper Saddle River (1969). No. Monograph
33. Zienkiewicz, O.C., Chang, C.T., Bettess, P.: Drained, undrained, consolidating and dynamic behaviour assumptions in soils. Geotechnique **30**(4), 385–395 (1980)
34. Staubach, P., Machaček, J.: Influence of relative acceleration in saturated sand: analytical approach and simulation of vibratory pile driving tests. Comput. Geotech. **112**, 173–184 (2019)
35. Brezzi, F., Fortin, M.: Mixed and Hybrid Finite Element Methods, vol. 15. Springer, Heidelberg (2012)
36. Hilber, H.M., Hughes, T.J., Taylor, R.L.: Improved numerical dissipation for time integration algorithms in structural dynamics. Earthq. Eng. Struct. Dyn. **5**(3), 283–292 (1977)
37. Schnabel, P.B.: SHAKE: a computer program for earthquake response analysis of horizontally layered sites. EERC Report 72-12. University of California, Berkeley (1972)
38. Bardet, J., Ichii, K., Lin, C.: "EERA," a computer program for equivalent-linear earthquake site response analyses of layered soil deposits. University of Southern California, Department of Civil Engineering, User's Manual (2000)
39. Robinson, D., Dhu, T., Schneider, J.: SUA: a computer program to compute regolith site-response and estimate uncertainty for probabilistic seismic hazard analyses. Comput. Geosci. **32**(1), 109–123 (2006)
40. Mejia, L., Dawson, E.: Earthquake deconvolution for FLAC. In: Proceedings of the 4th International FLAC Symposium on FLAC and Numerical Modeling in Geomechanics, Madrid, Spain, pp. 211–219. Citeseer (2006)
41. Mánica, M., Ovando, E., Botero, E.: Assessment of damping models in FLAC. Comput. Geotech. **59**, 12–20 (2014)
42. Lysmer, J., Kuhlemeyer, R.L.: Finite dynamic model for infinite media. J. Eng. Mech. Div. **95**(4), 859–878 (1969)

43. Fuentes, W.: Contributions in mechanical modelling of fill materials. Veröffentlichung des Instituts für Bodenmechanik und Felsmechanik am Karlsruher Institut für Technologie (KIT), Heft 179 (2014)
44. Fuentes, W., Triantafyllidis, T.: ISA model: a constitutive model for soils with yield surface in the intergranular strain space. Int. J. Numer. Anal. Methods Geomech. **39**(11), 1235–1254 (2015)
45. Fuentes, W., Tafili, M., Triantafyllidis, T.: An ISA-plasticity-based model for viscous and non-viscous clays. Acta Geotechnica **13**(2), 367–386 (2017)
46. Wichtmann, T.: Soil behaviour under cyclic loading-experimental observations, constitutive description and applications. Th. Triantafyllidis, Veröffentlichung des Instituts für Bodenmechanik und Felsmechanik am Karlsruher Institut für Technologie (KIT), Heft 181 (2016)
47. Wichtmann, T., Fuentes, W., Triantafyllidis, T.: Inspection of three sophisticated constitutive models based on monotonic and cyclic tests on fine sand: hypoplasticity vs. Sanisand vs. ISA. Soil Dyn. Earthq. Eng. **124**, 172–183 (2019)
48. Günther, H.: Numerische Untersuchung zur Bodenverflüssigung bei Tagebau-Restseeböschungen unter Erdbebeneinwirkung. Master's thesis. Institut für Bodenmechanik und Felsmechanik am Karlsruher Institut für Technologie (KIT) (2015)
49. Niemunis, A.: Homepage A. Niemunis (2016). http://www.pg.gda.pl/aniem/
50. Weber, B.: Bodenverstärkung in der südlichen Niederrheinischen Bucht. Ph.D. thesis. Universität zu Köln (2007)
51. Wichtmann, T., Steller, K., Triantafyllidis, T., Back, M., Dahmen, D.: An experimental parametric study on the liquefaction resistance of sands in spreader dumps of lignite opencast mines. Soil Dyn. Earthq. Eng. **122**, 290–309 (2019)
52. Bauer, E.: Calibration of a comprehensive hypoplastic model for granular materials. Soils Found. **36**(1), 13–26 (1996)
53. Boulanger, R.W., Idriss, I.M.: Liquefaction susceptibility criteria for silts and clays. J. Geotech. Geoenviron. Eng. **132**(11), 1413–1426 (2006)
54. Ramirez, J., Barrero, A.R., Chen, L., Dashti, S., Ghofrani, A., Taiebat, M., Arduino, P.: Site response in a layered liquefiable deposit: evaluation of different numerical tools and methodologies with centrifuge experimental results. J. Geotech. Geoenviron. Eng. **144**(10), 04018073 (2018)
55. Ghofrani, A., Arduino, P.: Prediction of LEAP centrifuge test results using a pressure-dependent bounding surface constitutive model. Soil Dyn. Earthq. Eng. **113**, 758–770 (2018)
56. Zeghal, M., Goswami, N., Kutter, B.L., Manzari, M.T., Abdoun, T., Arduino, P., Armstrong, R., Beaty, M., Chen, Y.-M., Ghofrani, A., Haigh, S., Hung, W.-Y., Iai, S., Kokkali, P., Lee, C.-J., Madabhushi, G., Tobita, T., Ueda, K., Zhou, Y.-G., Ziotopoulou, K.: Stress-strain response of the LEAP-2015 centrifuge tests and numerical predictions. Soil Dyn. Earthq. Eng. **113**, 804–818 (2018)
57. Abe, K., Soga, K., Bandara, S.: Material point method for coupled hydromechanical problems. J. Geotech. Geoenviron. Eng. **140**(3), 04013033 (2013)
58. Bandara, S., Soga, K.: Coupling of soil deformation and pore fluid flow using material point method. Comput. Geotech. **63**, 199–214 (2015)
59. Aubram, D., Savidis, S.A., Rackwitz, F.: Theory and numerical modeling of geomechanical multi-material flow. In: Triantafyllidis, T. (ed.) Holistic Simulation of Geotechnical Installation Processes, pp. 187–229. Springer (2016)
60. Arias, A.: A measure of earthquake intensity. In: Hansen, R. (ed.) Seismic Design for Nuclear Power Plants. MIT Press, Cambridge, Massachusetts (1970)

Infrastructure Projects and Monitoring Concepts

Prediction and Control of Ground Deformations for Mechanized Tunneling in Clays with Mixed Face Conditions

Andrew J. Whittle(✉) and Vasiliki Founta

Massachusetts Institute of Technology, Cambridge, USA
ajwhittl@mit.edu

Abstract. This paper describes a new methodology for predicting and controlling ground movements for a tunnel that traverses an interface between two clay layers of contrasting shear strength and stiffness. We use non-linear finite element analyses to investigate effects of face pressure on tunnel face stability and steady state ground surface deformations. This leads to a series of design charts linking soil properties (undrained shear strength and stiffness), stratigraphy (layer interface), tunnel cover depth and mechanized control parameters (face and grout pressure). The methodology has been validated for a recent case study where EPB tunnels traverse an interface between stiff clay (Old Alluvium) and soft, Marine clays. The proposed methodology successfully uses the measured face pressures to predict ground movements for these mixed face conditions. Further generalization of the method is now needed to represent mixed face conditions with contrasting permeability.

1 Introduction

Closed-face Earth Pressure Balance (EPB) tunnel boring machines have been very effective in reducing risks associated with urban tunneling projects. The EPB design controls face pressures within a sealed cutter-head chamber that is filled with the excavated soil. As the machine advances the soil is mixed with conditioning agents to produce a low shear strength paste (undrained shear strength, s_u = 20–30 kPa; Mair 2008) that can flow through the screw conveyor, Fig. 1a. Tunnel excavation inevitably generates sources of ground movements due to stress relief at the excavated face, overcutting and ploughing of the shield, complex boundary conditions around the tail void and deformations of the segmental lining system. The EPB operation aims to control ground deformations through the face pressure (and machine advance rate) and through grouting (and grout pressures) applied within the tail void. While the performance of EPB machines has been very impressive, especially in homogenous ground conditions (e.g., Ieronymaki et al. 2018), the control of ground movements is particularly difficult when the TBM encounters an interface between materials of contrasting shear strength, stiffness and permeability. These mixed face conditions can result in excessive settlements (including the creation sinkholes; Shirlaw 2008), due to over-excavation at the face, water inflow, abnormal or excessive equipment wear and time delays during construction (Zhao et al. 2007).

T. Triantafyllidis (Ed.): *Recent Developments of Soil Mechanics and Geotechnics in Theory and Practice*, LNACM 91, pp. 267–280, 2020.
https://doi.org/10.1007/978-3-030-28516-6_14

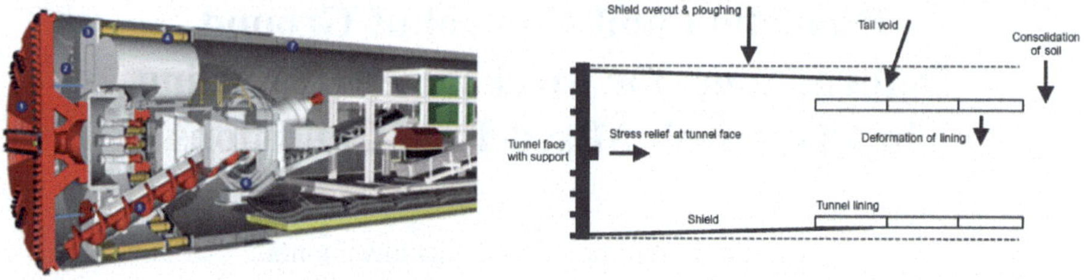

1) Cutterhead; 2) Face chamber; 3) Bulkhead; 4) Thrust cylinders; 5) Screw conveyor; 6) Segment erector; 7) Segmental lining

a) Design of EPB machine

b) Sources of tunneling-induced ground movements

Fig. 1. Closed-face EPB shield tunneling (adapted from Möller 2006)

The goal of this study was to develop a simplified framework for predicting the ground movements as a function of face pressures for tunneling across an interface between clay layers of contrasting strength and stiffness. We have developed a semi-empirical method based on non-linear finite element analyses and illustrate its application from a recent tunneling project. The analysis framework provides a predictive framework that can be used to establish target face pressures during EPB operations.

2 Methodology

In prior studies we have investigated predictions of EPB tunnel-induced ground movements using a variety of soil models and representations of the ground-TBM interactions (Founta et al. 2013; Founta 2017). These studies have shown that relatively simple representations of the tunnel advancement process through incremental stepping of the boundary conditions (Möller 2006) achieve comparable predictions of ground movements to more complex representations of the TBM as a separate structural body. Similarly, the zone of soil plasticity occurs in the near-field of the TBM for well-executed mechanized tunneling such that relatively simple soil models can be used to estimate far field ground deformations (Ieronymaki et al. 2016). The current analyses use the linearly-elastic perfectly plastic (EPP) model to represent soil behavior and assume that ground movements caused by closed-face EPB tunnel construction can be estimated by considering 4 independent sources of settlement: (1) Face pressures; (2) Grout pressures within the tail void (the grout sets between the assembled segmental lining and the surrounding soil); (3) Overcutting due to the oversize cutterhead and tapered diameter of the shield that are used to reduce friction; and (4) Net buoyancy between the weight of the soil excavated within the tunnel cavity and the weight of the tunnel boring machine.

3 Face Stability

In order to use control settlements induced by tunneling, it is first necessary to establish the face pressure needed to ensure stability of the tunnel face. Sloan (2013) conducted 3D numerical limit analyses (NLA; upper and lower bounds) for the undrained stability of a lined-tunnel in a homogeneous clay with an unsupported heading of length, P. The results in Fig. 2a, show that the computed face pressures, σ_f/s_u, required for stability tend to underestimate the measured data from centrifuge model tests (Mair 1979), but the measured data are close to the computed Lower Bound results. Tschuchnigg et al. (2015) have subsequently done detailed comparisons that show very good agreement between numerical limit analyses (rigorous solutions for rigid plasticity) and stability analyses obtained by stress reduction methods (SRM) that are widely used in displacement-based finite element software. The SRM method, considers an initial equilibrium state and then progressively reduces the Mohr-Coulomb (c', ϕ') shear strength parameters. The SRM method resolves any stress points that violate the strength criterion (using standard stress point algorithms that redistribute the stresses) at any stage of the analysis until reaching a state where equilibrium can no longer be achieved. The factor of safety for the SRM is defined by:

$$FS = \frac{tan\phi'}{tan\phi'_{mob}} = \frac{c'}{c'_{mob}} = \frac{s_u}{s_{umob}} \tag{1}$$

where subscript *mob* refers to the mobilized shear strength parameters, and s_u is the undrained shear strength of the clay.

The current methodology uses the SRM (*phi-c* method; Brinkgreve and Bakker 1991) to compute face stability within the finite element program Plaxis 3D (Brinkgreve et al. 2012). These analyses are conducted through a simple two-step process that creates a tunnel heading within a large FE model of the ground: (i) Tunnel excavation is simulated by simultaneous de-activation of soil elements within the tunnel cavity and activation of the plate-elements for the tunnel lining. The tunnel face is equilibrated at a constant face pressure representing average total stresses at the springline (depth, H); (ii) stability is evaluated directly through strength reduction.

Figure 2a shows that the current SRM analyses match closely the Lower Bound (LB) results presented by Sloan (2013). The mechanisms of failure vary with tunnel cover depth (C/D) for the EPB tunnel boring machine (where P/D = 0). Figure 2b shows further comparisons of the face stability number computed by SRM with results from prior limit analyses. The results show that stability number converges to N = 13 for C/D $\geq$ 3.5 corresponding to a failure mechanism for a deep tunnel that does not extend up to the ground surface.

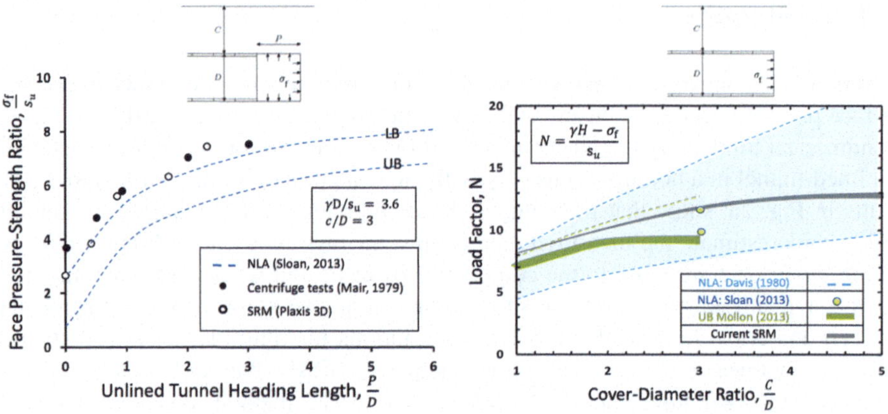

a) Effect of heading length, P

b) Effect of cover-diameter ratio

Fig. 2. Analysis of undrained face stability of lined tunnels in homogeneous clay

Strength reduction analyses can be readily extended to consider the face stability for mixed face conditions. For example, Fig. 3 illustrates the computed failure mechanisms for a tunnel heading that encounters two clay units of contrasting undrained shear strength. Here we consider a system comprising a soft clay (layer T) overlying a stronger clay (layer B) with undrained strength ratio, $s_{uB}/s_{uT} \geq 1$. The interface between the two layers bisects the tunnel face and is characterized by the embedment ratio, $-1 \leq D_T/D \leq 1$. The strength reduction method assumes the same factor of safety in both clay units. Figure 3 illustrates the face pressure required for stability for a shallow tunnel ($C/D = 1$) at a prescribed undrained strength ratio $s_{uB}/s_{uT} = 5$. The results are presented as functions of the embedment ratio, D_T/D, and strength of the upper clay unit, $s_{uT}/\gamma H$ (where H is the depth to the springline). For a given embedment depth, face pressure is only required when the strength of the upper clay unit is below a threshold value which increases with D_T/D. When the tunnel is fully embedded in the upper layer ($D_T/D = 1$) the face is stable for $\sigma_f/\gamma H = 0$ and $s_{uT}/\gamma H > 0.12$. In contrast, when the face is fully embedded in the lower clay ($D_T/D = 0.0$) the threshold strength of the upper unit is $s_{uT}/\gamma H = 0.04$, if the face pressure is increased to $\sigma_f/\gamma H = 0.5$ then the face is stable for $s_{uT}/\gamma H > 0.02$.

These results reflect differences in the face instability mechanisms associated with the two-layer system. Figure 4 illustrates the effects of the relative strength ratio, s_{uB}/s_{uT}, for a heading with C/D = 2 and D_T/D = 0.25. It is very interesting to observe the transition from a failure mechanism that extends across the full face of the tunnel and up to the ground surface (homogeneous case, s_{uB}/s_{uT} = 1), to a failure mechanism where the larger incremental deformations ($|\Delta u|/|\Delta u_{max}| > 0.75$) occur only within the upper soft layer (s_{uB}/s_{uT} = 2), or the majority of the mechanism ($|\Delta u|/|\Delta u_{max}| > 0.05$) is enclosed within the upper layer (s_{uB}/s_{uT} = 10).

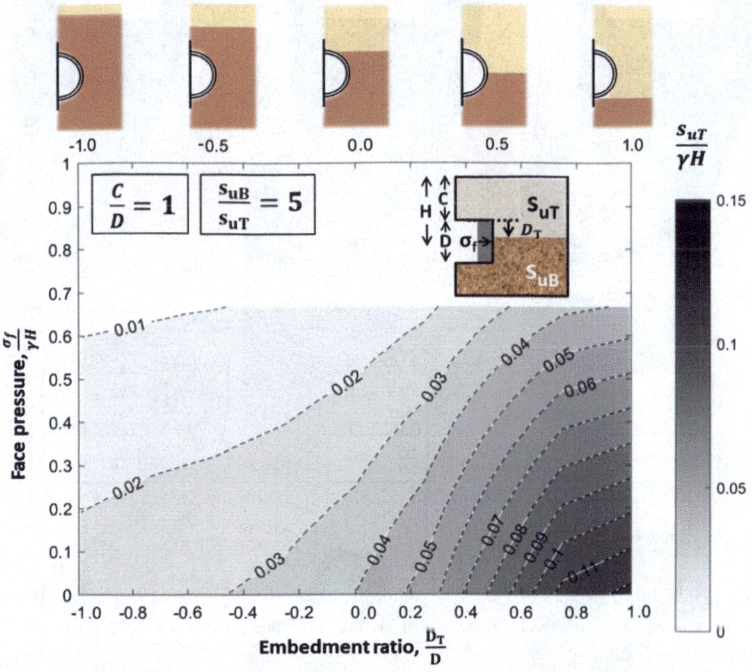

Fig. 3. Critical face pressure required to achieve stability for tunnel heading intersecting clay layers of contrasting strength computed by SRM

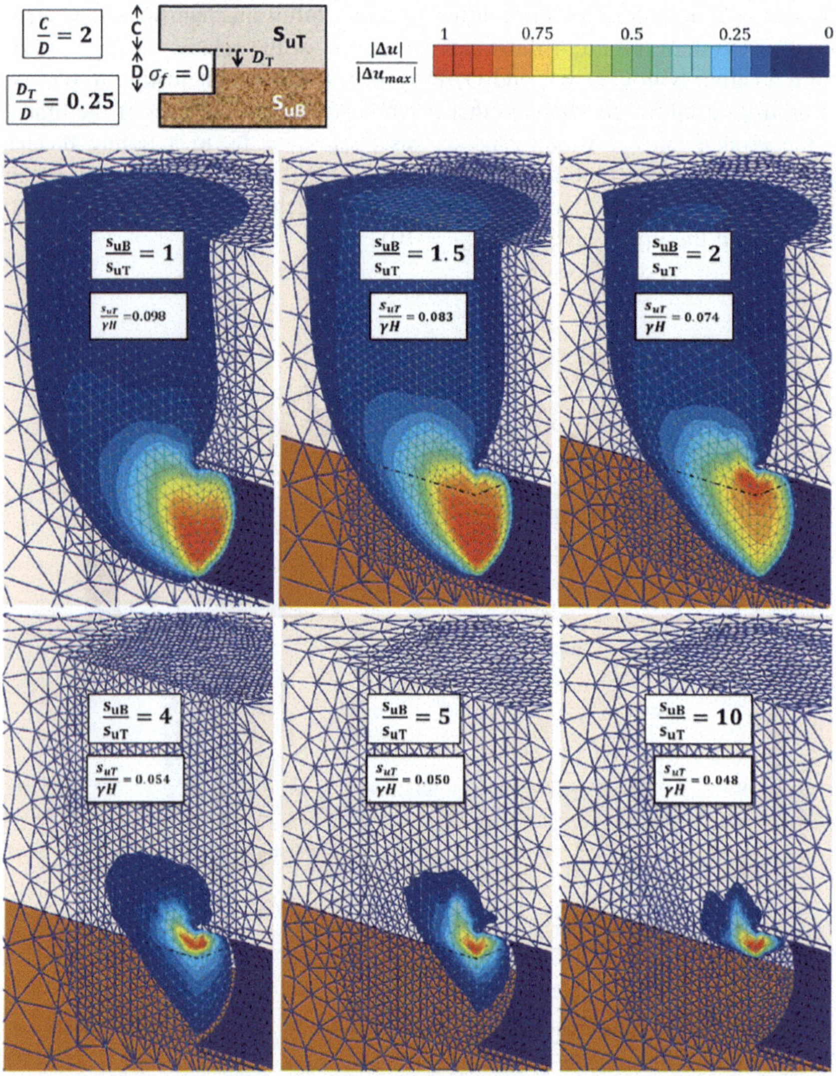

Fig. 4. Effect of relative strength ratio on collapse mechanisms for mixed face tunnel in clay

4 Ground Deformations Due to Face Pressures

The calculation of ground movements due to face pressures is accomplished using the same FE model used for face stability. The undrained shear behavior of the soil mass is represented by the EPP soil model (characterized by two parameters, G, s_u). We compute the net ground deformations associated with steady state advance of the TBM at a fixed face pressure. In practice, this is achieved by switching the frame of reference and integrating spatially the incremental movements associated with a single round of advance for the tunnel heading. The initial state involves equilibration of the lined tunnel (with fixed radius) with the heading located in the mid-section of the FE ground model (the analyses consider a reference tunnel with 7 m diameter and a ground model with length 270 m). Far-field boundaries are located such that there are effectively zero incremental movements at each step of the advancing tunnel heading. Net deformations (surface settlements, u_z) are then computed by integrating the incremental movements (δu_z) over the length of the FE model for each step. Figure 5 shows that this process typically converges within 4–5 rounds of tunnel advance and hence, enables very efficient calculation of the net settlement. The effects of the shield conicity and tail void grouting are not considered in this analysis but are analyzed separately.

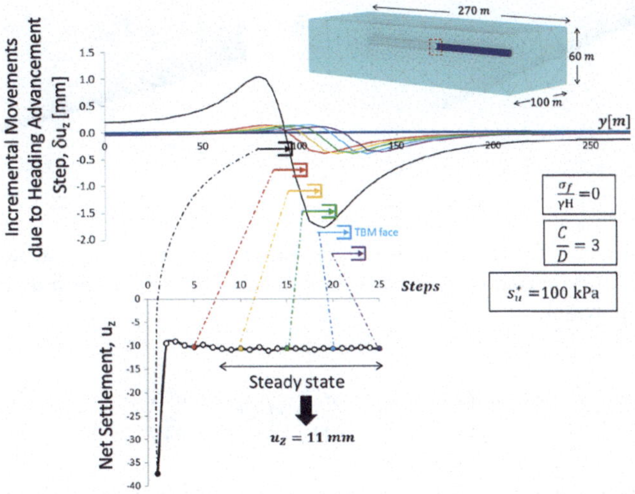

Fig. 5. Calculation of net settlement from incremental advances of heading at specified face pressure

Figures 6a and b illustrate typical predictions of the computed normalized net settlements for tunnels in homogeneous clays as functions of the face pressure and undrained shear strength at two cover ratios, C/D. The computed settlements are unbounded settlements for undrained strengths approaching critical values (s_u^c) obtained from the prior face stability calculations and converge to a range of residual deformations at high values of $s_u/\gamma H$. This behavior is well characterized by a rect-angular hyperbola:

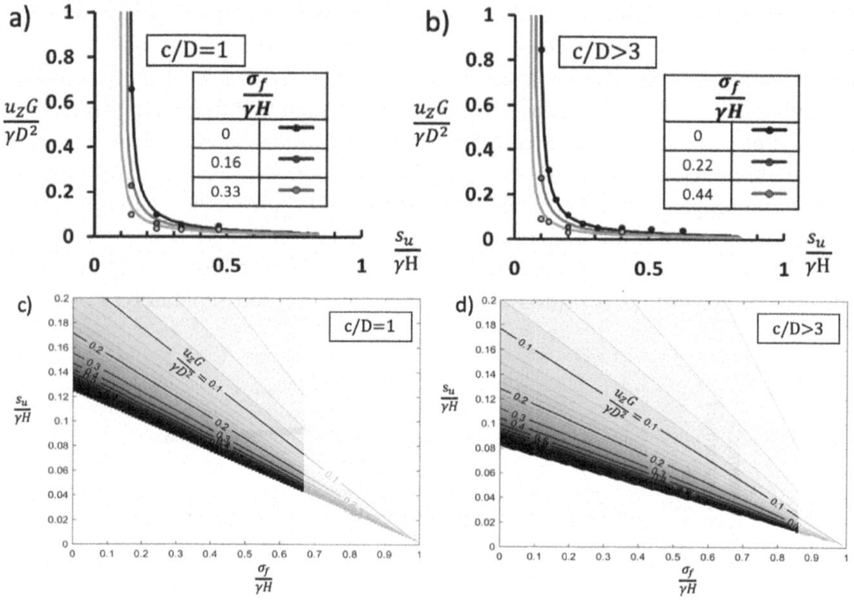

Fig. 6. Net settlements for tunnels in homogeneous clay (a), (b) computed results and curve fitting of data; (c), (d) derived design charts for estimating net settlement from undrained shear strength and face pressure

$$\frac{u_Z G}{\gamma D^2} = a\left(1 - \frac{\sigma_f}{\gamma \mathrm{H}}\right) / \left(\frac{s_u}{\gamma \mathrm{H}} - \frac{s_u^C}{\gamma \mathrm{H}}\right) \tag{2}$$

where the constant $a = a_{hom} = 0.012$, controls the residual net deformations and is obtained by least square fitting of the computed results.

Based on the known face stability results, Eq. 2 can be used to generate design charts that enable predictions of net settlements for specified combinations of (s_u and σ_f) as shown in Figs. 6c and d.

The same computational procedure can be used for 2-layer mixed face conditions with a constant interface embedment ratio, D_T/D. Figure 7 shows the computed net settlements as functions of the embedment ratio for a tunnel with C/D = 3 and undrained strength ratio, $s_{uT}/s_{uB} = 2$. These results highlight the relationship between the undrained shear strength and the critical face pressure (where uncontrolled settlements). The remaining contours can be used to guide the selection of face pressure and hence, control the predicted surface settlements, u_z.

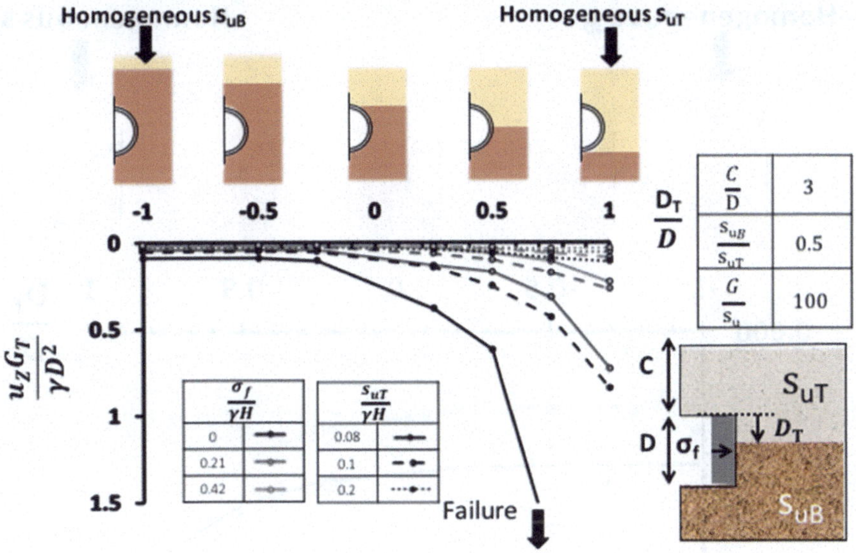

Fig. 7. Effect of tunnel embedment on computed settlements for two-clay layer mixed face conditions

A similar procedure can be used to relate face stability and surface settlements for mixed face conditions. Figure 7 summarizes the effects of the embedment ratio, D_T/D, on the computed surface settlements for a two clay layer system with $s_{uB}/s_{uT} = 2$ and $C/D = 3$ and selected values of the face pressure and undrained shear strength of the overlying clay ($\sigma_f/\gamma H$, $s_{uT}/\gamma H$). The surface settlements increase with the embedment ratio and failure occurs for the case (0.0, 0.08) at an embedment ratio, $D_T/D \approx 0.7$.

For given face pressure, the surface settlements can again be related to the undrained shear strength of the overlying clay (s_{uT} and the critical face pressure, $s_u^c/\gamma H$ as previously reported in Fig. 3) through a rectangular hyperbola function Eq. 2, where the constant is now a function of the embedment and undrained shear strength ratios, *a* (D_T/D, s_{uT}/s_{uB}).

Figure 8 shows the best fit values of the parameter *a* for the two-layer case (using the least squares method). The overall response is well-characterized by a logistic dose response curve (e.g., Joseph and Yang 2010) with input parameters as follows:

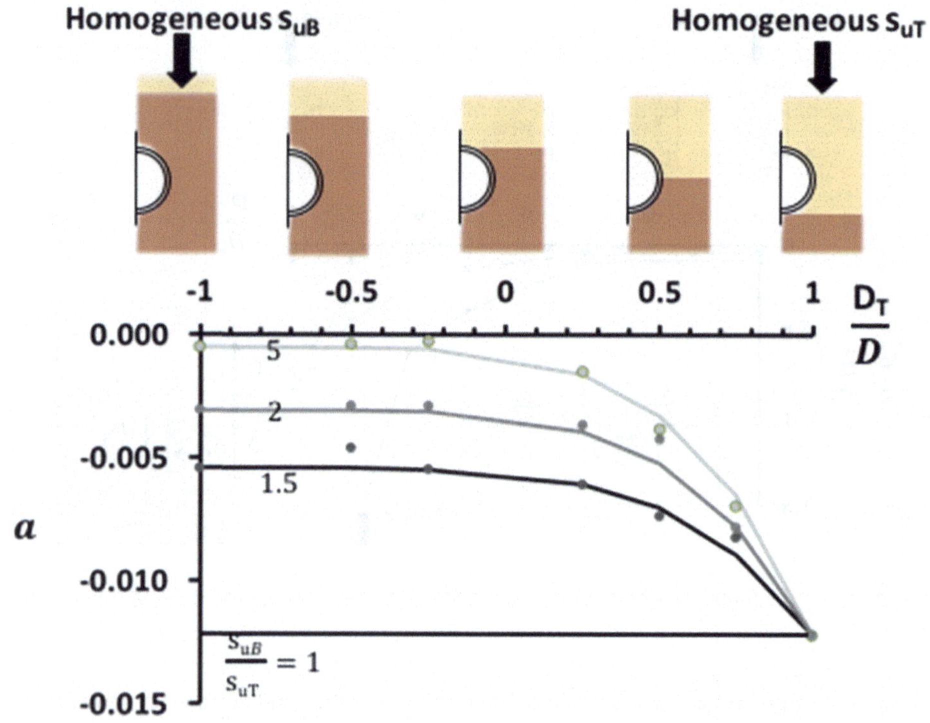

Fig. 8. Empirical interpretation of constant a in Eq. 2 for tunnels with mixed face conditions. Computed results in symbols, fitted logistic dose response function in solid lines

$$a = a_{hom}\left\{\left(1 - \left(\frac{s_{uT}}{s_{uB}}\right)^2\right)\left[\left(\frac{\frac{D_T}{D}+1}{2}\right)^5 - 1\right] + 1\right\} \tag{3}$$

Based on these results it is now possible to predict the surface settlements as a function of the embedment ratio, reference undrained strength ($s_{uT}/\gamma H$) and face pressure ($\sigma_f/\gamma H$). Figure 9 illustrates the resulting design charts. These results show how the face pressure must be increased in order to control surface settlements as the tunnel transits through the mixed face condition between the two clay layers.

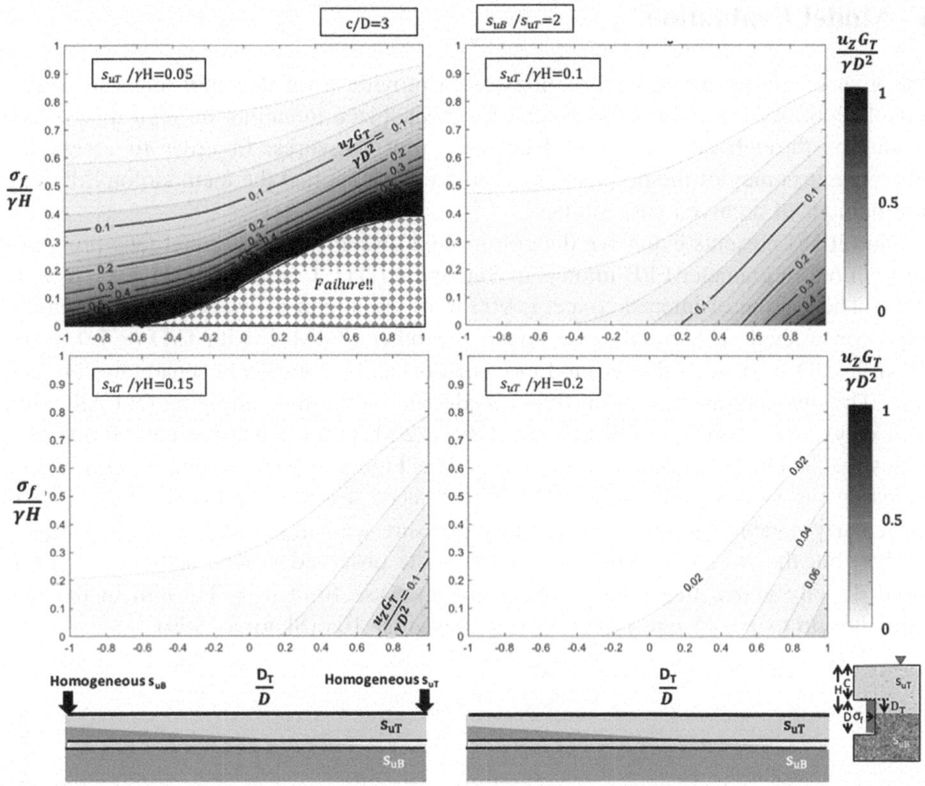

Fig. 9. Proposed design charts for predicting the effects of face pressure on the control of surface settlements of a tunnel heading for a two-layer, mixed face clay condition

5 Analyses of Other Sources of Ground Deformations

As indicated in the introductory section we have considered three other potential sources of ground deformations associated with EPB tunneling: The effects of grouting within the tail void are simulated by an independent analytical model following a methodology very similar to that used for face pressures. We first consider stability conditions for the tail void i.e., the critical undrained shear strength for an unsupported annulus (zero grout pressure over one-round length of segmental lining behind the shield) and then compute surface settlements as a function of the grout pressure that is actually applied within this void. The relationship between the surface settlement and grout pressure is similarly represented by a rectangular hyperbolic function that is calibrated to the computed results for each ground condition.

Effects of overcutting and buoyancy can be estimated from the known cavity contraction associated geometry of the TBM shield and cutterhead and from the ratio of the TBM to excavated soil weight, respectively. Far field ground deformations due to these effects can be well-estimated from elastic theory using analytical solutions (Pinto and Whittle 2014).

6 Model Evaluation

The goal of the proposed methodology is to provide a reliable tool that can enable reliable control of ground deformations for mechanized tunneling through mixed face conditions through adjustment of face and grout pressures. In order to assess the predictive accuracy of the proposed method we have applied the methodology directly to a number of archived case studies.

Su (2015) presents extensive data on ground settlements and tunnel face pressures for twin 6.6 m diameter EPB tunnels in Singapore (DTL Contract C933), Figs. 10a, b, and c. The section of interest concerns 900 m of the alignment where the twin tunnels were constructed sequentially one above the other (TBM 1 with C/D $\approx 5.0 \pm 03$; TBM 2, C/D = 3), such that ground movements can be considered separately for each bore. The tunnels transition from full embedment within the underlying Old Alluvium (stiff clay) layer with $s_u = 250$ kPa and $G/s_u = 200$; into a much weaker Marine clay with $s_u \approx 40$ kPa (i.e., $s_{uB}/s_{uT} = 6.25$), Fig. 10c. Figure 10b shows that the contractors increased the face pressure as each TBM advanced towards the marine clay. This is particularly notable for TBM2 where face pressure was increased from $\sigma_f = 2.5$ bar–4.5 bar, but this was not sufficient to control the observed ground settlements (measured directly above the tunnel crown) that increase from $u_z = 3$–5 mm in the Old Alluvium up to 20–25 mm in the Marine clay, Fig. 10a (Sharma 1999).

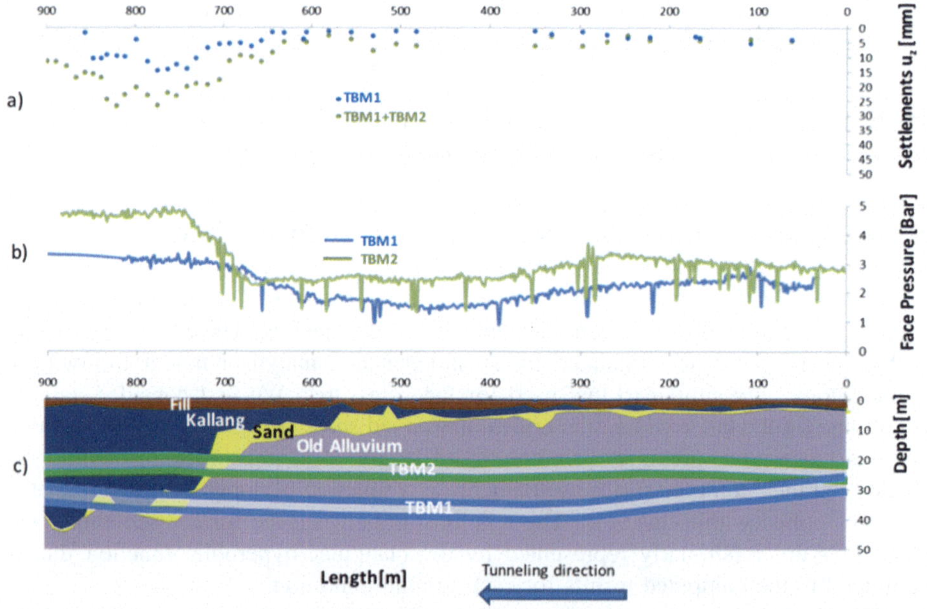

Fig. 10. Surface settlements, face pressures and stratigraphy for EPB tunnel construction at DTL C933 (after Su 2015)

The deformations at each location were predicted by superimposing the calculated displacements due to the TBM contraction and those due to the applied face pressure, Fig. 11. For the purpose of this comparison, we decided to ignore the effect of the grout pressure as high grout pressures were applied over this segment (close or equal to the geostatic stresses), resulting in minimal disruption of the surface.

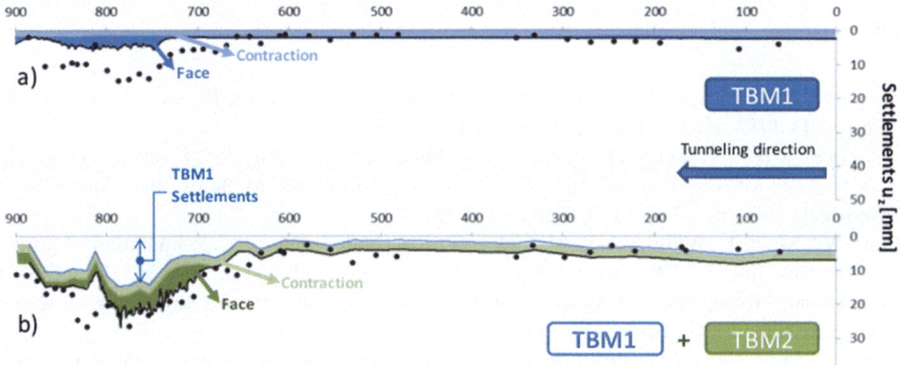

Fig. 11. Comparison of predicted and measured surface settlements due to EPB tunnel construction, DTL C933

Overall, there is a good agreement between the predictions and the measured surface settlements, Figs. 10a and b. For TBM1, there is excellent agreement for the segments where the EPB is embedded within the stiff old alluvium formation (0 to 700 m). However, the analysis underestimates settlements as the heading progresses into the Marine clay (Fig. 11a). This result may by caused by the presence of pockets of sand (unit F1). The performance of the shallower bore, TBM2, is much more reliably estimated (Fig. 11b; only one pocket of sand at around 700 m) showing the potential of the proposed analyses to represent realistically the ground movements.

7 Conclusions

This paper has demonstrated the use of computational models based on 3D finite element analyses to predict the relationship between tunnel geometry (C/D, DT/D), control parameters (face pressure, $\sigma_f/\gamma H$ and grout pressure), soil properties ($s_u/\gamma H$, s_{uB}/s_{uT}) and surface settlements ($u_z G/\gamma D^2$) for EPB tunnels. The current analyses have focused on a two-layer clay system, but the methodology employed in this work is general and can be applied to more complex ground profiles. The key assumption is that control of face pressures ensures that tunnel construction induces relatively small deformations in the surrounding ground (while face stability conditions are accurately estimated by SRM). The proposed methodology offers a tool that can be used to assess the control of ground movements based on a series of relatively simple numerical analyses. Initial evaluations for a recent case study in Singapore show that the proposed framework can provide credible predictions of performance and hence, could be used to design the face pressures for mixed face conditions.

Acknowledgements. This research was supported in part by Ferrovial-Agroman through the MIT-Ferrovial research Program and by the MIT-Singapore Alliance for Research and Technology (SMART) through the Center for Environmental Sensing and Modeling. The second Author is also grateful for support for the Onassis Foundation. We would also like to thank Dr Thiri Su (LTA) for her assistance in interpreting data from the DTL C933 project.

References

Brinkgreve, R.B.J., Engin, E., Swolfs, W., Waterman, D., Chesaru, A., Bonnier, P., Galavi, V.: Plaxis 3D 2012, Users Manual. Plaxis, Delft (2012)

Brinkgreve, R.B.J., Bakker, H.L.: Non-linear finite element analysis of safety factors. In: Proceedings of the International Conference on Computer Methods and Advances in Geomechanics, pp. 1117–1122. Balkema (1991)

Founta, V., Ninic, J., Whittle, A.J., Meschke, G., Stascheit, J.: Numerical simulation of ground movements due to EPB tunnelling in clay. In: Proceedings of the 3rd EURO:TUN Conference: Computational Methods in Tunneling and Subsurface Engineering, Bochum, Germany (2013, in print)

Founta, V.: Prediction of instability and ground movements during tunnel construction in non-homogeneous conditions. Ph.D thesis. Massachusetts Institute of Technology, Cambridge, MA (2017)

Ieronymaki, E.S., Whittle, A.J., Simic, D.: Interpretation of greenfield ground movements caused by mechanized tunnel construction. ASCE J. Geotech. Geoenviron. Eng. (2016). https://doi.org/10.1061/(ASCE)GT.1943-5606.0001632

Ieronymaki, E., Whittle, A.J., Einstein, H.H.: Comparative study of the effects of three tunneling methods on ground movements in stiff clay. Tunn. Undergr. Space Technol. **74**, 167–177 (2018)

Joseph, D.D., Yang, B.H.: Friction factor correlations for laminar, transition and turbulent flow in smooth pipes. Physica D **239**(14), 1318–1328 (2010)

Mair, R.J.: Centrifuge modeling of tunnel construction in soft clay. Ph.D thesis. Cambridge University (1979)

Mair, R.J.: Tunnelling and geotechnics: new horizons. Géotechnique **58**(9), 695–736 (2008)

Möller, S.C.: Tunnel induced settlements and structural forces in linings. Ph.D thesis. Institute of Geotechnical Engineering, University of Stuttgart (2006)

Pinto, F., Whittle, A.J.: Ground movements due to shallow tunnels in soft ground: 1. Analytical solutions. ASCE J. Geotech. Geoenviron. Eng. **140**(4), 0401.3040 (2014)

Sharma, J.S., Chu, J., Zhao, J.: Geological and geotechnical features of Singapore: an overview. Tunn. Undergr. Space Technol. **14**(4), 419–431 (1999)

Shirlaw, J.N.: Mixed face conditions and the risk of loss of face in Singapore. In: International Conference on Deep Excavations (ICDE 2008), Singapore, p. 7 (2008)

Sloan, S.W.: Geotechnical stability analysis. Géotechnique **63**(7), 531–572 (2013)

Su, T.: Study on ground behavior associated with tunneling In mixed-face soil condition. Ph.D thesis. National University of Singapore, Singapore (2015)

Tschuchnigg, F., Schweiger, H.F., Sloan, S.W., Lyamin, A.V., Raissakis, I.: Comparison of finite-element limit analysis and strength reduction techniques. Géotechnique **65**(4), 249–257 (2015)

Zhao, J., Gong, Q.M., Eisensten, Z.: Tunnelling through a frequently changing and mixed ground: A case history in Singapore. Tunn. Undergr. Space Technol. **22**(4), 388–400 (2007)

Two Neighbouring Tunnels in Saturated Soil Under Blast Loading

Vladimir A. Osinov(✉) and Stylianos Chrisopoulos

Institute of Soil Mechanics and Rock Mechanics,
Karlsruhe Institute of Technology, 76128 Karlsruhe, Germany
{vladimir.osinov,stylianos.chrisopoulos}@kit.edu

Abstract. The paper deals with the numerical modelling of the blast-induced deformation of two neighbouring shallow tunnels and the surrounding soil. The deformation is caused by an explosion inside one of the tunnels. The explosion is simulated by a short-term pressure load of moderate amplitude (8 MPa) applied to the tunnel lining. The lining of both tunnels is circular with an inner diameter of 9.6 m and consists of concrete segments (tubbings) assumed to be linearly elastic. The tunnels are located at a depth of 17 m in fully saturated soil. The effective stresses in the soil are described by a hypoplasticity model. The modelling incorporates pore water cavitation at zero absolute pore pressure. The dynamic problem is solved in a two-dimensional plane-strain formulation with the finite-element program Abaqus/Standard. The transient deformation of the tunnel lining and the soil is analysed in detail. In particular, the solution reveals the emergence of large cavitation zones in the soil during the dynamic deformation.

1 Introduction

The problem of an explosion in a tunnel is a dynamic soil-structure interaction problem whose solution is determined by both the design of the structure (the tunnel lining) and the mechanical properties of the surrounding medium (soil, rock). The ground surface also plays an important role for shallow tunnels and should be included in the formulation of the problem. The modelling is usually aimed at finding the deformation and possible damage of the tunnel lining and neighbouring structures, e.g. an adjacent tunnel or a construction above the tunnel. If the tunnel is situated in saturated soil, an important question is that of the expected residual deformations and permanent stress changes in the soil. If the effective pressure in the soil is reduced to zero, the soil turns into a liquefaction state. Saturated soils can liquefy not only under cyclic shear deformation typical of earthquake-induced loading but also after large-amplitude compression-extension cycles [1,2].

This paper presents results of the numerical modelling of the dynamic deformation of two neighbouring tunnels and the surrounding saturated soil induced by an explosion inside one of the tunnels. Similar problems with a single tunnel

T. Triantafyllidis (Ed.): *Recent Developments of Soil Mechanics and Geotechnics in Theory and Practice*, LNACM 91, pp. 281–296, 2020.
https://doi.org/10.1007/978-3-030-28516-6_15

have been studied in [3,4]. The explosion is simulated by a short-term pressure load of moderate amplitude (8 MPa) applied to the tunnel lining. The lining of both tunnels is circular and consists of concrete segments (tubbings) assumed to be linearly elastic. The constitutive description of the soil is based on the effective-stress principle. The effective stresses are described by a hypoplasticity model for granular soils. The choice of a constitutive model for soil in problems with blast-induced deformation is dictated by the magnitude of the applied pressure. An increase in the effective pressure in a fully saturated soil under undrained conditions is a small fraction of the total pressure applied to the soil. For the loading considered in this paper, the transient effective-stress changes in the soil do not exceed a few hundred kilopascals and lie within the validity range of the hypoplasticity model. The constitutive description of the pore water includes cavitation with zero stiffness at zero absolute pore pressure and linearly elastic behaviour for a nonzero pressure.

The problem is solved with the finite-element program Abaqus/Standard in a two-dimensional plane-strain formulation. Although a real explosion originates from a point source and requires a three-dimensional formulation, an advantage of the two-dimensional formulation is that it enables us to solve the problem with fine spatial discretization. As pointed out in [3,4], problems with blast-induced soil deformation require finer spatial discretization than problems with quasi-static or earthquake-induced loading for the same geotechnical structure. Close inspection of the literature reveals that this requirement is often disregarded and the computational mesh is too coarse, especially for three-dimensional problems (for literature on the numerical modelling of blast-induced soil-structure interaction, see e.g. references in [5,6]). The necessary degree of spatial discretization depends on the spatial variation of the solution. Previous studies [3,4] as well as theoretical estimations show that, in order to obtain reliable numerical solutions, the grid spacing in a blast-induces soil-tunnel interaction problem should be as small as several centimetres. The present two-dimensional formulation allows us to obtain accurate solutions on a fine mesh and to analyse in detail the deformation of the tunnel lining and the soil.

2 Constitutive Description of the Soil

The total stress tensor in the soil is considered as the sum of the effective-stress tensor $\boldsymbol{\sigma}$ (compressive stresses are taken to be negative) and the isotropic tensor $-p_f\mathbf{I}$, where p_f is the pore pressure (positive for compression) and $\mathbf{I}$ is the unit tensor. The effective stress is described by a hypoplasticity model with intergranular strain proposed in [7] for cohesionless granular soils with rate independent behaviour. The model is an extension of an earlier version of hypoplasticity [8] and includes the intergranular-strain tensor as a new state variable in addition to the stress tensor and the void ratio. The constitutive equation gives the effective-stress rate $\dot{\boldsymbol{\sigma}}$ as a function of the strain rate $\dot{\boldsymbol{\varepsilon}}$, the effective-stress tensor $\boldsymbol{\sigma}$, the intergranular-strain tensor $\boldsymbol{\delta}$ and the void ratio e:

$$\dot{\boldsymbol{\sigma}} = \mathbf{H}\left(\dot{\boldsymbol{\varepsilon}}, \boldsymbol{\sigma}, \boldsymbol{\delta}, e\right). \tag{1}$$

The evolution of the intergranular-strain tensor is governed by the equation

$$\dot{\boldsymbol{\delta}} = \mathbf{F}(\dot{\boldsymbol{\varepsilon}}, \boldsymbol{\delta}). \tag{2}$$

For a detailed description of the functions **H** and **F** in (1), (2), see the original paper [7] or [9,10]. The constitutive parameters used in the present study are given in Tables 1 and 2.

Table 1. Constitutive parameters of Karlsruhe sand [11]

φ_c [°]	h_s [MPa]	e_{c0}	e_{d0}	e_{i0}	α	β	n
30	5800	0.84	0.53	1.0	0.13	1.0	0.28

Table 2. Additional constitutive parameters of the extended hypoplasticity model with intergranular strain [10]

R	m_R	m_T	β_r	χ
4×10^{-5}	5.0	5.0	0.05	1.5

The constitutive equation (1) is corrected at small stresses as described in [4]. The correction consists in the multiplication of the function **H** by a factor which depends on the mean stress. The correction has an effect on the constitutive response only at small stresses (below 1 kPa in absolute value). This small-stress correction, originally proposed in [12] for a different constitutive model, is used in order to avoid numerical problems caused by zero effective stress when the soil undergoes expansion. In the problem considered in this paper, large tensile deformations in the vertical direction are produced in the upper layer of dry soil after the reflection of the pressure wave from the free surface. With the correction factor introduced in (1), the mean stress in tension approaches zero asymptotically.

Assuming locally undrained conditions (zero soil permeability) and incompressible solid phase, the constitutive equation for the pore pressure reads

$$\dot{p_f} = -\frac{K_f}{n}\,\mathrm{tr}\,\dot{\boldsymbol{\varepsilon}}, \tag{3}$$

where K_f is the pore fluid bulk modulus, n is the porosity, and the deformation rate is negative for compression. In fully saturated soil, K_f is equal to the bulk modulus of pure water $K_w = 2.2\,\mathrm{GPa}$. If the soil undergoes tensile deformation and the absolute pore pressure (including the atmospheric pressure) becomes sufficiently low, the pore water begins to cavitate and the pore fluid bulk modulus abruptly falls to a nearly zero value. We assume that the pore water cannot withstand tensile stresses and that cavitation therefore begins at zero absolute pore pressure. If p_f^0 is the initial pore pressure in a soil element and ε_f is the current volumetric strain of the pore fluid (zero at p_f^0), the pore fluid bulk modulus is

$$K_f = \begin{cases} K_w & \text{if} \quad \varepsilon_f < (p_f^0 + p_a)/K_w, \\ 0 & \text{if} \quad \varepsilon_f \geq (p_f^0 + p_a)/K_w, \end{cases} \tag{4}$$

where p_a is the atmospheric pressure (100 kPa). The volumetric strain of the pore fluid, ε_f, is equal to the volumetric strain of the soil, $\operatorname{tr}\boldsymbol{\varepsilon}$, divided by the porosity. Equation (4) produces a piecewise-linear strain-pressure dependence for the pore fluid as shown by the solid line in Fig. 1. The compression from a cavitated state with $p_f = -p_a$ is assumed to follow the same piecewise-linear curve.

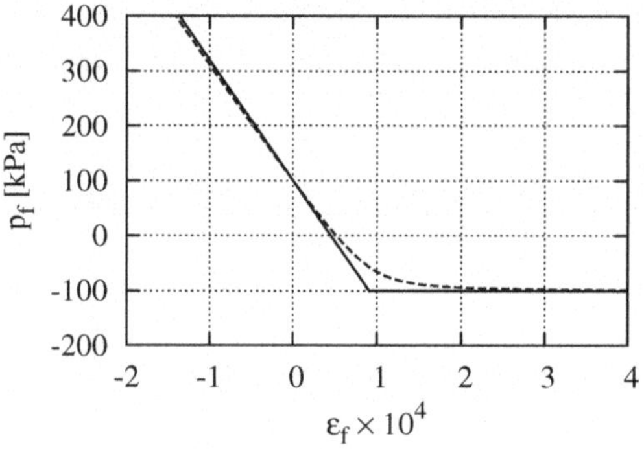

Fig. 1. Pore fluid pressure as a function of the pore fluid volumetric strain with $p_f^0 = 100$ kPa. Solid line: the function obtained with (4), dashed line: the smooth approximation

Figure 2 shows an example of uniaxial (oedometric) extension of saturated soil calculated with the hypoplasticity model for the soil skeleton. The kink on the total-stress curve shown by the solid line corresponds to the beginning of pore water cavitation according to (4). The total-stress curve for the compression from a cavitated state with $p_f = -p_a$ also has a kink at the point that corresponds to the closure of the voids in the pore water or, in terms of relation (4), at the point where $\varepsilon_f = (p_f^0 + p_a)/K_w$. The kink on the strain-stress curve due to pore water cavitation introduces a strong nonlinearity in the overall constitutive response of the soil in compression-extension in addition to the comparatively weak nonlinearity in the effective-stress response, see the curve for the effective stress in Fig. 2.

The strong nonlinearity in the compression from a cavitated to a fully saturated state is caused by a sharp increase in the soil stiffness at the moment when the cavitation voids in the pore water disappear. This increase has a substantial qualitative influence on the solutions of a dynamic problem. The increasing stiffness in combination with a high strain rate may lead to the formation of discontinuities (shock fronts) [13]. Examples related to a similar problem (an explosion in a tunnel) can be found in [3]. The shock fronts in [3] arise because of a small amount (few volume percent) of free gas in the soil and, as a consequence, the sharply increasing soil stiffness in compression. In the present study,

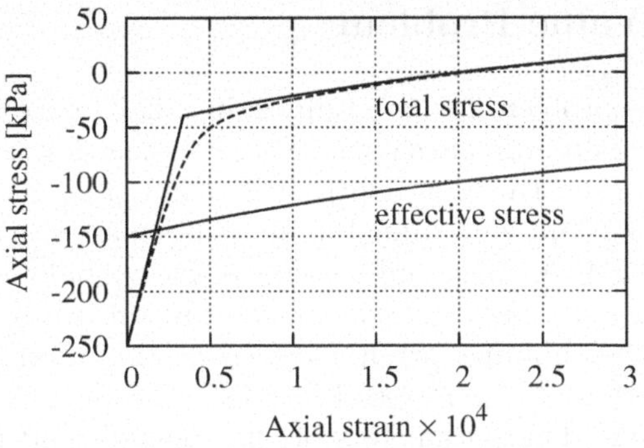

Fig. 2. Uniaxial extension of saturated soil calculated with the hypoplasticity model with a hydrostatic initial effective stress of -150 kPa, an initial pore pressure of 100 kPa and a void ratio of 0.6. Dashed line: the smooth approximation of pore water cavitation

along with the cavitation-related mechanism of shock formation, a shock front will also arise in the upper layer of dry soil where the mean stress will first be reduced to a nearly zero value during the wave reflection and then gradually restored during the subsequent compression. The increasing stiffness in compression will produce a shock front.

The formation and propagation of shock fronts make the numerical solution of a dynamic problem rather difficult, especially with commercial programs originally not intended to deal with discontinuous solutions. Besides the fact that problems with shock fronts require fine spatial discretization, they also require special means such as, for instance, viscous stresses. Otherwise the solution may be completely spoiled by spurious numerical oscillations. Viscous stresses with constant viscosity may not suffice: the oscillations may be eliminated, but at the cost of unrealistically high damping and the excessive smearing of the shock profile because of the too high viscosity. A more complicated approach may be needed to obtain satisfactory solutions.

In the present study, spurious oscillations in the numerical solutions are reduced by combining a smooth approximation of (4) and viscous stresses. First, the piecewise-linear strain-pressure relation for the pore pressure is approximated by a smooth curve as shown in Fig. 1 by the dashed line. The approximation is obtained by using the bulk modulus of a mixture of water and free gas instead of the modulus determined by (4). The smooth curve in Fig. 1 corresponds to the compression-extension of a mixture with an initial degree of saturation of 0.99999 at a pressure of 100 kPa. Second, viscous stresses are introduced in addition to the constitutive stresses (1) and the pore pressure. The values of the viscosity coefficients which produce satisfactory results have been found by trial and error. Constant viscosity coefficients turned out to be insufficient to achieve the desired effect for cavitation-induced shocks. The viscosity is made variable with a sharp increase during the transition of the pore fluid bulk modulus from zero to 2.2 GPa. For more detail, see [4].

3 Boundary-Value Problem

The boundary-value problem for two tunnels in a half-space is solved as a two-dimensional plane-strain problem shown in Fig. 3. The water table and the tunnel centres are located, respectively, at depths of 2 m and 17 m. The distance between the tunnel centres is 25 m. The tunnel lining is circular with an inner radius of 4.8 m. The lining is 0.45 m thick and consists of individual concrete tubbings as shown in Fig. 3. The tubbings and the concrete invert are modelled as linearly elastic materials. The tubbings are not bolted at the interfaces. The forces at the concrete-concrete and concrete-soil interfaces result from dry friction with a given friction angle. The parameters of the concrete and the friction angles are given in Table 3. The concrete-soil friction angle is taken to be zero because the contact algorithms available in Abaqus does not allow the user to model nonzero friction between concrete and saturated soil. The maximum shear stress at the contact surface is determined by the effective normal stress in the soil, whereas Abaqus uses the total normal stress when calculating the shear stress. The contact algorithms available in Abaqus can allow or forbid separation in the normal direction, depending on the user's choice. The problem is solved with separation for concrete-concrete interfaces and without separation for concrete-soil interfaces.

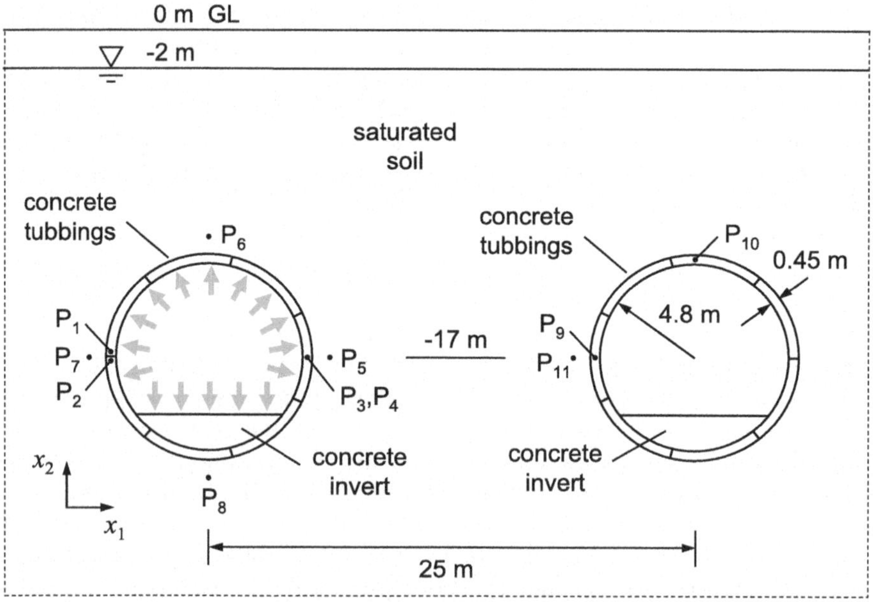

Fig. 3. Two tunnels in the half-space

Table 3. The concrete parameters and the friction angles

Young modulus [GPa]	40
Poisson ratio	0.2
Density [kg/m^3]	2400
Concrete-concrete friction angle [°]	45
Concrete-soil friction angle [°]	0

The initial stresses in the soil and in the tunnel lining are in static equilibrium with the gravity and are hydrostatic in the far field. The initial void ratio e is homogeneous and equal to 0.6. This corresponds to a relative density of 0.77 estimated as $(e_{c0} - e)/(e_{c0} - e_{d0})$, see Table 1. An explosion inside the left tunnel is simulated by a given pressure applied to the inner surface of the tunnel. The boundary condition for all points of the surface is shown in Fig. 4. The size of the computational domain is 250 m × 150 m. Transparent boundary conditions at the outer boundary are not used in the calculations. The type of boundary conditions at the outer boundary plays no role in this case, as the numerical solutions are valid only until the waves reflected from the outer boundary reach the domain of interest around the tunnels. The present computational domain enables us to model up to 120 ms.

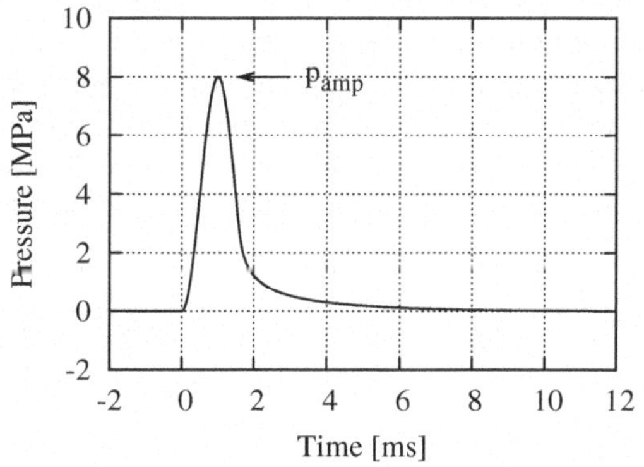

Fig. 4. Pressure applied to the inner surface of the tunnel

The boundary-value problem is solved with the finite-element program Abaqus/Standard with the implicit Hilber-Hughes-Taylor time integration scheme using the 4-node bilinear quadrilateral elements CPE4 with selectively reduced integration. The element size varies from about 6 cm in the tubbings and in the soil near the tunnels to 1 m near the outer boundary of the domain. The finite-element mesh near the lining is shown in Fig. 5. The time increment is 10^{-5} s.

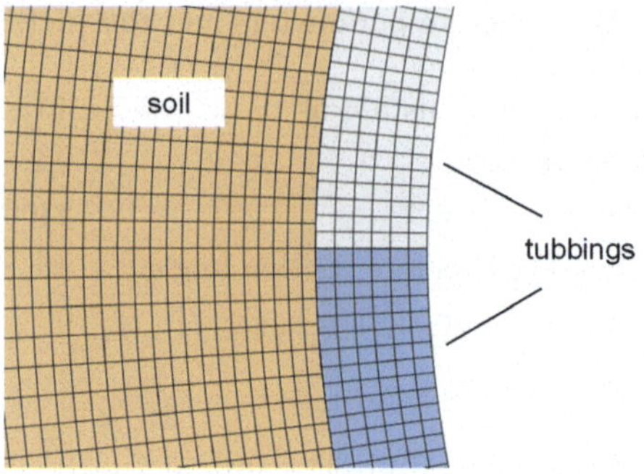

Fig. 5. Finite-element mesh near the lining

4 Deformation of the Lining

The pressure pulse shown in Fig. 4 is applied to the inner surface of the left tunnel. The motion of the tubbings can be traced by considering the displacements of two points P_1 and P_2 located in two neighbouring tubbings of the left tunnel near the contact surface between them as shown in Fig. 3. The displacement components at these points are shown in Fig. 6. The applied pressure causes the tubbings to move in the radial direction, see the curve for the horizontal displacement component u_1. Since the tubbings are not connected at the interface, their radial displacements result in the loss of contact between them at the beginning of the motion (in 1 ms), see two curves for the vertical component u_2. The maximum gap between the tubbings in this separation phase reaches 6 mm. The two curves for the vertical displacements also reveal high-frequency oscillations of the tubbings in the circumferential direction during the separation phase. The oscillations will be seen better from the stress curves below.

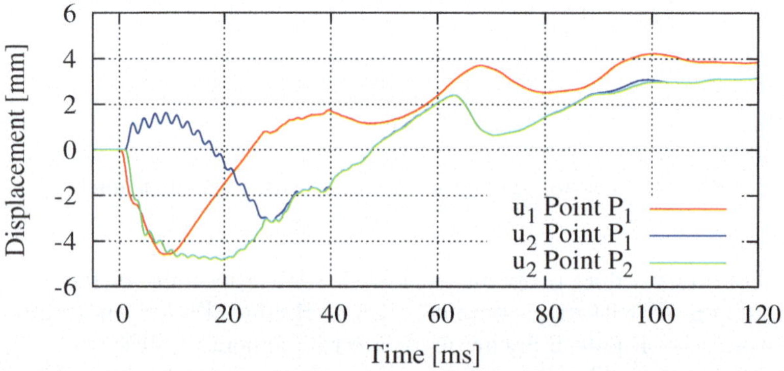

Fig. 6. Displacement components in the tubbings at points P_1 and P_2

The soil reaction decelerates the radial motion of the tubbings and then accelerates them in the opposite direction to the tunnel centre until the contact between them is restored again. As seen from Fig. 6, the separation phase lasts 27 ms. The duration of the separation phase may differ by few milliseconds for different pairs of tubbings. The maximum radial displacement of the tubbings during the separation phase is 4.5 mm, see the curve for u_1 in Fig. 6.

In order to trace the stresses induced in the tunnel lining during its deformation, consider two points P_3 and P_4 located, respectively, near the free surface and near the concrete-soil interface in the middle part of the tubbing, see Fig. 3. The circumferential stress component at points P_3 and P_4 is shown in Fig. 7. The curves in the figure reveal high-frequency oscillations of the tubbing in the circumferential direction during the separation phase mentioned above. The oscillations arise in the form of a standing wave with a frequency of 480 Hz. The frequency depends on the tubbing length in the circumferential direction and would be higher for shorter tubbings as is the case, for instance, in [4]. The maximum tensile and compressive stresses during the oscillations are, respectively, 15 and 12 MPa. These values are greater than the pressure applied to the lining (8 MPa). Since the oscillations occur in the form of a standing wave, the stress amplitude varies along the tubbing: it is maximal in the middle part and vanishes near the ends.

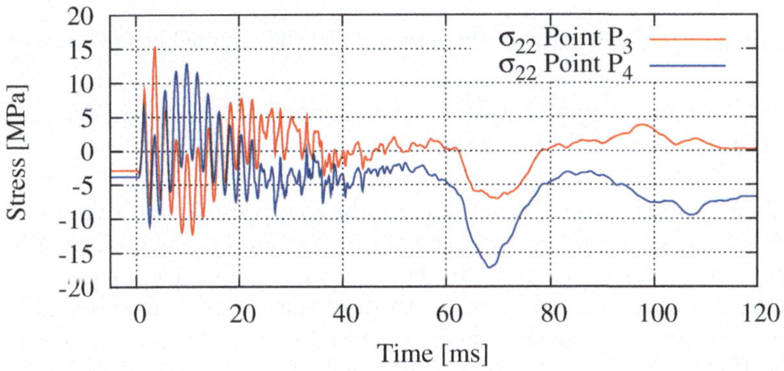

Fig. 7. Circumferential stresses in the lining of the left tunnel at points P_3 and P_4 located, respectively, near the free surface and near the concrete-soil interface

The closure of the gaps between different pairs of tubbings at the end of their separation phases is not simultaneous. The tubbings collisions during the nonsimultaneous closure induce irregular oscillations of the circumferential stress in the interval between 25 and 50 ms observed in Fig. 7. In the time between 60 and 80 ms, the curves in Fig. 7 show a relatively slow loading-unloading cycle with a high compressive stress between 7 and 17 MPa. This indicates that the whole lining is subjected to an external pressure. The cause of this pressure will be evident from the analysis of the soil deformation in the next Section.

The pressure pulse applied to the lining of the left tunnel is transmitted to the soil and transforms into a pressure wave. The wave propagates in the soil and compresses the lining of the right tunnel leading to an increase in the compressive circumferential stresses in the tubbings and a few oscillations of the lining as a whole. This is seen from Fig. 8 which shows the circumferential stresses at points P_9 and P_{10} in the tubbings of the right tunnel. The maximum increase in the compressive stresses is about 7 MPa. Neither tensile stresses nor separation of the tubbings are observed in the lining of the right tunnel.

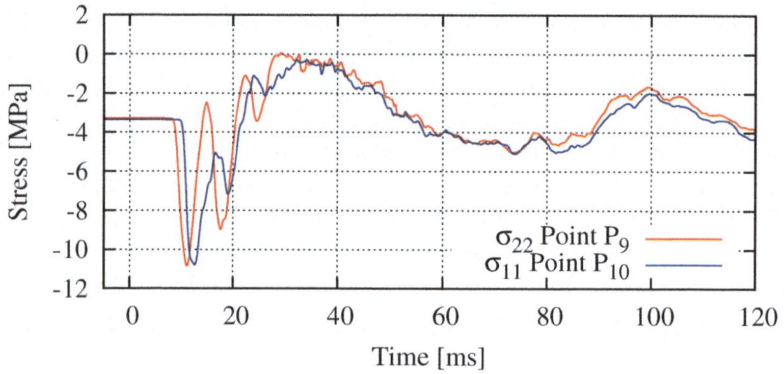

Fig. 8. Circumferential stresses in the lining of the right tunnel at points P_9 and P_{10}

5 Deformation of the Soil

The loading applied to the tunnel lining induces a complicated dynamic process in which the soil is subjected to both compressive and tensile deformation. The tensile deformation turns out to be large enough to reduce the absolute pore pressure to zero and thus to give rise to pore water cavitation. Recall that the transition to cavitation in this study is approximated by a smooth curve, see Fig. 1, so that zero absolute pressure is reached asymptotically and the onset of cavitation becomes undefined. However, for the analysis of solutions it is convenient to distinguish between cavitated and noncavitated water. In order to be able to identify a cavitation zone, we say that the pore water is cavitated if the pore pressure p_f is below -99 kPa.

The pore pressure distribution in the soil at different times is shown in Figs. 9, 10. The cavitation begins in three distinct places: near the upper half of the lining of the left tunnel, beneath the upper layer of dry soil and near the right tunnel. The cavitation around the left tunnel begins after 7 ms, yet before the top tubbings reach their maximum radial displacement. The cavitation beneath the dry soil layer arises at nearly the same time when the pressure wave (red colour in Fig. 9 for $t = 11$ ms) reaches this layer. If the half-space were linearly elastic, the reflected wave would have a tensile stress of the same order of magnitude as in the incident pressure wave. The interaction of a high-pressure wave with the ground surface in saturated soil is more complicated and has much in common with the reflection of high-pressure waves from the free surface in water. In both cases, tensile stresses are limited to 100 kPa (zero effective stress in the soil skeleton and zero absolute pore pressure assumed as a threshold for pore water cavitation). As a consequence, the interaction of the blast-induced pressure wave with the ground surface in the soil results in the formation of a cavitation zone instead of a reflected wave with high tensile stresses. The third cavitation zone near the right tunnel arises at a later time when the pressure wave passes around the tunnel and interacts with the tunnel lining, see Fig. 9 for $t = 11$ ms. The interaction leads to the wave reflection with large tensile deformations and pore water cavitation. The mechanism is similar to that of the reflection from the ground surface.

The three emerging cavitation zones grow in size, coalesce and form one big cavitation zone as seen in Fig. 9 for $t = 16$ and 26 ms. As the upper tubbings of the left tunnel come into contact with each other at the end of the separation phase, they stop the radial motion of the adjacent soil. The cavitated pore water in the soil is compressed to a noncavitated state, and the cavitation boundary moves upwards from the lining into the soil. At the same time, the upper boundary of the cavitation zone moves downwards, leading to the shrinkage of the cavitation zone as well. The shrinkage of the cavitation zone is a complicated process with highly nonlinear soil behaviour in compression near the boundary of the zone, see Figs. 1, 2 for the one-dimensional uniaxial case. In particular, Fig. 10 for $t = 56$ ms reveals singularities with high pressure gradients at the ends of the narrow cavitation zone at the final stage of its shrinkage.

The closure of the cavitation zone produces a spreading high-pressure zone with pressures of up to 1.5 MPa and two pressure waves, one of which propagates upwards and the other one downwards, see Fig. 10 for $t = 62$ ms. The wave that propagates upwards is reflected from the upper soil layer in the same manner as the primary blast-induced pressure wave and produces a new smaller cavitation zone, Fig. 10 for $t = 70$ ms. The wave that propagates downwards passes around the tunnel, deforms the tunnel lining and increases the circumferential stresses in the tubbings in the time between 60 and 80 ms as we have seen in Fig. 7.

The time dependence of the volumetric strain and the mean effective stress in the soil is shown in Figs. 11, 12 at four points P_5, P_6, P_7, P_8 around the left tunnel and at point P_{11} near the right tunnel. The points are located at a distance of 1 m from the tunnel, see Fig. 3. The blast-induced pressure wave produces a short compression-extension cycle followed by a longer extension-compression cycle. The latter coincides with the cavitation phase, except for point P_8 where no cavitation is observed. An analysis of the individual strain components reveals that the principal axes of the strain tensor rotate, making the deformation path rather complicated. Figure 12 shows that the mean effective stress in the soil is substantially reduced. The residual effective stress is difficult to predict, as it depends, besides the loading amplitude, on many factors such as the constitutive behaviour of the skeleton, the pore fluid compressibility and the

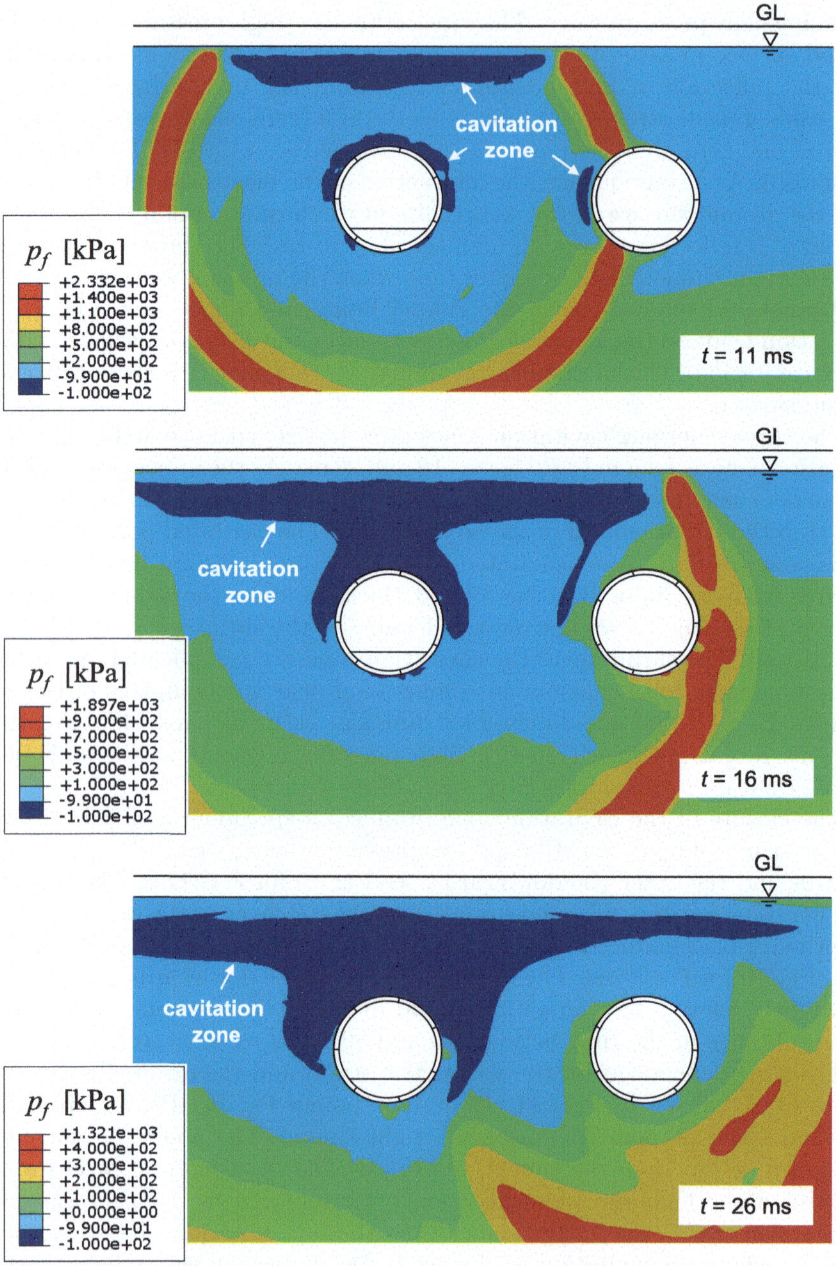

Fig. 9. Pore pressure in the soil at different times

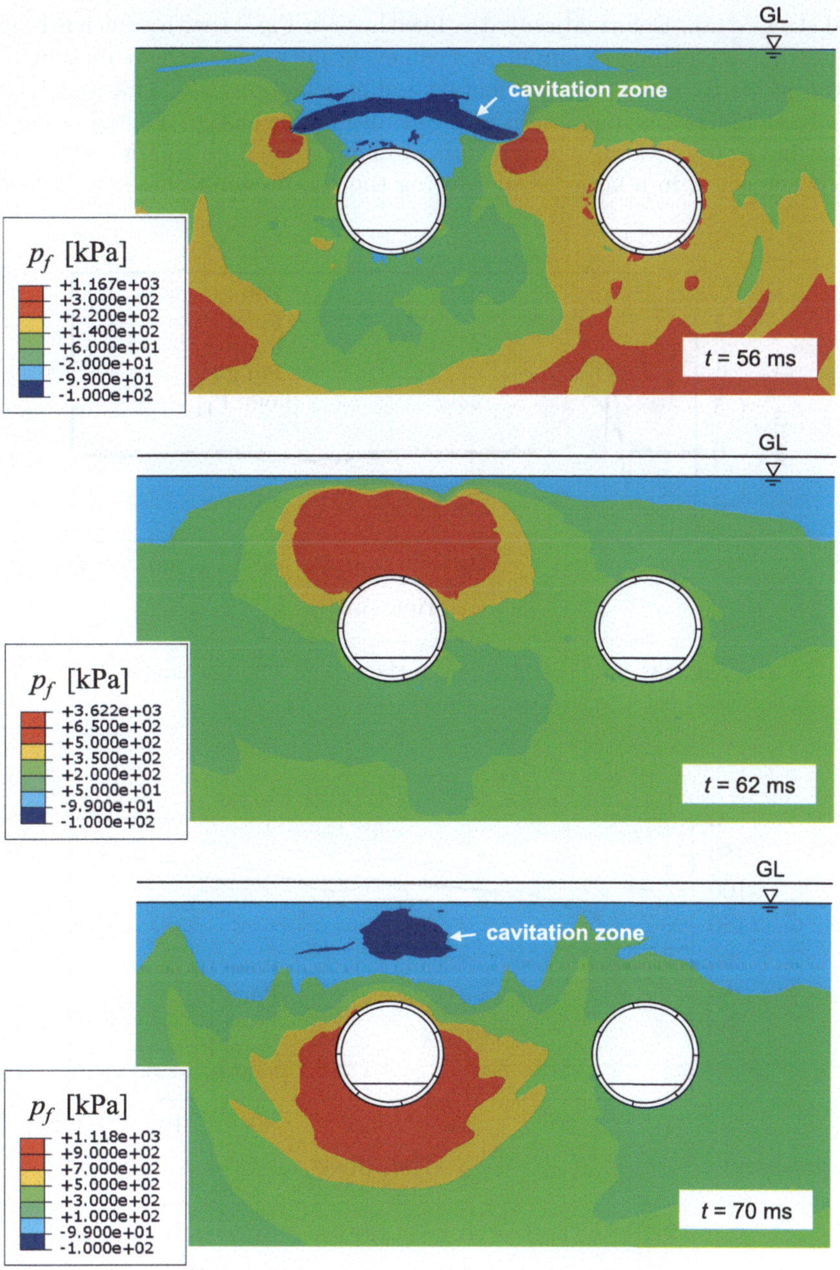

Fig. 10. Pore pressure in the soil at different times

initial stress state. Deformation paths like those in Fig. 11 with sufficiently large compression and extension amplitudes can eventually lead to the reduction of the effective stress to zero and to momentary soil liquefaction [1,2]. This may happen not only at higher loading amplitudes but also in the case of a small amount of free gas in the pore water, which results in higher pore fluid compressibility and, as a consequence, in a larger strain during the compression phase [3].

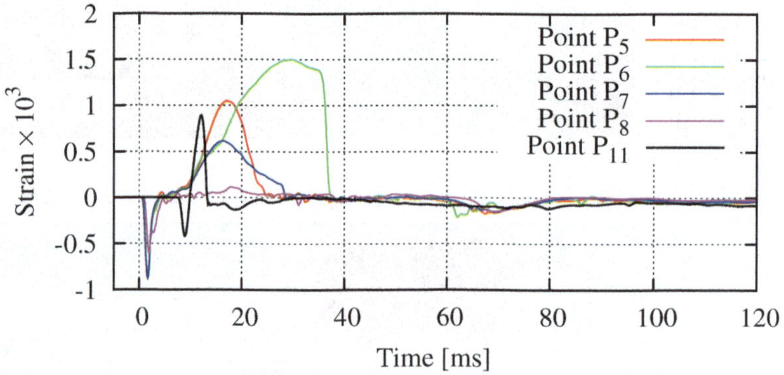

Fig. 11. Volumetric strain in the soil at different points

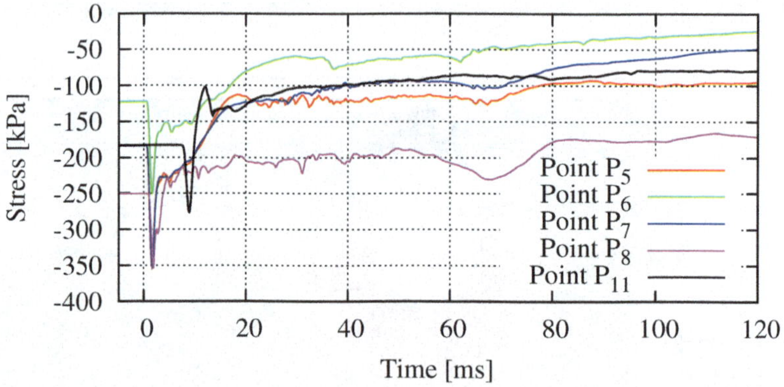

Fig. 12. Mean effective stress in the soil at different points

The motion of the soil does not end after 120 ms covered by the present numerical solution. This is seen in Fig. 13 which shows the vertical displacement components at three points in the soil above the left tunnel. The first two points lie in the upper layer of dry soil.

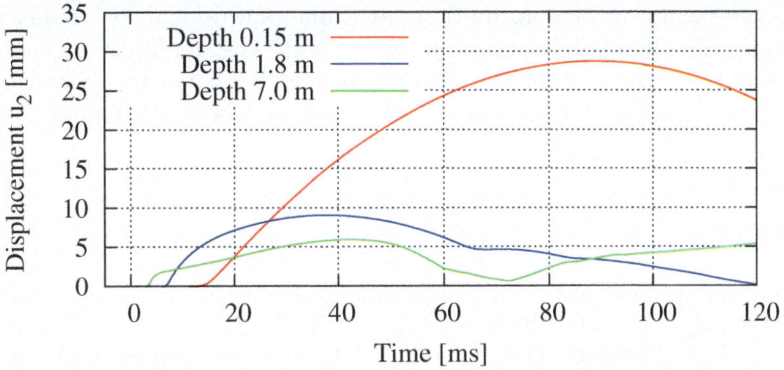

Fig. 13. Vertical displacement of the soil above the left tunnel

In order to assess the influence of the adjacent tunnel, the present problem has also been solved with a single tunnel. A comparison of the two solutions in the vicinity of the tunnel up to a few metres from the lining shows that the influence of the adjacent tunnel is insignificant.

6 Conclusion

The main features of the blast-induced deformation of the tunnel lining and the soil can be summarized as follows. The tubbings of the tunnel subjected to blasting lose contact with each other after 1 ms and are in a separation phase during the next 25–28 ms (Fig. 6). The maximum radial displacement and the maximum gap between the tubbings in the separation phase are of the order of several millimetres. The tubbings exhibit high-frequency oscillations (480 Hz) during the separation phase, with the maximum tensile and compressive oscillating stresses being by 50–80% greater than the blast pressure applied to the lining (Fig. 7).

The numerical solution shows an important role of pore water cavitation in the deformation process. The soil expansion following the blast-induced pressure wave is large enough to give rise to pore water cavitation (Fig. 9, $t = 11$ ms). A big cavitation zone forms after 15 ms between the tunnels and the ground surface (Fig. 9, $t = 16$ and 26 ms). The volumetric soil stiffness sharply increases in the transition from a cavitated state back to full saturation, making the strain-pressure relation highly nonlinear (Figs. 1, 2). As a consequence, the shrinkage of the cavitation zone is accompanied by high gradients (shock fronts) and singularities at the boundaries of the cavitation zone (Fig. 10, $t = 56$ ms). The ultimate closure of the cavitation zone produces a high-pressure domain which in turn induces two pressure waves (Fig. 10, $t = 62$ and 70 ms).

The complexity of the deformation paths makes it difficult to predict permanent changes in the effective pressure in the soil. The qualitative conclusion is that the effective pressure in the vicinity of both tunnels may be considerably reduced (Fig. 12). The largest decrease is observed above the tunnel subjected to blasting.

References

1. Fragaszy, R.J., Voss, M.E.: Undrained compression behavior of sand. J. Geot. Eng. ASCE **112**(3), 334–347 (1986)
2. Bolton, J.M., Durnford, D.S., Charlie, W.A.: One-dimensional shock and quasi-static liquefaction of silt and sand. J. Geot. Eng. ASCE **120**(10), 1874–1889 (1994)
3. Osinov, V.A.: Blast-induced waves in soil around a tunnel. Arch. Appl. Mech. **81**, 543–559 (2011)
4. Osinov, V.A., Chrisopoulos, S., Triantafyllidis. Th.: Numerical analysis of the tunnel-soil interaction caused by an explosion in the tunnel. Soil Dyn. Earthq. Eng. **122**, 318-326 (2019)
5. Feldgun, V.R., Karinski, Y.S., Yankelevsky, D.Z.: The effect of an explosion in a tunnel on a neighboring buried structure. Tunn. Undergr. Space Technol. **44**, 42–55 (2014)
6. Lu, G., Fall, M.: A coupled chemo-viscoplastic cap model for simulating the behavior of hydrating cemented tailings backfill under blast loading. Int. J. Numer. Anal. Meth. Geomech. **40**, 1123–1149 (2016)
7. Niemunis, A., Herle, I.: Hypoplastic model for cohesionless soils with elastic strain range. Mech. Cohesive-frict. Mater. **2**(4), 279–299 (1997)
8. von Wolffersdorff, P.A.: A hypoplastic relation for granular materials with a predefined limit state surface. Mech. Cohesive-frict. Mater. **1**(3), 251–271 (1996)
9. Osinov, V.A.: Large-strain dynamic cavity expansion in a granular material. J. Eng. Math. **52**, 185–198 (2005)
10. Osinov, V.A., Gudehus, G.: Dynamics of hypoplastic materials: theory and numerical implementation. In: Hutter, K., Kirchner, N. (eds.) Dynamic Response of Granular and Porous Materials under Large and Catastrophic Deformations, pp. 265–284. Springer, Berlin (2003)
11. Herle, I., Gudehus, G.: Determination of parameters of a hypoplastic constitutive model from properties of grain assemblies. Mech. Cohesive-frict. Mater. **4**, 461–486 (1999)
12. Osinov, V.A., Chrisopoulos, S., Grandas-Tavera, C.: Vibration-induced stress changes in saturated soil: a high-cycle problem. In: Triantafyllidis, T. (ed.) Holistic Simulation of Geotechnical Installation Processes. Benchmarks and Simulations, pp. 69–84. Springer, Cham (2016)
13. Osinov, V.A.: Longitudinal shock waves in soil (this volume)

In Situ Stress Assessment Based on Width and Depth of Brittle Borehole Breakouts

Eleni Gerolymatou(✉) and Alexandros Petalas

Chalmers University of Technology, Gothenburg, Sweden
eleni.gerolymatou@chalmers.se, petalas@chalmers.se

Abstract. Borehole breakouts, as well as breakouts in tunnels and shafts, are a common occurrence, especially under high in situ stresses or stress states with high deviatoric component. Though they can pose a risk to stability, often they are of use, especially in deep boreholes, as they can help to determine to a certain extent the primary in situ stress. Observations have shown that while their depth evolves, their width remains constant. Currently the width only is used in conjunction with the Kirsch analytical solution to establish a linear relationship between the two in plane principal primary stress components. The stress state cannot be fully determined since one equation is available (failure criterion) for two unknowns. A recently proposed numerical tool based on conformal mapping is used in this work to simulate the formation of shear breakouts and investigate the feasibility of the determination of both principal primary in situ stress components, by making use of both the depth and the width of the breakout. Concluding, recommendations are provided for the use of the proposed methodology and limitations of its applicability are discussed.

1 Introduction

The *in situ* stress state and its knowledge is significant for a variety of geotechnical applications including tunneling, mining and reservoir engineering. In the case of construction in rock it is particularly important for applications at deep depths, where the stresses are high enough for failure to take place. Such failure takes place as a result of stress redistribution induced by man-made interventions. Its prediction is crucial for deep reservoirs in applications such as geothermal energy and oil and gas recovery, where the contact to the host rock is necessary and the lining of the borehole is usually not present or partial.

Failure may be local or global. In the case of boreholes, global failure translates to the total loss of the borehole, while local failure leads to the formation of borehole breakouts, where only a certain part of the borehole wall fails. The local failure results in a new geometry for the borehole, which is stable, at least to a certain extent. A significant number of experiments are available from the literature, showing a strong dependence of the shape of the breakouts on the

T. Triantafyllidis (Ed.): *Recent Developments of Soil Mechanics and Geotechnics in Theory and Practice*, LNACM 91, pp. 297–319, 2020.
https://doi.org/10.1007/978-3-030-28516-6_16

primary stress state, see for example [1]. This shape, along with the observed mode of failure, can provide information on the direction and the magnitude of the *in situ* stress state [2]. Tension induced breakouts are aligned with the direction of the maximum in plane principal stress, while shear or compression induced ones are aligned with the direction of the minimum in plane principal stress. It has also been observed both *in situ* and in the laboratory that large boreholes fail at significantly smaller loads than smaller boreholes [3,4]. Small boreholes fail at stresses that are significantly higher than the ones expected from the theoretical stress distribution and uniaxial or triaxial tests performed in the laboratory.

The material response can be ductile or brittle. The simulation of breakout formation is challenging in both cases, in the first case because of the softening of the material and in the second case because of the loss of continuity. In the ductile regime, failure is characterized by zones of localized deformation that cannot be modeled by simple constitutive models in the sense of Noll [5]. Nonlocal [6], micropolar [7] or higher gradient [8] models can be used to simulate the formation of breakouts in materials exhibiting softening. In the brittle regime both continuum and discrete numerical methods have been proposed to simulate borehole breakouts. Among others, [9,10] used the boundary element method to simulate the development of borehole breakouts in brittle rocks and recently [11] proposed a methodology using the finite element method. Simulations using discrete elements have also attempted to provide an answer [12]. In this case, the limited number of particles that can be used poses a limit to the applicability, since size dependence that has been observed in situ, numerically and in the lab, plays a significant role. In terms of analytical methods, conformal mapping has been used in [13] to evaluate the stress field around underground openings. However, the shape of the opening has always been considered known and constant and the provided solutions are only valid for a specific form of the mapping.

In the present work a recently developed semi-analytical approach based on conformal mapping [14] is utilized for the assessment of the in situ stress state using both the width and the depth of borehole breakouts. The manuscript is organized as follows. In Sect. 2 a brief discussion about the characteristics of borehole breakouts is presented. In Sect. 3 the utilized semi-analytical method with conformal mapping is outlined, its extension to include scale effects is presented and its validation against finite element simulations and experimental data from the literature is highlighted. In Sect. 4 the methodology for the assessment of the in situ stress state is discussed and Sect. 5 presents the obtained results. A final discussion for the possibilities and the limitations of the proposed methodology is presented in the final section of conclusions.

2 Breakout Formation and Arrest

As already mentioned in Sect. 1, the geometry of the failure of the borehole wall varies strongly with the mode of failure. Tensile failure takes place when the ratio of the maximum to the minimum in-plane principal stress is large enough

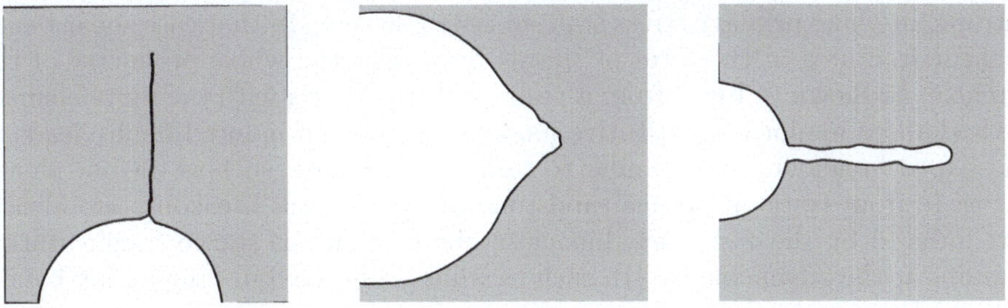

(a) Tensile crack induced by hydraulic pressure traced after [15].

(b) Shear induced breakout traced after [1].

(c) Compaction induced breakout traced after [1].

Fig. 1. Breakout shapes depending on failure mode.

to lead to tensile stresses at the borehole circumference that are larger than the tensile strength of the rock, or when the internal pressure in the borehole is large enough. The failure presents itself in the form of a crack oriented parallel to the direction of the maximum in-plane principal stress, as shown in Fig. 1a after hydraulic fracturing tests on sandstone blocks [15]. As a rule, such tensile cracks are not denoted as breakouts, but they belong in the same category from a physical and mechanical point of view.

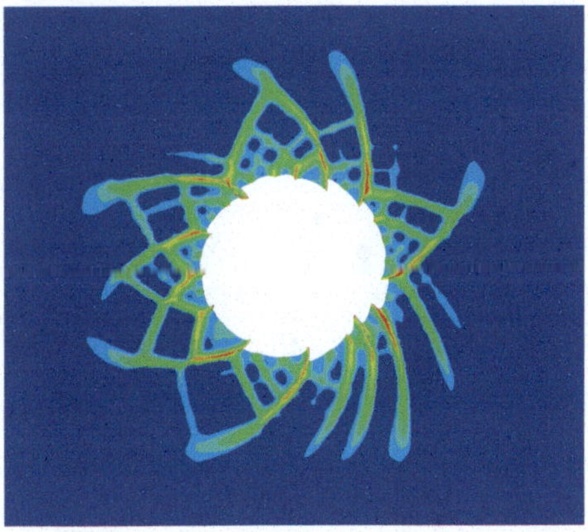

Fig. 2. Characteristic spirals of localized deformation forming as a result of shear induced failure of the rock around a borehole. Numerical simulation with the finite element method and a nonlocal constitutive model

Shear or splitting failure produces so called 'dog-ear' or 'cusp' breakouts, similar to what is shown in Fig. 1b from tests on a porous sandstone after [1]. It is mostly this type of borehole breakout that is used for the assessment of the *in situ* primary stress state. Shear induced failure may also take place under isotropic in-plane primary stress state in isotropic rock. In that case shear bands or shear fractures in the form of spirals form over the whole perimeter of the borehole, as shown in Fig. 2 from a numerical simulation with the finite element method and a nonlocal constitutive model. Compaction induced failure leads to slit shaped breakouts, very similar to compaction bands, such as the one shown in Fig. 1c from tests on porous sandstone after [1]. Such breakouts are also in part induced by shear stresses, but close observation has shown fragmentation of grains at the advancing tip [1], such as what is observed in compaction bands. This mode has only received more attention in the last years.

Failure may be ductile or brittle. In the first case the material response may be hardening or softening, but continuity is preserved. The damaged zone continues carrying some amount of load, even if not to the same extent as it did before yield. In the case of brittle failure continuity is lost. The damaged material is removed from the borehole wall, leaving a gap behind. In reality the distinction is not perfectly clear cut. Closer observations using methods, such as scanning electron microscope imaging or thin sections, have shown damaged areas in the vicinity of the breakout [16,17] in cases where the general image was that of brittle failure. This indicates the likely presence of a process zone also in the case of what macroscopically appears to be brittle failure.

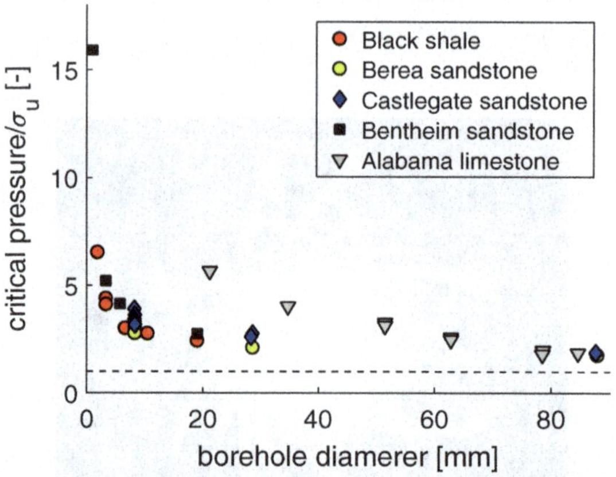

Fig. 3. Compilation of experimental results on size effects from different publications. The stresses are normalized with the uniaxial strength of each material. The theoretical description that ignores size effects predicts that breakouts start forming at a normalized critical pressure of 0.5 (dashed line). The data sources are: Black shale [4], Berea and Castlegate sandstone [18], Bentheim sandstone [19], Alabama limestone [10]

Scale effects are commonly present in rock mechanics. It has been observed that the initiation of breakout formation is strongly sensitive to scale. Large perforations start showing damage at loads that are in agreement with theoretical predictions using the solution of Kirsch. Smaller perforations on the other hand only start exhibiting failure at much higher loads, often several times greater than the ones theoretically predicted [3,4]. Based on elasticity theory, the initiation of breakouts should take place when the external pressure is equal to one half of the uniaxial strength. As shown in Fig. 3, breakout initiation is observed at significantly higher loads than those predicted by the elastic solution, depending on the material and on the size of the perforation.

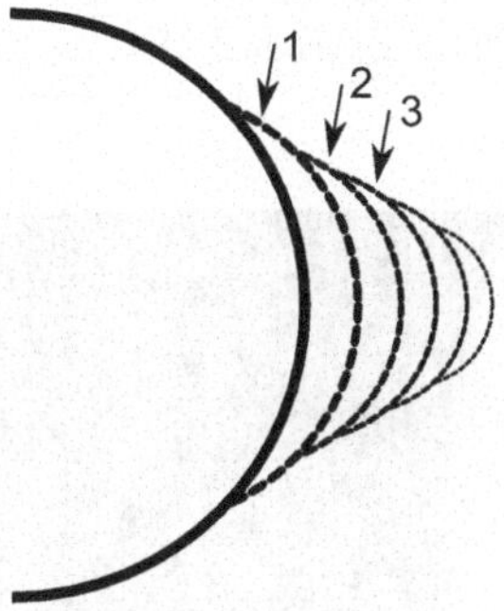

Fig. 4. Progressive failure propagation forming a dog-ear breakout: the section 1 is removed first, the other sections follow in succession, until the final shape is formed

It has been observed by several authors, such as in [9,10], that the propagation of the breakouts takes place in such a way that their width remains constant, while the depth gradually increases. The procedure is sketched in Fig. 4. Successive zones of material are detached either by spalling or by the formation of shear bands. Each successively detached domain is less wide than the previous one, leading to an overall shape that becomes more narrow with increasing depth. The procedure continues until the arrest of the failure propagation.

The reasons for the arrest of dog-ear of cusp breakouts are not completely clear. The arrest may be purely linked to shape variations that lead in turn to stress redistribution. This is however not likely to be the only reason. It has been postulated that the phenomenon is linked to a plastic zone [20] or a process zone [4]. The effect is the same as that of the formation of a plastic radius around a circular perforation: the stresses are locally reduced and the elastic domain is protected. Alternatively, the arrest has been attributed to scale effects, see for example [10], which may in turn be linked to the formation of a process zone.

3 Solution of the Direct Problem

The solution of the direct problem is the evaluation of the shape of the resulting breakout, when the *in situ* stress state is known. For this part a recently developed semi-analytical method [14] that is based on conformal mapping is modified and used. This involves several limitations and assumptions. The first assumption made is that the stress state may be described as plane strain. The out of plane stress can be taken into account when the Poisson ratio is known.

A further assumption is that the stress state is coaxial to the borehole axis. This is not the case in most of the *in situ* applications. However, while a non-coaxial primary stress state can be accommodated by the method used, it is not deemed necessary for the present work. For laboratory experiments the stress state is in the vast majority of cases coaxial to the borehole.

A final limiting point of the present approach is the assumption that failure of the rock is purely brittle. This assumption ignores the presence of a process zone or of a plastified area of the material that is in a position to carry load.

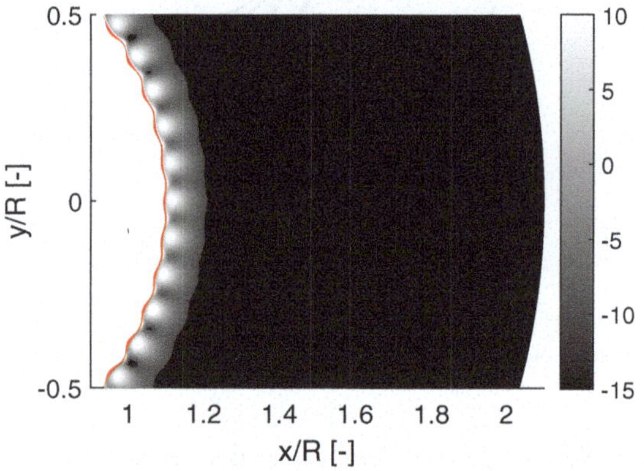

Fig. 5. Propagation and intensification of spurious oscillations, as a result of stress concentrations around a non smooth boundary. R stands for the borehole radius and f is the value of the yield function. As may be observed, small oscillations of the boundary lead to large local stress concentrations

3.1 Method Outline

For the sake of clarity, the method used for the direct problem is summarized in the pseudocode below.

Program Code

```
initialize the borehole boundary
discretize the domain
while (new surface > Tol)
  evaluate the stress state using conformal mapping
  evaluate the yield function
  remove the area formed that yieldeds
  discretize the new borehole boundary
end
```

The code developed in [14] was modified for the present article. The use of a series expansion for the description of the stress functions leads to small oscillations of the evaluated stresses. These tend to become more pronounced as the breakout becomes larger. The reason is that the oscillations of the boundary lead to high stress concentrations, as one would expect from the theory of elasticity. An example is shown in Fig. 5. A vertical stress of 60 MPa and a horizontal one of 30 MPa were used with a Mohr-Coulomb failure criterion with a cohesion of 15 MPa and a friction angle of 55°.

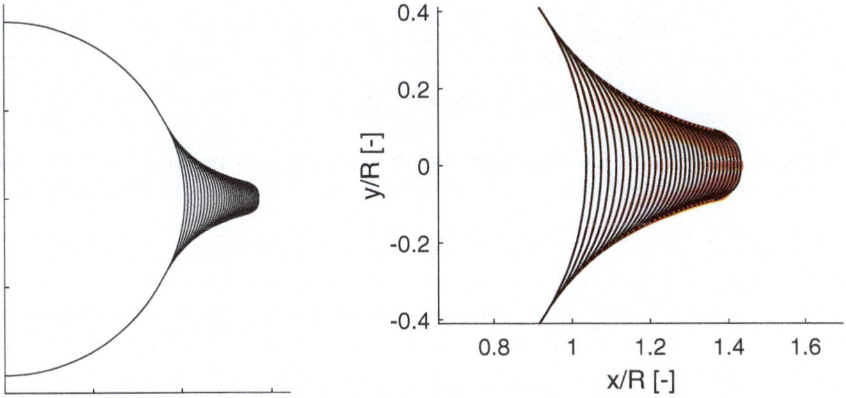

(a) Breakout propagation. (b) Interpolation.

Fig. 6. Example of breakout propagation prediction.

To resolve the issue, which is a numerical one, in the previously used version a smoothing of the boundary was used with an averaging technique. The degree of smoothing was however found to affect the results to some degree. It was chosen here therefore to substitute the smoothing with a polynomial interpolation. An example of the results of the successive iterations is shown in Fig. 6a. The formed breakout is of the cusp type. The corresponding polynomial interpolation is shown in Fig. 6b for a polynomial degree of 5. The interpolation is marked in black, while the original points are marked in red. The agreement between original data and interpolation is good. The results remained the same for polynomial degrees ranging from 3 to 8. A polynomial interpolation of the fifth order is used for subsequent simulations in the present work, unless otherwise stated.

It was found that using the new approach developed here the breakouts in many cases tend to advance indefinitely as far as the depth is concerned, even if the width remained constant. This phenomenon is known from previous works using methods based on elasticity [10, 20, 21]. The discrepancy with experimental results is related to the reasons for breakout arrest that were discussed in Sect. 2, namely local plastic response of the host rock, the formation of a process zone or scale effects. It is worth noting that all may be different aspects of the same phenomenon.

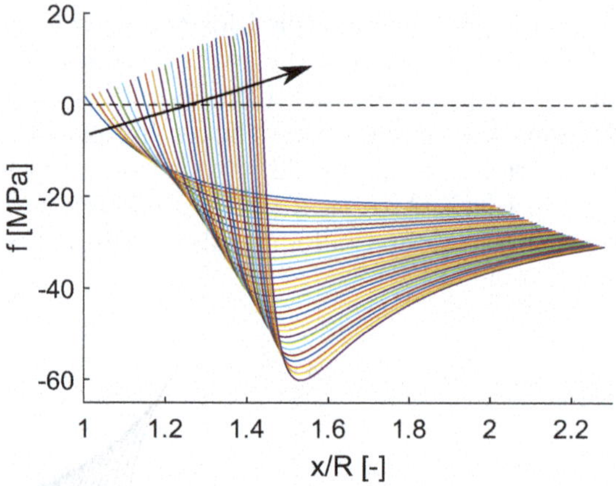

Fig. 7. Evolution of the value of the yield function with successive iterations. The values were measured along the line $y = 0$ for positive values of the horizontal coordinate

For a more thorough explanation of the mechanism leading to continuing breakout propagation, the value of the yield function was plotted as a function of the normalized horizontal coordinate, over a line on the x-axis, for each successive iteration of breakout formation. The result is shown in Fig. 7. The arrow indicates the direction of increasing iterations. As may be seen, in the beginning the curve is monotonic, with a gradual decrease. As the breakout propagates, the maximum value of the yield function becomes larger, while its gradient also becomes larger. For later iterations the gradient becomes very pronounced and a local minimum appears. The value of the yield surface changes drastically within a very small area. A plastic rather than brittle response of this area would lead to a very different response and probably arrest the advance.

3.2 Scale Effects

As the failure propagates and the breakout becomes deeper, the height of the area that fails becomes smaller. This is akin to the buckling of pillars of ever-decreasing height: the smaller height is linked to size effects, meaning that the load required to reach failure becomes progressively higher. Such scale effects have been observed also in laboratory tests on boreholes, as shown for example in Fig. 3.

To incorporate scale effects the results on Alabama limestone by [10] were used. The normalized tangential stress at which failure was first observed is plotted over the perforation radius in Fig. 8. The stress is normalized with the uniaxial strength of the material. The fit is given by the equation

$$y = max(-1.3 \ln (x/85), 1) \tag{1}$$

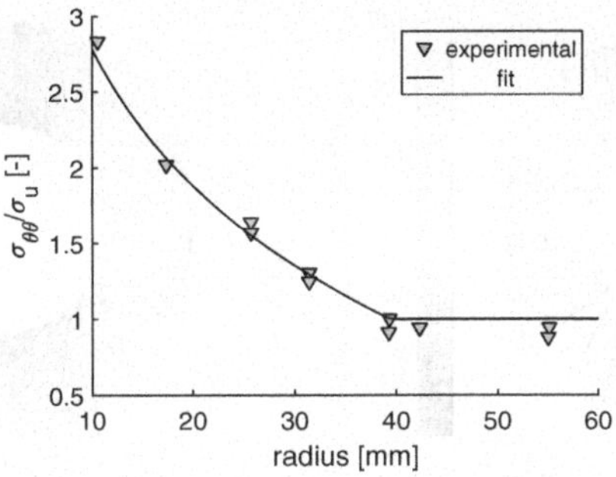

Fig. 8. Normalized tangential stress at failure as a function of the borehole radius. Data from Alabama limestone after [10]. The fit is given by Eq. (1)

It should be noted that this is only one of the many options available. It must also be remarked that the cut-off at a value equal to unity is not relevant for the present case, as all experimental points correspond to radii of 11 mm or smaller.

3.3 Stress State Estimation: FEM *vs* Conformal Mapping

In this section a comparison of the computed stress state around the borehole by the conformal mapping method proposed in [14] and finite element analysis is presented. Three stages of the breakout development are simulated in order to compute the stress redistribution around the borehole. The first is when the borehole is drilled just before the breakout appears. For this stage the conformal mapping retrieves the exact analytical solution of Kirsch, as shown in [14]. However, for the second and third stage, when the breakout is halfway and fully developed respectively, an exact analytical solution does not exist and this is why the results of the conformal mapping method are validated against the FEM simulations.

For the FEM simulations the commercial software PLAXIS 2D is utilized. The geometry for the three different FEM simulations (immediately after the drilling of the borehole, when half the depth of the breakout is developed and when the breakout is fully developed) is input-based on the geometry resulting from the semi-analytical method with conformal mapping, after the removal of the area where yielding took place and the new borehole boundary was discretized as explained in the pseudocode of Sect. 3.1. This means that the elastic solution for the stress state around the borehole and the breakout is directly compared for the two methods based on the same problem geometry.

In the FEM simulations the problem is analyzed as plane strain. However, since in conformal mapping the effect of the out-of-plane stress is not taken into account in this work, a linear elastic constitutive law is used with zero Poisson's ratio ($\nu = 0$). The Young's modulus is set to 15 GPa, even though it does not influence the solution for the major and minor principal stress components. A rectangular domain is used with the geometry of the borehole and borehole

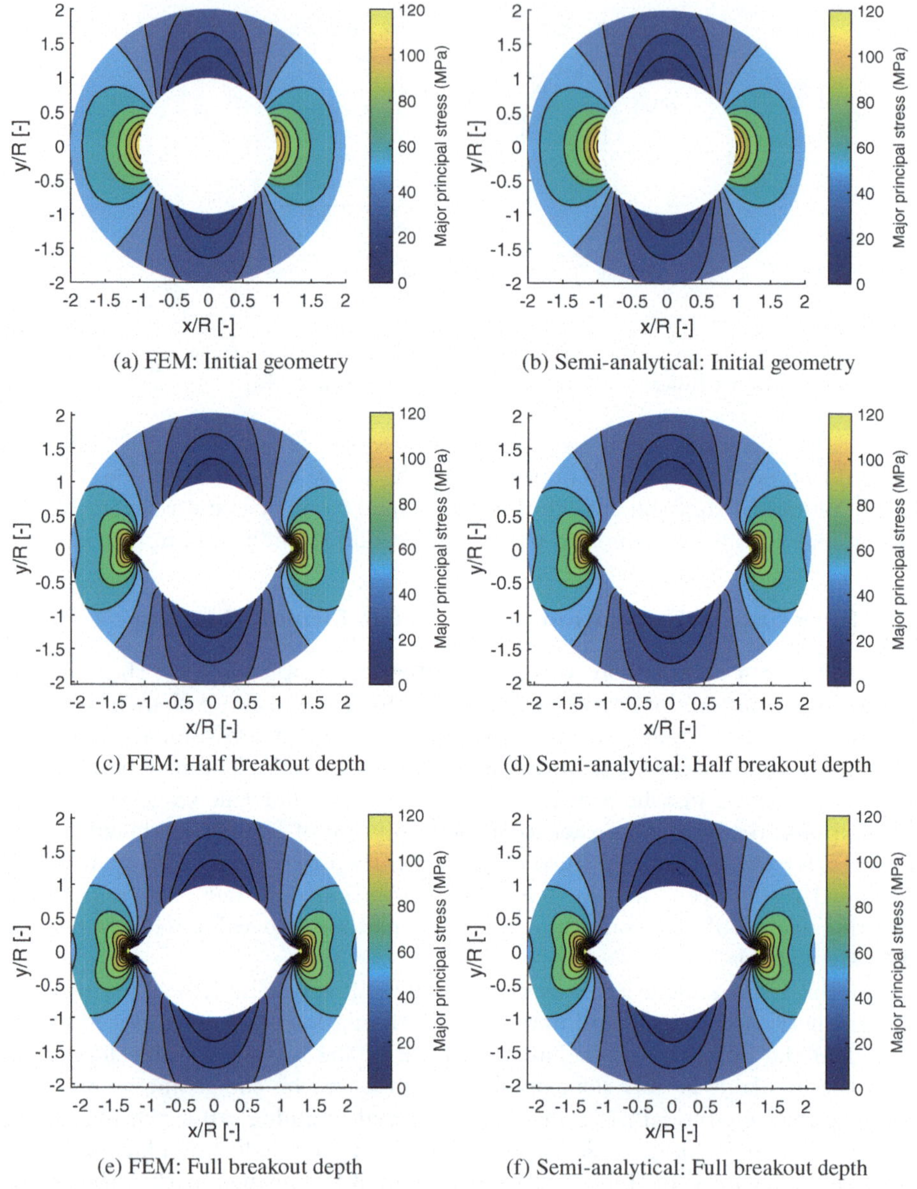

Fig. 9. Comparison of stress calculation based on the finite element method and the proposed semi-analytical method with conformal mapping [14].

breakout in the center. Zero displacement boundaries are placed 100 radii away from the borehole, both in the vertical and the horizontal axis, to ensure that the boundary has no effect on the solution (infinite medium). The mesh is refined around the borehole until the solution for the two principal stress components is not influenced by the discretization. 15-noded triangular elements are used in Plaxis 2D.

The results of the simulations as far as the major principal stress component is concerned are presented in Fig. 9a–f, for the three stages. The results of the conformal mapping method are in very good agreement with the FEM simulations and give confidence for the applicability of this semi-analytical method both for solution of the direct problem and for the *in-situ* stress state assessment that is presented in the sequel.

3.4 Comparison to Experiments

The semi-analytical method based on conformal mapping is used to simulate experiments performed in the literature on borehole breakout development. The Mohr-Coulomb failure criterion is used to pinpoint the areas that yield during the breakout development. It requires two material properties, i.e. the friction angle ϕ and cohesion c that are calibrated based on experimental results. It reads

$$f = q - \sin(\phi)p - 2c\cos(\phi) \tag{2}$$

where in terms of principal stresses

$$p = \frac{\sigma_1 + \sigma_2}{2} \tag{3}$$

and

$$q = \frac{\sigma_1 - \sigma_2}{2} \tag{4}$$

ϕ is the friction angle and c is the cohesion of the material. The simulated experiments were performed on Alabama limestone and their results are published in [10]. For the friction angle a value of 18° was used in accordance to the measurements by [10]. The cohesion was calculated at 14 MPa using the value of the uniaxial compressive strength provided by [10] and the Mohr-Coulomb failure criterion for minimum principal stress equal to zero.

The size effect (as explained in Sect. 3.2) is incorporated in the cohesion used in the Mohr-Coulomb failure criterion. For each successive iteration a polynomial fit of the second degree was fitted to the new part of the borehole boundary. The curvature of this polynomial-corresponding on approximation to the curvature of the tip of the breakout-was used to evaluate the equivalent radius. The equivalent radius was subsequently used to evaluate the strength for the next iteration by modifying the cohesion in accordance with the experimental data on Alabama limestone and the fit given in Eq. (1). An example of the results is shown in Fig. 10. As expected, the width remains constant, while the depth increases up to the point of arrest. The failure surfaces of the successive iterations, delimited

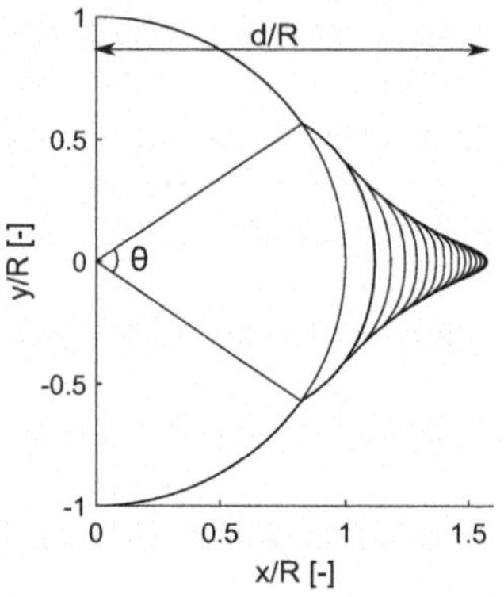

Fig. 10. Example of the final shape of the borehole breakout, showing the intermediate shapes for all iterations. As may be observed, the width remains constant, while the depth increases. The overall shape is slightly reminiscent of a cusp.

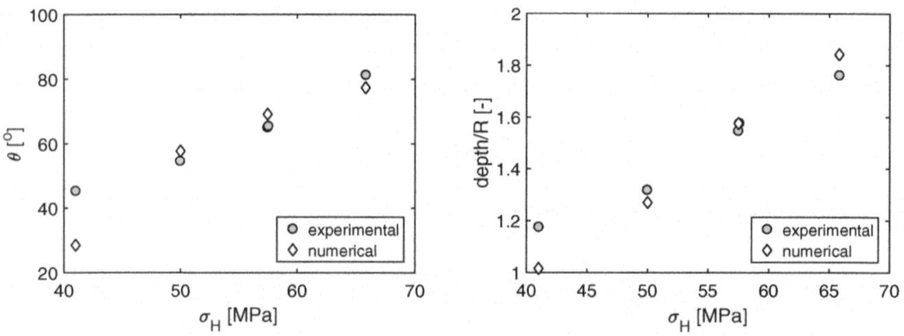

(a) Aperture angle.

(b) Normalized depth.

Fig. 11. Numerical simulation of borehole breakout development and comparison to experimental results in Alabama limestone. Experimental data after Herrick and Haimson [10].

in the figure with black lines, become gradually smaller. The overall shape is slightly reminiscent of cusp borehole breakouts, though not far from dog-ear shaped ones.

The first test series by Herrick and Haimson [10] was used for the comparison. Prismatic specimens with a side length of 13 cm in the horizontal and 17 cm in the vertical direction were loaded with the principal stresses in the horizontal and vertical directions. The maximum and the minimum principal stresses were horizontal. A borehole with a radius of 1.1 cm was drilled under load in the vertical direction and the breakout dimensions measured. The aperture angle and the normalized depth were registered, see Fig. 10. For the first test series the minimum principal stress was equal to 14 MPa, the vertical stress was equal to 21 MPa and the maximum principal stress varied between 40 and 67 MPa.

The comparison of the numerical to the experimental results is shown in Fig. 11. It should be remarked that, since the breakout width does not increase, the aperture is determined by the Kirsch solution and, of course, the failure criterion. The agreement between numerical and experimental results is rather good, both concerning the aperture angle and the normalized depth, with the exception of the point corresponding to the smallest maximum principal stress. In this case both breakout angle and depth are underestimated. The agreement between experimental and numerical results concerning the depth is particularly good, considering that, since the borehole radius was equal to 1.1 cm, a step of 0.1 in the normalized depth corresponds to a little more than one mm.

4 Numerical Method for Stress Assessment

The current common practice for the use of borehole breakouts for the assessment of the *in situ* stress state is outlined in Fig. 12. The Mohr-Coulomb failure criterion with the parameters used for the comparison to the experimental results was utilized. The vertical stress was assumed to be known and equal to 80 MPa. Commonly in practice the vertical stress is evaluated from the overburden. It is reasoned that the *in situ* stress state cannot be beyond the yield locus of the material. For different assumptions concerning which of the three principal stresses is the maximum and which is the minimum principal stress, different limiting loci are attained. From them it results that the primary *in situ* stress must be within the polygon marked in blue in Fig. 12. From the Kirsch solution

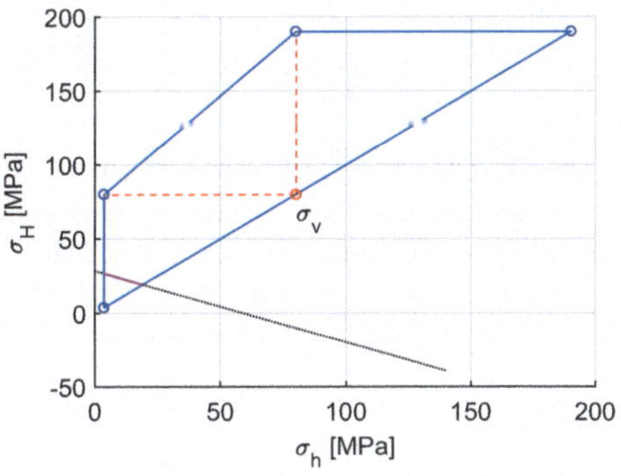

Fig. 12. Example of the standard use of borehole breakouts for the estimation of the *in situ* stress state. Based on global failure, the stress state is limited to the interior of the blue polygon. Based on the width of the borehole breakout it is limited to the magenta line.

for the circular cross-section of the borehole under the assumption of plane strain state, it results that at the wall of the borehole

$$\sigma_H = \frac{\sigma_{\Theta\Theta} - (1 - 2cos(2\Theta))\sigma_h}{1 + 2\cos(2\Theta)} \tag{5}$$

where Θ is the angular coordinate. Setting 2Θ equal to the borehole aperture θ results in

$$\sigma_H = \frac{\sigma_u - (1 - 2cos(\theta))\sigma_h}{1 + 2\cos(\theta)} \tag{6}$$

where σ_u is the uniaxial compressive strength of the material, when ignoring the out of plane stress. The above equation for an aperture of 80° results in the black dotted line of the diagram in Fig. 12. Combining with the polygon, it results that the stress state must lie on the magenta line of the same diagram. A more detailed description may be found in [2]. In what follows it will be assessed whether the depth of the breakout can be used in conjunction with its width to further limit the domain of possible stress states, using the conformal mapping methodology presented in Sect. 3 for the solution of the direct problem.

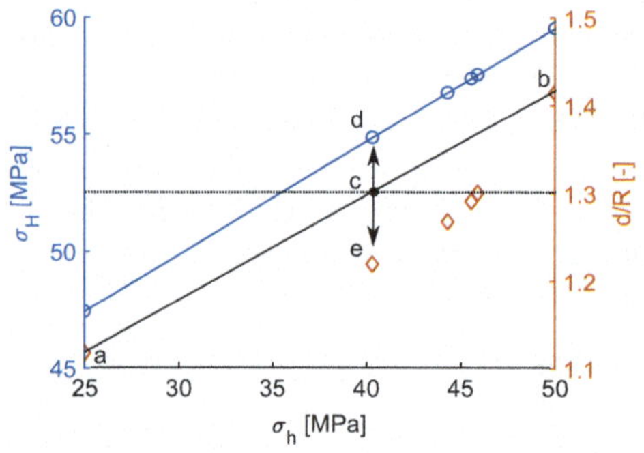

Fig. 13. Example of the method for the estimation of the *in situ* stress state. Points a and b are the depths corresponding to the minimum and maximum estimate of the minimum principal stress. The dotted line is the target depth and point c results from linear interpolation. The corresponding maximum principal stress is at point d and the breakout depth at point e. Points e and b are used for the next steps in lieu of points a and b

A modified version of the iterative bisection method in conjunction with the conformal mapping algorithm are used to estimate the *in situ* minimum σ_h and maximum σ_H principal horizontal stress components, based on the measured width and depth of the borehole breakout and the uniaxial strength σ_u of the material. The workflow of the proposed numerical procedure is as follows. An initial guess for the minimum and maximum values of σ_h is provided by the user, denoted also as left and right guess respectively. These two initial guess values can be estimated based on the admissible range as explained for the example of

Fig. 12 (e.g. the magenta line). Based on the aperture width and the uniaxial strength of the material, σ_H is determined by Eq. (6) for the two initial guess values of σ_h. Then the conformal mapping algorithm is used to solve the direct problem (see Sect. 3) evaluating the depth of the breakouts for the two pairs of σ_h and σ_H. The resulting breakout depths are compared with the target depth value (input borehole breakout depth) and a new interval for the minimum principal stress σ_h is evaluated using the method described in the next paragraph. The procedure is repeated until a stress state is reached for which the estimated depth is within tolerance of the target breakout depth value.

In the bisection method, where two initial guess points are also provided, the middle point is used to evaluate the interval in which the solution is to be found for the next iteration, bisecting thus the domain in which the solution is sought. In the proposed numerical procedure, a linear relationship between the minimum *in situ* in plane stress and the breakout depth is assumed in order to evaluate the new guess of the minimum principal stress σ_h for the next iteration, instead of the middle point. In order to clarify the numerical procedure a graphical example is given in Fig. 13. In this example the minimum or left estimate for the minimum *in situ* principal stress equals 25 MPa and the maximum or right estimate for the minimum *in situ* principal stress equals 50 MPa for the first iteration. The corresponding values of the maximum *in situ* stress σ_H are calculated by Eq. (6) and determine the two blue points with x-coordinates 25 MPa and 50 MPa. The calculated corresponding normalized depth d/R is about 1.1, located at point a, (lower than the input 1.3) and about 1.4, located at point b, (higher than the input 1.3). Since convergence is not achieved, a third, middle value is selected for the minimum principal stress σ_h by using the intersection of the line between points a and b with the line expressing the target normalized depth, found at point c in the example. For this new value the maximum principal stress σ_H and the normalized breakout depth d/R are estimated, corresponding to points d and e respectively. From the three evaluated normalized depths, the new interval for the solution is gained. For the new interval the procedure is repeated until a stress state is reached corresponding to a depth within precision of the target one. The code is outlined in more detail below.

Program Code

```
provide the following
  left bracket for the minimum stress shl
  right bracket for the minimum stress shr
  breakout width theta
  breakout depth d
evaluate
  left bracket maximum stress sHl
  left bracket depth dl
  right bracket maximum stress sHr
  right bracket depth dr
initialize estimated depth dm to zero
while (abs(d-dm) > Tol)
   evaluate the new minimum stress guess shm using
     q=(dr-d)/(dr-dl)
     shm=shl+q(shr-shl)
  evaluate the new minimum stress guess sHm
  evaluate the new depth guess dm
  if (d in [dl dm])
    shr=shm
    sHr=sHm
    dr=dm
  else
    shl=shm
    sHl=sHm
    dl=dm
  end
end
return shm, sHm, dm
```

5 Results

As an example the case shown in Fig. 12 is used. The width of the breakout is set to 80°, while the initial estimates for the minimum and the maximum value of the *in situ* in plane minimum principal stress are 9 MPa and 18.5 MPa respectively. The order of polynomial interpolation for the discretization of the borehole boundary after the removal of the yielded area (see discussion in Sect. 3.1) was set to 3 instead of 5, as this was found to be more stable in the area of low minimum and high maximum principal stresses and close to the area of total failure. The borehole radius was set to 100 mm, because a 20 cm diameter is common in practice.

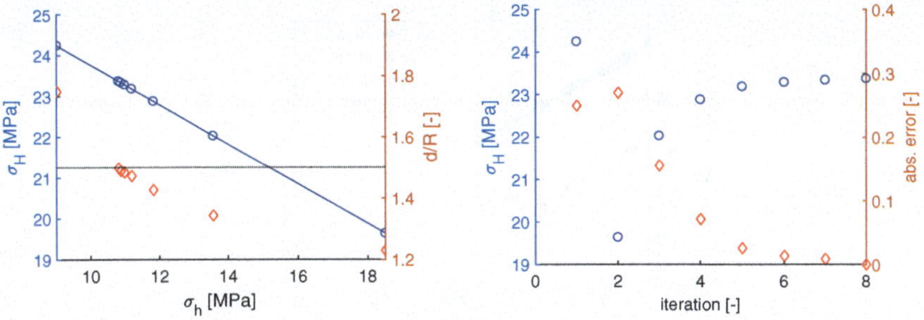

(a) Results. (b) Convergence.

Fig. 14. Target normalized depth 1.5

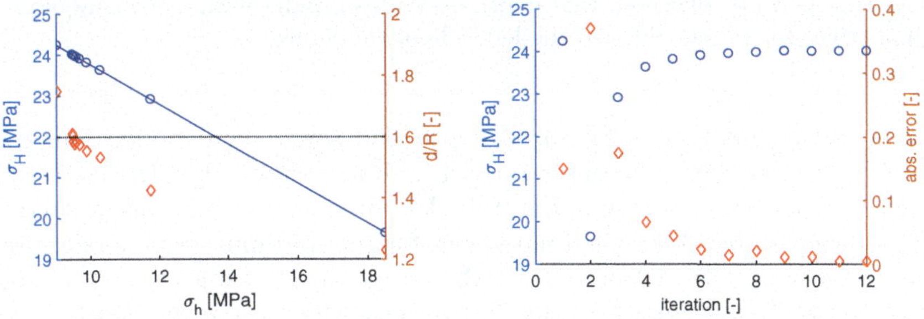

(a) Results. (b) Convergence.

Fig. 15. Target normalized depth 1.6

Two different target depths were tested. The normalized target depth was set equal to 1.5 in the first and 1.6 in the second case. The results are shown in Figs. 14 and 15 respectively. In Figs. 14a and 15a respectively the estimate of the minimum in situ in plane principal stress is shown on the horizontal axis. The left vertical axis corresponds to the estimate of the maximum in situ in plane principal stress with the respective data marked in blue, while the right vertical axis corresponds to the normalized depth with the respective data marked in red. In Figs. 14b and 15b on the left vertical axis the estimate of the maximum in situ in plane principal stress and on the right vertical axis the absolute error of the normalized depth are shown as a function of the number of iterations.

For the first case 8 iterations were required for an accuracy of 10^{-3}. This includes the initial guesses as iterations. The final values for the minimum and maximum primary stresses were 10.80 MPa and 23.37 MPa respectively, while the corresponding normalized depth was equal to 1.499. It can be seen that the target value is approached monotonically from one side, when ignoring the first iteration. The value of both maximum principal stress and normalized breakout depth changes rapidly in the first iterations and the increments decrease as the number of iterations increases.

The behavior is very similar in the second case. The area corresponding to small minimum principal stresses is however characterized by a very strong

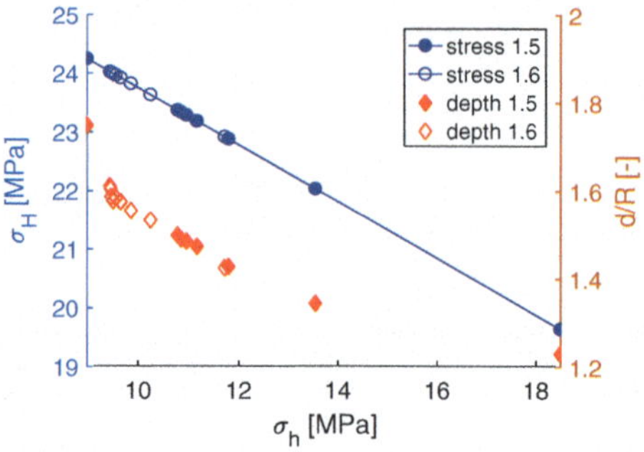

Fig. 16. Superposition of the results for the two test cases. Full symbols correspond to a target normalized depth of 1.5, while open symbols correspond to a target normalized depth of 1.6. It can be observed that depth decreases rapidly with increasing minimum principal stress for points close to the left side of the figure

variation of the breakout normalized depth and some slight oscillations in the values for the depth may be observed in Fig. 15a. Despite this, the convergence remains satisfactory, as shown in Fig. 15b. A number of 12 iterations was required for an accuracy of $5 \cdot 10^{-3}$. The final values for the minimum and maximum primary stresses were 9.48 MPa and 24.01 MPa respectively, while the corresponding normalized depth was equal to 1.605. In this case the normalized depth varies in a much stronger manner than the stresses: a normalized depth of 1.578, evaluated for the eighth iteration, corresponds to a minimum and a maximum primary stress of 9.53 MPa and 23.99 MPa respectively.

The results for both cases are plotted together in Fig. 16. This highlights the variation of the normalized depth of the breakout with the stresses. For minimum *in situ* in plane principal stresses between 9 MPa and 10 MPa, the normalized breakout depth varies roughly between 1.8 and 1.6, while for minimum principal stresses between 10 MPa and 18 MPa the normalized depth varies roughly between 1.6 and 1.2, showing a much slower variation.

The breakout shape, including the successive iterations to evaluate it, is shown for the initial guesses, which are the same for the two cases. It can be seen that while the width is the same, the depth is very different. However, at the same time it should be observed that for the borehole shape marked in blue the depth close to the area of the borehole wall that remains sound is very small. Another interesting observation, particularly obvious in the larger breakout, is that the breakouts are dog-ear shaped. For the small scale simulations the breakouts are instead cusp shaped, as shown in Fig. 10. This is very likely a scale effect.

On the whole the performance of the procedure for the evaluation of the primary stress state using the shape of the borehole breakouts may be deemed satisfactory. In the next section the feasibility of the stress assessment is discussed on the basis of experimental results.

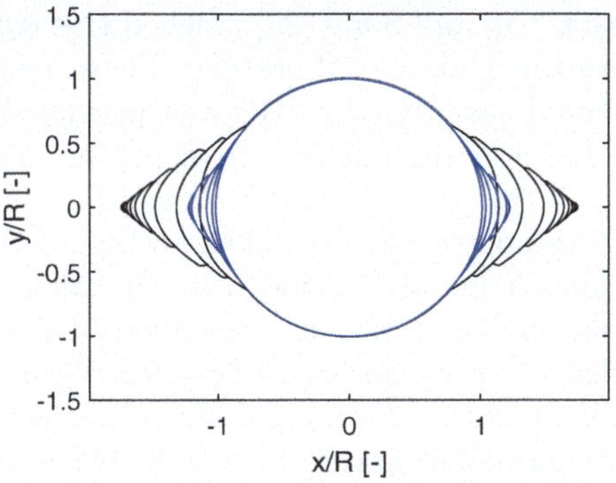

Fig. 17. Breakout shape for the initial guesses. It can be seen that while the width is the same, the depth is very different. Moreover, in contrast to the small scale simulation, the breakouts are dog-ear shaped instead of cusp shaped

6 Discussion on the Feasibility of Stress Assessment

Though the opinion voiced in [22] that the shape of the borehole breakouts depends solely on the material properties and not on the *in situ* stress state has been shown through experimental results not to reflect reality, it is as yet unclear whether the depth and the width of borehole breakouts are independent from each other. It has been postulated [10,23] that, while they both depend on the stress state, they vary in the same manner with it. This would mean that each time either the depth or the width of the borehole breakout may be used for the stress assessment, as the use of the one renders the use of the other automatically redundant. This in turn means that one relationship for the stresses may be gained from the shape of the breakouts, rather than two, as done in the previous section.

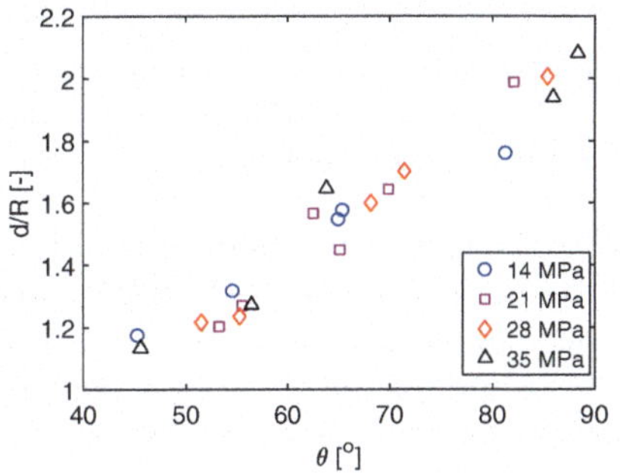

(a) Results on Alabama limestone by [10].

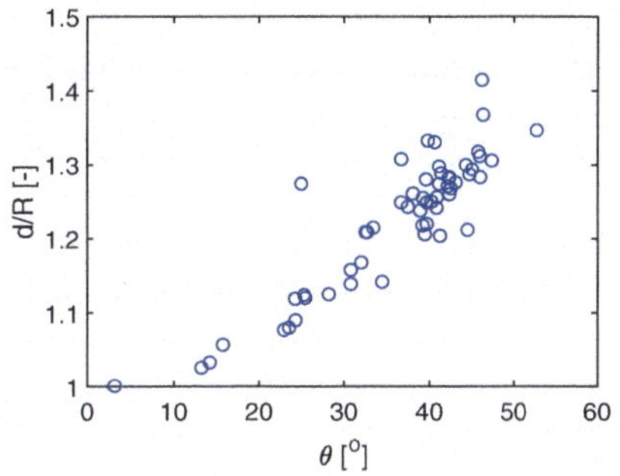

(b) Results on Westerly granite by [23].

Fig. 18. Borehole breakout normalized depth as a function of the breakout width

Herrick and Haimson [10] and Song [23] reached this conclusion on the basis of results of tests performed in the laboratory. These results are presented in Fig. 18. The experimental program by [10] was previously discussed in some detail in Sect. 3.4. In Fig. 18a the normalized depth is plotted as a function of the aperture angle for all tested stress states. It can be observed that the width of the distribution of the points, which is close to linear, is rather narrow. This in turn implies that knowledge of the width is equivalent to knowledge of the depth and no additional information may be gained by the use of both.

Song [23] performed a large number of tests on Westerly granite, that are plotted in Fig. 18b. The maximum principal stress was acting in the horizontal direction, with the borehole cored in the vertical direction. Its values lay between 140 and 240 MPa. The minimum principal horizontal stress and the vertical stress were significantly smaller and ranged between 20 and 50 MPa and 20 and 120 MPa respectively. Once more, the results lie on a relatively narrow band, which is similar to the one observed in Fig. 18a, especially when taking into account the difference of the limits of the axes for the two graphs. It should be remarked that the results by [23] correspond to significantly smaller aperture angles and depths and that the minimum horizontal principal stress was in all cases significantly smaller than the maximum one.

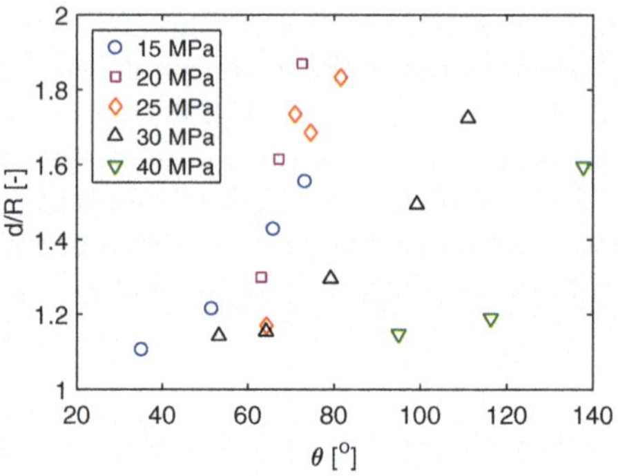

Fig. 19. Borehole breakout normalized depth as a function of the breakout width for different far field stress states by [1]. The vertical stress for the different sets marked in the legend was equal to 30, 40, 40, 40, 50 MPa from top to bottom

In contrast to the findings discussed above, the results by Haimson and Lee [1] on Tablerock sandstone do not show a one to one relationship between breakout aperture width and normalized depth, as may be seen in Fig. 19. Even in this case it should however be remarked that the results of the three first test series are located in a rather narrow domain. The differences observed indicate that in some cases at least different information is included in the two quantities describing breakout shape. It remains unclear however, why this variation was not present in the previous cases. In the case of [23] it may attributed to the small variation of the minimum principal stress, when compared to the maximum principal stress, though this is only an assumption.

The numerical results, as shown for example in Figs. 17 or 16, show a significant variation of the normalized depth with constant breakout width. The discrepancy may be due to the selection of the stress states tested in the laboratory or to the fact that a failure criterion not incorporating the out of plane stress was used.

7 Conclusions

In this work, a recently proposed semi-analytical method for breakout geometry estimation [14] is used for the assessment of the *in situ* stress state in brittle rocks. The semi-analytical method is based on conformal mapping for the calculation of the stress state redistribution during breakout development and the Mohr-Coulomb criterion is used to determine the material failure.

The algorithm is extended in this work to include experimentally observed scale effects on the strength of the material around the borehole. A logarithmic relationship between the borehole radius and the normalized tangential stress at failure is used, based on fitting experimental results on Alabama limestone from the literature [10], to modify the yield criterion and account for scale dependent strength. This improved the capability of the algorithm to simulate the arrest of the breakout development with higher accuracy, when the modified yield criterion was used in combination with a more accurate higher order polynomial approximation for the boundary of the area failing in each iteration.

A two step validation process of the proposed semi-analytical method with conformal mapping is presented in this work for the first time. First, the simulation of the stress redistribution is compared with results from finite element analysis for different stages during breakout development. The results of the conformal mapping method are in good agreement with the results from finite element analysis and this shows that the methods can be applied with confidence to simulate both the direct and the inverse problem when the *in situ* stress state is unknown. The second validation process is the comparison of the simulations with experimental results from the literature on Alabama limestone [10]. The agreement of the simulations with the results is good, both for the predicted aperture width and borehole breakout depth, especially for higher levels of maximum in plane principal stress. The algorithm underestimates the breakout angle and depth for the case of the smallest maximum in plane principal stress.

The semi-analytical method based on conformal mapping is utilized in conjunction with a modified version of the iterative bisection method to assess the *in situ* stress state by solving the inverse problem. Both the width and the depth of borehole breakout are used as input. The Kirsch solution provides one relationship for the two stress components; the solution depends on the breakout width and the uniaxial strength of the material. The conformal mapping algorithm is used for the second relationship, by simulating the borehole depth based on the estimate of the minimum *in situ* in plane principal stress. A solution is obtained when the predicted depth equals (within a tolerance) the input one. The iterative methodology shows good convergence characteristics and two examples are illustrated in Sect. 5, where the *in situ* stress state is estimated based on the proposed methodology. It is concluded that the proposed method can be used for the *in situ* stress state assessment and the first results show great potential for future use in engineering applications.

For the assessment of the components of the *in situ* in plane stress state on the basis of the geometry of borehole breakouts it is necessary that the depth and the width of borehole breakouts are independent from each other. This is necessary so that two stress relationships based on the geometry of the breakout can be retrieved. The experimental results from the literature are contradictory on this aspect as it was thoroughly discussed in Sect. 6. It seems therefore likely that the accuracy of the suggested methodology, as well as of any other methodology estimating both principal stresses, may depend on the sensitivity of the breakout shape of the specific material to the stress state.

The proposed methodology has several limitations and assumptions that should be improved or further investigated in future works. A strong assumption in the proposed algorithm is the restriction to two dimensions of the stress state and the failure criterion. This is not a characteristic of the method. The selection for the restriction to two dimensions was made because it reflects the current state of practice. A further strong assumption is the assumption of coaxiality between borehole axis and *in situ* stress. This will be raised in the future. Finally, in its present state the approach requires the knowledge of scale effects from laboratory tests. On approximation and in the absence of suitable data, a power law can be used.

Despite these limitations and assumptions, the proposed methodology for *in situ* stress state assessment has good potential to be used as a practical tool in several engineering applications including geothermal energy and/or oil and gas recovery.

Acknowledgements. The first author would like to acknowledge the support of BeFo under the auspices of project No. 408. The second author would like to acknowledge the support by the Energy Area of Advance Initiative of Chalmers University of Technology.

References

1. Haimson, B., Lee, H.: Borehole breakouts and compaction bands in two high-porosity sandstones. Int. J. Rock Mech. Min. Sci. **41**, 287–301 (2004)
2. Zoback, M.D., Barton, C.A., Brudy, M., Castillo, D.A., Finkbeiner, T., Grollimund, B.R., Moos, D.B., Peska, P., Ward, C.D., Wiprut, D.J.: Determination of stress orientation and magnitude in deep wells. Int. J. Rock Mech. Min. Sci. **40**, 1049–1076 (2003)
3. Cuss, R.J., Rutter, E.H., Holloway, R.F.: Experimental observations of the mechanics of borehole failure in porous sandstone. Int. J. Rock Mech. Min. Sci. **40**, 747–761 (2003)
4. Meier, T., Rybacki, E., Reinicke, A., Dresen, G.: Influence of borehole diameter on the formation of borehole breakouts in black shale. Int. J. Rock Mech. Min. Sci. **62**, 74–85 (2013)

5. Noll, W.: A new mathematical theory of simple materials. Arch. Ration. Mech. Anal. **48**, 1–50 (1972)
6. Crook, T., Willson, S., Jian, G.Y., Owen, R.: Computational modelling of the localized deformation associated with borehole breakout in quasi-brittle materials. J. Petrol. Sci. Eng. **38**, 177–181 (2003)
7. Papamichos, E.: Borehole failure analysis in a sandstone under anisotropic stresses. Int. J. Numer. Anal. Meth. Geomech. **34**, 581–603 (2010)
8. Zervos, A., Papanastasiou, P., Vardoulakis, I.: Modelling of localisation and scale effect in thick-walled cylinders with gradient elastoplasticity. Int. J. Solids Struct. **38**, 5081–5095 (2001)
9. Zheng, Z., Kemeny, J., Cook, N.G.W.: Analysis of borehole breakouts. J. Geophys. Res. **94**, 7171–7182 (1989)
10. Herrick, C.G., Haimson, B.C.: Modeling of episodic failure leading to borehole breakouts in Alabama limestone. In: 1st North American Rock Mechanics Symposium, ARMA–1994–0217, Austin, Texas. American Rock Mechanics Association (1994)
11. Zhang, H., Yin, S., Aadnoy, B.S.: Poroelastic modeling of borehole breakouts for in-situ stress determination by finite element method. J. Petrol. Sci. Eng. **162**, 674–684 (2018)
12. Lee, H., Moon, T., Haimson, B.C.: Borehole breakouts induced in Arkosic sandstones and a discrete element analysis. Rock Mech. Rock Eng. **49**, 1369–1388 (2016)
13. Exadaktylos, G.E., Stavropoulou, M.C.: A closed-form elastic solution for stresses and displacements around tunnels. Int. J. Rock Mech. Min. Sci. **39**, 905–916 (2002)
14. Gerolymatou, E.: A novel tool for simulating brittle borehole breakouts. Comput. Geotech. **107**, 80–88 (2019)
15. Stanchits, S., Surdi, A., Gathogo, P.: Onset of hydraulic fracture initiation monitored by acoustic emission and volumetric deformation measurements. Rock Mech. Rock Eng. **47**, 1521–1532 (2014)
16. Lee, M., Haimson, B.: Laboratory study of borehole breakouts in Lac du Bonnet granite: a case of extensile failure mechanism. Int. J. Rock Mech. Min. Sci. **30**, 1039–1045 (1993)
17. Haimson, B.: Micromechanisms of borehole instability leading to breakouts in rocks. Int. J. Rock Mech. Min. Sci. **44**, 157–173 (2007)
18. van der Hoek, P.G.: Prediction of different types of cavity failure using bifurcation theory. In: DC Rocks 2001, The 38th U.S. Symposium on Rock Mechanics (USRMS), ARMA–01–0045, Washington, D.C. American Rock Mechanics Association (2001)
19. Dresen, G., Stanchits, S., Rybacki, E.: Borehole breakout evolution through acoustic emission location analysis. Int. J. Rock Mech. Min. Sci. **47**, 426–435 (2003)
20. Cheatham, J.B.: A new hypothesis to explain stability of borehole breakouts. Int. J. Rock Mech. Min. Sci. Geomech. Abstr. **30**, 1095–1101 (1993)
21. van der Hoek, P.G., Smit, D.-J., Khodaverdian, M.: Material-dependent size effect of hollow cylinder stability: theory and experiment. In: 1st North American Rock Mechanics Symposium, ARMA–1994–0411, Austin, Texas. American Rock Mechanics Association (1994)
22. Gough, D.I., Bell, J.S.: Stress orientations from borehole wall fractures with examples from Colorado, east Texas, and northern Canada. Can. J. Earth Sci. **19**, 1358–1370 (1982)
23. Song, I.: Borehole breakouts and core disking in Westerly granite: mechanisms of formation and relationship to in situ stress. Ph.D. thesis. The University of Wisconsin, Madison (1998)

Foundation Engineering: Challenges and Achievements

Geotechnical Challenges for the Numerical Prediction of the Settlement Behaviour of Foundations in Rosenheim's Seeton

Roberto Cudmani, Daniel Rebstock, and Joshua Schorr(✉)

Zentrum Geotechnik, Technical University of Munich, Munich, Germany
joshua.schorr@tum.de

Abstract. The Rosenheimer Basin in the alpine region near the German-Austrian border is a deep ancient glacial lake, which has been filled with fine-grained sediments over the course of the past 10.000 to 100.000 years. Due to the rapid economic growth in the region over the last few decades, the high population density and the increasing utilisation of infrastructure in the region a significant demand for development exists, particularly for structures with large loads. The design and construction of such structures represents a significant challenge for engineers due to the sensitive soft, fine-grained lacustrine sediments. This article focusses on the conception and design of a cable-stayed bridge pylon in Rosenheim. The foundation consists of bored piles connected to a pile cap, with additional displacement piles and vertical drains for soil improvement. As the stiffness and strength of the sensitive lacustrine sediments are strongly reduced due to the disturbances caused by pile installation, displacement piles in combination with vertical drains around the piles are prescribed in order to re-consolidate the soil ("soil healing") and increase the shaft friction after the pile installation. In order to take into account the effect of the soil disturbance and the healing effect of the soil improvements on the foundation behaviour realistically, high quality pile loading tests were carried out. Based on the results of laboratory and field tests, and the simulation of the single pile loading tests using a visco-hypoplasticity constitutive model, a 3D Finite-Element Model was developed and calibrated to predict the time-dependent behaviour of the mixed foundation. The numerical prediction shows that the serviceability requirements of the foundation can be fulfilled and the scheduled underpinning of the superstructure to compensate the foundation settlements will likely not be required prior to 50 years of operation. An extensive monitoring program will be implemented during the construction to validate and, if necessary, to adjust the numerical model to realistically predict the long-term deformation behaviour of the bridge foundations.

1 Introduction

As part of the new western by-pass road B15, Westtangente Rosenheim, two bridges line have been planned with a total length of around 670 m to cross the Mangfall River and the Mangfall canal, the industrial area Aicherpark and a railway. The bridges consist of a cable-stayed bridge with a maximum span between the pylons of 100 m, and a 480 m long, multi-span beam bridge with spans varying between 20 m and 31 m (Figs. 1 and 2).

T. Triantafyllidis (Ed.): *Recent Developments of Soil Mechanics and Geotechnics in Theory and Practice*, LNACM 91, pp. 323–346, 2020.
https://doi.org/10.1007/978-3-030-28516-6_17

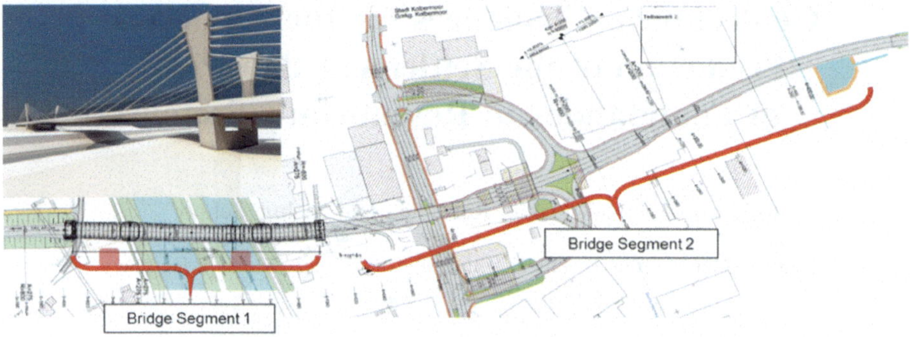

Fig. 1. Bridge structures crossing over the Mangfall river, Mangfall canal, the Aicherpark industrial estate and the Holzkirchen - Rosenheim railway line: Bridge Segment 1 (Mangfall bridge): Ingenieurbüro Grassl GmbH (Structural design) | Reinhart + Partner Architekten und Stadtplaner (Visualisation); Bridge Segment 2 (Aicherpark bridge): SSF Ingenieure AG (Structural design).

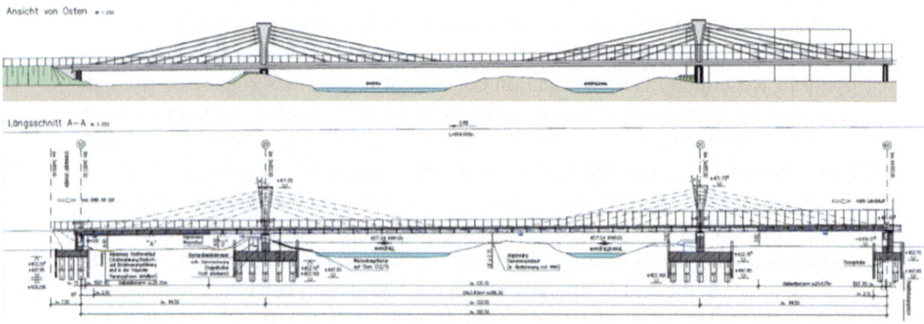

Fig. 2. Bridge Segment 1: Cable-stayed bridge over the Mangfall river and Mangfall canals (side view); Ingenieurbüro Grassl GmbH (Structural design) | Reinhart + Partner Architekten und Stadtplaner (Visualisation).

A particular challenge for the realisation of the project is the subsoil, which consists of quaternary, sensitive, very soft, low plastic and fine-grained lacustrine deposits, the so-called "Seeton" of the Rosenheim sedimentary basin. Due to the thickness of the lacustrine clayey sandy silt layer, down to depths of approx. 150 m in the vicinity of the bridge construction site, floating foundations are required to transfer the bridge loads to the subsoil. Due to the viscous behaviour and low bearing capacity, fulfilment of the geotechnical limit and serviceability states for this soil is a difficult task, especially for the foundation of the pylon. The relatively high loads necessitate the use of bored piles, despite the fact that the installation of the piles cause a considerable disturbance to the subsoil, considerably affecting their bearing behaviour and reducing their bearing capacity.

The development of an innovative floating foundation to account for the complex subsoil conditions in Rosenheim, with focus on the three dimensional Finite Element Analysis (FEA) of the foundation behaviour is presented in this article. Firstly, the main conclusions of the comprehensive soil investigations that were carried out to assess the mechanical time-dependent behaviour of the clay are described. Subsequently, the additional requirements for the piling and concreting works, which were specified to reduce the disturbance of the sensitive soil as well as the results of three high-quality static pile loading tests are described. These results are used to calibrate the numerical model of the foundation and to validate the application of the FEA to simulate its bearing behaviour. Using the pylon foundation as example, the strategy for the fulfilment of the geotechnical design requirements for the limit and the serviceability state of the floating foundation will be discussed. The observation method will be applied in order to deal with uncertainties related to the settlement prediction over the lifespan of the described bridges in the lacustrine sediments.

With this aim, in addition to the FEA of the foundation behaviour, a comprehensive monitoring program will be implemented, during both the construction and operation of the bridge. The observation method allows for the employment of remediation works, such as the possibility to lift the bridge superstructure, if the admissible settlements of the foundation, as per the structural design are exceeded. With the help of the FEA model, the time-dependent foundation settlement can be assessed and the appropriate time for lifting of the bridge can be scheduled.

2 Geological and Geotechnical Subsoil Conditions

The area around Rosenheim lies in the catchment area of the Inn River. In the last ice age about 100.000 to 1.000.000 years ago, the Ur-Inn was a huge glacial valley. It consisted of a solid ice layer at the base of the Alps surrounded by moraine, which was pushed by the glacier. After the Ice Age, the moraine dammed the meltwater of the glacier to form the Rosenheimer Lake (surface area of around 420 km^2), which over time, was continuously filled with fluviatile sediments transported by the Inn. Depending on the grain size and the flow velocity the fine-grained soils consisting of silt, clays and fine sands with thicknesses of up to 300 m were deposited in the ancient lake. About 12,000 years ago, the moraine barrier north of the village "Wasserburg am Inn" was broken through resulting in a complete loss of the water contained in the lake. Overlying the lacustrine clay or "Seeton" is a cover layer of alluvial sediments consisting of gravel, sand, and silt mixtures of varying thicknesses is typically encountered.

At the location of the construction site, the Seeton, with a thickness of 150 m, is covered by such a cover layer with a thickness of 6 m to 8 m and consisting of silty gravel. According to geotechnical investigations, the lacustrine sediments at the site consist mainly of silt and clayey silt interbedded with very fine sand layers, which can seldom be identified by a visual inspection. According to the results of the soil classification tests, the Seeton can be classified predominantly as a low plastic clayey-silt with a very soft to soft consistency (liquid limit $w_L = 0{,}38$, plastic limit $w_P = 0{,}20$, natural water content $w_n = 0{,}30$ to 0,38). Typically, the Seeton has a very small permeability in the vertical direction ($k_v \leq 10^{-8}$ m/s) due to the presence of the fine-grained layers, while the permeability in the horizontal direction, which is controlled by the permeability of the interbedded sand layers, is up to two orders of magnitude larger.

The investigation of the subsoil in the vicinity of the pylons, especially in the area of the pile loading tests, consisted of a total of 3 boreholes, 15 CPT and pressiometer tests using the cone-pressiometer testing device (CPTM), with final depths of up to

70 m. Furthermore, 31 CPT tests were carried out up to a depth of 70 m in locations corresponding to the piers of the multi-span bridge. Additionally, soil samples were retrieved with a tube sampler for the determination of the index properties and the investigation of the soil mechanical behaviour of the Seeton in the laboratory.

The shear strength c_u derived from the cone penetration resistance q_c by means of the common empirical relationship described in Fig. 3 shows a linear increase with depth which is typical for normal consolidated clays. These c_u values were confirmed by the results of the undrained triaxial compression tests carried out on the "disturbed" samples retrieved from the boreholes. The relationship $\Delta c_u / \Delta \sigma'_0 = 0.1$ (σ'_0: in situ vertical effective stress) corresponds to the lower boundary of the range of the data encountered in the literature for soft clays (e.g. Hansbo 1957), which typically increase from about 0.1 to 0.4 for a limit liquid of $w_L = 0.25$ to 0.8. These interbedded thin sand layers are most probably the cause for the sudden increases in the cone resistance, as repeatedly observed in Fig. 3.

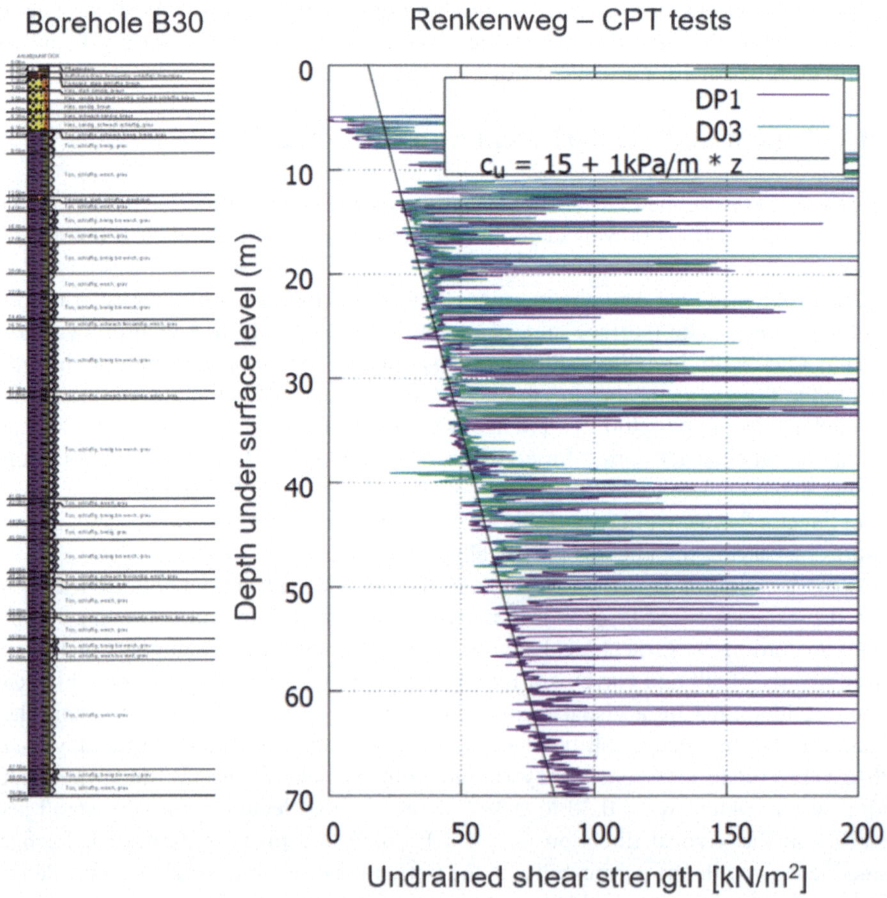

Fig. 3. Example of the determination of the undrained shear strength from the cone penetration resistance $c_u = (q_c - \sigma_{v0})/N_k$ with $N_k = 15$.

The mechanical behaviour of the Seeton is rate dependent. The viscosity index (Leinenkugel 1976) determined in the laboratory according to Krieg (2000) was found to be $I_v = 0.03$. Therefore, the time-dependent soil behaviour and the loading rate must be considered for a realistic assessment of the deformations of the bridge foundations.

3 First and Second Pile Loading Test Campaign

In the original geotechnical report of the project, in which the authors of the article were not involved, a floating pile foundation was recommended to transfer the loads of the pylon and the piers to the ground. For the foundation design, a characteristic value of the skin friction of $q_s = 30$ kN/m^2 and a negligible end bearing $q_b = 0$ was recommended. Furthermore, it was assumed that an increase of skin friction of $q_s = 60$ kN/m^2 could be realizable by post-grouting of the pile shaft. It was stated that the skin friction q_s applied in the design must be confirmed by pile loading tests (PLT).

With this aim, two pile loading campaigns were carried out between October 2015 and September 2016 - without the participation of the authors. In the first campaign, three bored piles were subjected to tension loads and in the second campaign, eight bored piles were installed in two test fields and subjected to compression loads. The test piles had a diameter of 1.2 m and lengths of between 17.5 m and 27.5 m (1st campaign) and 30.5 m (2nd campaign). Due to the thickness and the planned location of the pile caps below the ground surface, the most part of the cover layer will be excavated and hence, the bearing behaviour of the piles will be determined predominantly by the shaft and end bearing resistance in the Seeton. For this reason, measures to eliminate or at least to reduce the skin friction in the upper gravel layer during the pile testing had to be carried out. The measures used to achieve this consisted of welding a thin steel tube to the reinforcement in the area of the upper gravel layer in order to create a gap between this tube and the temporary casing used to drill the borehole. It was intended, that after withdrawal of the temporary casing, the gravel would close the annular gap and become looser, leading to a reduction of the friction angle and the horizontal stresses, resulting in a reduction in the skin friction.

In both, the first and second PLT campaigns, the skin friction in the Seeton layer assumed for the design could not be confirmed. The evaluation also showed that the highest skin friction was mobilised in the cover layer, where the skin friction should have been eliminated. The reason for the ineffectiveness of the measures to eliminate skin friction became apparent once the pile heads were uncovered: the annular gap between the steel tube and the gravel layer had been filled with concrete. Instead of a reduction, this led to an increase of the skin friction in the cover layer.

With the aim of improving the bearing capacity, in four of the eight tested piles, additional measures were tested consisting of (post) grouting and (post) installation of vertical drains around the piles. A systematic improvement of the pile bearing capacity with respect to the piles without any treatment was not observed. The measured bearing capacity, defined as the pile resistance for a displacement equivalent to 10% of the pile diameter, varied between 2,000 kN and 4,800 kN. After subtracting the skin friction in the cover layer and the end bearing, the skin friction in the Seeton varied between 0 and 25 kN/m^2, with a mean value below $q_s = 10$ kN/m^2. Despite the relative homogenous soil conditions encountered at the site, the large variations of the measured pile skin friction meant that it was not possible to extrapolate the PLT data to the design of the bridge foundations. Therefore, the pile loading test results did not fulfill an essential condition for the experimental determination of the pile bearing capacity, according to the applicable standards and recommended guidelines published by the geotechnical community. Even assuming reproducibility of the PLT results, adopting a characteristic skin friction $q_s = 10$ kN/m^2 for the foundation design would have led to pile lengths of more than 100 m making construction of the bridge extremely difficult from a technical, and most certainly infeasible from an economical point of view.

Ultimately, it was concluded that neither the evaluation of the skin friction and the load bearing behaviour of the tested piles nor the extrapolation of the skin friction to the foundation piles were reliable based on the results of the two PLT campaigns. For this reason, the execution of a third PLT campaign was recommended to finally assess the technical and economic feasibility of a floating pile foundation of the bridge.

4 Third Pile Testing Campaign

4.1 Concept Und Pile Instrumentation

The third PPB campaign was conceived by the authors in cooperation with the technical department of BAUER Spezialtiefbau GmbH and involved three test piles (P09, P10 and P11) and six reaction piles (R19 to R24) with a nominal diameter of D = 1.2 m. The length of the test piles were 30 m (P09 and P10) and 60 m (P11). Four reaction piles with D = 1.2 m and a length of 40 m were used to load the piles. The new test field was located adjacent to the previous test field at Aicherpark, such that the three existing reaction piles (R10, R13 and R16) from the second PPB campaign could be reused (Fig. 4).

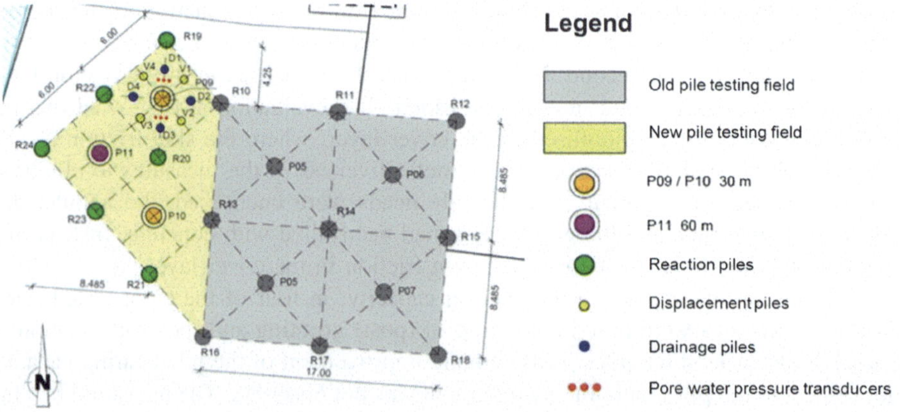

Fig. 4. Third static pile testing campaign – Test field and pile layout, including the reaction piles.

In order to evaluate the mobilized pile resistance over the depth, the piles were instrumented with concrete strain gauges (BVG) and extensometers (EXT). These two independent systems provided a redundancy in the determination of the normal forces, affording a certain level of quality control over the strain measurements. Three Geokon "Sister Bars" were installed at each of the BVG cross-sections. Each pile has two BVG cross-sections in the upper gravel layer, where the skin friction has been eliminated, for calibration purposes. The extensometer used was the retrievable "Model 1300" from Geokon (Fig. 5). The extensometers are pneumatically anchored in steel pipes that were installed in the piles. A vibrating wire displacement transducer is mounted below each anchor point. The piles P09 and P10 were each equipped with two extensometer strings with four and five anchoring points, respectively. For pile P11, the two short extensometer strands were converted into a long strand with eight anchor points.

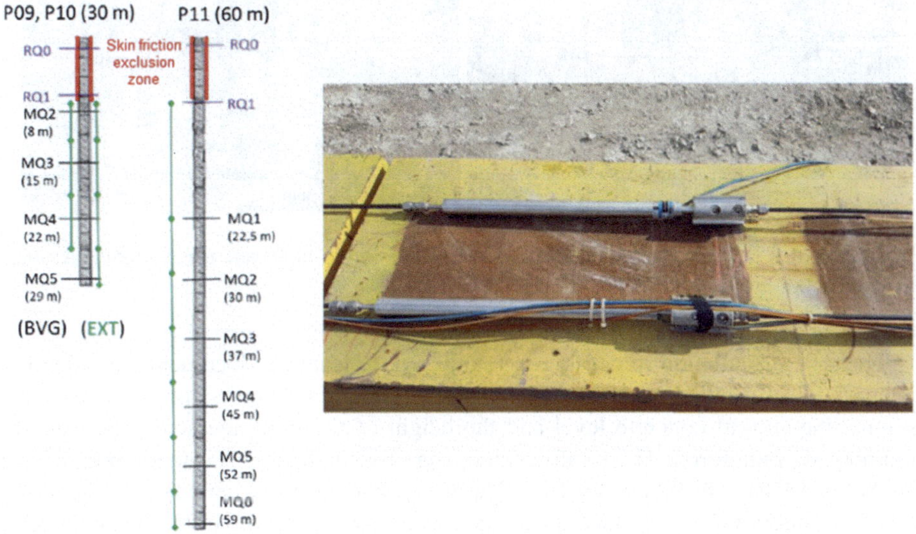

Fig. 5. Instrumentation of the test piles – left: locations of the instrumented cross sections with strain gages (BVG); and right: retrievable extensometer (EXT) with pneumatically expandable anchors and vibrating wire strain gauges.

4.2 Technical Specification for Pile Installation

The three test piles and the six new reaction piles were installed between Jan. 10th and Feb. 7th 2017. The drilling rig used was a BG 46 with an additional casing oscillator (Fig. 6, left). In order to reduce the disturbance to the sensitive Seeton as much as possible, different technical requirements were specified for the piling works. These included; drilling under water load to prevent hydraulic failure, pre-advancing the casing to a depth $\geq 2 \cdot D_s \approx 2.4$ m below the bottom of the bore-head (also when the planned final depth is reached) as well as the limitation of the drilling and withdrawal speed of the drilling bucket. In order to stabilise the base of the borehole after reaching the final depth, an additional 0.5 m thick gravel layer was deposited at the base of the piles and compacted with the drill bucket prior to the installation of the reinforcement.

Fig. 6. BAUER BG46 with: casing oscillator and water inflow (left) and ring auger to eliminate the skin friction in the cover layer (right).

The pile installation process was subjected to careful documentation. During concreting, the installed or removed casing and tremie pipe lengths, the actual concrete volume, the current concrete level and the height of the reinforcement cage were all continuously monitored. The reinforcement cage for the test piles settled about 65 cm for P09 and P10, and 90 cm for P11. The major part of these deformations occurred just after withdrawal of the first liner segment to above the bottom of the borehole, after pouring the first concrete load. The settlement resulted from the gap left by the casing below the bottom of the borehole and the consolidation of the Seeton under the weight of the reinforcement cage.

The exclusion of the skin friction in the area of the gravelly surface layer was carried out with a new and innovative system. It consisted of drilling a borehole with a diameter of 2.0 to a depth of about 7 m below the ground level, with an approx. embedment of 0.5 m into the Seeton and filling the borehole with plastic concrete. Next, a special ring auger (Fig. 6, right) was used to drill a hole of greater than 1.2 m in diameter. This meant that following the normal installation of the pile, an annular gap between the pile and the plastic concrete is created, ensuring the complete elimination of the skin friction in the cover layer as the remaining plastic concrete ring supports the surrounding soil and the annular gap is filled with water.

4.3 "Healing" the Seeton After Pile Installation

Following the installation of test pile P09 (see site plan in Fig. 4) four full displacement piles V1 to V4 (CMC: Controlled Modulus Columns, System Menard GmbH), with a diameter of 0.4 m and a length of 28 m below ground level, were installed with the aim of compensating for the disturbance of the Seeton due to the pile installation. The displacement auger which was used, is shown in Fig. 7.

Fig. 7. Displacement auger used to install the CMC piles.

In the plan view, the CMCs were installed symmetrically around the pile, within a relative short distance from the pile shaft, in order to induce a re-consolidation of the Seeton and thereby increase the skin friction. In addition, four vertical drains D1 to D4 were installed prior to the installation of the CMCs with the aim of accelerating consolidation. The consolidation process was monitored by pore pressure transducers installed in and around the pile.

4.4 Loading System and Instrumentation of the Pile Head

A reaction system consisting of two steel beams with a span of 8.5 m was used to apply the test loads. As mentioned before, the reaction loads were transferred to the four reaction piles by two GEWI bars per reaction pile (Fig. 8). The reaction system was designed for a capacity of 11 MN. The load was applied by means of a hydraulic jack, which allowed a variable oil intake and a SDA 1100 hydraulic cylinder (max. load of 11,000 kN at a nominal pressure of 692 bar). The load was kept constant by means of an electronic contact pressure gauge (accuracy of ±0.5 bar). The force was recorded by an electronic load cell. The vertical pile head deformations were recorded using three digital displacement transducers. Two additional electric displacement transducers were used to check the horizontal displacement of the pile head. In addition, the tensile forces in the GEWI bars were recorded by means of electronic load cells in order to check that the load was evenly distributed. The entire test data were recorded and stored digitally and could be evaluated and displayed in real time. To check the functionality and the results of the main measuring system; the settlements of the pile heads, the measurement bridges, the reaction piles and a fixed point (outside the area of influence of the pile loading apparatus) were recorded manually using a Leica DNA03 digital dumpy level.

Fig. 8. Pile loading test (pile P10 is depicted) – load transfer beams and loading setup, vehicle containing the measurement equipment and a surveyor's dumpy level.

4.5 Execution of the Pile Loading Tests

The pile loading tests took place between March and May, 2017 and consisted of two phases for all three piles. In the first, load-controlled phase (CLT: constant load test), the piles were loaded in increments of approx. 250 kN, at the start of the test and approx. 100 kN, towards the end of the test. During these loading stages, the creep coefficients were determined with the following relationship:

$$k_s = \frac{(S_2 - S_1)}{\log t_2 - \log t_1}$$

Where s is the Settlement of pile head and t is the time since begin of the load stage. The coefficients were determined in intervals of 20 min. According to the recommendations of the German geotechnical society, EA-Piles (2012), the load was only increased when the creep coefficient became constant or began to decrease with time and the velocity of the pile head displacement was lower than 0.005 mm/min. Aside from the determination of the load-settlement behaviour of the pile, the aim of the load-controlled test phase was to investigate the time-dependent developments of the settlements under constant load and the possibility of a creep failure at high loading levels. Unstable behaviour, indicated by an increase in the settlement velocity and the creep coefficient with time, was not observed up to the maximum load of the load-controlled phase (load leading to a pile head settlement of 0.1 D). To the contrary, despite the fact that quite large creep coefficients of up to 20 mm were recorded near the maximum CLT increments, for long observation times, the creep coefficient remained either constant or decreased with time (Fig. 9). The load-controlled test phase was carried out until the limit displacement of 0.1•D ≈ 12 cm according to EA-Piles (2012), was reached. An overview of the load-controlled test phases for the three test piles can be seen in Table 1.

Table 1. Overview of the load controlled phases

		P09 (L = 30 m, with soil improvement)	P10 (L = 30 m, without soil improvement)	P11 (L = 60 m, without soil improvement)
Max. Test force controlled phases	[kN]	3257	1990	6509
Number of loading steps	[–]	20	12	26
Total testing time	[h]	~328	~205,5	~274,5
	[d]	~13,5	~8,5	~11,5

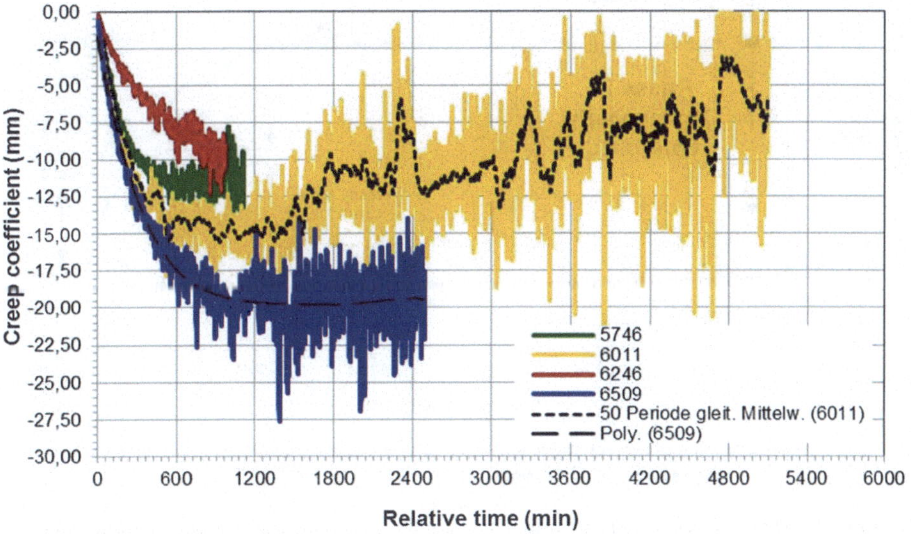

Fig. 9. Pile test P11 – Variation of the creep coefficient during loading stage time for the last four loading stages.

The load-controlled test phase was followed by a second, velocity-controlled phase (CRT: constant rate test) that included three "velocity jump tests". The initial velocity at the pile head of approx. 0.6 mm/min was increased by the factors 3.5, 0.5 and 7 in the jump tests. The velocity-controlled test phase was finished once the maximum displacement range of the hydraulic cylinder of 25 cm was reached. From the jump tests, the pseudo-viscosity index could be determined Krieg and Goldscheider (1998):

$$I_v = \frac{\ln(R_1(u)/R_2(u))}{\ln(\dot{u}_1/\dot{u}_2)} = 0.027 - 0.032$$

Where $R_i(u)$ is the pile resistance that would develop for the velocity $\dot{u}_i$ of the pile head settlement at a vertical displacement u_i. In Fig. 10, the results of the CRT and the CLT phases are presented for the pile P09. The results show that the load-settlement response from the CRT phase can be used to derive the load-settlement response for the CLT. Indeed, the red curve, which is the load-settlement calculated assuming a constant loading velocity of 0.005 mm/min at the pile head, agrees with the green curve, which is the load-settlement curve derived from the CLT when the velocity 0.005 mm/min is used as criterion to increase the load.

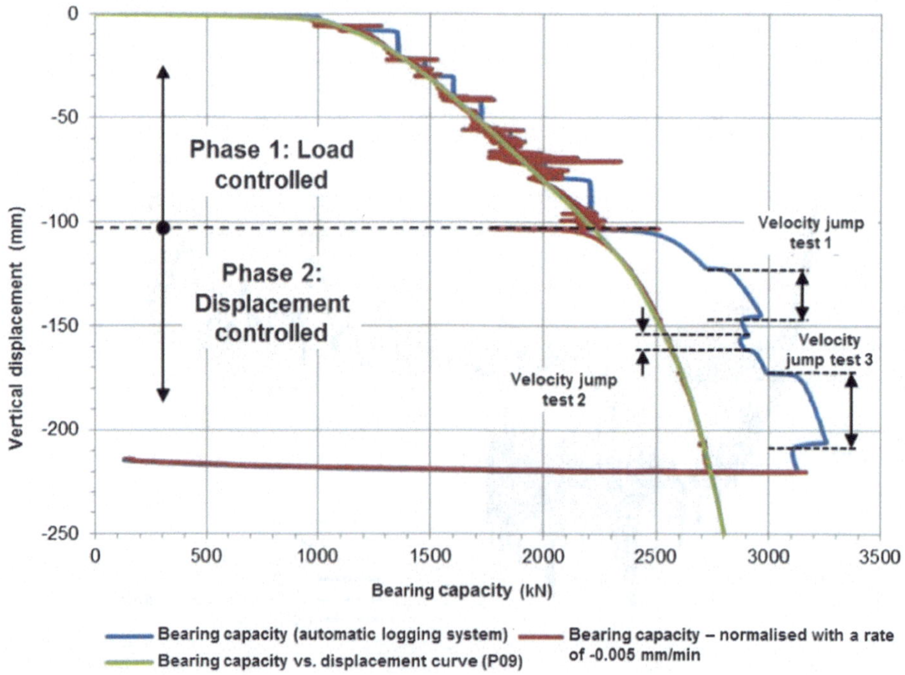

Fig. 10. Pile load vs. pile head settlement for the CLT and CRT phases for pile P09.

4.6 Results of the PLT

The comparison of the results of the three test piles in Fig. 11 shows, as expected, that the 60 m pile reached the highest bearing capacity of 6.5 MN at the displacement of approx. 0.1 D. The bearing capacities of the 30 m long piles were 3.2 MN (P09) and 2.2 MN (P10). This indicates that the pile capacity could be increased by approximately 30% by the proposed soil improvement. The CRT phase shows that even at pile displacement larger than 0.1 D, significant additional pile resistances can be still mobilised at expenses of an increase in the deformation velocity. Since the base resistance is similar for the three piles, the differences in the pile capacities originate in the skin friction, as seen in Fig. 12.

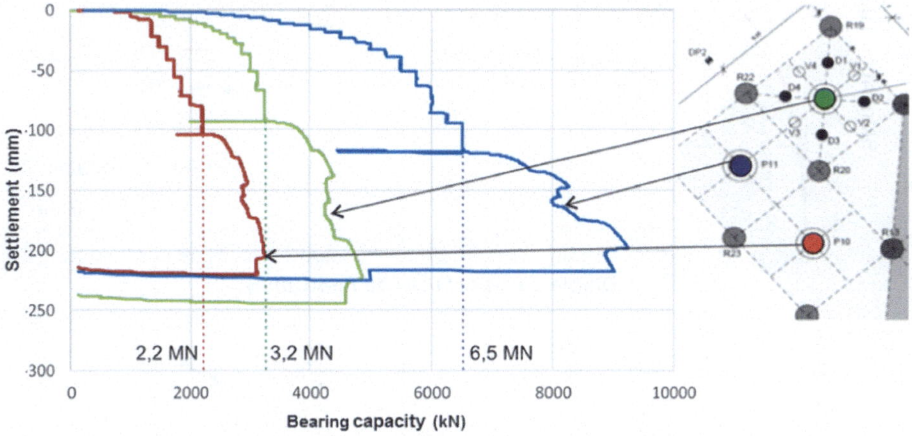

Fig. 11. Comparison of the bearing capacity-settlement lines of the test piles.

From the evaluation of the individual measurement cross-sections, the (depth-dependent) skin friction and the end bearing pressure were determined (Fig. 12). The skin friction reaches its maximum value at pile head displacements of 15 mm to 30 mm. The end bearing of all three piles increases almost linearly with the pile displacement to about 1,000 kN (P09 and P11) or 900 kN (P10) and does not appears to be fully mobilised at the end of the load-controlled phase. Despite a depth of almost double the piles P09 and P10, the mobilised tip resistance of pile P11 is still about 1,000 kN, compared to a similar value of 900 kN for P10 with a length of only 30 m. This result indicates that the end bearing is mainly controlled by the installation-related disturbance of the sensitive Seeton.

The "healing effect" of the soil improvement can be seen in the evaluation of the skin friction of the two 30 m piles with recorded values of about 25 kN/m^2 for P09, compared to 15 kN/m^2 for P10. Due to the soil improvement, the end bearing behaviour of P09 is lightly stiffer (with a displacement of 900 kN at 85 mm instead of 100 mm). The average skin friction is 25 kN/m^2 for P11 (60 m) and is thus higher than 15 kN/m^2 for P10. The fact that the mean skin friction of the 60 m pile is greater than the value of the 30 m pile is consistent with the increase of the shear strength with the depth indicated by the results of the field and laboratory tests.

Taking into account safety factors relating to the statistical scattering of the data from EA-Piles, characteristic values of the pile resistances could be reliably derived from the results of the third pile testing campaign.

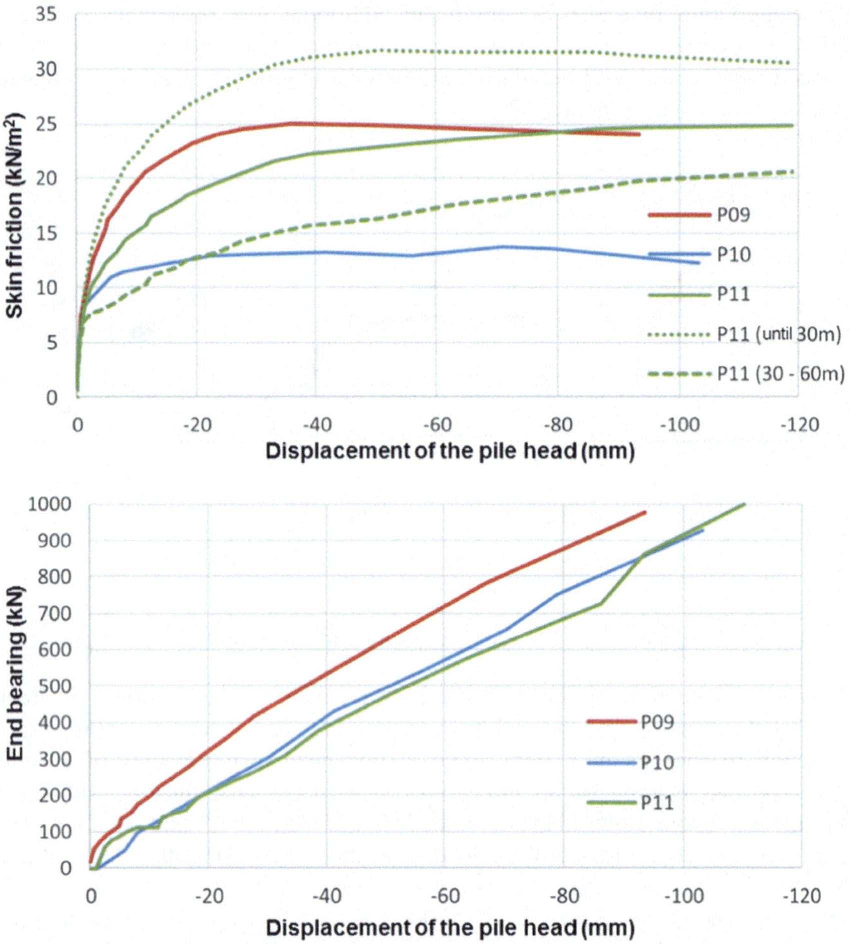

Fig. 12. Comparison of the mean skin friction (top) and the end bearing (bottom) over the head displacement of the three test piles in the CLT phase.

5 Foundation Design

On the basis of the test load results, two technically feasible alternatives for the foundation of the pylons (design vertical load $E_{v,d} \approx 80$ MN, characteristic vertical permanent load $G_k \approx 49$ MN, characteristic vertical traffic load $Q_k \approx 8$ MN) were identified. These options are limited due to the fact that the dimensions of the pile cap (15 m × 15 m) cannot be increased due to site constraints. The first foundation alternative is a classical pile foundation. In order to reduce the group effect of the piles and to avoid excessive disturbance of the subsoil during pile construction, 14 piles with a diameter of 1.2 m, a centre-to-centre distance of 4.5 m and a length of 65 m are specified. Under the characteristic load, a settlement of the single piles of about 8 mm can be estimated from the results of the pile tests. The settlement of the pile group for the permanent loads were estimated to be 3 cm to 4 cm using a simplified approach based on analytical solutions to the theory of elasticity with the program ELPLA 10.

The second alternative is a mixed foundation (Fig. 13), in which the loads are transferred through the piles, the pile cap and the soil improvement. In this foundation type, the bored piles are installed first, followed by the vertical drains and finally the unreinforced displacement piles, following the same sequence as for test pile P09. The displacement piles cause a consolidation of the subsoil and lead to an increase of the effective stress of the soil around the pile, which in turn increases the skin friction of the piles. The vertical drains enable a rapid dissipation of the excess pore water pressure generated by the soil displacement and thus a rapid consolidation of the Seeton. The length of the displacement piles is limited due to the method of installation and available equipment to approx. 50 m.

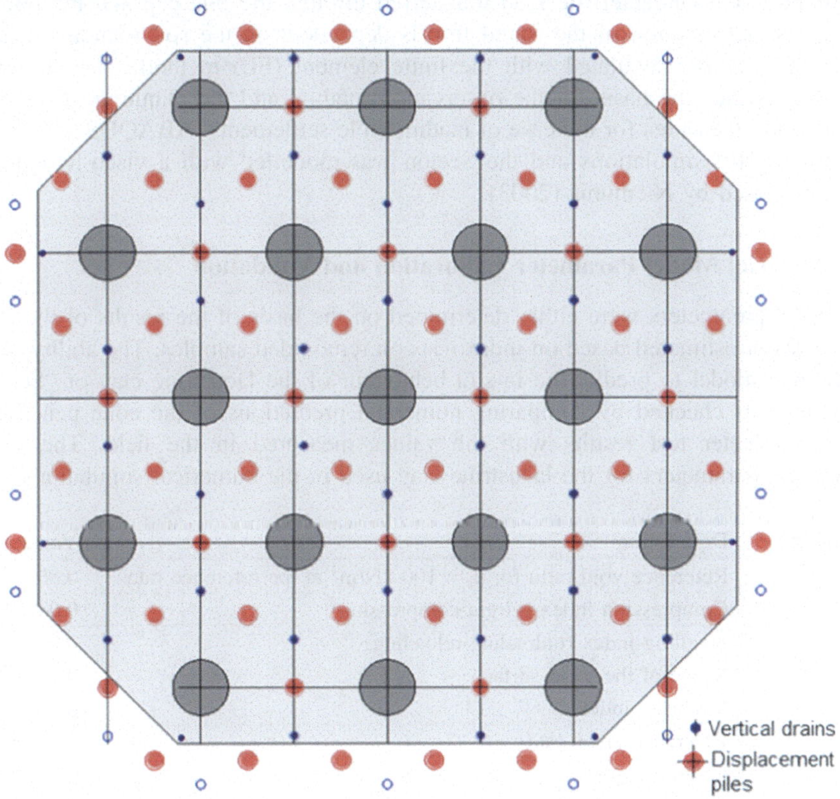

Fig. 13. Plan view of the foundation and layout of bored piles, displacement piles and vertical drains for the second alternative.

The single pile was designed in accordance with the DIN EN 1997-1:2014 and the DIN 1054:2010 for the portion of the load to be carried by the piles. Additionally, the ultimate limit state check of the bearing capacity of the pile group was performed for the total load, replacing the pile group by a single 'replacement pile' in accordance with

DIN EN 1997-1 and DIN 1054. The displacement piles are only intended to improve the subsoil and therefore no separate limit state verifications need to be performed.

From a geotechnical point of view, the mixed foundation was regarded as the more robust foundation type. The shorter piles reduce the risk of a disturbance to the subsoil during construction. In addition, the foundation behaviour is less dependent on the load bearing behaviour of the individual piles than the pure bored pile foundation variation. The displacement piles improve and homogenise the subsoil and counteract the installation-related disturbances.

6 Prediction of the Deformation Behaviour of the Mixed Foundation

The proportional characteristic load transferred through the pile cap and the piles as well as the deformation of the foundation is dependent on the soil-foundation interaction and can be calculated with the finite element (FE) method. The numerical modelling is thus the basis for the observation method and the planning of potential rehabilitation measures for the case of inadmissible settlements. ABAQUS v2017 was used for the FE simulations and the Seeton was modelled with a visco-hypoplastic model proposed by Niemunis (2003).

6.1 Material Model Parameter Calibration and Validation

The model parameters were either determined on the basis of the results of the laboratory tests or estimated based on index tests on remoulded samples. The ability of the constitutive model to predict the in-situ behaviour of the lacustrine clay or "Seeton" parameter was checked by comparing numerical predictions of the cone penetration and pressiometer test results with the values measured in the field. The visco-hypoplastic parameters for the lacustrine clay used in the numerical simulations are:

Parameters	Description	Value
e_r	Reference void ratio for p_r = 100 kN/m^2 at the reference rate	0.86
λ	Compression index (virgin compression)	0.04
κ	Swelling index (unloading-reloading)	0.01
β_R	Shape of the yield surface	0.5
I_v	Viscosity index	0.03
D_r	Reference creep rate	1e−7 m/s
φ'_c	Friction angle	30°
m_2	Intergranular strain parameters	5.0
m_5		5.0
R_{max}		1E−4
β_x		0.2
χ		1.5

The numerical and experimental results of oedometric compression tests and undrained triaxial compression tests are compared in Figs. 14 and 15. In Fig. 16, the relationship between the radial pressure and the normalised radial displacement r/r_0 (r_0, r: initial and current radius of the cavity), as derived from the pressiometer tests, are compared with the numerical solution of the cylindrical cavity expansion problem under undrained conditions for depths of 10 m–20 m and 30 m–40 m. In the cavity expansion model, the mechanical behaviour of the Seeton is simulated by the visco-hypoplastic model with the model parameter given above. The initial state of stress of the soil is defined by the vertical stress σ_z resulting from the overburden, the radial and annular stresses:

$$\sigma_{\theta,r}/\sigma_z = K_0 = 1 - \sin\varphi'.$$

In addition, an overconsolidation ratio OCR = 1.5 was assumed for the in-situ state, which has been determined with the constitutive model for an estimated age of the Seeton of 10.000 years. Furthermore, the expected creep settlements of the Seeton for free-field conditions in the next 100 years starting with OCR = 1.5 was analysed. The calculated settlement of a few centimetres during this period of time was considered realistic, as large subsidences of the ground surface have not been observed in the region of Rosenheim so far. In Fig. 16 only the numerical results for the upper and

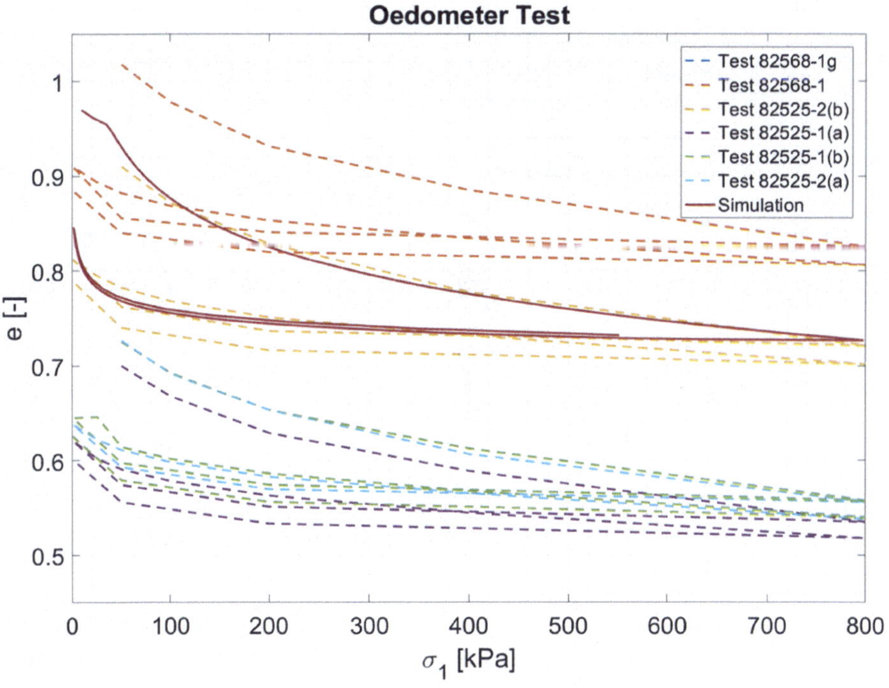

Fig. 14. Oedometric compression: void ratio versus effective vertical stress from laboratory tests and numerical element test simulations with the visco-hypoplastic constitutive model.

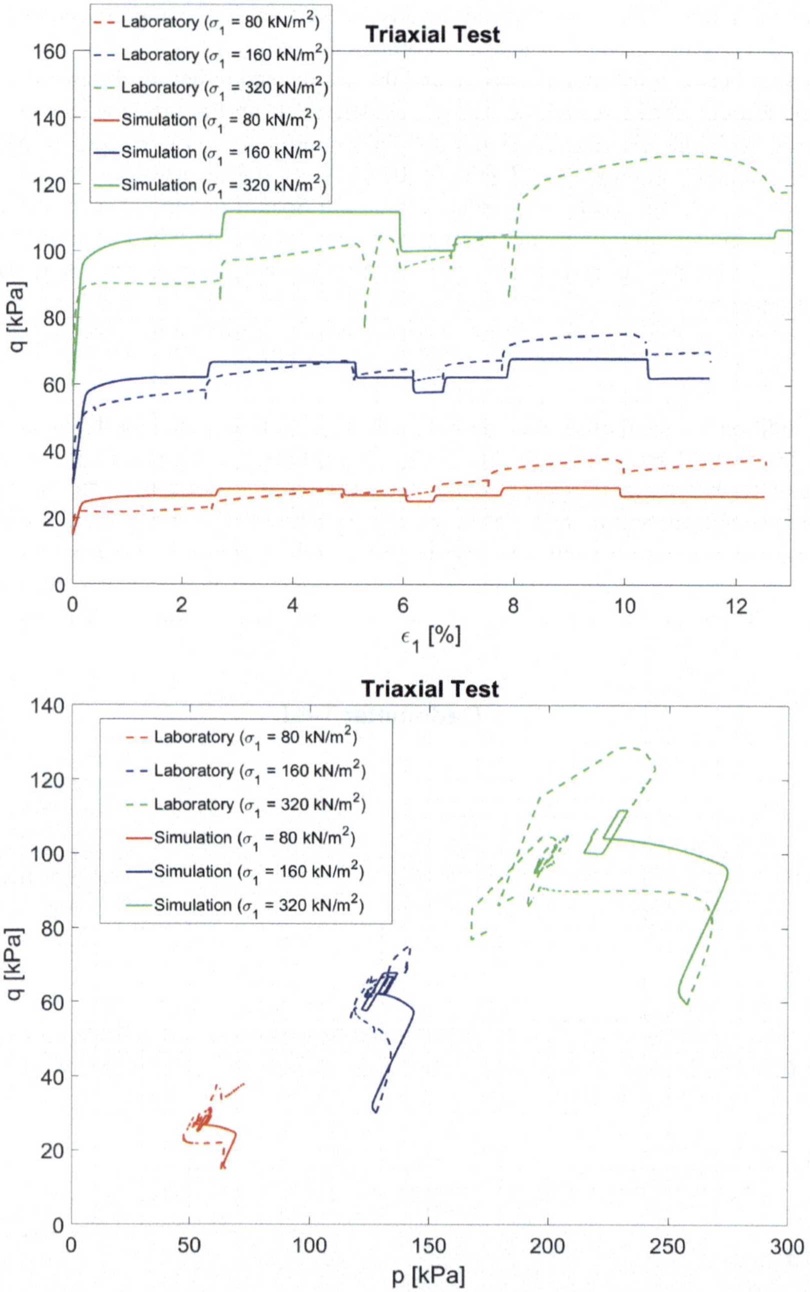

Fig. 15. Undrained triaxial compression: deviatoric stresses versus axial strain (top), deviatoric stresses versus mean stress (bottom) from laboratory tests and numerical element test simulations with the visco-hypoplastic constitutive model.

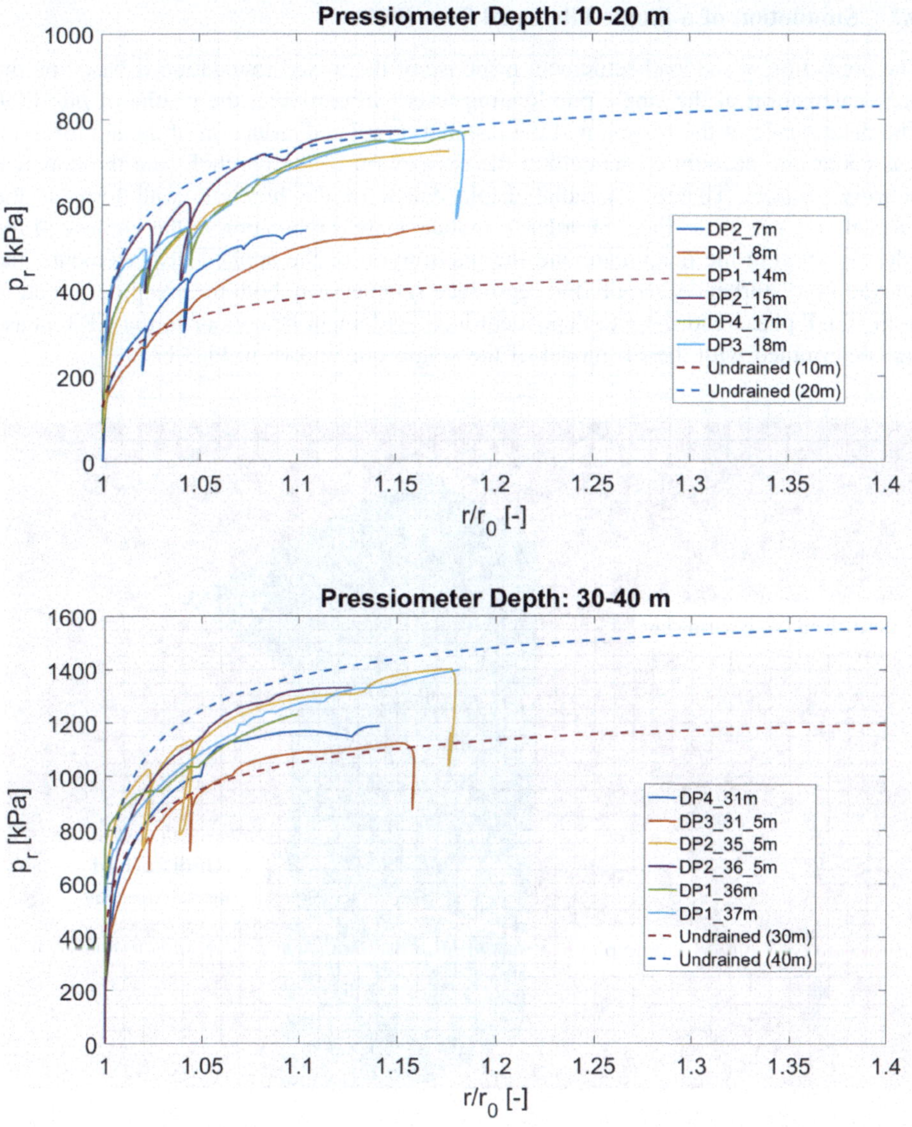

Fig. 16. Comparison of experimental pressure-expansion curves obtained with the cone-pressiometer testing device (CPTM) with the prediction using a cylindrical cavity expansion model for the depth intervals 10–20 m (top) and 30–40 m (bottom).

lower values of the considered depth range have been plotted. As it can be seen, both the laboratory and pressiometer tests can be realistically simulated by the constitutive model.

6.2 Simulation of a Single Pile Load Test (P09)

The prediction of the load-settlement response of the mixed foundation is based on the back-calculation of the single pile loading tests - in particular the results of pile P09. The disturbance of the subsoil and the installation-related reduction of the skin friction was taken into account by modelling the excavation of the borehole and the concrete pouring process. Thereby, a radial displacement of the borehole wall towards the vertical axis was prescribed, in order to induce a stress relaxation in the vicinity of the pile. By varying the magnitude and the distribution of the applied displacements, the relaxation (disturbance) required to reproduce satisfactorily both the creep deformation in the CLT phase and the rate-dependent load-settlement behaviour in the CRT phase was determined with the FE-model of the single pile shown in Fig. 17.

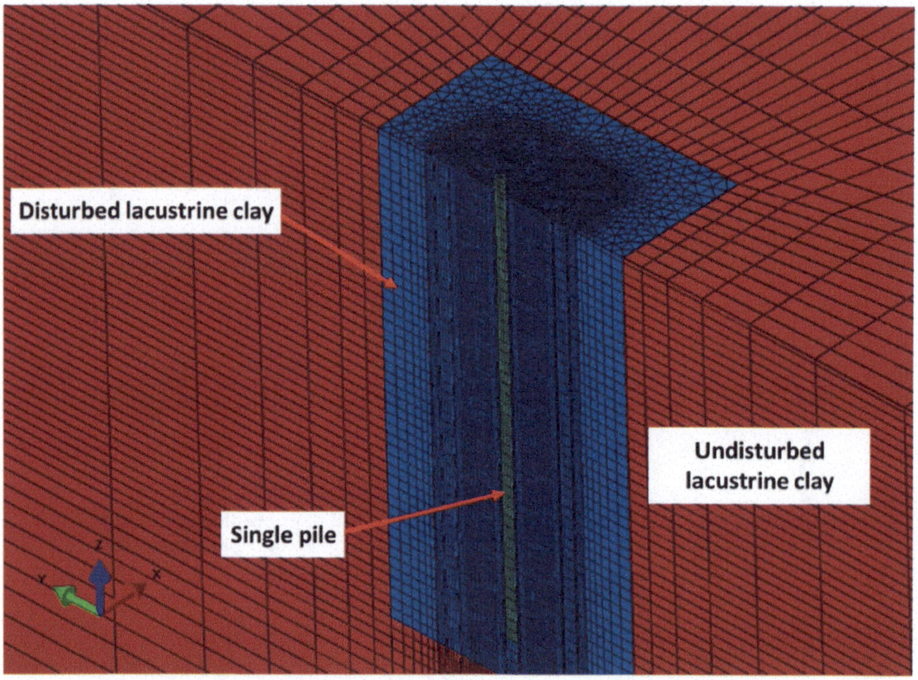

Fig. 17. 3D FE-Model of the single pile with soil improvement (P09) with the same mesh discretisation as the pile group model.

The result of the simulation of the time-dependent behaviour of the pile P09 during the two test phases along with the results of the PLT using the 3D FE-model of the single pile are shown in Fig. 18. A comparison of the numerical and experimental determined pile head displacements for four constant load stages is depicted in Fig. 19. As it can be seen in both figures, the proposed technique to account for pile installation in the numerical simulation enables the back-analysis of the time-dependent behaviour of the single pile realistically, although no relative displacement between the soil and the pile can occur in the numerical model.

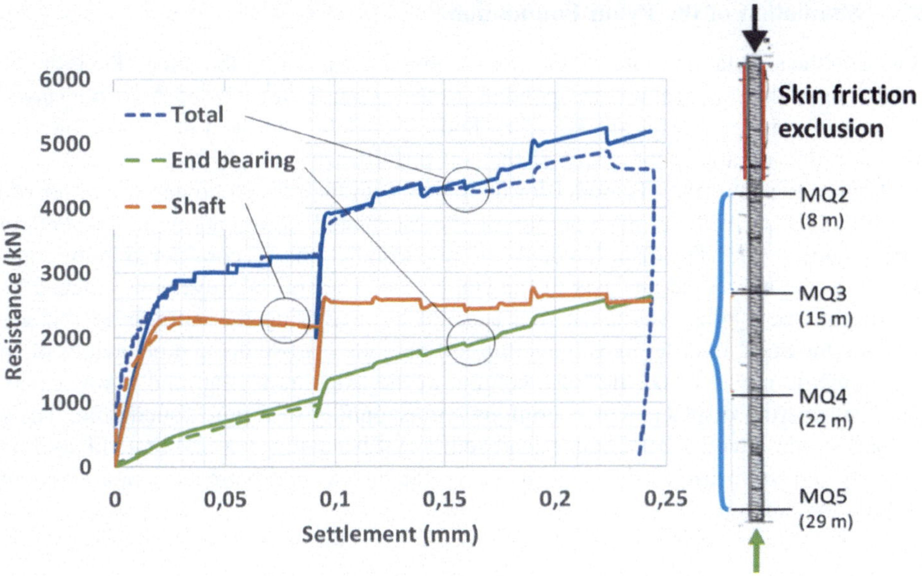

Fig. 18. Comparison of the results of the pile test for P09 (dashed lines) with the results of the FE modelling (continuous lines): pile load (black), the shaft resistance (blue) and end bearing resistance (green) as a function of the settlements at the pile head.

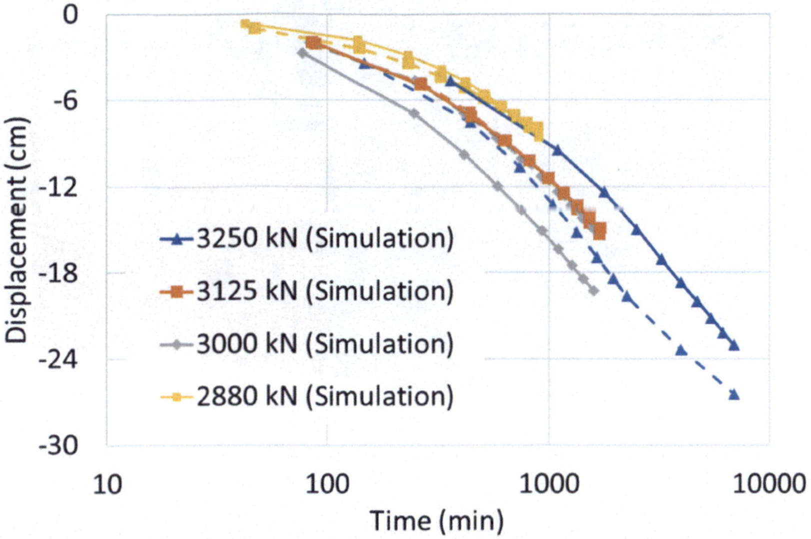

Fig. 19. Settlement of the pile head versus the relative time for the last four load increments: Comparison of results of the pile loading test (dashed lines) and the numerical simulations for P09 (continuous lines).

6.3 Simulation of the Pylon Foundation

Based on the simulation of the single pile loading test, and using the same 3D mesh, the behaviour of the pile group corresponding to the pylon foundation of the cable stayed bridge was simulated using the same method of relaxation of the soil surrounding the pile, in order to capture the effect of the soil disturbance.

Figure 20 shows the FE model for the pylon with the pile group and the improved subsoil (left), as well as the contour of the calculated deformations at the end of construction (right). The disturbance due to pile installation was considered in the same way as in the numerical analysis of the single pile. The soil improvement induced by the displacement piles was taken into account by increasing the stiffness of the soil between the piles. In order to estimate this stiffness the load-settlement behaviour of an axisymmetric model consisting of a column in the axis and soil around it was calculated considering displacement conditions corresponding to a rigid foundation. Then, the soil stiffness with a comparable load-settlement behaviour was calibrated based on the same axisymmetric model with the same dimensions and boundary conditions but without the column.

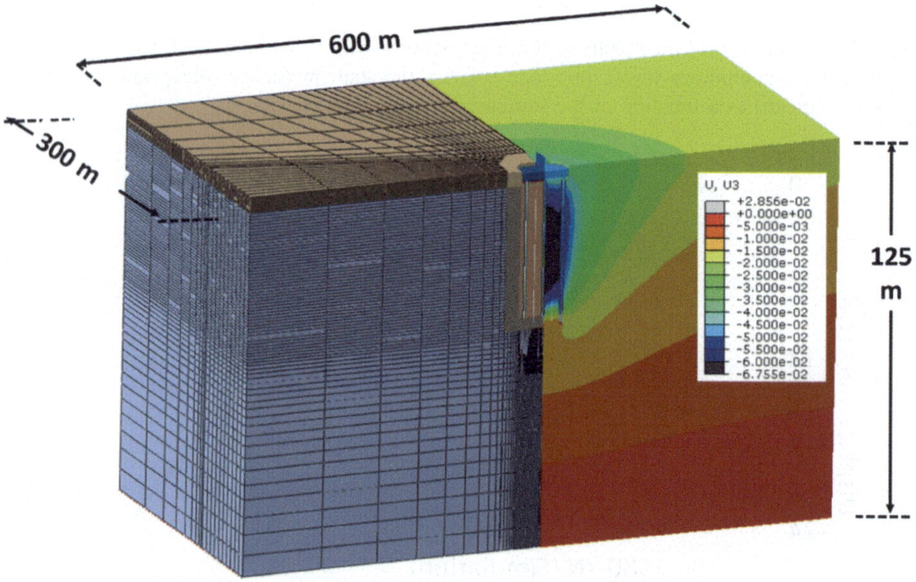

Fig. 20. FE model Pylon view of the mesh (left) and the resulting displacement after 40 years in operation (right).

The FE model of the pylon predicts a maximum foundation settlement of about 6–7 cm after 40 years of operation. The large concentration of the soil deformation in the area of the pile group and the significantly lower settlements of the surrounding area is clearly visible. The development of the foundation settlement predicted by the model for an operation time of 40 years is presented in Fig. 21. A monitoring program will be implemented to track the soil and the foundation deformations during construction.

Based on the data from the monitoring, the FE-model and the prediction will be validated and if required, adjusted and improved.

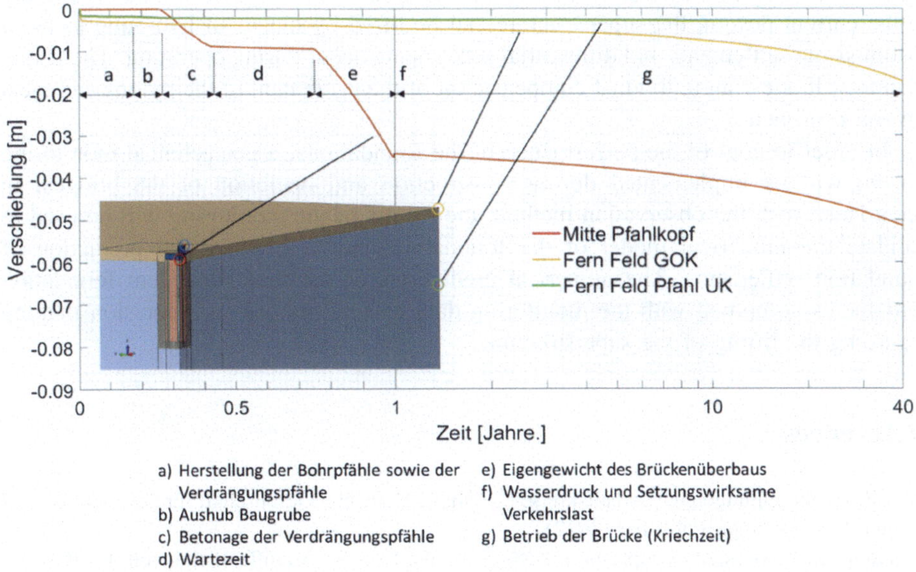

Fig. 21. Evolution of the vertical displacements of the centre of the pylon foundation. For easier visualization, the time axis is linear until in the end of construction (1 year) and logarithmic thereafter (from 1 to 40 years).

7 Final Remarks

Geotechnical solutions in sensitive fine-grained soils, such as the Rosenheim Seeton, require special demands on the investigation of the subsoil, the planning and the construction. The comprehensive subsoil investigations, including laboratory and field tests, enabled the geotechnical characterisation and parameter calibration necessary for modelling the time-dependent mechanical behaviour of the Seeton. The results of the 3rd pile testing campaign illustrate the pronounced sensitivity of the fine-grained soil and highlight the special precautions that must be taken during pile construction to reliably, and reproducibly achieve the pile resistance required to transfer the bridge loads to the subsoils.

A mixed pile-raft foundation was proposed as an alternative to a classical piled foundation, which was considered infeasible from a technical and economical point of view. In order to counteract the effect of soil disturbances due to pile construction and to improve the stiffness of the soil between the bored piles, vertical drains and displacements piles in this sequence, will be installed after the piles.

The time-dependent load-settlement behaviour of a single pile, also including the displacement piles and the vertical drains was investigated by means of high quality pile loading tests. Based on the back-calculation of the single pile tests, a 3D FE Model of the mixed foundation, which considers the time-dependent behaviour of the Seeton and the load settlement behaviour of the single pile, was developed and used to predict time-dependent behaviour of the foundation.

In order to mitigate the geotechnical risks associated with the Seeton and according to the current design, the superstructure can be lifted by means of hydraulic jacks, if inadmissible settlements and differential settlements occur during operation. The lifting system will allow an individual compensation of the settlement of the pylons and piers several centimetres.

In order to control the performance of the foundations, a geotechnical monitoring system will be implemented during construction and operation of the bridges. In accordance with the observation method, the results of the monitoring will be used to validate the numerical model of the foundation and to improve the prediction of foundation settlements. The numerical predictions of the time-dependent foundation settlements combined with the monitoring data will be the basis for decision making regarding the lifting of the superstructure.

References

EA-Piles, Empfehlungen des Arbeitskreises "Pfähle". Deutsche Gesellschaft für Geotechnik, 2nd edn. Ernst & Sohn Publisher (2012)

Niemunis, A.: Extended hypoplastic models for soils, Heft 34. Veröffentlichungen des IGB der Ruhr-Universität Bochum (2003)

Krieg, S.: Viskoses Bodenverhalten von Mudden, Seeton und Klei, Heft 150. Veröffentlichungen des IBF, Karlsruhe Institute of Technologie (2000)

Krieg, S., Goldscheider, M.: Bodenviskosität und ihr Einfluss auf das Tragverhalten von Pfählen. Bautechnik 75, vol. 10, pp. 806–820 (1998)

Leinenkugel, H.J.: Deformations- und Festigkeitsverhalten bindiger Erdstoffe. Experimentelle Ergebnisse und ihre physikalische Deutung, Heft 66. Veröffentlichungen des IBF, Karlsruhe Institute of Technologie (1976)

Hansbo, S.: A new approach to the determination of the shear strength of clay by the fall-cone test. In: Proceedings of the Royal Swedish Geotechnical Institute, vol. 14 (1957)

Diaphragm Wall Technique – Planning, Execution and Development over the Last 65 Years

Gebhard Dausch(✉)

BAUER Spezialtiefbau GmbH, Schrobenhausen, Germany
Gebhard.dausch@bauer.de

Abstract. The development of diaphragm wall technology in Europe began at the beginning of the 1950s with the work of Christian Veder and Hans Lorenz. The advantages of this special civil engineering method quickly became apparent in the construction of inner-city excavation pits and infra-structural measures in the form of low-deformation, water-impermeable reinforced concrete walls, as well as in the use of sealing walls in earth dam and landfill construction.

In the beginning, the focus was on the investigation of the supporting effect of bentonite suspension in the open slot, but soon the technical developments of the excavation tools followed. After setbacks in the quality of the walls, the pioneers of the method tried to identify the sources of error and to avoid them by consistently adhering to the self-imposed specifications. To this day, the process is a speciality in terms of planning and execution and requires a great deal of experience on the part of those involved. The project-specific quality assurance of diaphragm wall construction sites is a decisive aspect for the safe and economical execution of the work.

1 Basics

A diaphragm wall usually has 2 functions. On the one hand, the static function serves to absorb the loads from the neighbouring buildings and the earth and water pressure. Secondly, it has a waterproofing function. This combination of functions is particularly evident in inner-city construction pits. From the upper edge of the diaphragm wall to a few metres below the construction pit floor, the wall has to assume the sealing function in addition to the static effects, and deeper down it actually only serves to shut off the water. Trough pits with a sealing base are constructed to avoid lowering the groundwater level. This is achieved either by incorporating the walls in a soil layer with low water permeability or, for example, by means of a grouting or jet grouting slab.

With the diaphragm wall construction procedure, a construction pit enclosure can be planned directly at the property boundary for optimum utilisation of the area. Compared to a sheet pile wall, there are hardly any vibrations or noise disturbances in the direct vicinity. The depth of the wall is freely selectable under consideration of the boundary conditions (number of basements, load, geology, etc.) and is only limited to a limited extent by technical equipment restrictions. In the 70 s, wall depths of 25 to 30 m were already a special feature. In recent years, diaphragm walls for inner-city

T. Triantafyllidis (Ed.): *Recent Developments of Soil Mechanics and Geotechnics in Theory and Practice*, LNACM 91, pp. 347–354, 2020.
https://doi.org/10.1007/978-3-030-28516-6_18

excavation pits with a depth of 40 to 50 m have been constructed. In Hamburg, a diaphragm wall with a depth of up to 72 m was constructed in 2017 at the Olympus Campus pit.

The wall thicknesses have increased in recent years. Today, wall thicknesses of 1.0 m with deep excavation pits are no longer a special feature. Walls with a thickness of 1.5 m are planned for the construction pits for the new metro main line in Munich.

2 Development of Equipment and Tools for Diaphragm Wall Execution

The equipment and tools used in Milan subway construction in the 1950s can no longer be compared with those used today. Christian Veder patented a process for the production of a watertight wall, which was characterized by the stringing together of uncased bores, which were produced with a free-fall chisel in a flushing stream from clay suspension. This continuous wall surface was then concreted using the tremie concrete method (Fig. 1).

Fig. 1. Diaphragm wall works in Milan from 1956 (Quelle: Wikipedia)

In the early 1960s, the first excavation methods with continuous soil excavation by means of indirect flushing with bentonite suspension were developed in Japan by the companies Tone Boring and Okumura. Soletanche brought a further development of this excavation technology onto the market in the 1970s. Menck & Hambrock started building the first series-produced diaphragm wall grabs in 1966. At the end of the

1960s, STEIN began working with diaphragm walls and initially designed its well-known rope grab for its own use. Later, STEIN also built the grabs for its competitors (Fig. 2).

Fig. 2. Construction site with 3 cutters for the construction of a shaft in the UK (Quelle: BAUER)

BAUER is the only manufacturer of diaphragm wall cutters in Germany. BAUER has been building its various cutters for the world market since 1984. The use of these cutters enables them to be integrated into rock formations and to reach great depths with minimal deviation in verticality. The cutter wheels can be adapted to the geological conditions with an appropriate choice of loosening tools. Verticality is continuously measured by means of an integrated inclinometer and counter-controlled by flap controls in the event of deviations from the nominal position. An additional control of the position of the cutter is carried out at BAUER with the specially developed "Cutter Inclination System", CIS for short. This system uses a tachymeter to record the above-ground position of the retaining ropes of the cutter three-dimensionally and extrapolates it to the cutter body, thus determining the position of the cutter.

With the development of the hydraulic diaphragm wall grabs, the mechanical rope grabs were partially replaced. The hydraulic grab does not require such high demands on the machine operator and can optionally correct the vertical position of the trench during excavation by means of control plates. In France, SOLETANCHE had already adapted hydraulic grabs to Kelly bars in the 1970s and thus reached depths of between 50 and 60 m. In 1993, BAUER developed a hydraulic grab with a width of 35 cm for use with sealing walls.

The mechanical grab achieves its high closing forces by the closing cable guide via deflection pulleys. For operation with the mechanical rope grab, a rope excavator requires free fall winches. As a rule, the grab is lowered to the bottom of the excavation, lifted up again a little and then dropped onto the bottom. This strengthens the penetration of the grab teeth into the soil and increases the digging capacity. In contrast to this, the hydraulic grab is simply placed on the excavation base when open and the grab shells are closed hydraulically. The "chiselling" or "stabbing" of the grab is not carried out with the hydraulic grabs.

The first rope crawler crane for diaphragm wall work, which were used area-wide in Germany for use with mechanical rope grabs, were MENCK cranes. After the insolvency of the MENCK company in 1978, LIEBHERR began building crawler cranes with free fall winches.

3 Supporting Fluids

The use of bentonite suspension was a major mystery in the early years of diaphragm wall technology. Only the consequent research of the properties and modes of action of thixotropic clay suspensions by Hans Lorenz, Fritz Weiss and Hanno Müller-Kirchenbauer enabled the well-founded handling and calculation of the stability of the "open, suspension-supported trench". With the publication of the dissertation by Fritz Weiss in 1964, the total stability of the panel and the single grain in the diaphragm wall could be calculated for the first time. Fritz Weiss introduced the concept of flow limit to understand the suspension properties, and his pendulum device for determining the shear strength was initiated by standards. With the development of the ball harp (german: Kugelharfe) in 1974, Paul von Soos brought a measuring instrument onto the market that can still be found on all diaphragm wall construction sites today.

Two phenomena are described with the mode of action of a bentonite suspension. In cohesive and fine sandy soils a filter cake is formed. The water in the suspension is partially filtered out at the interface of the soil layer and the hydrostatic pressure acts on that of the suspension and is transferred to the soil. In the case of coarse sand and gravel without sufficient fines, the bentonite suspension penetrates the soil to a certain depth. Although there is no filter cake formation, stabilisation is nevertheless effective on the soil. This stabilizing effect is described by the term flow limit (shear strength of the suspension).

The use of polymers as additives for a hybrid suspension has become increasingly common in recent years. For example, a bentonite suspension is enriched with the addition of polymers especially for special applications or properties. In Germany, the use of a pure polymer suspension is not as common as in Asia, for example. In France, diaphragm walls with pure polymer suspensions were already successfully manufactured in the 1990s. BAUER switched from a bentonite suspension to a pure polymer suspension at a milling construction site in the Middle East because the milling wheels bonded due to the geological conditions and this phenomenon no longer occurred due to the change in the supporting fluid.

In March 2019, the European Foundation Engineering Federation (EFFC), in cooperation with the American Foundation Engineering Federation (DFI), published a guideline entitled "Guide to support fluids for deep foundations", describing the source materials for support fluids and explaining how to handle them on the construction site with practical test options.

4 Formation of Joints

The formation of joints is a fundamental issue in the construction of diaphragm walls. The selected joint system must be known as early as the planning and design of the diaphragm wall reinforcement. The different types of joint formation have an influence on the geometry of the panels and the reinforcement steel cages.

In the last 20 years, the use of round joint tubes to limit the panel has been reduced in favour of flat joint elements. The depth of use of the recoverable formwork elements is 35 to 40 m for standard tasks. For deeper trenches, lost joint elements are used.

Prefabricated formwork elements are a somewhat more expensive but high-quality alternative among the lost formwork elements. Due to the stable design of the concrete elements, there are actually no distortions or solder deviations. However, they are very heavy and their use is limited by depth.

The proper and complete removal of the circulated concrete and bentonite residues around the formwork element is very important for a watertight execution.

For cuttered diaphragm wall lamellas, as a rule no stop-ends are required. The cutter cuts the previously concreted primary panel and the concrete of the secondary panel is seamlessly joined to the existing panel.

5 Execution

The essential success factor for proper execution is the competent and experienced team on the construction site.

In addition to knowledge of the geological composition of the soil, the on-site team also needs information on the groundwater situation. Careful adjustment of the bentonite suspension and storage in silos or reservoirs is one of the basics. In the event of a sudden loss in the suspension level in the trench, the team must immediately take countermeasures to prevent the trench from sinking further and to prevent it from collapsing.

The working platform should be sufficiently high in relation to the groundwater level in accordance with the boundary conditions. In this way, the stability of the excavation wall can be ensured by a sufficient pressure gradient.

Panel layout, joint formation and reinforcement design must be planned and determined in advance before execution. In the case of installations in the reinforcement cage, e.g. anchor ducts, care must be taken to ensure a flow-optimised shape. The rising concrete should be able to completely enclose the internals.

For the economical construction of a diaphragm wall, the execution sequence is also an important point. Unfortunately, malfunctions repeatedly lead to deviations from the ideal sequence. This has a strong influence on the daily output and logistics on the construction site (provision of reinforcement cages).

As early as the end of the 1960s, those responsible were intensively concerned with the damage that had occurred and developed instructions for action, which were included in the first diaphragm wall standard. Since then, numerous publications and lectures have reported on the proper execution

6 Quality Assurance

The diaphragm wall standard DIN EN 1538 and other regulations describe what has to be done on construction sites in terms of quality assurance. The topics exceeding the requirements of the standard are regulated in the contract and in the quality assurance plan of the construction site. After the collapse of the city archives in Cologne, the subject of quality assurance has become an even higher priority.

During the work preparation before excavation, special attention is paid to the geometric position of the wall, the building materials used and the reinforcement cages. During wall construction, the following partial steps are monitored and controlled: the excavation process, adherence to verticality and torsion, cleaning of the slots, joint formation, installation of reinforcement and concrete installation.

7 Outlook

Further developments in diaphragm wall technology will include not only developments in equipment but also the disposal of bentonite and polymer suspensions, wider use of polymer suspensions, quality improvement in joint formation, control of concrete paving and quality.

Over the next few years, we will be able to observe how diaphragm wall construction units are increasingly being equipped with data acquisition from production and various sensors for monitoring and quality assurance. Construction management will be able to concentrate on digital process management alongside execution technology.

8 Chronological Development of the Diaphragm Wall Technique

1912 Patent of the company Carl Brandt, Düsseldorf: Process for the manufacture of pillars, piles and the like for civil engineering purposes using the thick flushing process commonly used in mining

1950 Hans Lorenz: On the use of thixotropic liquids in foundation engineering

1950 Christian Veder: Patent for a process for producing watertight walls by stringing together uncased boreholes and using a thick flush stream

1956 Diaphragm wall work in Milan carried out by the company ICOS-Veder

1959 First diaphragm wall projects in Berlin and Munich carried out by the company Pollensky & Zoellner

1961 Development of the technology for the regeneration of bentonite suspension on a construction site in Belgrade/Petuelstraße, Munich

1962 Laboratory tests at Pollensky & Zoellner to investigate the support of earth walls by liquids

1964 Dissertation by Fritz Weiß: Theory of liquid-supported earthwalls and technology of supporting liquids. Introduction of the flow limit. Development of the pendulum device for determining the flow limit

1964 TONE BORING company from Japan develops a tool for the continuous conveyance of soil by means of 3-cone chisels which rotate around their vertical axis.
1965 OKUMARA company from Japan develops the first diaphragm wall milling machine with horizontal milling wheels, which are moved by electrically driven chains
1966 First set of rules by Fritz Weiß for the Munich Federal Railway Authorities for the safe and damage-free manufacture of diaphragm walls
1966 First serially manufactured diaphragm wall grab from Menck & Hambrock, nominal thickness 60 cm and 80 cm with grab width of 2.6 m
1967 First training course conducted by Schönebecker Brunnenfilter GmbH: "Workshop on diaphragm wall construction", from 20 to 22 September 1967 in Munich in the meeting room of Munich Central Station with 126 participants
1969 SoOLETANCHE company builds first diaphragm wall with prefabricated parts
1969 The STEIN company builds the first diaphragm wall grab (K121) with oil-bearing rope pulleys and uses it on the Kaufhof construction site in Munich
1970 Start of production of single-phase diaphragm walls in Europe
1971 Foundation of the NABau "Working Committee II diaphragm walls" with 32 employees under the direction of Dr.-Ing. Bub, Institute for Civil Engineering, Berlin for the development of the later standards DIN 4126 and DIN 4127
1972 Calculation method for the "stability calculation of the open trench" by Hanno Müller-Kirchenbauer with introduction of the stagnation gradient
1973 First single-phase sealing wall in Germany at the Iffezheim barrage by the KELLER company
1975 SOLETANCHE company develops a slotted wall grab on a 60 m long Kelly bar
1975 SOLETANCHE company develops a cutter with horizontally mounted counter-rotating, hydraulically driven cutting wheels
1978 Development of the ball harp for determining the flow limit by Paul von Soos
1981 Publication of ATV 18313 "Schlitzwandarbeiten" (diaphragm wall works)
1982 CASAGRANDE launches a chain-driven cutting machine on the market
1982 First "earth concrete wall" manufactured in Germany at the Igelsbachsperre by BAUER company
1984 First sealing wall with adjusted HDE film at the Malsch landfill by the ZÜBLIN company
1984 Publication of the preliminary standards DIN 4126 "Diaphragm walls" and DIN 4127 "Clays for diaphragm walls"
1984 First German diaphragm wall cutter with direct hydraulic gear from BAUER, Schrobenhausen, Germany
1986 DIN 4126: "Diaphragm walls", version August 1986 and DIN 4127 "Diaphragm wall clays", version August 1986
1989 First cutter for rock works (rock strength up to 150 MPa), BAUER, Schrobenhausen
1989 First single-phase slurry wall with cutter in Ottmaringer Valley by Bilfinger + Berger

1990	Recommendations of the Working Committee "Ufereinfassungen": E 144 Application and design of diaphragm walls and E 156 Application and manufacture of sealing trench walls and sealing narrow walls
1990	DVWK Merkblatt 215 "Dichtungselemente im Wasserbau"
1991	Mini cutter MBC 30 (4.5 × 5 × 5 m) from BAUER with winding device for 150 m depth
1993	Construction of a BC 50 milling machine for off-shore work for diamond search at a depth of 150 m off the Atlantic coast of South Africa
1993	Hydraulic single rope grab for sealing walls with d = 35 cm from BAUER, Schrobenhausen, Germany
2000	First European standard for diaphragm wall work: EN 1538 "Diaphragm walls", version July 2000
2004	Theodoros Triantafyllidis, Planning and construction of a special foundation, Part 1: diaphragm wall and slurry wall technology, Ernst & Sohn Verlag
2005	Sealing wall for the Peribonka dam: Use of a cutter for rock strengths of up to 200 MPa with almost vertical rock flanks and a depth of 120 m. Carried out by BAUER Spezialtiefbau
2009	Use of a CBC25/MBC 50 low-headroom cutter at the Yeleh Water Power Station in China for the construction of a slurry wall from a 6.0 m wide × 6.5 m high tunnel: depth up to 75 m, width 1 m
2012	Center Hill Damm, USA: Thickest 2-phase slurry wall with d = 2.25 m manufactured with a cutter BC50. Depth up to 64 m (BAUER)
2012	Experimental field Gualdo Forli-Cesena for milling depth of 250 m (TREVI)
2015	Revised European standard EN 1538 "Diaphragm walls"
2016	EFFC/DFI Best Practice Guide to Tremie Concrete for Deep Foundations, 1st Edition
2017	The ZÜBLIN company uses the first rope grab with a nominal thickness of 2.0 m (MFS Maschinenfabrik GmbH & Co KG) for the EMSCHER BA60 project in Oberhausen
2018	EFFC/DFI Guide to Tremie Concrete for Deep Foundations, 2nd Edition
2019	EFFC/DFI Guide to Support Fluids for Deep Foundations
2019	Star-Orion South Diamond Project, Canada: Diamond exploration with a 3.2 m × 1.5 m milling machine for 250 m depth (Rio Tinto)

Stability and Large Deformations of Slender Structures Supported by Soil Materials

Matthias Baeßler[1(✉)], Pablo Cuellar[1], Falk Lüddecke[2], and Albrecht Victor[2]

[1] Bundesanstalt Für Materialforschung Und-Prüfung (BAM), Berlin, Germany
Matthias.baessler@bam.de
[2] Jörss-Blunck-Ordemann GmbH, Hamburg, Germany

Abstract. The stability and geometric nonlinearities of slender structures are a major topic in structural design. While this topic is most relevant in the field of Structural Engineering, e.g. for slender steel or concrete structures, only few applications take the role of soil-structure-interaction explicitly into account. The focus of this paper is placed on the impact of soil support and its modelling for the buckling analysis based on examples both for pile foundations and for railway track stability. The general interaction between steel design and the geotechnical input is addressed.

1 Introduction: Structural Stability and Soil-Structure-Interaction

Based on the early work of *Leonhard Euler* for critical loads of axially compressed columns the investigation of the stability of slender constructions belongs to the standard design criteria to be proved. For some types of structures, these investigations of structural stability are combined with considerations on the subsoils or soil-like materials contribution. To name a few, technical questions include the global buckling of pile like foundations, but also local buckling phenomena of shell-like foundations such as in suction buckets or monopiles. Very similar to global buckling considerations of piles embedded in soft soil (see e.g. Vogt et al. 2005) is the lateral stability of ballast track originating from significant axial forces occurring along the rail. (Ballast will be addressed as a soil material in the following chapters.) To complete the list of topics, the buckling of silo structures for granular materials is specifically interesting and has been intensively investigated in the past e.g. in the DFG Collaborative Research Centre SFB 219 in Karlsruhe University. However, piles and track will be the main focus in this paper. With respect to the piles the paper considers especially the view of the steel designer.

T. Triantafyllidis (Ed.): *Recent Developments of Soil Mechanics and Geotechnics in Theory and Practice*, LNACM 91, pp. 355–369, 2020.
https://doi.org/10.1007/978-3-030-28516-6_19

For any structural member contacting adjacent soil material the latter can be both: a significant loading and a supporting element. The interaction to the subsoil is often modelled using springs. The characteristics of theses springs can be linear or nonlinear. While it is undisputed that soil seldomly behaves linear, the role of the calculation procedure for the structural stability has a decisive impact on the need for a linearization of the stiffness. For the most common numerical linear buckling analysis a full linearization of stress-deformation behavior will be necessary.

However, if for the steel design material and geometrical nonlinearities as well as imperfections are considered, soil behavior in any complexity can be modelled as well, see e.g. Gottschalk (2017) who investigated the buckling of suction buckets considering also nonlinear springs as well as Hübner (2007) who investigated buckling of embedded tubular piles using also a Hypoplastic model for the derivation of springs or lately Bakroon et al. (2018), considering pile tip buckling and deformations using a full ALE concept. In this context the paper discusses a few examples on stability and large deformations of slender structures supported by soil materials.

In the load-deformation diagram buckling will show for the occurrence of a horizontal tangent or a bifurcation point in the solution path. In the example from Fig. 1, it can be assumed that in the case of large deflections the cross-section is plasticized, and a plastic hinge is formed in the middle. A limit load occurs below the critical load at which no load increase is possible. The load-deformation diagram results from the consideration of geometrical and material nonlinear behavior. The limiting load is dependent on the amplitude of the initial disturbance $w_{0,i}$. The soft transition results from a successive plasticizing process of the cross-section.

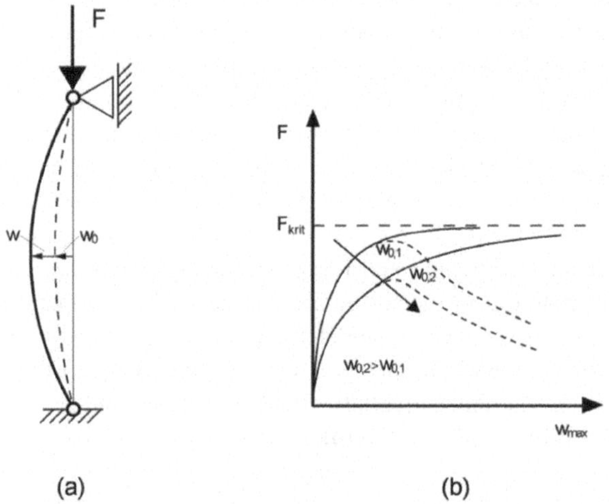

Fig. 1. Limit loads for an imperfect rod including material plasticity

A similar behavior can be observed when a nonlinear spring reaches the plasticity limit (Fig. 2) and is described in the next chapter for the lateral stability of a track model.

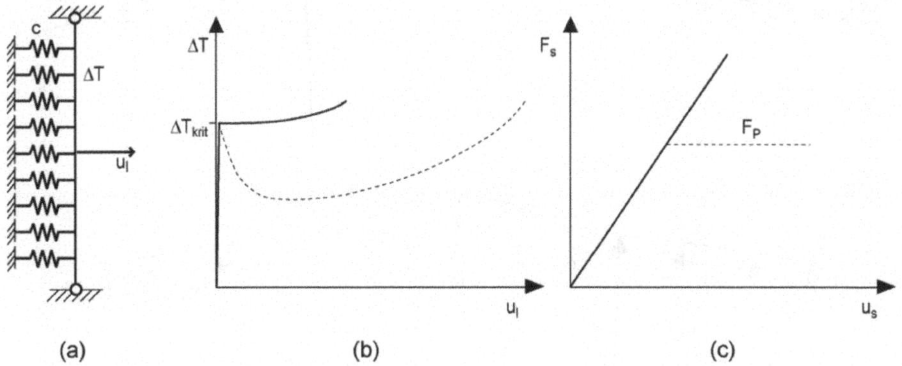

Fig. 2. (a) model of axially constrained rod, (b) critical temperature for linear (solid line) and elastic-plastic (dotted line) support, (c) characterizing function of the considered stiffness, see Van (1997).

2 Stability of Railway Track

The continuously welded rail track is prone to lateral shift and track buckling in case of large track irregularities and external loading. The geometric non-linearity here results from the pressure-loaded track with track irregularities, while the nonlinear material behavior is primarily due to the sublinear increase of the lateral bedding resistance with the lateral load force until a limit resistance is reached. Figure 3 shows an example of a one-sided distortion of the track and the scheme for its modelling in a FE environment while Fig. 4 adds the definitions for the misalignment. Some fundamental comprehensive work for the understanding and modelling of this design situation has been done by Van (1997) and Kish et al. (1998).

The lateral resistance (QVW from German Querverschiebewiderstand) depends on the specific track construction, especially the type of sleeper but also largely on the maintenance conditions (see Fig. 5 and Van 1997). For common B70 sleepers lateral resistance should be in the range of 8 kN (Chatkeo 1985), as assumed here.

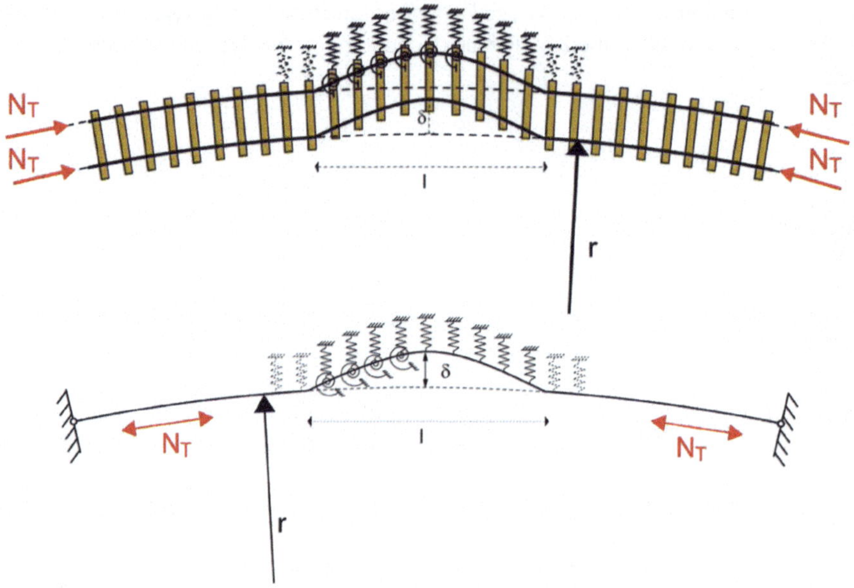

Fig. 3. (Top) an example of a one-sided distortion of the track and (bottom) the scheme for its modelling in a FE environment

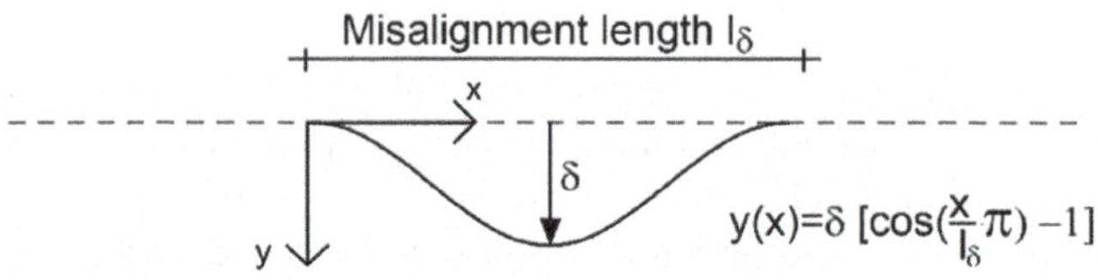

Fig. 4. Definitions for the initial lateral misalignment: Misalignment shape, misalignment length l and misalignment amplitude δ

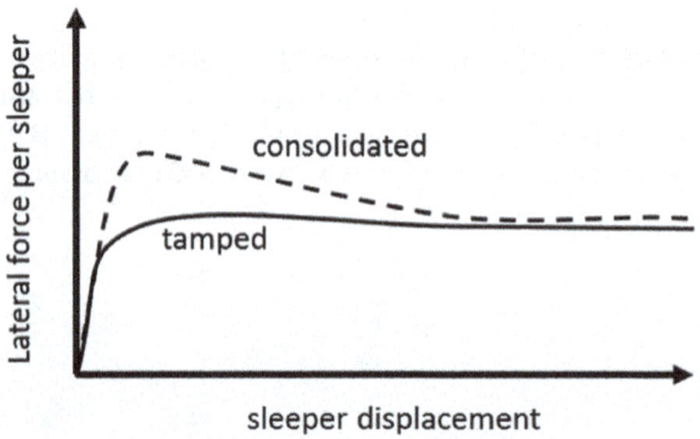

Fig. 5. Lateral resistance of individual sleepers: consolidated state (dashed line) and after tamping (solid line)

The transverse displacement resistance profile QVW1 is calculated from the values for a B70 sleeper, which is suitable for a good (prior to track maintenance) to averaged (after track maintenance) track. The resistance approaches asymptotically the peak resistance of 8.1 kN for all approaches. QVW2 is a bilinear curve, QVW3 and QVW4 result from the fact that QVW1 in direction can be stretched or compressed by a factor of 2. The assumptions for lateral resistance are summarized in Fig. 6.

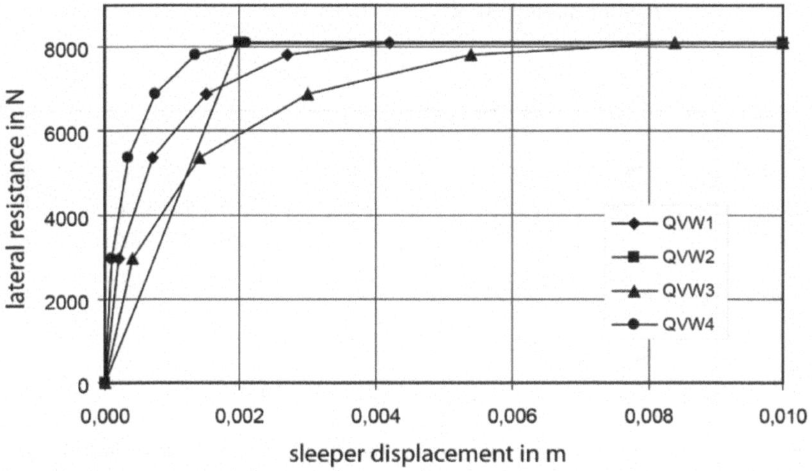

Fig. 6. Sleeper resistance with different approaches for load displacement curvature

The overall model consists of the elements shown in Figs. 3 and 4 and is further explained below.

- Lateral bedding: For the description of lateral bedding, springs are used located at each rail support point. The characteristic curves are simplified as non-linear elastic
- Rails: In reality, the rails are usually continuously welded. For the Model it is sufficient to model a section. The rail will be 48 meters long and constrained at its ends. The rails are modelled as Bernoulli beam. Nonlinear geometric effects are taken into account. The two rails are merged. Investigations by Van (1997) show that this results in a negligible deviation.
- Track geometry: The track geometry is determined by the specification of the curve radius and the track direction error (shape, length and amplitude), see Fig. 4.

For a set of parameters, a study was conducted to better understand the critical lateral track stability limit with respect to geometric parameters and imperfections, the curvature of the lateral track resistance based on a steady increase of axial compression in the rail due to temperature increase (Baeßler 2008). Table 1 summarizes the standard parameters for the model.

Table 1. Standard parameters of the mechanical model

Parameter	Value	Dimension
Radius r	5 000	M
Track length l_G	48	M
Sleeper distance	0.6	M
Torsional stiffness c_T	170 000	Nm/rad per m
Amplitude of misalignment δ	0.005	M
Length of misalignment $1_δ$	10.0	M

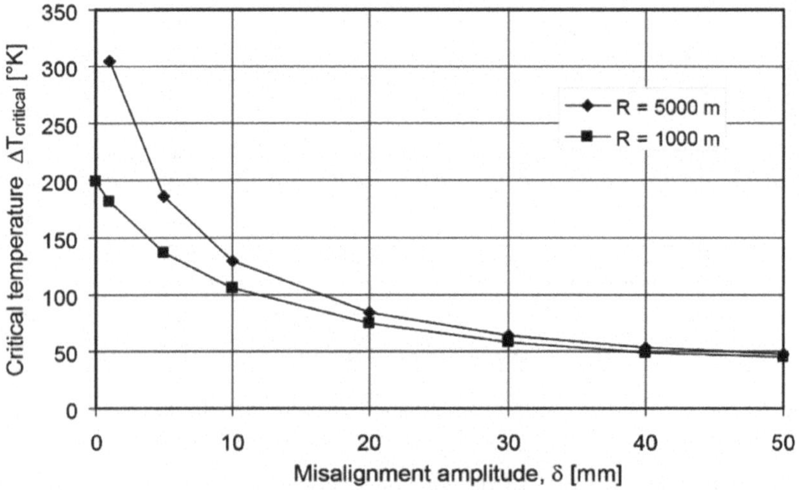

Fig. 7. Critical temperature for track buckling depending on misalignment amplitude δ and track radius.

For the track stability analysis, the critical temperature is important when buckling is observed. In principal rail temperature and axial loading of the rail are connected by thermal expansion and the longitudinal constraining of the rail. Thereby, a critical track buckling load is associated with a critical temperature ΔT. Track radius and misalignment amplitude have been varied in the investigation. As can be expected any deviation from a straight track e.g. narrow radius or extend of misalignment is highly important and reduces the critical temperature, see Figs. 7. The misalignment should be limited as much as possible.

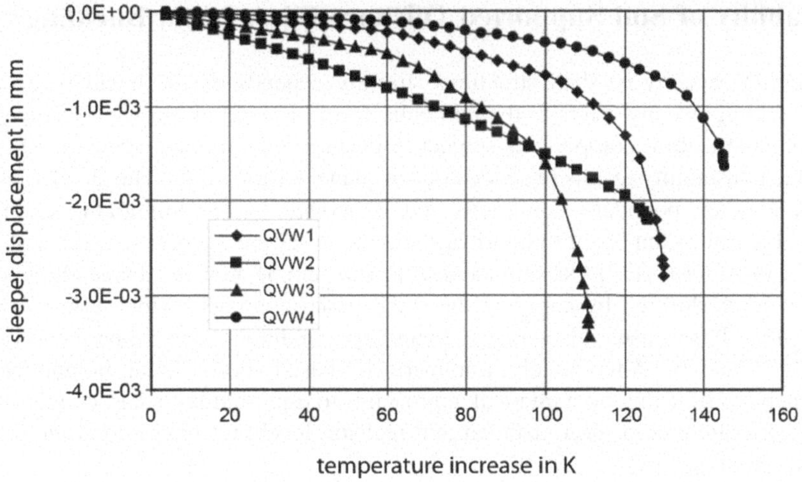

Fig. 8. Sleeper displacement vs. temperature increase for different lateral resistance (QVW)

Furthermore, the general importance of the soil support for the lateral stability of the track is obvious, see Fig. 8. However, as can be found from the parameter study the peak resistance is not sufficient for verification but also the resilience is decisive. A softer spring with the same plasticity limit decreases the critical temperature by more than 30% in the study.

3 Stability of Slender Piles in Soft Soil: Global Buckling

Slender piles embedded in soft soil are mechanically very similar to track: a beam like structure, axially loaded and supported by soil material.

From a general point of view, the problem of pile buckling concerns the stability of the slender pile, either before or during installation in the ground and thereafter during its application as a structural foundation element. In this context, the term global buckling can usually be used to describe the transverse deformation of the pile profile (e.g. Euler's buckling) without significant deformation of the cross section. For piles, fundamental investigations on the buckling of slender piles in soft soils are described in Vogt et al. (2005). For an elastic bedding the analytical solution of *Engesser* for single bars is given there. It is stated that an undrained shear strength c_u of 10 to 15 kPa has been assumed to be sufficient to prevent buckling. However, the aforementioned work describes that also the low stiffness for soft soils can be decisive.

Global buckling goes beyond the single column as e.g. specifically in Offshore Structures like Jackets with piles, where the buckling verification of the global structure includes the piles.

4 Stability of Soil Supported Offshore Piles: Local Buckling

The bearing capacity of shell structures strongly depends on the buckling behavior. Local buckling usually refers to the particular form of buckling for shell elements and thin-walled structural components subject to in-plane compressive stresses, where the mechanical instability results in sudden out-of-plane deformations. The slenderness, the type of loading, boundary conditions and imperfections are important determining factors. Typical examples of cylindrical shell structures are support structures of Wind Energy Converter (WEC) like tubular steel towers and Monopiles. Typically, buckling analyses are performed by applying the stress-based approach, which refers to a few documented base cases of loading, boundary conditions and imperfections (e.g. DIN EN 1993-1-6). Alternatively, a numerically based approach can be applied. The main intention of using the numerical approach is to represent boundary conditions and stress distributions at a more detailed and realistic level and finally to come to more economic designs.

In particular the stress-based approach does not provide documented reference cases for lateral supported tubular slender piles. A very common solution is to neglect the effect of the soil and to accept a lower bearing capacity due to buckling.

One option to consider more complex load situations and boundary conditions is to evaluate the overall slenderness λ_{ov}, where F_{Rpl} is the plastic buckling load calculated by a material non-linear analysis (MNA) and F_{Rcr} is the elastic buckling load, calculated by a linear buckling analysis (LBA).

$$\lambda_{ov} = \sqrt{\frac{F_{Rpl}}{F_{Rcr}}} = \sqrt{\frac{r_{Rpl}}{r_{Rcr}}}$$

Alternatively, the load factors r can be used to determine the overall slenderness λ_{ov}. For an axially loaded pile the overall slenderness is comparable to the slenderness λ_x. High values for λ indicate a low buckling resistance. In a first step, the result of these two buckling analyses (LBA, MNA) is the ideal buckling load. The effect of imperfection is considered in a second step by tabulated values which refer to certain tolerance class and determine the real buckling load respectively the real buckling resistance of the structure.

Imperfections can be also considered in the numerical analysis. A geometrically and materially non-linear analysis including imperfections (GMNIA) must be performed therefor. A challenging task is to cover all possible imperfections and to find the lowest real buckling resistance. Typically, eigenform related imperfections cause the lowest buckling resistance. Therefore, numerous buckling eigenforms must be analyzed to find the design driving case.

The potential of the numerical approach shall be described for a simplified case. An embedded Monopile of 8 m diameter is 80 m long and over 35 m laterally supported by soil, see Fig. 9.

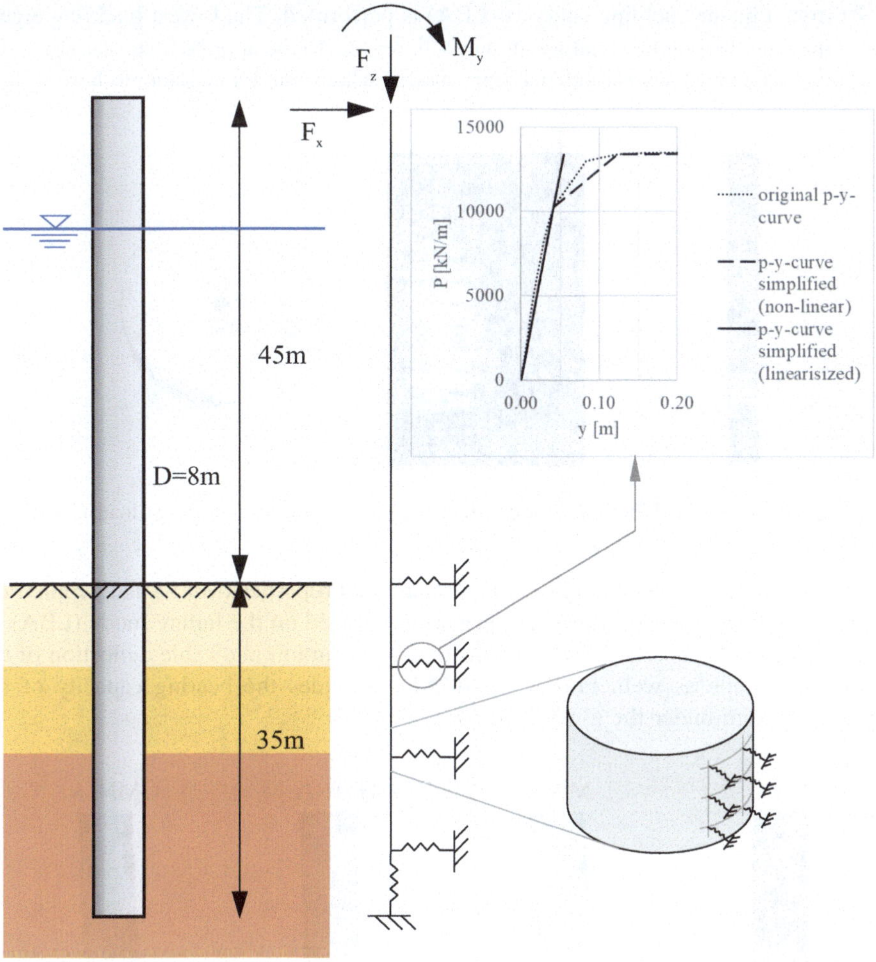

Fig. 9. Monopile – boundary conditions and model simplifications

The soil profile consists of a loose sand layer (0.5 m) on top and several stiff and very stiff clay layer which alternate with sand layer down to the pile tip level (35 m). The soil is represented by linear Winkler springs, which have been derived from non-linear p-y-curves. Linearization may not be applicable, e.g. for weak soil layer. In these cases deformations of the steel structure exceed the limit, where linear soil resistance can be assumed. Then non-linear Winkler springs must be applied, which increases the numerical complexity. T-z curves which usually represent the skin friction of the pile have not been applied here. They would decrease the axial load in the pile towards the pile tip. The t-z effect on buckling near the seabed is considered negligible. The structure is loaded by a vertical and a horizontal load and a bending moment on the top. The magnitude of imperfections has been chosen to be in line with the quality class B of DIN EN 1993-1-6.

At first, a linear buckling analysis (LBA) is performed. The lowest buckling eigenmodes indicate the potential failure modes with respect to the applied load case (Fig. 10). Other load cases may lead to other failure modes which are not considered here.

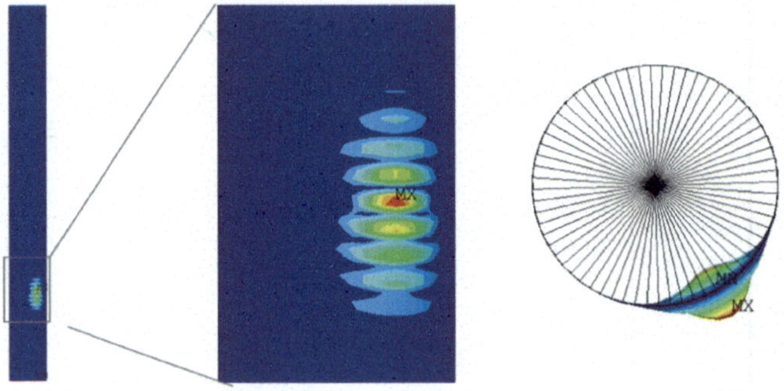

Fig. 10. Local buckling eigenmode (deflections) referring to the lowest load factor

Further analyses considering non-linear material properties (MNA) and geometrical non-linear deformations (GMNA) are performed. Based on the failure mode (LBA) an imperfect geometry is created which considers a maximum allowable deflection of the pre-buckled shape as well. Finally, a GMNIA provides the bearing capacity of the imperfect system under the given load case.

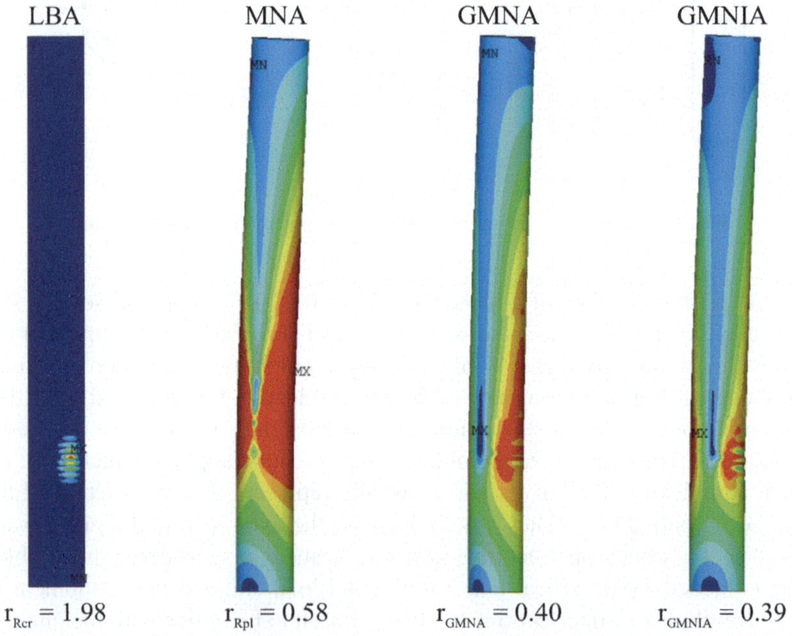

Fig. 11. Load factors r for different types of analyses

The load factors r shown in Fig. 11 describe the level of the buckling load with respect to the applied set of loads in the analyses (LBA, MNA, GMNA and GMNIA). The load factors respectively the buckling load decrease with the introduction of materially and geometrically non-linear characteristics and finally the imperfections. The overall slenderness λ_{ov} as the result of an LBA and MNA can be compared to the slenderness λ_x of the stress-based approach. One reason for the difference is the lateral resistance of the soil, which has been considered in the numerical model only.

Numerical approach	Stress-based approach
$\lambda_{ov} = \sqrt{\frac{r_{Rpl}}{r_{Rcr}}} = \sqrt{\frac{0.58}{1.98}} = 0.54$	$\lambda_x = \sqrt{\frac{f_{y,k}}{\sigma_{x,Rcr}}} = \sqrt{\frac{355}{1750}} = 0.45$

The numerical approach shown for one single situation (load case, imperfection) must be applied on numerous combinations including collapse affine and construction related imperfections as well, which is not shown here. The load factor $r_{GMNIA} = 0.39$ applied on the loads of the investigated load set describes the lowest elastic-plastic buckling loading of the structure. In comparison to this the load factor of the stress-based method is about 0.24 and does not consider any lateral resistance of the soil layer. The GMNIA approach allows for a consideration of the soil support and shows potential reserves in design.

Finally, the result of the GMNIA should be verified on known problems to confirm the reliability of this approach. Very often this verification is not possible due to missing comparative data. It is also recommended to evaluate different magnitudes of imperfections to confirm that the lowest buckling load factor has been calculated.

The numerical approach allows to include complex boundary conditions like soil embedment. The effect of soil embedment has been shown only for one load set and one shape of imperfection. For a detailed analysis numerous load sets and imperfections must be investigated. The potential of this approach with respect to the stress-based method has been indicated. It is expected that a more economic design is achievable by using the numerical and more complex approach instead of relying on the stress-based concept.

5 Pile Tip Buckling of Tubular Steel Piles

Damages to the pile, particularly localized at the pile tip, often arise during any of the many phases of pile installation. The topic is addressed for an offshore application.

For instance, the operations involving a high risk for pile tip damage include the general lifting and handling on deck, the lowering into the pile-sleeve cone, the transitory pile-in-sleeve state and the pile driving itself whenever the pile tip impacts on any obstacles within the subsoil (e.g. large boulders, stiff soil strata, etc.) (Aldridge et al. 2005). Since it is often not easy to discern from literature if shell buckling was involved in investigations and reported failures (or if it was only a propagation of initial damage or imperfection) the term “pile tip buckling” is partly used as an umbrella term for pile tip deformations and their propagation during installation. Buckling of the lower end of the pile is mainly driven by the pile tip resistance. However, for the case of large pile diameters as currently used for monopile foundations, the assumption of a tip resistance that is equally distributed all over the tip surface becomes increasingly questionable. In this respect, the scientific literature shows that the buckling of cylindrical shells depends largely on the boundary conditions of the support. Typically, these supports (foundations) are designed to provide a defined boundary condition which is then considered in the buckling analysis and design of the superstructure shell. For piles, this boundary condition depends on the (geo) mechanical response of the foundation and would have to be provided by a specific geotechnical analysis considering the local soil conditions. In practice however, this requirement is often circumvented by adopting simple assumptions which allow a safe but also economic design of the pile.

One option to include Pile Tip Buckling scenarios is to vary the pile tip resistance, see Fig. 12.

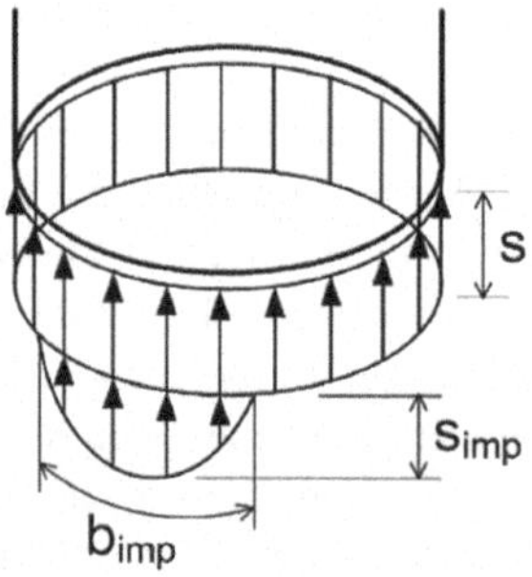

Fig. 12. Equally-distributed tip resistance s and increased tip resistance b_{imp} to consider an unequal distribution of tip resistance

However, one of the key open aspects to be further investigated is the buckle propagation during pile driving and the consequent collapse of the cross section, as outlined in (BUMP 2018) and widely reported for various offshore platforms (see e.g. Alm et al. 2004; Erbrich et al. 2011; Senders et al. 2013). Such phenomena are characterized by the progressive accumulation of large pile deformations (see Fig. 13, as opposed to the sudden failure which is typical for the shell stability buckling problems. While the latter are usually analysed by means of GMNIA approaches the former phenomena must be treated explicitly as a transient process instead.

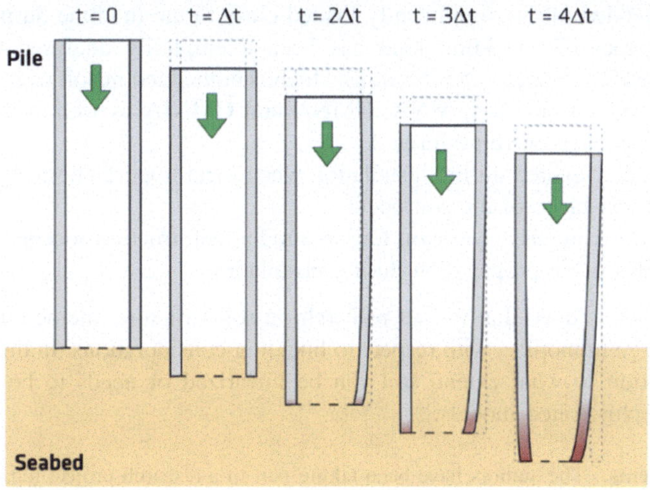

Fig. 13. Pile buckling propagation during dynamic driving

Currently no design rules for Pile Tip Buckling are available except for rough prescriptions for the wall thickness as the API-criterion which might eventually be very costly for larger Monopiles.

6 Conclusions/Summary

Stability and large deformations of slender structures are a major topic in structural design and can be decisively influenced by structure-soil-interaction. The paper addresses several subtopics where structural stability depends on the soil support. Firstly, the investigation of lateral stability of an axially compressed ballasted railway track is summarized. The rail itself is an extremely slender structure. So, its lateral stability under relevant compression loading is completely dependent on a stable lateral support of the ballast. Ballast support is modelled as local springs. Neither the mechanical ultimate strength of the sleeper ballast interaction nor the initial stiffness is sufficient to fully describe the lateral shift until buckling of the track. The use of nonlinear springs improves the results and is still numerically feasible.

Due to its similarity considering the mechanical system slender piles in soft soil have been addressed afterwards addressing the work of *Vogt et al. (*2005*)*. The problem of global buckling for pile structures is also very relevant for complete foundation structures as for offshore jackets, usually implying simplification as linear springs for the pile soil interaction.

In contrast to global buckling local buckling deals with local buckling phenomena of tubular shells including large radial deformations. Local buckling has to be considered for all tubular members of the steel structure. Considering a Monopile several scenarios have to be addressed that are relevantly influenced by the structure-soil-interaction:

(a) The embedded pile is significantly loaded close to the mudline during operation. In this paper the buckling load has been exemplarily analysed for an 80 m Monopile (D = 8 m, t = 55–65 mm) which is embedded in soil over 35 m. Based on one load set an LBA, MNA, GMNA and GMNIA as well as a stress-based buckling analysis were performed.
(b) Buckling is triggered during installation when local imperfections in the structure or in the resistance of the soil occur
(c) Pile Tip Buckling as a synonym for large radial deformations occurs as a transient process of buckle propagation during installation.

It must be concluded that further research on soil-structure-interaction would help to optimize large monopiles with respect to buckling considerations. In this respect it is an open question to what extend soil can be linearized or needs to be incorporated using more sophisticated models.

Acknowledgments. The authors have been taking part in a research project regarding pile and pile tip buckling (Carbon Trust OWA BUMP). While the results shown are not a specific part of this project the general opportunity to investigate pile buckling as well as the fruitful discussions with the Technical Working Group of the Carbon Trust Offshore Wind Accelerator and the group of Prof. Rackwitz (TU Berlin) is very much appreciated.

References

Vogt, N., Vogt, S., Kellner, C.: Buckling of slender piles in soft soils (in German), die Bautechnik **82**(12), 889–901 (2005)

Gottschalk, M.: Zur Beultragfähigkeit von Suction Buckets, Dissertation, Leibniz University Hanover (2017)

Hübner, A.: Tubular Piles – Buckling Design in a Complex Situation, Dissertation, University Fridericiana Karlsruhe (2007)

Bakroon, M., Daryaei, R., Aubram, D., Rackwitz, F.: Numerical evaluation of buckling in steel pipe piles during vibratory installation. Soil Dyn. Earthquake Eng. (2018). https://doi.org/10.1016/j.soildyn.2018.08.003

Van, M.A.: Stability of continuous welded rail track, Dissertation Tu Delft (1997)

Kish, A., Samavedam, G., Wormley, D.: Fundamentals of track lateral shift for high high-speed rail applications. In: ERRI Interactive Conference on Cost Effectiveness and Safety Aspects of Railway Track, Paris (1998)

Chatkeo, Y.: Die Stabilität des Eisenbahngleises im Bogen mit engen Halbmessern bei hohen Axialdruckkräften, Mitteilungen des Prüfamtes für Bau von Landesverkehrswegen der Technischen Universität München Heft 46, München (1985)

Baeßler, M.: Lageveränderungen des Schottergleises durch zyklische und dynamische Beanspruchungen, Dissertation, TU Berlin (2008). opus.kobv.de/tuberlin/volltexte/2008/1934/pdf/baessler_matthias.pdf

Baeßler, M., Cuéllar, P., Rücker, W.: The Lateral stability of ballasted tracks on vibrating bridge decks. IJRT 3(2) (2014)

DIN EN 1993-1-6:2010-12: "Design of Steel Structures – Part 1-6: Strength and stability of shell structures", Eurocode 3, December 2010

Alm, T., Snell, R.O., Hampson, K.M., Olaussen, A.: Design and Installation of the Valhall Piggyback Structures, OTC-16294-MS. In: Proceedings of Offshore Technology Conference, 3–6 May 2004, Houston, Texas (USA) (2004)

Erbrich, C.T., Barbosa-Cruz, E., Barbour, R.: Soil-pile interaction during extrusion of an initially deformed pile. In: Gourvenec, S., White, D. (eds.) Frontiers in Offshore Geotechnics II, pp. 489–494. Taylor & Francis Group, London (2011)

Senders, M., Banimahd, M., Zhang, T., Lane, A.: Piled foundations on the north west shelf. Australian Geomechanics 48(4) December 2013

Baeßler, et al.: BUMP Buckling Assessment of Monopiles; Report unpublished for the Carbon Trust Offshore Wind Accelerator (2018)

Aldridge, T.R., Carrington, T.M., Kee, N.R.: Propagation of pile tip damage during installation. In: Gourvenec, Cassidy (eds.) Frontiers in Offshore Geotechnics: ISFOG 2005, pp. 823-827, Taylor & Francis Group, London (2005)

Excavation Pits: Calculation Methods

Achim Hettler[1(✉)] and Theodoros Triantafyllidis[2]

[1] TU-Dortmund, Lehrstuhl Baugrund-Grundbau, Dortmund, Germany
achim@a-hettler.de
[2] Institute for Soil Mechanics and Rock Mechanics (IBF),
Karlsruhe Institute of Technology (KIT), Karlsruhe, Germany
Theodoros.Triantafyllidis@kit.edu

Abstract. Three methods in engineering practice are mainly implemented to investigate the behavior of excavation walls. In the majority of the cases beam models with classical supports seem to be sufficient.

For modelling more accurate the deformation behavior of the foot of the wall it may be worth to improve the prediction of the deformation behavior using a subgrade reaction model. In cases of more complex pit geometries and soil conditions a finite element analysis may be more appropriate.

All these three methods of calculation are shortly described in the paper and compared with each other.

With the consideration of bound theorems an attempt is made in the paper to discuss the safety issues. All the presented three methods can be used to calculate the wall deformations. It is demonstrated that even a finite element analysis has limitations in cases of deformation predictions induced due to geotechnical installation processes in the vicinity of the wall. Data from field records may be used to estimate the order of magnitude of wall deformations due to installation processes. As an example of using field data in the wall deformation prediction and FEM in a recent research project the vibroinstallation of uplift piles near to the wall has been used.

The numerical results show quite satisfactorily that the new developed model may serve as a basis for the prediction of wall deformations due to some installation processes.

1 Introduction

In practice, there are essentially three methods available for the calculation of excavation pits - beam model, modulus of subgrade reaction method, finite element method - which are presented and compared with each other in the article. When verifying the ultimate limit state, the question arises about the safety of the individual methods. By means of simplified considerations, it can be shown that the calculated systems can generally be classified as approximately safe. For most standard cases realistic results for the serviceability state are achieved with the available methods. However, with some exceptions, the limits of the procedures are reached when deformations due to installation processes are to be predicted. The current state of technology and science is discussed on the basis of examples.

T. Triantafyllidis (Ed.): *Recent Developments of Soil Mechanics and Geotechnics in Theory and Practice*, LNACM 91, pp. 370–382, 2020.
https://doi.org/10.1007/978-3-030-28516-6_20

2 Design Analysis

Typically, wall constructions are calculated and verified using classical framework models. A special feature are the supports of the bracings and anchors and the earth support.

In the case of unsupported walls, the simplification of Blum has proven successful (Fig. 1). This makes it possible to convert the highly statically indeterminate fixed support in the ground into a statically determined model with two unknowns. The earth resistance E_{ph}, or the mobilised earth resistance E'_{ph} as well as the equivalent force C_h can be determined directly from the two available equilibrium conditions. It should be noted that in the Blum model a vertical tangent is assumed at the fixed support point. Thus, a comparison with the modulus of subgrade reaction method and with finite element calculations yields the highest fixing effect.

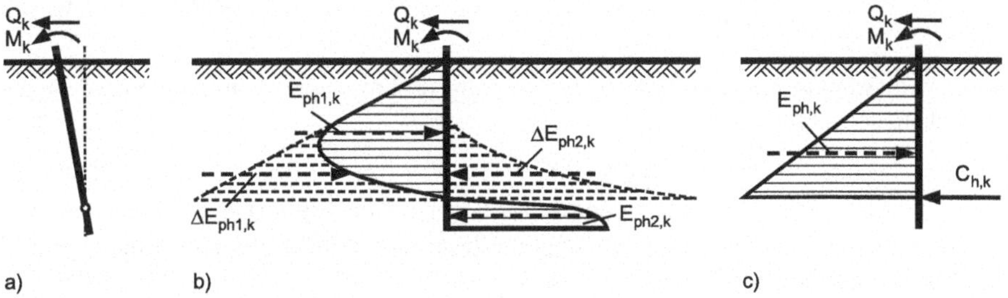

Fig. 1. Blum's simplification [17]: a) wall movement b) supposed earth pressure c) simplified earth pressure distribution.

In general, in the case of one or more supported excavation pit walls with support in the ground, a distinction must be made with regard to the static system between

- free supported walls (Fig. 2a),
- partial fixed walls (Fig. 2b),
- fixed walls (Fig. 2c).

All three systems assume a fixed support at the actual wall embedment length or at the theoretical point of bearing.

With a free support according to Fig. 2a, the wall embedment length is only supported by ground reactions $\sigma_{h,k}$ on the excavation pit side; the support force at the wall base must result from the calculation to $C_{h,k} = 0$. In the case of partial fixed support (Fig. 2b), the ground reaction $\sigma_{h,k}$ and the equivalent force $C_{h,k}$ generate a reversing moment, which reduces the inclination of the bending line at the wall base in comparison with a free support. If the back-turning moment is so large that at the theoretical bearing point C, where the equivalent force $C_{h,k}$ is applied, a vertical tangent of the bending line is reached, this is referred to as a full soil mechanical fixing or a

Blum fixed support. Their special feature is that the fixing moment $M_{C,k} = 0$ at the fixed support must result from the calculation (Fig. 2c). Depending on the degree of utilization of the soil reaction, the wall base can be free supported, partially fixed or fully fixed for the same depth of embedment.

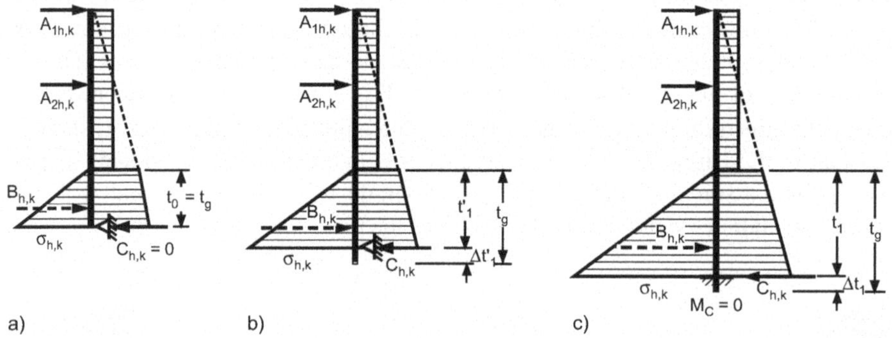

Fig. 2. Earth support for supported excavation walls: (a) Free support, (b) Partial fixed support, (c) Fixed support [11]

- The smallest possible embedment depth t_0 for a free support is obtained when the design value of the soil reactions with the boundary conditions according to Fig. 2a is 100% utilized.
- The smallest possible embedment length t_1 up to the theoretical bearing point with a fixed soil mechanical support is obtained if the design value of the soil reactions with the boundary conditions according to Fig. 2c is 100% utilized.

The support points of struts and anchors are generally assumed to be freely rotatable and immovable. Under these conditions, wall structures such as beams in building construction can be calculated. For details see *Hettler/Triantafyllidis/Weißenbach* [11].

3 Extension with Modulus of Subgrade Reaction Method

If the displacements of the wall base are to be calculated more precisely than with the beam model with non yielding support, the modulus of subgrade reaction method can be used. Here, the soil reaction is replaced by springs that are usually not coupled to each other. The advantage of this method is that the magnitude of the soil reaction and the displacements can be mapped realistically. At the same time, the effort for the static calculation is limited and the influence of different parameters can be clearly recorded. The main difficulty of the method lies in determining the spring characteristics in such a way that the calculated displacements and bending moments come as close as possible to the reality. The following points, among others, must be taken into consideration:

- The actual relationship between displacement and soil reaction is non-linear.
- Depending on the type of wall of design, e.g. parallel movement, rotation about the foot or rotation about the top of the wall, other spring characteristic curves result.
- Due to the arching effect, earth pressure redistributions occur and the spring characteristic curves are actually coupled with each other.
- The initial earth pressure condition has a great influence on the spring characteristics and must be taken into consideration.

In recent years, the working group for excavations from the German Geotechnical Society (DGGT) has also dealt in detail with the modulus of subgrade reaction method and issued Recommendation EB 102 [5, 9]. According to EB 102, paragraph 1, the method may be used for verifying the embedment depth, for determining the internal forces and also for verifying the serviceability limit state.

The approach of the initial stress condition plays an important role. Systematic investigations show, see e.g. Besler [2] and Hettler/Vega-Ortiz/Mumme [13], that the wall toe displacements become much too large without consideration the preloading arising from the weight of the excavated volume of the pit. According to EB 102 [5], the initial stress state may be the earth pressure at rest, calculated from the ground surface (Fig. 3). The idea is that during the excavation, the original at-rest earth pressure is maintained. In order to meet the requirement of the maximum possible limit state, the initial stress condition is limited from the excavation pit floor to the passive earth pressure.

Below the depth z_e the mobilisation of the soil reactions is modelled by subgrade reaction springs. On the earth side, the actions are applied like in the case of non yielding support of the beam.

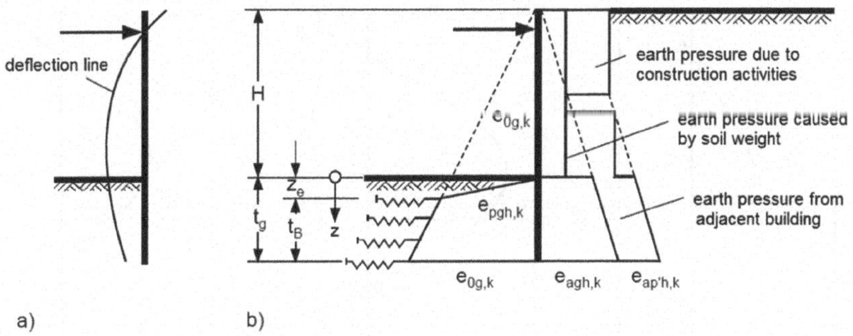

Fig. 3. Subgrade reactive for non-cohesive soil without zero displacement point: (a) wall deformation, (b) pressure diagram [11]

If a large embedment depth and flexible walls cause the wall to rotate backwards with a zero displacement point (Fig. 4a), it is obvious to apply the at-rest earth pressure below the zero displacement point on the earth side as well (Fig. 4b).

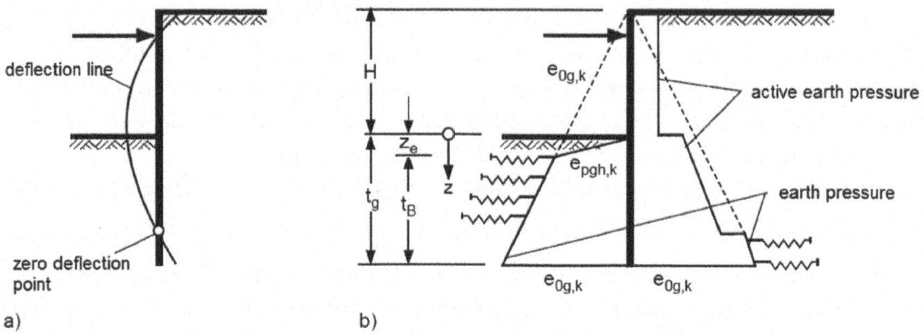

Fig. 4. Subgrade reaction model for non-cohesive soil with zero displacement point: (a) wall deformation, (b) pressure diagram [11]

As shown in Fig. 4a, no vertical tangent can occur at the zero displacement point according to the modulus of subgrade reaction method. In this respect, the fixing effect is less than with the Blum model.

The movement of the wall foot depends very much on the bending stiffness of the wall, the flexibility of the earth support and the embedment depth. Figure 5a, for example, shows the bending line of a simply supported, relatively rigid wall. As the embedment depth increases and the wall becomes more flexible, the deformation pattern changes (Fig. 5b). A backward rotation occurs and the wall toe movement consists of a parallel displacement and a rotation about the toe of the wall. In the case of long, flexible walls, a bending line is obtained as shown in Fig. 5(c).

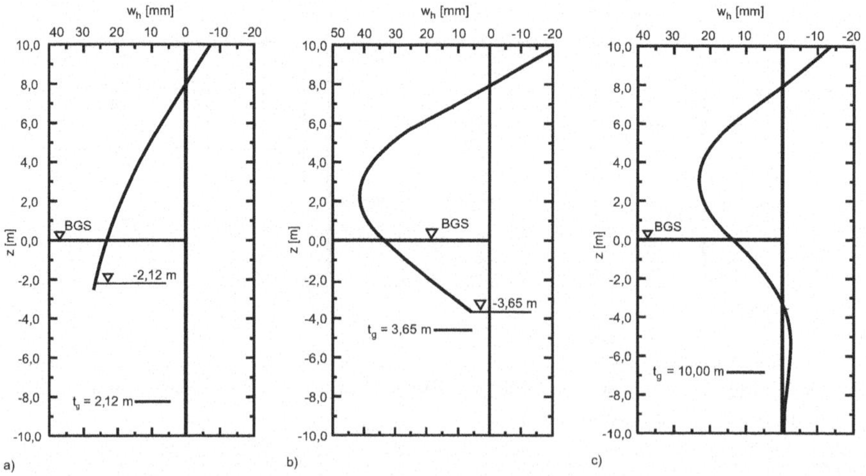

Fig. 5. Different forms of wall movement: (a) relatively stiff wall, (b) flexible wall with reverse rotation, (c) long wall [11].

For further details, including the determination of the subgrade reaction modulus, see *Hettler/Triantafyllidis/Weißenbach* [11].

4 Finite Element Calculations

Numerical calculations using the finite element method (FEM) are much more complex than simple beam models or extended models using the modulus of subgrade reaction method. Following the recommendations of the Working Group for Excavations (EAB), FEM is suitable in the following cases:

- Excavation walls with supporting conditions for which a reliable determination of size and distribution of the earth pressure is not possible, e.g. with strongly deforming walls;
- Construction with difficult geometric dimensions, e.g. recessed or protruding corners, which do not permit a reliable distribution of the earth pressure with conventional assumptions;
- staggered excavation walls with a berm width, which do not allow a reliable determination of the size and distribution of the earth pressure with conventional assumptions;
- Excavation wall designs where a realistic assessment of the effects of excavation, prop or anchor pre-stressing on the earth pressure redistribution and the displacements of the excavation wall is required;
- Constructions, where a realistic recording of the seepage flow and the associated water pressures is required;
- excavation pits next to buildings, pipes, other utility installations or traffic areas;
- exceptionally deep excavations.

In simple cases, classical methods and FEM provide similar results, as the following example in Fig. 6 shows for a wall that is simply propped with a free earth support. The classical beam model with non yielding support, the extended beam model with subgrade reaction and FEM calculations based on hypoplasticity soil model are compared.

In the example in Fig. 6, the support in the FEM calculation was modelled as non yielding. This explains the higher support force compared to the classical model and the increased earth pressures in the support area (Fig. 6a). Apart from that, there is a very good agreement between the active earth pressure assumed according to EAB and the FEM calculation. This also applies to the moment distribution (Fig. 6c). If the wall foot is modelled with non-linear subgrade reaction, the soil reactions in front of the wall in the classical beam also agree practically with the FEM result (Fig. 6a). The wall deformations require a separate discussion (Fig. 6b). Here, stiffness plays a significant role for small strains. In principle, the deformations of the entire terrain caused by the excavation of the excavation pit are added in the FE calculations.

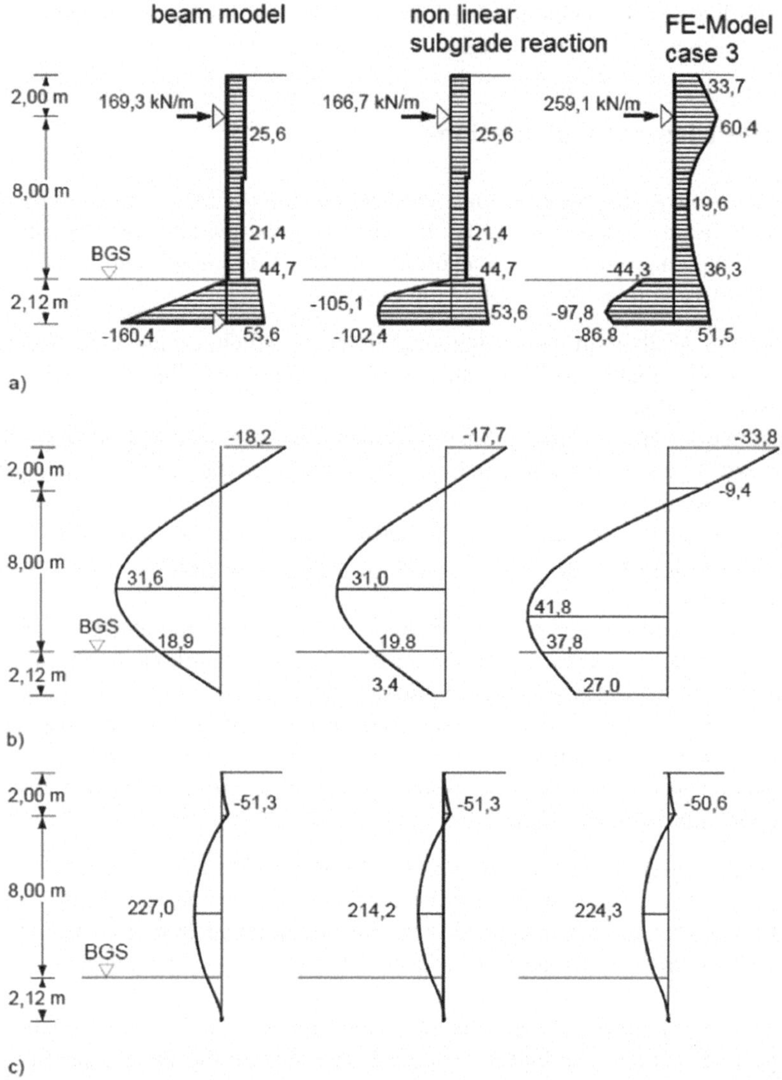

Fig. 6. Comparison of the classical calculation with the FEM for a sheet pile wall with free earth support at the embedment depth t = 2.12 m: (a) Stress distribution [kN/m^2], (b) Bending line [mm], (c) Moment line [kN m/m] [11].

As this example shows, in the standard cases covered by the EAB, the FEM is unlikely to provide any advantages and improvements in the determination of the earth pressure distribution. The FEM can, however, be of importance if the range of validity of the classical earth pressure theory and the empirical basis of the EAB are abandoned. This applies in particular to complex geometry or flexible anchors and flexible walls.

5 Safety Considerations

One of the most difficult tasks is the question of the safety factors of systems.

Brinch Hansen/Lundgren [3] investigate different limit states for an anchored sheet pile wall (Fig. 7)

- Failure with rotation about the top of the wall point
- Rotation about the top of the wall with yielding joint in the middle of the wall and flexible wall toe
- Rotation about the top of the wall with yielding joint in the middle of the wall and fixed at the toe of the wall
- Rotation around the head point with yielding joints in the middle of the wall and in the embedment part of the wall.

and explain: "One could perhaps imagine that it would be necessary - just as for failure mechanisms in the ground - to examine all (or at least some) types of limit state in order to find the critical one. But this is not the case. On the contrary, it turns out that one can design a structure for any chosen type of limit state and thus obtain a safe construction. It can be shown that a logically dimensioned construction subject to earth pressure can actually fail in no other way than assumed in its calculation. This has to do with the fact that the deforming part is relieving stresses and the movement that has begun comes to an end again."

Brinch Hansen/Lundgren [3] justify the explanations with plausibility considerations.

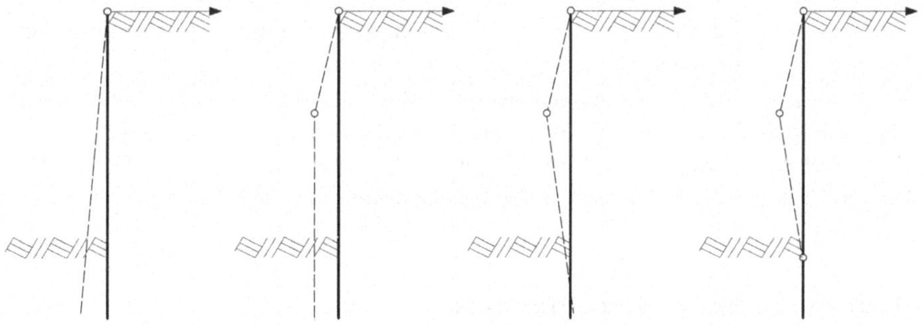

Fig. 7. Types of limit states for an anchored sheet pile wall [3]

From today's point of view, these theories can also be theoretically substantiated with the help of plasticity theory. Collapse loads can be determined within the theory of ideal plastic materials by the analysis of kinematic mechanisms and static stress fields, and upper and lower bounds are obtained see *Drucker/Greenberg/Prager* [4] *Koiter* [14] *and Gudehus* [8].

Put simply, the lower bound theorem is: If a static stress field can be found within the soil that is in equilibrium with the dead weight and the external forces and does not violate the boundary condition at any point, then the external forces form a lower bound for the collapse loads.

The upper bound theorem says: If there is a failure mechanism, for which the work done by the external forces and by the forces from own weight is smaller than the work from plastic deformations in shear zones and shear bands, then the external loads can be classified as the upper bound for the collapse loads, cf. also *Atkinson* [1] and *Powrie* [15].

Strictly speaking, the bound theorems apply only to ideal plastic materials. An important assumption is, among other things, the so-called normality rule, which is generally not fulfilled for soils. Nevertheless, the static and kinematic methods have proved their validity in many applications, especially in earth pressure, and the error made seems to be relatively small. A detailed description and application to soils can be found at *Goldscheider* [6]. As *Goldscheider* can show, the essential assumption for the proof is not the so-called normality condition, but Drucker's postulate.

Transferred to excavation walls, one could classify the calculations on the basis of correctly determined earth pressure distributions in the sense of the static theorem as a solution lying on the safe side, and only one solution suffices. With the realistic load diagrams, the design is further restricted in the EAB. By specifying a load diagram in the calculation and corresponding prestressing of anchors or struts in the design, a more economical design in comparison to any earth pressure approach should normally be possible.

All these considerations assume ductile behavior. In the event of sudden failure, e.g. brittle fracture or loss of a prop caused by collision with an excavator, additional safety measures are required. For this reason, EAB [5] requires increased partial safety factors in the following cases:

- Verification of stability deep slip surfaces in accordance with EB 44, paragraph 10 (Section 7.3) for excavations adjacent to structures,
- Verification of the design against overall failure of terrain in accordance with EB 45, paragraph 7 (section 7.4) for excavation pits adjacent to structures,
- Design of struts according to EB 52, paragraph 14 (section 13.7)
- Design of anchorages for walls in full excavation condition.

6 Determination of Deformations

The regulations of the EAB [5] ensure that, in the case of at least medium-dense non-cohesive soil and at stiff cohesive soil, the displacements of the earth support of a multiple-supported wall are small and correspond in magnitude to the movements and deformations of the remaining excavation pit wall. As a rule, this eliminates the need for special investigations into the size of the deformations and displacements.

A separate serviceability limit state verification may be required in particular

- in the case of construction pits, next to very high buildings, badly founded buildings or buildings in poor structural condition,
- for construction pits with very little or no distance to an existing building,

- for excavation pits alongside structures with high ground water level
- in pits next to structures founded in soft cohesive soil,
- in excavation pits, in proximity to structures which particularly high demands on their deformation, e.g. because of the sensitivity of machines,
- for excavation pits with an anchorage steeper than 35°.

Two cases have to be distinguished when proving serviceability:

- If the deformations of the wall are to be recorded more precisely, but the effects of the deformations on the environment are rather subordinate, the accuracy of the deformation prognoses can be increased by improving the static system, e.g. by recording the flexibility of the anchors, taking into account the pre-deformations in the various construction stages and considering the amount of subgrade reaction in the ground.
- If both the deformations of the wall and those of the surrounding soil are to be determined, numerical investigations, e.g. using the finite element method with consideration of the initial stress state, are required. For further details see *Hettler/Triantafyllidis/Weißenbach* [11].

Even when highly complex FE models are used, the limits are reached when deformations caused by geotechnical installation processes are to be predicted (*Hettler and Triantafyllidis* [12]). Only in isolated cases installation processes have been investigated in the scientific field, e.g. *Grabe and Mahutka* [7] or *Triantafyllidis* [16]. In practice, on the other hand, it is often necessary to include in the deformation prognosis the installation effects of e.g.

- excavation pit walls
- grouted anchors
- Bearing and sealing injection soles
- Basement anchorages

to be taken into account. As long as reliable forecast models are not yet available, it is necessary to carry out an estimation on the basis of measurements in comparable cases. In a contribution to the Baugrundtagung 2012 conference [10] some examples were compiled, see also *Hettler/Triantafyllidis/Weißenbach* [11].

From a research perspective, a DFG research group in recent years has succeeded in producing a complete prognosis model at least for the case of excavations with back-anchored concrete floors and vibrated piles.

The deformations of excavation pit walls due to dynamic installation methods, such as the vibration of piles next to a diaphragm wall [18], can be considered as cyclic creep with the HCA model [19]. The vibration amplitude is set by measurements with the distance to the vibration position of the pile and the time for a penetration of the pile is converted into a number of cycles. *Grandas/Vogelsang/Triantafyllidis* [20] used the sequence of pile installation and the frequency of the vibrator used (34 Hz) as well as the measurements of the particle velocities to determine the geometric radiation of wave attenuation in order to estimate the strain amplitude in all Gaussian points in the FE mesh.

With the HCA model and the simulation of the sequence of pile installation, the deformation curve of the D-wall could be reconstructed for the two published measuring positions of the diaphragm wall at Potsdamer Platz [18] (see Fig. 8), which could not be explained previously by local condensation assumptions and reduction of the passive resistance caused by the vibration processes. Since this model is also used promisingly for other dynamic installation processes, e.g. deep vibratory compaction [21], new possibilities for the estimation of shoring wall deformations due to dynamic effects close the shoring walls open up with the help of the HCA model in combination with FEM analysis.

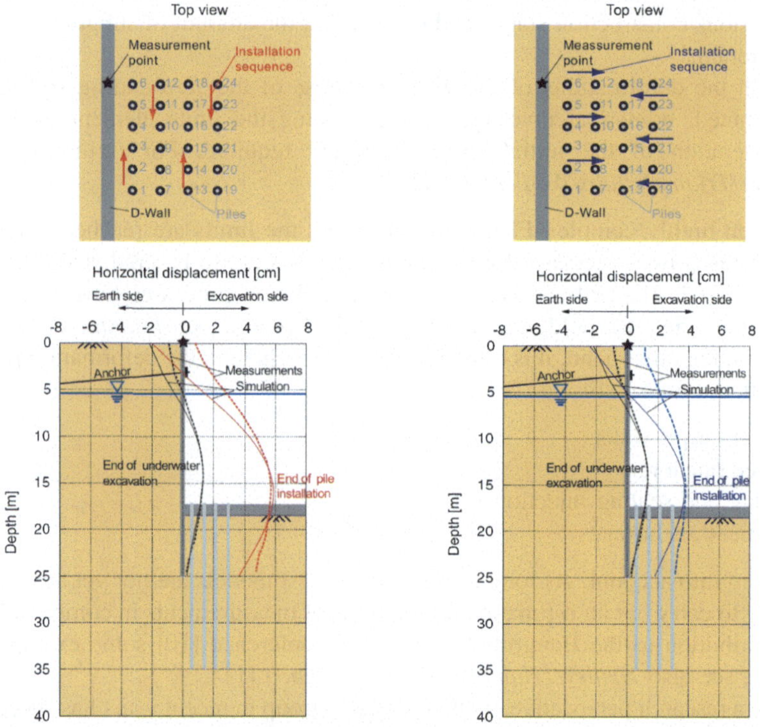

Fig. 8. Simulation of excavation wall deformations with the HCA model, using the example of measurements at Potsdamer Platz in [20], left: Pile construction parallel to the wall, right: Pile construction perpendicular to the wall

7 Further Information

Although a great deal of knowledge about the calculation and construction of excavation pit walls has been gathered over the last few decades, damages also occur again and again - sometimes despite careful planning and execution. It is striking that water effects are often identified as the main cause of damage, e.g.

- ground water drowdown
- ground water under high pressure or artesian conditions
- removing the ground when dewatering the excavation pit
- defects in sealing soles or diaphragm wall joints
- Anisotropy in water permeability
- unstable filter structure

An extensive collection of case histories with damages can be found in *Lutz Wichter* [22].

References

1. Atkinson, J.: The Mechanics of Soils and Foundations, 2nd edn. CRC Press/Taylor and Francis Group, Boca Raton (2007)
2. Besler, D.: Wirklichkeitsnahe Erfassung der Fußauflagerung und des Verformungsverhaltens von gestützten Baugrubenwänden. Schriftenreihe des Lehrstuhls Baugrund-Grundbau der Universität Dortmund, Dortmund, no. 22 (1998)
3. Brinch Hansen, J., Lundgren, H.: Hauptprobleme der Bodenmechanik. Springer, Heidelberg (1960)
4. Drucker, D.C., Prager, W., Greenberg, H.J.: Extended limit design theorems for continuous media. Quart. Appl. Math. **9**(4), 381–389 (1952)
5. Empfehlungen des Arbeitskreises "Baugruben" EAB, 5 Auflage, Ernst und Sohn, 2 Korrigierter Nachdruck (2017)
6. Goldscheider, M.: Gültigkeitsgrenzen des statischen Kollapstheorems der Plastomechanik für Reibungsböden. Geotechnik **36**(4), 243–263 (2013)
7. Grabe, J., Mahutka, J.: Finite-Elemente-Analyse zur Vibrationsrammung von Pfählen. Bautechnik **82**, 632–640 (2005)
8. Gudehus, G.: Lower and upper bounds for stability or earth-retaining structures. In: Proceedings of the 5th European Conference on Soil Mechanics and Foundation Engineering, Madrid, vol. 1, pp. 21–28 (1972)
9. Hettler, A.: Empfehlung EB 102 des Arbeitskreises "Baugruben" der DGGT zur Anwendung des Bettungsmodulverfahrens. Bautechnik **88**(5), 640–645 (2011)
10. Hettler, A., Borchert, K.-M.: Herstellbedingte Verformungen bei tiefen Baugruben. DGGT Baugrundtagung München (2010)
11. Hettler, A., Triantafyllidis, Th., Weißenbach, A.: Baugruben, 3 Auflage. Ernst und Sohn (2018)
12. Hettler, A., Triantafyllidis, Th.: Deformations of deep excavation walls induced by construction processes. In: Proceedings of the 17th ICSMGE, Alexandria (2009)
13. Hettler, A., Vega-Ortiz, S., Mumme, B.: Berechnung von Baugrubenwänden mit verschiedenen Methoden: Trägermodell, nichtlineare Bettung, Finite-Elemente-Methode. Bautechnik **83**(1), 35–45 (2006)
14. Koiter, W.T.: General theorems for elastic-plastic solids. In: Sneddon, I.N., Hill, R. (eds.) Progress in Solid Mechanics, Chap. IV, pp. 166–221. North-Holland Publishing Company, Amsterdam (1960)
15. Powrie, W.: Soil Mechanics – Concepts and Applications, 3rd edn. CRC Press/Taylor and Francis Group, Boca Raton (2013)
16. Triantafyllidis, T.: Optimierung der Herstellung von Verbauwänden im Hinblick auf Verformungen vorhandener Bauwerke. Abschlussbericht zum BMBF-Forschungsvorha-ben, Förderkennzeichen: 19 W 2086A, Ruhr-Universität, Bochum (2007)

17. Weißenbach, A.: Baugruben, Teil III: Berechnungsverfahren, unveränderter Nachdruck 2001. Ernst & Sohn, Berlin/München/Düsseldorf (1977)
18. Triantafyllidis, Th.: Neue Erkenntnisse aus Messungen an tiefen Baugruben am Potsdamer Platz in Berlin. Bautechnik **75**(3), 133–154 (1998)
19. Niemunis, A., Wichtmann, T., Triantafyllidis, Th.: A high-cycle accumulation model for sand. Comput. Geotech. **32**(4), 245–263 (2005)
20. Grandas-Tavera, C., Vogelsang, J., Triantafyllidis, Th.: Simplified simulation of the installation of vibro-piles in water saturated soil. Soil Dyn. Earthq. Eng. **121**, 491–498 (2019)
21. Kimmig, I., Triantafyllidis, Th.: Abschätzung der Verdichtungswirkung bei einer Baugrundverbesserung mittels Rütteldruckverdichtung mit einem Akkumulationsmodell. Heftbeitrag 26, Darmstädter Geotechnik Kolloquium, pp. 44–57, 7 March 2019
22. Wichter, L.: Schäden an Baugruben und Stützkonstruktionen. Eigenverlag, Teichland im März (2019)

Traces of Prof. Triantafyllidis in Bochum

Diethard König[(✉)], Achim v. Blumenthal, and Meisam Goudarzy

Chair of Soil Mechanics, Foundation Engineering and Environmental Geotechnics, Ruhr-Universität Bochum, Bochum, Germany
diethard.koenig@rub.de

Abstract. From 1998 to 2008 Prof. Triantafyllidis was head of the chair of foundation engineering and soil mechanics of the Ruhr-Universität Bochum. During that time, he initiated a number of research projects and the topics of these initiatives came along with the scientific work of Prof. Triantafyllidis for the next decades. But also in Bochum these research activities created a base for further developments. Within this contribution a number of the research initiatives of Prof. Triantafyllidis are recalled and two examples are given, how these works are influencing the developments in Bochum up to know.

1 Introduction

After getting his diploma in civil engineering Prof. Triantafyllidis worked as a doctoral student and post doc for 10 years at the institute of soil and rock mechanics of Universität Fridericiana in Karlsruhe. During this time, he mainly worked on the field of soil dynamics, wave propagation and dynamic soil structure interaction and contributed to the application of elasto-dynamics to soil mechanic problems (Triantafyllidis 1984, Triantafyllidis 1989). One experimental component was the determination of soil stiffness depending on shear strain by resonant column tests and piezo elements. In the following years Prof. Triantafyllidis worked for the construction companies Bilfinger & Berger and Brückner Grundbau getting a deep insight to several types of special foundation engineering works as bored piles, bored pile walls and slurry walls and grouted anchors. Partly, he contributed to further developments of those techniques (Triantafyllidis et al. 1997).

At the end of the 1990th Prof. Triantafyllidis was head of the technical division of Brückner Grundbau GmbH and responsible for the deep construction pits "Potsdamer Platz" and "Lehrter Bahnhof" in Berlin. These challenging projects (Triantafyllidis 1998) with excavation depth of about 20 m, water pressures up to 170 kN/m^2 without embedding of the walls into a low permeable soil layer and large areas of up to 100.000 m^2 led to a new milestone in the design and execution of deep construction pits. A thorough design was necessary to fulfill the specification to minimize the effects of the construction activities on the ground water level. Slurry walls, bored pile walls and sheet pile walls were used supported by one layer of prestressed grouted anchors located just above the ground water level. Underwater concrete bases were installed to seal the bottom of the construction pits. These concrete bases were stabilized against uplift forces by vibro injection piles.

T. Triantafyllidis (Ed.): *Recent Developments of Soil Mechanics and Geotechnics in Theory and Practice*, LNACM 91, pp. 383–407, 2020.
https://doi.org/10.1007/978-3-030-28516-6_21

To get an insight into the behavior of such constructions in general, and into the interactions between the soil, the walls and the base and at least to the mechanisms causing deformations in detail, a construction pit was equipped with measurement technique. Specially, horizontal wall deformation, soil deformation behind the wall and below the base were measured during the excavation of the soil, the installation of the vibro injection piles, the placement of the underwater concrete base as well as during the pumping of the water. The surprising result was, that most of the deformations of the slurry wall were initiated at the toe of the wall during installation of the vibro injection piles. The magnitude of the horizontal deformations revealed a strong dependency on the sequence of installation of the vibro injection piles parallel or perpendicular to the wall. At another site pore water pressures and accelerations within the soil caused by the installation of the vibro injection piles were measured. A significant increase of pore pressures were detected up to a distance of about 3 m. Prof. Triantafyllidis analyzed the data and published the results of the measurements and his interpretation in Triantafyllidis (1998). During his presentation at Ruhr-Universität Bochum in the frame of his application for the position of the head of the chair of foundation engineering and soil mechanics he demonstrated the results and highlighted the question of impact of the construction procedure on the behavior of the construction itself in an impressive manner. However, the enthusiasm to get an insight into this problem holds on him up to now.

In the following section an overview is given of the research activities which have been initiated by Prof. Triantafyllidis during the period he was head of the chair for foundation engineering and soil mechanics at Ruhr-Universität Bochum from 1998 to 2008. Most of these initiatives have had a further impact to the research of the chair in the subsequent years up to today. By two examples, namely the investigations of the characteristics of the filter cake formed during slurry wall instalation and the studies on the small strain stiffness of granular materials, this impact is shown in detail in Sects. 3 and 4.

2 Research Activities in Bochum

Becoming the head of the chair of foundation engineering and soil mechanics at Ruhr-Universität Bochum in March 1998 Prof. Triantafyllidis was directly forced to enter the collaborative research center SFB 398 "Lifetime-Oriented Structural Design Concepts". This research network just prepared the application for the second founding period. Prof. Triantafyllidis put his focus on the subproject A8 "Influence of changes in soil structure on the lifetime of buildings". The former subproject of his precursor was related to the contribution of the soil to changes in the soil structure interaction mainly due to cyclic loading. The new project shifted the focus on the structural change in the soil due to cyclic loading and the development of an accumulation model to describe the increasing deformation with increasing number of cycles. Beneath other practical application the understanding of the structural changes in the soil and a setup of an accumulation model would also support the simulation of the observed deformation in front of the toe of a retaining wall due to installation of vibro injection piles.

Within the following years the existing triaxial devices which were already equipped for cyclic soil testing were systematically improved (Fig. 1) and by performing a large number of tests a data base was created for the development of an accumulation model for non-cohesive soils (Wichtmann 2005; Niemunis et al. 2005, Wichtmann et al. 2009, Niemunis et al. 2009). Later, when Prof. Triantafyllidis became Professor in Karlsruhe, these research topics were one start point to install the research group GEOTECH dealing with the holistic consideration of geotechnical installation processes, which brought him back to his afore mentioned motivation (Triantafyllidis 2017). The triaxial equipment developed at that time in Bochum was later used e.g. to study the liquefaction behavior of sandy soils (Rahemi 2017).

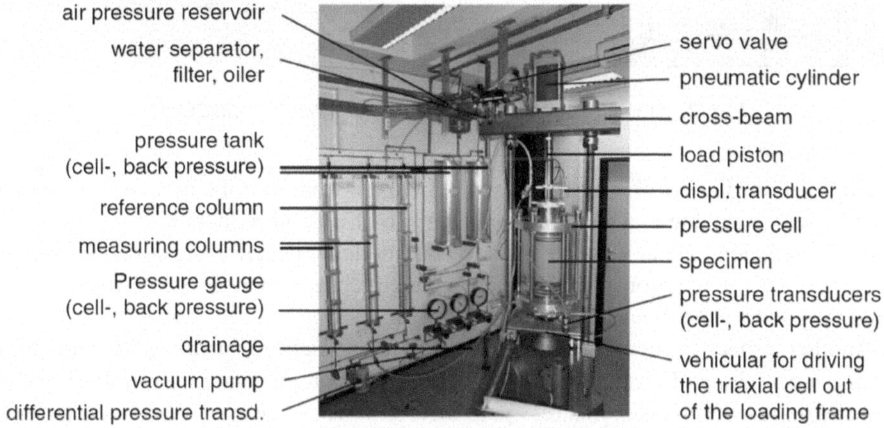

Fig. 1. Triaxial device for cyclic testing (Wichtmann 2005)

Directly related to his interest on the influence of the construction procedure on the deformation behavior of the structure was the DFG proposal which has been accepted in 2001 "Influence of the construction method on the deformations of retaining walls in cohesive soils". Within this project the influence of the construction method was investigated by numerical simulations (Schäfer 2004). Two questions raised during this work. The first one was linked to the pressure acting from the fresh concrete on the soil during concreting a single trench. The second one was related to the transfer of shear forces between the concrete of the already installed slurry wall and the soil taking into account the formation of a filter cake, which built up during the time the open trench is supported by a bentonite suspension and which remains partly after concreting the trench. In the frame of a BMBF project, which was realized in cooperation with Ed. Züblin AG, additional investigations on these questions were performed. Arwanitaki et al. (2007) and Arwanitaki (2009) showed clearly the strong influence of the soil particles brought into the bentonite suspension during the excavation of the trench on the characteristics of the filter cake which has been neglected before. This will be discussed more in detail in Sect. 3. Loreck (2008) performed centrifuge model tests and element tests investigating different mechanisms influencing the concrete pressure

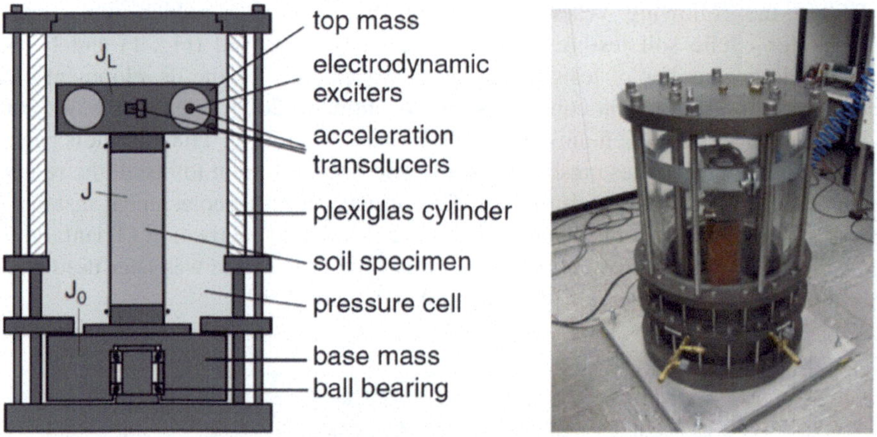

Fig. 2. Bochum resonant column device (Wichtmann 2005)

development during concreting a slurry wall. He showed, that the concrete pressure is reducing with time due to a combination from silo effect and reduction of the horizontal pressure coefficient. He developed an analytical model to describe the concrete pressure development as a function of the trench geometry, the concrete age at beginning of concreting and of the concreting process.

Another aspect in numerical modelling of construction processes in foundation engineering is the sufficient modelling of the soil behavior using constitutive models taking into account changes in stress path direction. These changes are connected with changes in soil stiffness and strain dependent soil stiffness. From his experience at Karlsruhe Prof. Triantafyllidis was familiar with the resonant column (RC) test to determine strain depending shear stiffness. The dynamic soil properties determined by RC device are also important to analyze the influence of dynamic impact caused by special foundation works like installation of vibro injection piles as discussed above. He forced the installation of a RC device (Fig. 2) in Bochum (Wichtmann et al. 2001). A DFG project was granted studying the influence of grain size distribution and fine content of granular soils on their dynamic properties (Wichtmann and Triantafyllidis 2009). Further research on the dynamic properties of soils are going on in Bochum up to now and will be presented in Sect. 4.

Also an early initiative of Prof. Triantafyllidis in Bochum was to investigate the influence of carbonic acid on the bearing capacity of grouted anchors. The bearing capacity is influenced due to limestone-dissolving environment causing a corrosion of the cement body of the anchor. Taking this into account the occurrence of carbon acid often leads to limitation in the use of grouted anchors on the construction site. In a number of cooperative research projects, pull out tests of anchor bodies exposed to water with excess carbonic acid were performed at different scales. The carbonic acid concentration, the exposure time and soil type were varied. The corroded zone of the anchor bodies was examined by chemical analysis. Dependency of the bearing capacity on concentration, time and soil type was detected (Hof et al. 2002; Hof 2004).

This research is carried forward by a cooperation between Ruhr-Universität Bochum and Federal Waterways Engineering and Research Institute (BAW) in Karlsruhe (Heidenreich et al. 2019).

A second collaborative research center was founded in cooperation with the Faculty of Geoscience SFB 526 "Rheology of the Earth". Within the subproject A1 the phenomenon of formation of shear bands in granular material under extension was studied. In particular, the geometry of shear band patterns defined by shear band spacing and inclination of shear bands was investigated. Boundary conditions (geometry, stress level) and material properties have been varied in a number of experiments (Fig. 3). The outcome was, that shear band spacing is connected to both geometry and material properties. In sand and sand dominated cohesive-frictional materials the softening gradient is a crucial factor on the resulting shear band distance (Wolf 2005; Röchter 2011).

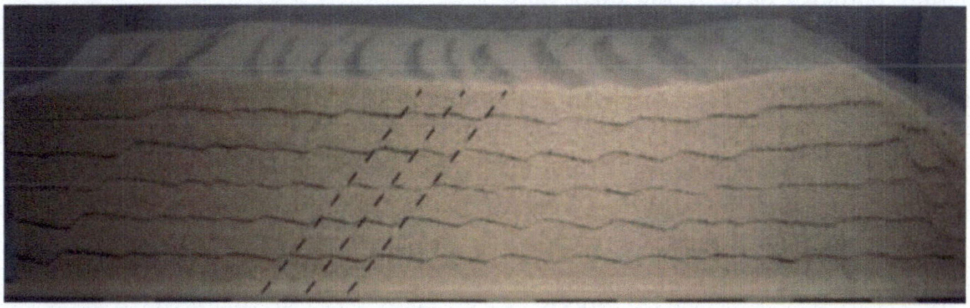

Fig. 3. Formation of shear bands in dense sand activated within the Bochum extension device (Wolf 2005)

In cooperation with the IKT the interaction between buried pipes and different types of filling materials was studied in large scale tests and modeled by finite element method (Arsic 2009).

3 Further Work in Bochum on Characteristics of Filter Cakes Formed During Slurry Wall Installation

3.1 Work of Arwanitaki (2009)

The friction between the soil and a retaining wall, e.g. a slurry wall, represents an input value for the design of the wall. The active earth pressure decreases and the passive earth pressure increases with an increase in the wall friction. From economical point of view this will improve the wall design significantly. In case of slurry walls a filter cake forms during the excavation of the trench supported by a bentonite slurry and remains after concreting between the soil and the wall. This causes the requirement that the wall friction angle has to be reduced in case of slurry walls compared to other types of walls like sheet pile walls. In the German regulations the wall friction angle of slurry walls is reduced to one half of the internal friction angle of the soil and in case of a period of more than 30 h between starting the excavation of the trench and beginning of concreting the wall, the friction angle has to be reduced to zero.

Arwanitaki et al. (2007) and Arwanitaki (2009) performed field studies on construction pits supported by slurry walls. The soils wherein the slurry walls have been installed were mostly sands with more or less content of fines, in one case sandy gravel. She has taken samples from the filter cake and of the nearby soil. She determined the grain size distributions of the filter cake and recognized that the filter cake is composed mainly of grains from silt and fine to medium sand with grain size diameter smaller than 2 mm. Assuming, that grain size of Bentonite is mainly below 0.001 mm she concluded, that the content of bentonite in the filter cake is between 5% and 10%, in case the nearby soil was a sandy gravel up to 17%, and that the remaining part of the grains originates from the nearby soil. This soil is loosened during the excavation process and some of its grains remain in the suspension. The suspension is loaded up with soil grains and these soil grains are filtrated at the walls of the trench like the bentonite particles forming the filter cake.

Arwanitaki (2009) created artificial filter cakes showing grain size distributions like the real filter cakes and performed shear tests on these samples. The results of this shear tests demonstrate, that shear strength of the filter cakes in silty to sandy soils is similar to the one of a silt. For more gravely soils the filter cake does not form a continues layer. In that case, the thickness of the filter cake is varying from a few millimeters to 30 mm and is regularly crossed by coarser grains leading to a direct contact between the soil and the slurry wall. Corresponding to that the surface of the slurry wall is extremely rough.

Based on this Arwanitaki (2009) recommended to set the wall friction angle between the soil and the slurry wall to 25° to 30°.

The field studies of Arwanitaki (2009) are related to four construction sites with slurry walls installed mainly in sandy soils. The influence of the time period in which the trench is stabilized by the bentonite suspension is discussed on the base of one laboratory test and theoretical analyses. The amount of bentonite in the filter cake is predicted by back analyses of the grain size distribution.

In further studies the findings of Arwanitaki (2009) are confirmed by determining the composition of the filter cake measuring the cation exchange capacity. In addition, the influence of the time period in which the trench is stabilized by the bentonite suspension on the composition and thickness of the filter cake is investigated. These works will be presented in the following.

3.2 Detection of Composition of Filter Cake Material

Arwanataki (2009) deduced the composition of the filter cake and the bentonite content from the grain size distribution of filter cake samples. For this, it is assumed that the grain size of the bentonite in the suspension and therefore in the filter cake is mainly below 0.001 mm. The aim of further research was to establish a method to determine the Bentonite content independent of grain size distribution. To get representative samples out of a thin filter cake from the site is often difficult and these samples have usually a small mass of a few grams. Such small amount of material is not enough to determine the grain size distribution e.g. by sedimentation but should be sufficient to employ the new method.

A typical parameter used by geologists and clay scientists to characterize the mineralogy of soils is the cation exchange capacity (CEC). CEC of sand and silts containing mostly quartz is very low, close to zero. The CEC of Bentonites is about 70 meq/100 g. The former mentioned works are showing that the filter cake is mostly composed of sand and silt with varying bentonite content. The idea is that with increasing bentonite content the CEC of the filter cake is also increasing and following Dohrmann et al. (2001) it is expected that this relation is linear. Therefore, knowing the CEC of the nearby soil as minimum value and of the bentonite used for mixing the fresh support slurry as maximum value from the measured CEC of the filter cake the bentonite content of the filter cake can be interpolated linearly.

For determining the CEC the Cu(II)-Triethylenetetramine method (Meier and Kahr 1999) is used as described by Dohrmann et al. (2001). Figure 4 shows the CEC measured on different soil-bentonite mixtures with changing content of bentonite starting from pure soil material ending up with pure bentonite (data from Scherbeck 2018). In case of mixture A soil material was Hostun sand, for B kaolin and for C local silt.

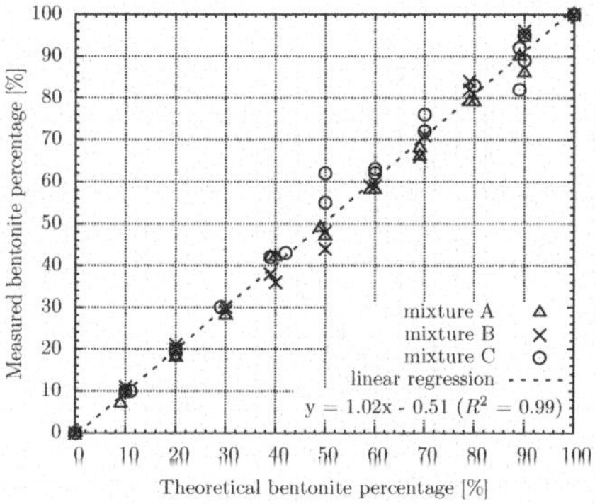

Fig. 4. Relation between predetermined (theoretical) bentonite content of mixtures of Hostun sand (A), kaolin (B) and a local silt (C) and Bentonite to the measured bentonite content by analyzing CEC of the pure materials and of the mixture

The bentonite content, which is predetermined by the mixing process of the mixtures, is plotted on the x-axis, on the y-axis the bentonite content, which is measured by analyzing the CEC of the pure soil material, of the bentonite and of the mixture. The results are grouped along the diagonal of the isometric scaled graph which confirms a linear relation between CEC and bentonite content.

Figure 5a demonstrates an example for determining the bentonite content of a filter cake material. The material has been taken from a construction site of a slurry wall in north of Germany. The soil profile is starting at ground surface with fillings followed by glacial loams overlaying glacial drift. IBECO B1 bentonite has been used for preparing the support slurry. The CEC of the soil has been determined to 7.25 meq/100 g and of

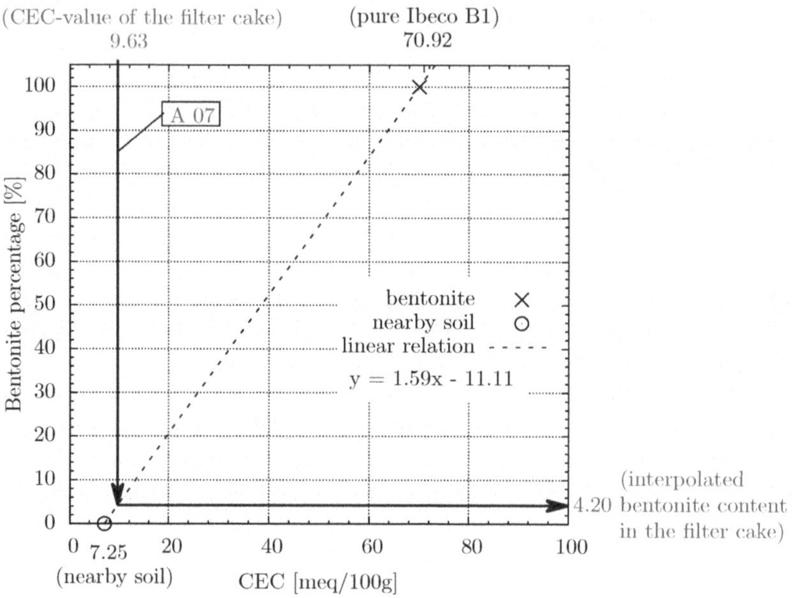

a)

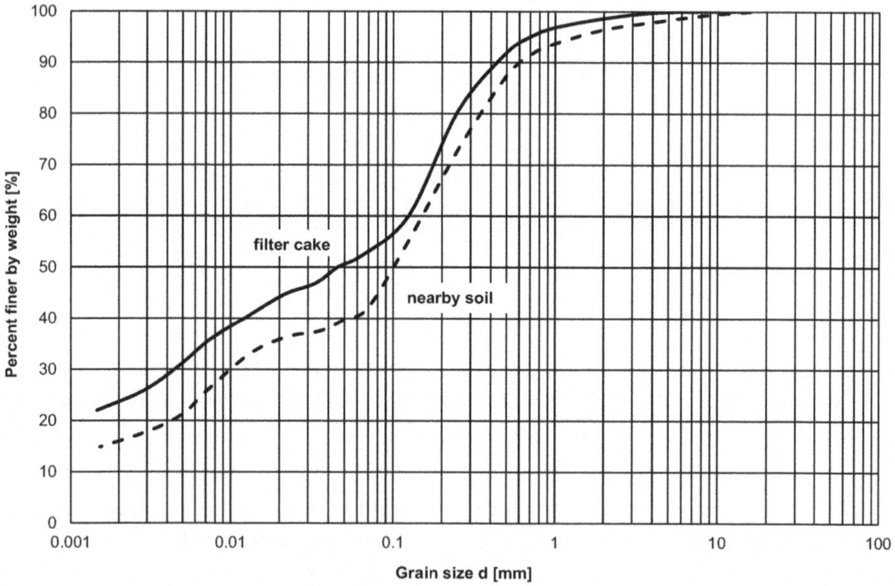

b)

Fig. 5. (a) Bentonite content of a filter cake material taken from a construction site in the north of Germany determined by CEC and (b) typical grain size distribution of the nearby soil and grain size distribution of the filter cake material the bentonite to 70,92 meq/100 g. The CEC of the filter cake material was found to be 9.63 meq/100 g (data from Birinci 2019). Assuming the linear distribution of CEC with bentonite content, as discussed above, the bentonite content is determined to 4.2% of the dry mass of the filter cake material.

This value is slightly lower than bentonite contents derived by Arwanitaki (2009) for slurry walls installed in sandy soils (5% to 10%) from grain size distributions. This small Bentonite content or the other way round the high content of grains of the nearby soil within the filter cake may be due to the grain size distribution of the nearby soil. This grain size distribution is dominated by fine sand and silt particles which are suspended in the slurry. As presented in Fig. 5b, in this case the grain size distribution of the filter cake material shows a content of grains smaller than 0.002 mm which is about 6% higher than the one of the typical nearby soil. This shift may be due to the bentonite particles which form together with the grains of the nearby soil the filter cake.

These results demonstrate that the CEC allows to determine the bentonite content of a filter cake in a time saving procedure specially in cases, where only a very limited mass of filter cake material is available.

3.3 Influence of Time from Trench Excavation to Concreting

König and Schroeder (2015) investigated the filter cake formed at two different slurry wall panels with a total depth of 24 m. The time between starting the excavation and beginning of concreting (slurry support period of the trench) was about 16 h for panel 28 and about 70 h for panel 30. Both panels were located close to each other with a distance of 10 m. Soil conditions were similar at both locations and characterized by a fill followed by a well graded gravel overlaying a sand marl. For supporting the excavated trench, a bentonite slurry was used.

During the excavation of the construction pit the surface texture and roughness of the slurry wall was inspected, the thickness of the filter cake was measured and samples of the filter cake were taken in different depths.

Whereas in the depth of the sand marl a continues filter cake was formed in the depth of the gravel no continuous filter cake was observed. In this part the wall was very rough and parts of a filter cake were found in pocket-like cavities with several centimeters of thickness. Nose-like elevations of the slurry wall surface were in direct contact with the soil. This was similar for both panels. The thickness of the filter cake in the depth of the sand marl was about 8 mm (panel 28) and 10 to 16 mm (panel 30). This increased thickness of the filter cake of panel 30 may be due to the longer time period in which the trench was supported by the bentonite slurry (slurry support period of the trench).

Figure 6 shows the grain size distributions of the filter cakes taken in a depth of 5 m, 6.5 m and 9 m for panel 28 (solid thin lines) and panel 30 (solid bold lines). The dashed lines represent the grain size distributions of the nearby soils. It is obvious, that the filter cakes show for all depths and both panels similar grain size distributions. These distributions are similar to that one of the sand marl found in a depth of about 9 m. A reason for this is that during the excavation finer soil particles from the sand marl are distributed in the bentonite slurry over the whole height of the trench and are separated during the filtration process of the slurry at the walls of the trench in the same manner like bentonite particles. The suspension seems not able to hold grains of the well graded gravel with a grain size above 2 mm and due to that these grains are not found in the filter cake. These findings are in agreement with the results of Arwanitaki (2009). In addition, there is no significant difference in the grain size distribution of the two panels although the slurry support period was 16 h for panel 28 and 70 h for panel 30.

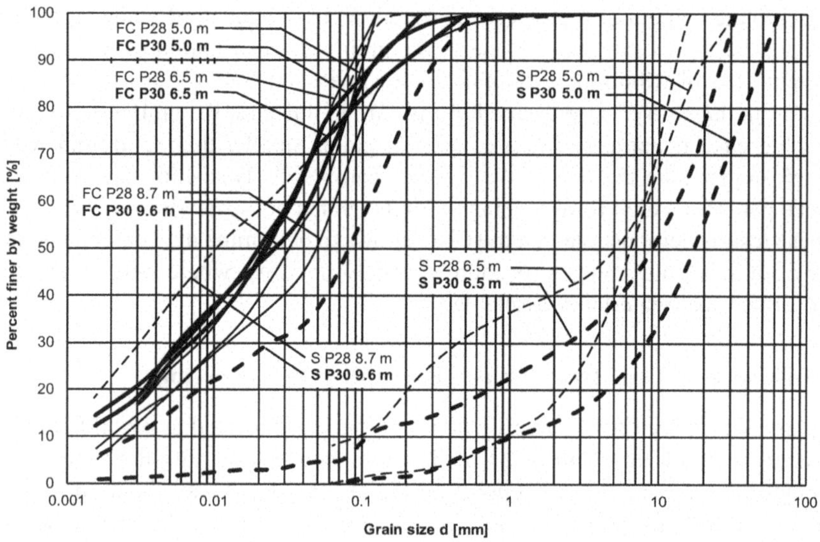

Fig. 6. Grain size distributions of filter cakes (FC, solid lines) and nearby soils (S, dashed lines) from panel 28 (P28, 16 h slurry support period, bold lines) and panel 30 (P30, 70 h slurry support period, thin lines) (data from König and Schroeder 2015)

To determine the bentonite content in the filter cakes, from remaining filter cake samples from three panels with different slurry support periods the CEC-values have been measured (Table 1). For this the remaining filter cake samples were divided in sub samples. The number of sub samples for each filter cake sample depended on the mass of the filter cake sample. It can be seen in Table 1, there is a slight variation of the CEC determined for the different sub samples. One reason for that is, that the sample mass used for the Cu(II)-Triethylenetetramine method is very small (1 g or less) and filter cake material is not homogenous on that scale. So the composition of the sub samples may vary.

The CEC of the gravel is assumed to be zero. The CEC of the bentonite (IBECO B1) has been determined to 70.92 meq/100 g as average value of 10 measurements within a range of 67.99 meq/100 g and 73.05 meq/100 g. According to Fig. 7 the bentonite content of the filter cake samples is determined and the data are summarized in Table 1 as upper and lower values taking into account the scattering of the CEC values of the different sub samples of the filter cake materials. The bentonite contents are varying between 10.52% and 19.11%. Arwanitaki (2009) is giving a value of 17% in case the nearby soil is a gravel. A significant and systematic influence of slurry support period of the trench cannot be detected. This supports the conclusion of König and Schroeder (2015) that the influence of slurry support period on the composition of the filter cake is minor. However, these findings have to be confirmed by additional investigations related to other construction sites.

Table 1. Parameters of three panels of one construction site and measured CEC values on sub samples of filter cake materials and upper and lower limit of bentonite content (determination of bold values is shown in Fig. 7)

Panel		28	31	30	30
Slurry support period	h	16	42.5	70	70
Sample No.		13_507	13_513	13_509	13_512
Depth	m	6.5	5.5	5.0	6.5
Nearby soil		Well graded gravel			
CEC					
Sub sample 1	meq/100 g	9.16	11.69	15.00	9.83
Sub sample 2	meq/100 g		11.10	14.02	10.79
Sub sample 3	meq/100 g		10.96	14.15	9.79
Sub sample 4	meq/100 g		11.25	13.47	9.52
Sub sample 5	meq/100 g		10.49		
Sub sample 6	meq/100 g		10.25		
Bentonite content					
Upper limit	%	**10.52**	14.24	**19.11**	12.92
Lower limit	%		12.13	16.86	11.05

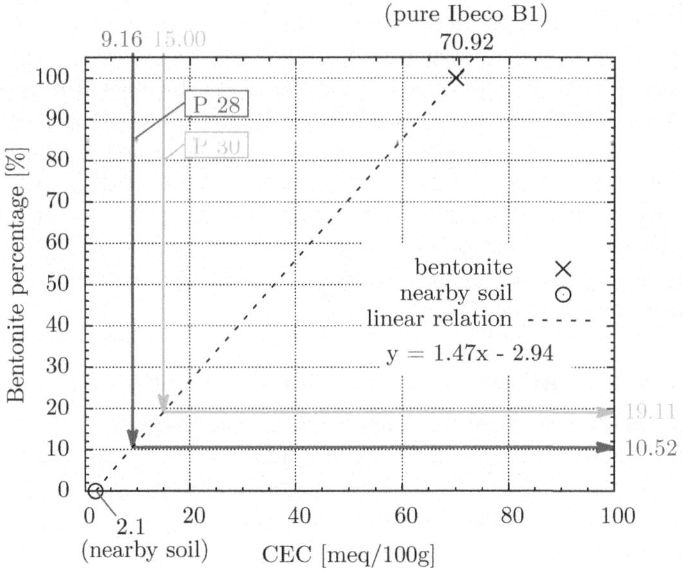

Fig. 7. Determination of bentonite content of filter cake materials taken from two different panels of one construction site (see bold values in Table 1)

4 Further Work in Bochum with RC Device

4.1 Bochum RC Device

The Bochum RC device installed under the guidance of Prof. Triantafyllidis (Fig. 2) is originally designed for applying isotropic stress conditions to the sample. Later on it has been retrofitted for applying anisotropic stress paths to the sample. In the following the original design of the Bochum RC device is described briefly and the modifications are presented. Several tests to calibrate and validate the modified RC device are shown and finally the influence of three different stress paths on the dynamic properties of a sand are discussed.

Resonant column technique is based on the oscillation, commonly torsional mode, of a cylindrical specimen to measure its resonant frequency. Torsional resonant column devices are divided into two groups, free-free and fixed-free resonant column devices which are applicable to determine the dynamic properties of soils for a wide range of strain from very small strain (shear strain, $\gamma < 10^{-6}$) to medium strain ($\gamma < 10^{-3}$). The bottom of the Bochum RC device is free in rotation, and torsional vibration is applied on the top of the specimen. Therefore, the device is considered as a free-free resonant column device. This device includes four parts, which are surrounded by a plexiglas cell. These parts are: (i) driving system, (ii) specimen, (iii) bottom mass, (iv) volume change measurement elements.

The driving system (i) is a cubic aluminum mass, which includes two mini-shakers for applying torsion and also two accelerometers to record signals (driving system is called actuator hereafter). The adopted mini-shakers (Brüel & Kjaer, Type 4810) are electro dynamic type with a permanent field magnet. A coil is flexibly suspended in the field of the permanent magnet. An alternating current signal is passed through the coil to produce a vibration. Two accelerometers are also mounted on the shaker (Fig. 8) and on a corner of the actuator for visualization of signals, which are called in this paper Exc. 1 and Res. 1, respectively. Exc. 1 is used to detect the applied force through the shaker, and Res. 1 to derive the displacement of the actuator during vibration. The specimen (ii) has a size of 10–15 and 20–30 cm in diameter and height, respectively, see Wichtmann et al. (2001) for more details. The bottom mass (iii) is a steel mass which is completely free in rotation. The volume change of the specimen is observed (iv) by the measurement of the radial deformation of the specimen using six noncontact displacement transducers and by measuring the vertical deformation by one additional noncontact displacement transducer.

4.2 Modification of RC Device to Apply Anisotropic Stress Paths

4.2.1 Instrumentation to Apply Anisotropic Stress State

The actuator of the Bochum resonant column device is a cubic part which includes the electrical equipment (Fig. 8). To apply an additional vertical stress inside of the specimen the actuator is loaded in vertical direction using a double acting pressure cylinder. The load of the cylinder is transferred with a loading bar, a hardened steel tip, to a smooth steel plate which is mounted at the central axis of the actuator and sample.

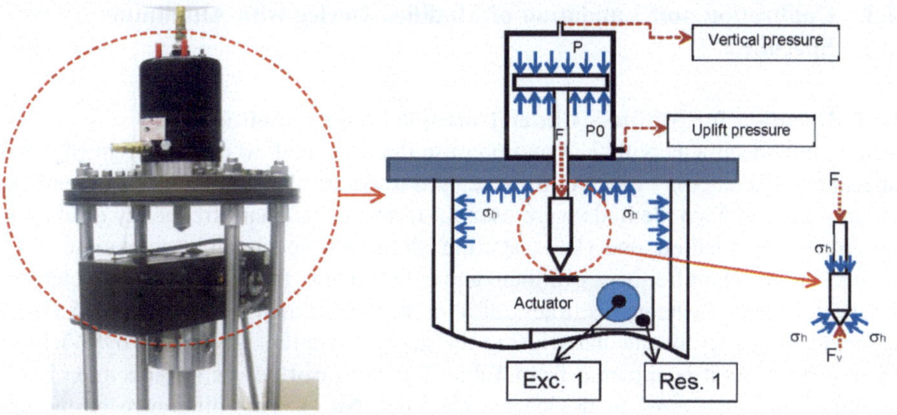

a) b)

Fig. 8. (a) The loading system for applying anisotropic loading on the top of the specimen in the Bochum resonant column device; (b) schematic sketch of the loading system and free body diagram of the pressures and forces that act on the loading bar

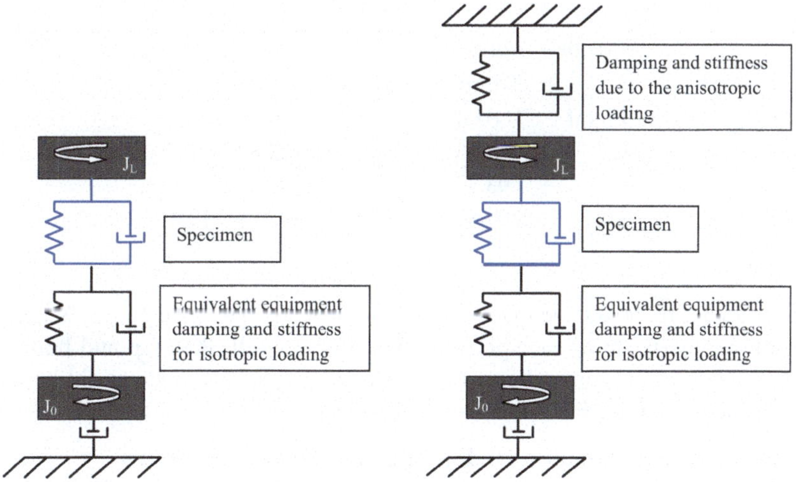

a) b)

Fig. 9. Spring-dashpot model of the Bochum resonant column device for: (a) isotropic stress state; (b) anisotropic stress state

Figure 9 shows spring-dashpot model of the modified device in comparison with the original device. This figure shows that the dynamic response of the device can be affected by the stiffness and damping at the interface between loading bar and actuator.

4.3 Calibration and Validation of Modified Device with Aluminum Specimen

(i) Calibration for Stiffness Under Isotropic Stress Conditions
Aluminum specimens with 2, 4 and 6 cm in diameter and 30 cm in height (defined as specimens No. 2, No. 4 and No. 6) were used to control the compliance of the resonant column device. Two methods were used to determine resonant frequency of aluminum specimens. In method one (RC-test), the aluminum specimen was mounted in RC device and resonant frequency of them and consequently their stiffness was determined. In second method (theory), stiffness of aluminum specimen was assumed 25 GPa and resonant frequency is calculated using the general relations given e.g. by Wichtmann et al. (2001). As it is apparent from Table 2, the error of device for specimen No. 2 is slightly more than error of device for specimen No. 4. This difference is due to the slenderness of the specimen because the diameter of specimen No. 2 is small in comparison to its length. Therefore, the results might be affected by flexural mode of vibration, i.e. bending of specimen during torsional vibration, due to the shape of actuator and size of specimen.

Table 2. Resonant frequency of aluminum specimens. Error is defined as theoretical frequency (Theory) minus measured frequency (RC test) related to theoretical frequency

Specimen	Frequency		Error
	Theory, G = 25 GPa	RC test	
d [cm]	f_r [Hz]	f_r [Hz]	[%]
2	23.43	22.82	2.6
4	93.45	92.41	1.11
6	209.86	182.95	12.8

Furthermore, the results show a significant error for specimen No. 6, which may be due to the mode of vibration. Specimen No. 6 is too stiff, thus top and bottom of the specimen rotate in the same phase. In this situation spurious mode might be measured instead of actual mode. For further calibration specimen No. 4 is selected.

(ii) Calibration for Damping Under Isotropic Stress Conditions
Aluminum specimen No. 4 was used to determine the damping of the Bochum resonant column device. Damping of aluminum is very small, approximately zero thus measured damping for aluminum specimen is related to the dissipation of energy in device (damping of device). Three methods were employed to measure the damping: energy method (Wichtmann et al. 2001), free vibration decay and bandwidth method (Stokoe et al. 1999). The results are presented in Fig. 10. As can be seen, at very small strain range ($\gamma < 10^{-6}$) measured damping by energy method and free vibration decay method is increasing with decreasing strains. For larger strain levels damping is determined to 0.0017 by all three methods. This value is also measured by bandwidth method for very small strains. The discrepancy between damping determined by energy method and free vibration decay method compared to bandwidth method is due to the

influence of noise at small strain range on the recorded signals. The energy method is based on the detected amplitudes of force of the shakers and displacements of the actuator by accelerometer measurements Exc. 1 and Res. 1. The accelerometer measurements may be effected by near field conditions, because at low amplitudes the effect of noise and near field effects are dominant. The free vibration decay method is based on the free oscillation of the sample which can be also affected by near field conditions at low amplitude of vibrations. The bandwidth method is only based on the detected frequencies. The impact of noise on frequencies and therefore on measured damping at very small strain range is low. Energy method is used in the experiments on soils presented in the following. Therefore, according to the observed results two calibration lines are defined for measured damping using energy method (Fig. 10). Firstly, at very small strain range, the amplitude of deformations is too small (oscillation with low amplitude). Thus, near field conditions or near field effects are dominant on the response of the specimen to the applied oscillation. On the other hand, at small strain range, damping is a constant value, which is related to the equipment damping. The equipment damping may be due to the connections and the quality of specimen.

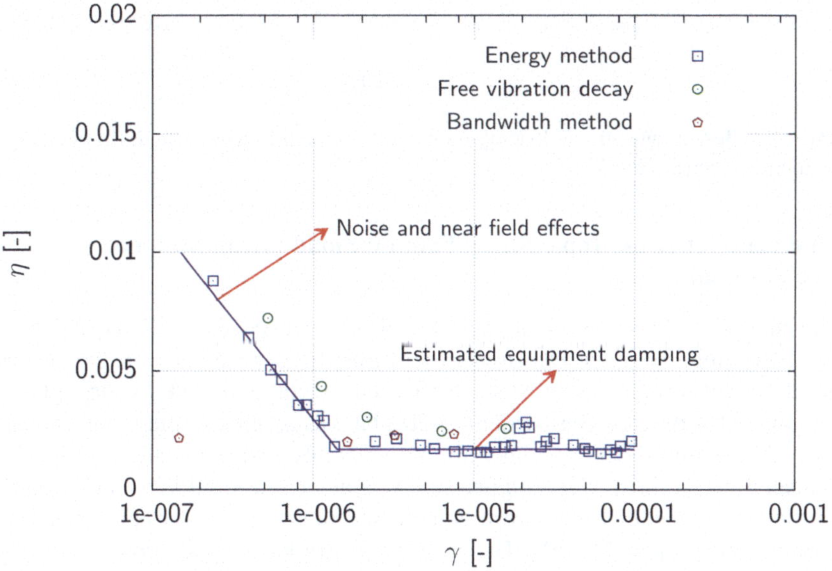

Fig. 10. Damping of the RC device using aluminum specimen No. 4 and different methods

(iii) Calibration for Stiffness and Damping Under Anisotropic Stress Conditions
Specimen No. 4 was installed in the resonant column device and vertical load was increased from 0 to 4000 N. It has to be noticed that for vertical load of zero contact between loading bar and actuator existed. The results show a small increment on the obtained resonant frequency with increase in the vertical load (Fig. 11a). The observed

increment (0.88%) in the measured resonant frequency may be due to the compliance of the aluminum specimen and connections. Damping of device was measured under increasing vertical load by energy method. Figure 11b shows that damping decreases slightly with an increase in the vertical load. This could be due to the fixity of connections of different parts. This means, by increasing the vertical load, connections of aluminum specimen and device may be fitted to each other more, therefore initial increasing in the stiffness and initial decreasing in the damping can be due to the compliance of connections. It is worthy to mention, that the small difference between damping under vertical load of zero and damping measured for isotropic stress state (Fig. 10) may be due to the influence of the contact between loading bar and actuator.

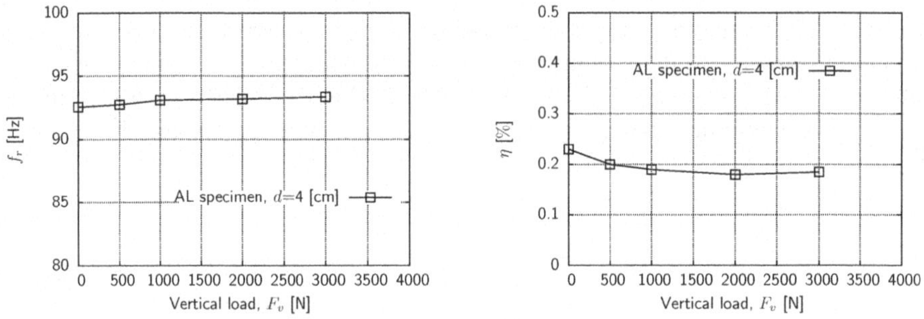

a) b)

Fig. 11. The effect of anisotropic loading on the: (a) resonant frequency and; (b) damping ratio of aluminum specimen No. 4

4.4 Influence of Anisotropic Stress Path on Small Strain Stiffness of Pure Sand

Dry Hostun sand (Flavigny et al. 1990) with G_S = 2.65, e_{max} = 1.04, e_{min} = 0.65, D_{50} = 0.355 mm (Fig. 12a) was used to perform RC experiments on the specimens subjected to anisotropic stress state. Specimens were prepared by dry pluviation method out of the device. The vacuum of 50 kPa was applied through bottom and top caps for stabilization of the specimen during assembly and preparation of the device. After assembling of the device, cell pressure was increased to 50 kPa and vacuum was reduced to zero. Then cell pressure was increased to set up an isotropic consolidation stress of e.g. $\sigma_v = \sigma_h$ = 200 kPa. The anisotropic stress state was applied through two different stress paths (Fig. 12b). In stress path I, cell pressure, σ_h, was kept constant and vertical stress, σ_v, was increased. Afterwards, RC experiments were performed for given stress states shown with circle symbols in Fig. 12b. In stress path II, $p = (\sigma_v + 2\sigma_h)/3$ was kept constant and equal to 200 kPa and $q = \sigma_v - \sigma_h$ was increased. Piezoelectric elements (shear plates, SP) were also installed in top and bottom caps of the modified resonant column device (Fig. 13a). This provides a possibility to perform RC and SP tests simultaneously. SP experiments were performed using sinusoidal wave with frequency of 10 kHz–20 kHz and peak to peak amplitude of 10 V (Fig. 13b). Note: resonant frequency for the adopted boundary conditions and material was a value between 30–65 Hz. However, as it was highlighted by Stokoe and Santamarina (2000) the effect of frequency on G_{max} in granular soils is negligible. Because of this reason, the results of RC and SP experiments were compared with each other for validation of the results.

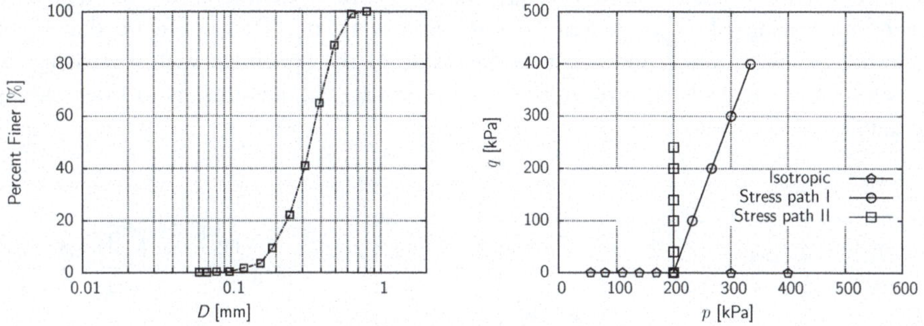

a) b)

Fig. 12. (a) grain size distribution of Hostun sand; (b) the adopted stress paths for experiments using the modified resonant column device

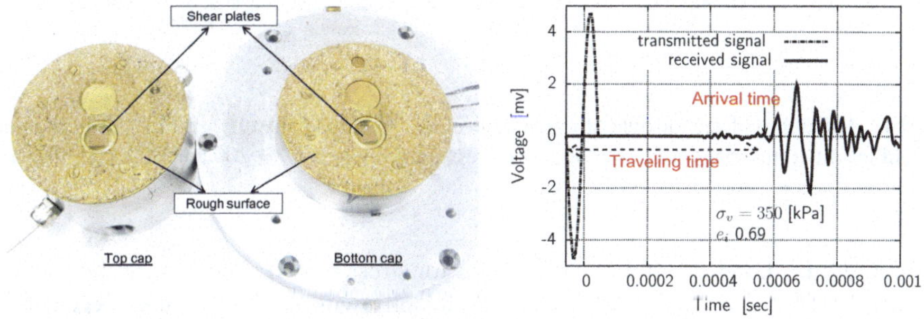

a) b)

Fig. 13. (a) Instrumentation of top and bottom caps of the Bochum resonant column device for performing shear wave velocity test using shear plates; (b) transmitted and received signals in experiment using the shear plates on the dense Hostun sand, subjected to the vertical and horizontal stresses of 350 and 200 kPa, respectively

(i) Maximum Shear Modulus, G_{max}

The maximum shear modulus, G_{max}, of the dense specimens with e = 0.71 subjected to stress paths I and II are shown in Fig. 14a and 14b. Firstly, these Figures show a good agreement between measured G_{max} using RC and SP experiments. It has to be mentioned that for more loose samples SP results are given lower G_{max} values compared to RC experiments which could be due to the weak contact between shear plates and sand particles in loose specimens.

For stress path I, Fig. 14a shows that G_{max} increases with an increase in the vertical stress, σ_v. The results show that the rate of increasing G_{max} with vertical stress decreases significantly, when σ_v/σ_h is more than 2.0, i.e. σ_v more than 400 and more than 600 kPa for specimens consolidated under isotropic stress state of 200 and 300 kPa, respectively. For stress path I, σ_h is constant and σ_v increases, triaxial compression stress path. Therefore, G_{max} increases due to the increasing of σ_v, but the rate of increasing of G_{max} decreases with increasing σ_v. This could be due to the increasing of shear stress and shear strain inside of the specimen with increasing σ_v. Therefore, the effect of anisotropic stress state on G_{max} along stress path I depends on the applied stress ratio.

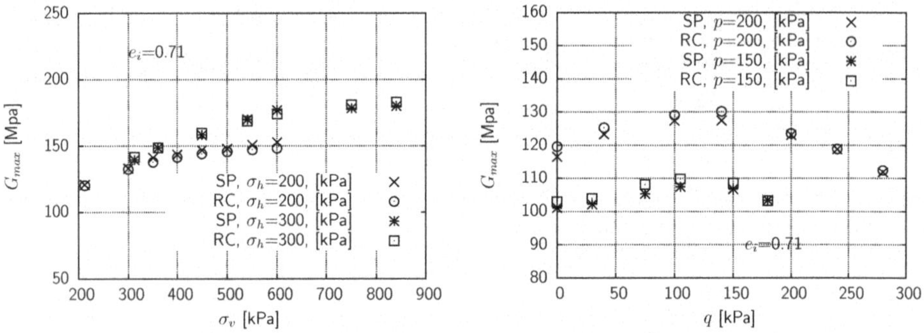

a) b)

Fig. 14. The effect of anisotropic stress state on G_{max} of dense Hostun sand using shear plates, SP, and modified resonant column along: (a) stress path I, when σ_h = 200 and 300 kPa; (b) stress path II, when p = 100, 150 and 200 kPa

Figure 14b shows the effect of q on G_{max} for stress path II. This Figure shows, G_{max} increases slightly with q and decreases significantly with increasing of q. To reach the target stress state for stress path II, σ_h was decreased and σ_v was increased from the initial state where $\sigma_v = \sigma_h$ = 200 kPa and 150 kPa, respectively. This will cause the buckling of force chains in the specimen from microscopic point of view and therefore, significant increasing of shear strain in the specimen from macroscopic point of view. This could be a reason for significant decreasing of G_{max} with anisotropic stress state along stress path II in comparison with stress path I.

Figure 15 shows the results for specimens with different density in normalized form. The above mentioned tendencies are independent from density. Furthermore, the results from RC experiment are compared with previous works using RC experiment (e.g. Hardin and Black 1966; Yu and Richart 1984) and also using bender elements (e.g. Ezaoui and Di Benedetto 2009 and Sadek et al. 2007). The results show also a good agreement between the current experiment using modified device and experiments from previous works.

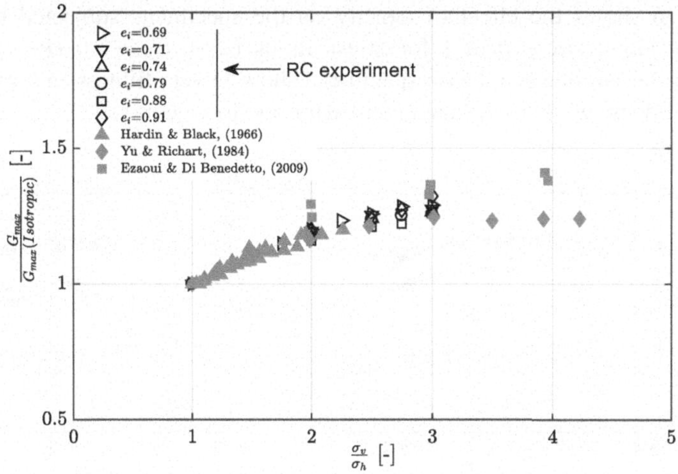

a)

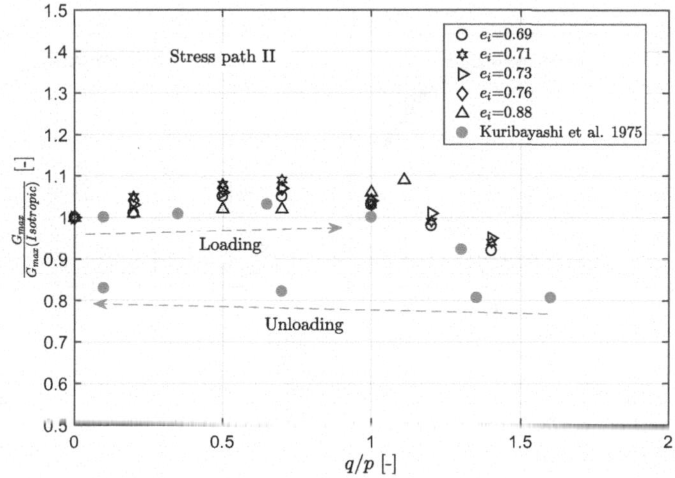

b)

Fig. 15. The effect of anisotropic stress state on G_{max} in comparison with previous works along: (a) stress path SP-I, when σ_h = 200; (b) stress path SP-II, when p = 200 kPa

(ii) Intermediate Strain Properties, G(γ) and η(γ)

RC experiments were also conducted to assess the effect of anisotropic stress states along various stress paths on intermediate strain properties, details of results have been presented in Goudarzy et al. (2018a, glass particles) and Goudarzy et al. (2019, sand). In this paper, for an example the effect of anisotropic stress state along stress path I and the influence of the density of specimens on the damping ratio and stiffness with respect to shear strain are presented. The effect of isotropic stress state on the modulus degradation and damping ratio of the adopted sand were presented in Goudarzy et al. (2018b), which will not be presented in this paper.

Figure 16 shows the effect of density on the specimens subjected to anisotropic stress state along stress path I for stress ratios of $\sigma_v/\sigma_h = 2.0$ (Fig. 16a) and 3.0 (Fig. 16b). As can be seen, loose specimens show lower stiffness in comparison with dense specimens, which is in agreement with previous works.

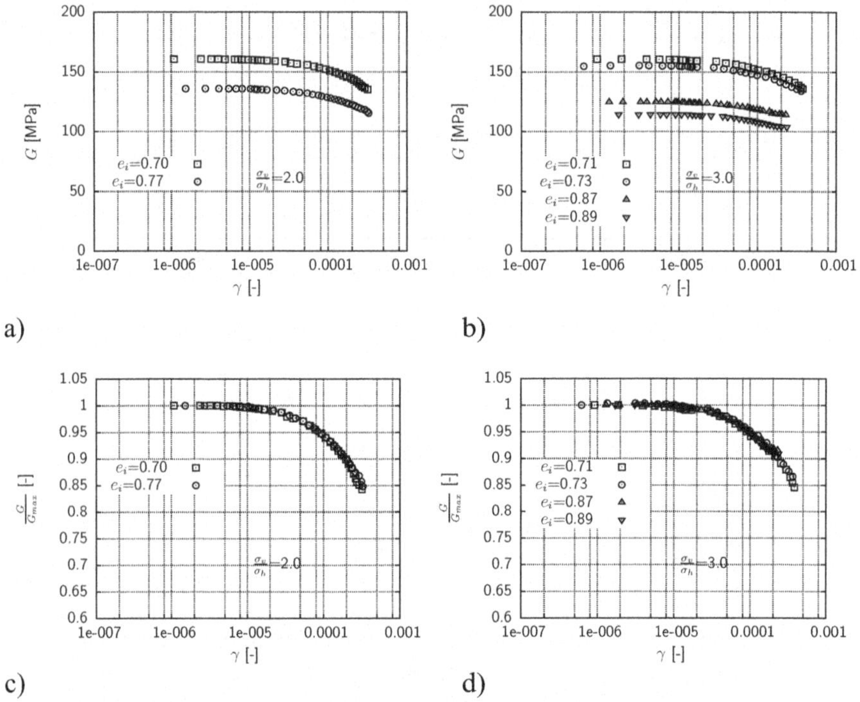

Fig. 16. The effect of density of the specimen on: (a) $G - \gamma$, when $\sigma_v/\sigma_h = 2.0$; (b) $G - \gamma$, when $\sigma_v/\sigma_h = 3.0$; (c) $G/G_{max} - \gamma$, when $\sigma_v\ \sigma_h = 2.0$; (d) $G/G_{max} - \gamma$, when $\sigma_v/\sigma_h = 3.0$ along stress path I for $\sigma_h = 200$

Figure 16c and 16d show the effect of density on $G/G_{max} - \gamma$ curves for stress ratio of 2.0 and 3.0, respectively. As it is apparent, the results show the independency of $G/G_{max} - \gamma$ curves to the density of specimens, which is in agreement with the previous observations on the effect of density on $G/G_{max} - \gamma$ for isotropic stress state (e.g. Tatsuoka et al. 1978; Kokusho 1980; Wichtmann and Triantafyllidis 2013; Goudarzy et al. 2018b). Figure 17 shows the effect of density on damping ratio for the specimens subjected to stress ratio 2.0 and 3.0 in Figs. 17a and 17b, respectively. The results show the independency of damping ratio to density of specimens subjected to anisotropic stress state which is also in agreement with previous mentioned works, but for isotropic stress states.

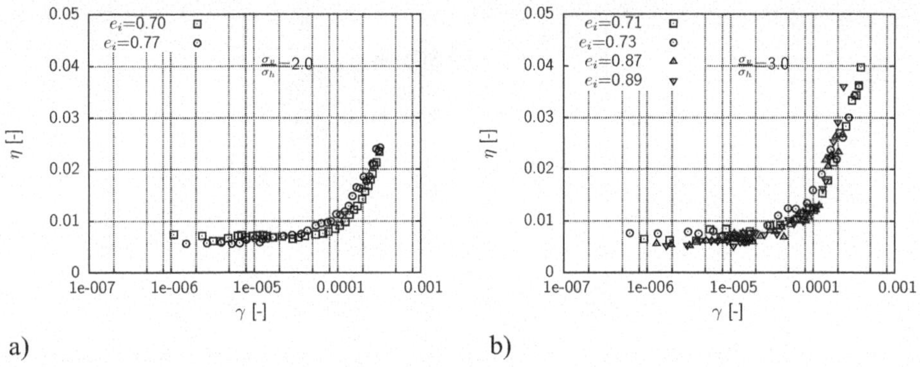

Fig. 17. The effect of density of the specimen on damping ratio, $\eta - \gamma$, when: (a) $\sigma_v/\sigma_h = 2.0$; (b) $\sigma_v/\sigma_h = 3.0$ along stress path I for $\sigma_h = 200$

Figure 18 shows the effect of stress state along stress path I on the strain dependency characteristics of the adopted sand with the initial density of $e_i = 0.7$. Figure 18a shows that G increases by increasing of stress ratio, but the rate of increasing decreases significantly after stress ratio 2.0. This could be due to the re-arrangement of particles or trend of specimen for going to dilation, which is followed by decreasing of stiffness for further anisotropic stress levels.

Figure 18b shows the effect of stress ratio along stress path I on $G/G_{max} - \gamma$ curves. This figure shows that all data are located in a narrow range. However, for ease of consideration, data for larger strains are scaled up in Fig. 18c. As can be seen, $G/G_{max} - \gamma$ curves slightly shifted above the data for isotropic stress state. However, with refereeing to Fig. 8a, this slightly increasing, with respect to isotropic stress state, could be due to the compliance of system rather than the effect of anisotropic stress state on the material. Figure 18d shows the effect of anisotropic stress state along stress path I on damping ratio, which is located in a narrow range. Again, for ease of consideration, data for larger strains are scaled up in Fig. 18e. As can be seen, $\eta - \gamma$ curve shows lower damping for anisotropic stress state in comparison with isotropic stress state. However, this slightly decreasing with respect to isotropic stress state could be due to the compliance of system (see Fig. 11b).

These findings are also in agreement with Tatsuoka et al. (1978), where they reported the insignificant effect of anisotropic stress state on damping ratio and G/G_{max} for specimens subjected to triaxial compression stress path. Furthermore, the results show that η_{min} is also independent of stress state, which is in agreement with Santamarina and Cascante (1996) for the same stress path.

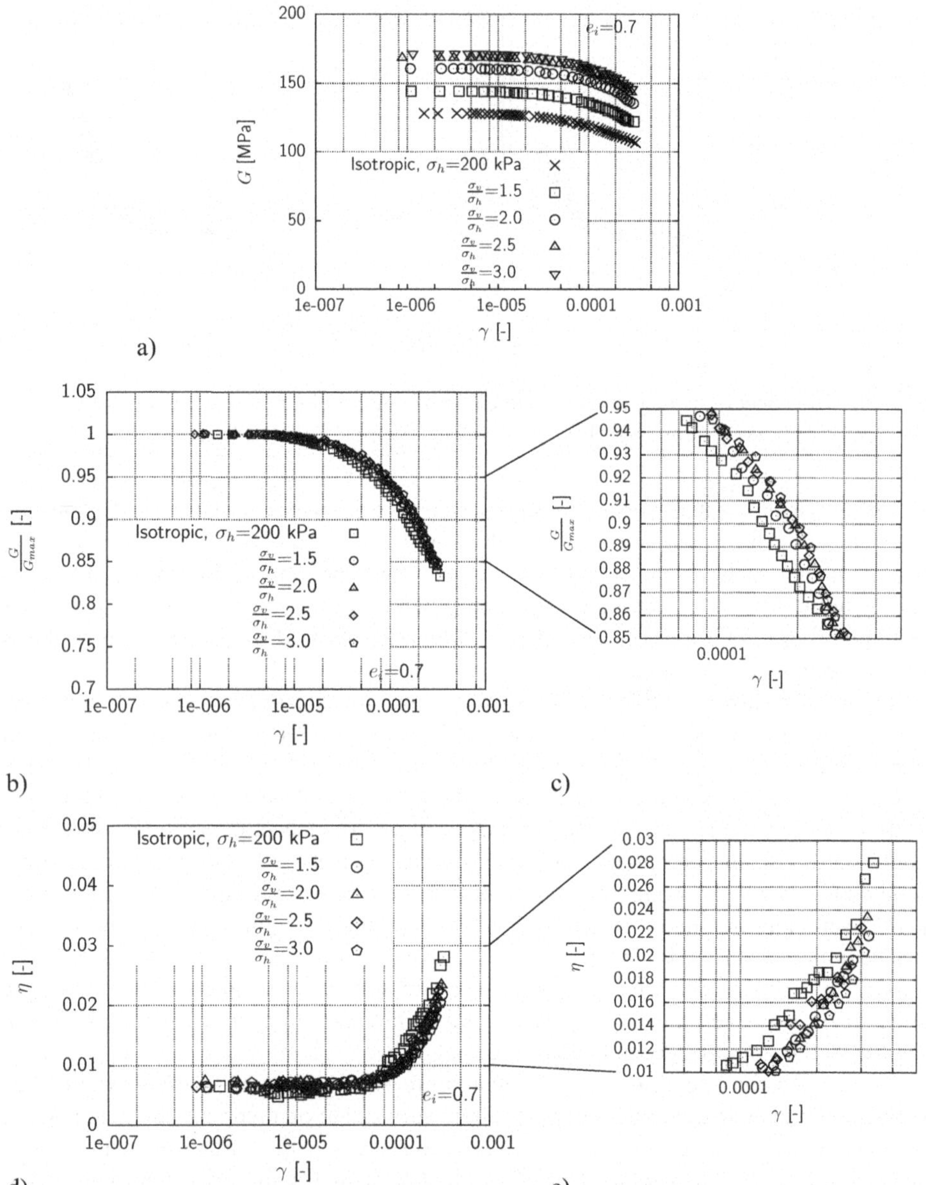

Fig. 18. The effect of anisotropic stress state along stress path SP-I on: (a) G − γ; (b) G G_{max} − γ; (c) η − γ of dense Hostun sand

5 Summary

Within this paper two research topics are highlighted which are followed by a number of scientists at the chair of soil mechanics, foundation engineering and environmental geotechnics at Ruhr-Universität Bochum. The activities on this research fields have been initiated by Prof. Triantafyllidis between 1998 and 2008 when he was head of the chair. His successor, Prof. Schanz, has had the farsightedness to enable and to support doctoral students and later post docs to continue these research activities beside an impressive number of new topics. The outcome has been a large number of high ranked publications giving impact to the scientific community and to practical applications. Many students have been trained on the laboratory equipment and have done field works on construction sites, which are now following their own research topics or are employed in the industry. The red line shown up in this paper from the first research topics of Prof. Triantafyllidis following his experience as engineer to his time as head of the chairs in Bochum and Karlsruhe may be simplifying and the development could be recount in another manner. However, the footprints of Prof. Triantafyllidis in Bochum are still visible and actual development shows, that they will have an impact also in the future.

References

Arsic, I.: Über die Bettung von Rohrleitungen in Flüssigböden. Schriftenreihe des Lehrstuhls für Grundbau, Boden- und Felsmechanik, Ruhr-Universität Bochum, no. 40 (2009)

Arwanitaki, A., König, D., Triantafyllidis, Th.: Zum Kontaktverhalten suspensionsgestützter Ortbetonwände und den anstehenden Baugrund. Bautechnik **84**(11), 781–972 (2007)

Arwanitaki, A.: Über das Kontaktverhalten einer Zweiphasen-Schlitzwand und nichtbindigen Böden. Schriftenreihe des Lehrstuhls für Grundbau, Boden-und Felsmechanik, Ruhr-Universität Bochum, no. 41 (2009)

Birinci, G.: Bestimmen der Zusammensetzung des Filterkuchens bei Schlitzwänden über die Kationenaustauschkapazität. BSc-thesis at the chair of soil mechanics, foundation engineering and environmental geotechnics, Ruhr-Universität Bochum (2019, unpublished)

Dohrmann, R., Ufer, K., Meyer, F.M.: Angewandte Tonmineralogie im Tunnelbau Möglichkeiten und Grenzen quantitativer Phasenanalyse. In: Schwaighofer, B., Müller, H.W., Schneider, J.F. (Hersg.) Baugeologisches Seminar, Mitteilungen Institut für Angewandte Geologie, Universität für Bodenkultur Wien, Reihe: Angewandte Geowissenschaften, no. 11 (2001)

Ezaoui, A., Di Benedetto, H.: Experimental measurements of the global anisotropic elastic behaviour of dry Hostun sand during triaxial tests, and effect of sample preparation. Geotechnique **59**, 621–635 (2009)

Flavigny, E., Dersuse, J., Palayer, B.: Le sable d'hostun rf. Geotechnique **53**, 67–70 (1990)

Goudarzy, M., König, D., Santamarina, J.C., Schanz, T.: Influence of anisotropic stress state on the intermediate strain behaviour of granular materials. Geotechnique **68**(3), 221–232 (2018a)

Goudarzy, M., König, D., Schanz, T.: Small and intermediate strain properties of sands containing fines. Soil Dyn. Earthq. Eng. **110**, 110–120 (2018b)

Goudarzy, M., König, D., Schanz, T.: Interpretation of small and intermediate strain characteristics of Hostun sand for various stress states. J. Soils Found. **58**, 1526–1537 (2019)

Hardin, B.O., Black, W.L.: Sand stiffness under various triaxial stresses. Soils Mech. Found. Div. **92**, 27–42 (1966)

Heidenreich, F., Herten, C., König, D.: Attack of lime-dissolving carbonic acid in laboratory and in situ tests on the load capacity of grouted anchors in sands. In: Proceedings of the XVII ECSMGE (2019, in print)

Hof, C., Triantafyllidis, Th., Schmidt-Döhl, F.: Erkenntnisse über den Einfluß von kalklösender Kohlensäure auf das Tragverhalten von Verpressankern. GEOLEX, **1**, 31–41 (2002)

Hof, C.: Über das Verpressankerverhalten unter kalklösenderm Kohlensäurenangriff. Schriftenreihe des Lehrstuhls für Grundbau, Boden-und Felsmechanik, Ruhr-Universität Bochum, no. 35 (2004)

König, D., Schroeder, Th.: Zusammensetzung des Filterkuchens an Schlitzwandlamellen mit kurzer und langer Standzeit. Bauingenieur **90**(2), 63–70 (2015)

Kokusho, T.: Cyclic triaxial test of dynamic soil properties for wide strain range. Soils Found. **20** (2), 45–59 (1980)

Kuribayashi, E., Iwasaki, T., Tatsuoka, F.: Effects of stress-strain conditions on dynamic properties of sands. In: Proceedings of the JSCE, no. 242, pp. 105–114 (1975)

Loreck, M.Ch.: Die Entwicklung des Frischbetondrucks bei der Herstellung von Schlitzwänden. Schriftenreihe des Lehrstuhls für Grundbau, Boden-und Felsmechanik, Ruhr-Universität Bochum, no. 39 (2008)

Meier, L.P., Kahr, G.: Determination of the cation exchange capacity (CEC) of clay minerals using the complexes of copper (II) Ion with triethylenetetramine and tetraethylenepentamine. Clays Clay Min. **47**, 386–388 (1999)

Niemunis, A., Wichtmann, T., Triantafyllidis, Th.: A high-cycle accumulation model for sand. Comput. Geotech. **32**(4), 245–263 (2005)

Niemunis, A., Wichtmann, T., Triantafyllidis, Th.: A high cycle model for soils. In: Stangenberg, F. et al. (ed.) Lifetime-Oriented Structural Design Concepts, Chap. 3.3.3., pp. 313–316. Springer, (2009). ISBN 978-3-642-01461-1

Rahemi, N.: Evaluation of liquefaction behavior of sandy soils using critical state soil mechanics and instability concept. Schriftenreihe des Lehrstuhls für Grundbau, Boden-und Felsmechanik der Ruhr-Universität Bochum, no. 64 (2017)

Röchter, L.: Systeme paralleler Scherbänder unter Extension im ebenen Verformungszustand. Schriftenreihe des Lehrstuhls für Grundbau, Boden-und Felsmechanik der Ruhr-Universität Bochum, no. 45 (2011)

Sadek, T., Lings, M., Dihoru, L., Wood, D.M.: Anisotropic stiffness of Hostun sand, pp. 147–152. Taylor and Francis Group (2007)

Santamarina, C., Cascante, G.: Stress anisotropy and wave propagation: a micromechanical view. Can. Geotech. J. **33**, 770–782 (1996)

Schäfer, R.: Einfluss der Herstellungsmethode auf das Verformungsverhalten von Schlitzwänden in weichen bindigen Böden. Schriftenreihe des Lehrstuhls für Grundbau, Boden-und Felsmechanik, Ruhr-Universität Bochum, no. 36 (2004)

Scherbeck, P.: Charakterisierung von beim Bau von Schlitzwänden entstandenem Filterkuchen. MSc-thesis at the chair of soil mechanics, foundation engineering and environmental geotechnics, Ruhr-Universität Bochum (2018, unpublished)

Stokoe, K.H., Darendeli, M.B., Andrus, R.D., Brown, L.T.: Dynamic soil properties: laboratory, field and correlation studies. In: Seco e Pinto. P.S. (ed.) Proceeding of the 2nd International Conference on Earthquake Geotechnical Engineering Conference, Lisbon, Portugal (1999)

Stokoe, K.H., Santamarina, J.C.: Seismic wave based testing in geo-technical engineering. In: International Conference on Geotechnical and Geological Engineering (GeoEng 2000), Melbourne, Australia, pp. 1490–1536 (2000)

Tatsuoka, F., Iwasaki, T., Takagi, Y.: Hysteretic damping of sands under cyclic loading and its relation to shear modulus. Soils Found. **18**, 26–39 (1978)

Triantafyllidis, Th.: Analytische Lösung des Problems der dynamischen Untergrundkopplung starrer Fundamente. Veröffentlichungen des Instituts für Bodenmechanik und Felsmechanik der Universität Fridericiana in Karlsruhe, no. 97 (1984)

Triantafyllidis, Th.: Halbraumlösungen zur Behandlung bodendynamischer Probleme mit der Randelementmethode. Veröffentlichungen des Instituts für Bodenmechanik und Felsmechanik der Universität Fridericiana in Karlsruhe, no. 116 (1989)

Triantafyllidis, Th., Ludwig, H.-Ch., Gipperich, Ch.: Verfahren zur Erstellung eines rückbaubaren Verpreßankers mit einem Verpreßkörper und einem oder mehreren angeschlossenen Zuggliedern. Europäisches Patentamt, Patentschrift Nr.: EP 0816570 A2, am 03 July 1997 (1997)

Triantafyllidis, Th.: Neue Erkenntnisse aus Messungen an tiefen Baugruben am Potsdamer Platz in Berlin. Bautechnik **75**(3), 133–154 (1998)

Triantafyllidis, Th. (ed.): Holistic Simulation of Geotechnical Installation Processes - Theoretical Results and Applications. Lecture Notes in Applied and Computational Mechanics, vol. 82. Springer (2017). ISBN 978-3-319-23159-1

Wolf, H.: Zur Scherfugenbänderung granularer Materialien unter Extensionsbeanspruchung. Schriftenreihe des Instituts für Grundbau und Bodenmechanik der Ruhr-Universität Bochum, no. 37 (2005)

Wichtmann, Th., Sonntag, Th., Triantafyllidis, Th.: Über das Erinnerungsvermögen von Sand unter zyklischer Belastung. Bautechnik **78**(12), 852–865 (2001)

Wichtmann, T.: Explicit accumulation model for non-cohesive soils under cyclic loading. Schriftenreihe des Instituts für Grundbau und Bodenmechanik der Ruhr-Universität Bochum, no. 38 (2005)

Wichtmann, T., Niemunis, A., Triantafyllidis, Th.: High-cycle laboratory tests on soils. In: Stangenberg, F. et al. (eds.) Lifetime-Oriented Structural Design Concepts, Chap. 3.2.2, pp. 198–207. Springer (2009). ISBN-978-3-642-01461-1

Wichtmann, T., Triantafyllidis, Th.: On the influence of the grain size distribution curve of quartz sand on the small strain shear modulus G_{max}. J. Geotech. Geoenviron. Eng. ASCE **135**(10), 1404–1418 (2009)

Wichtmann, T., Triantafyllidis, T.: Effect of uniformity coefficient of sands on G/Gmax and damping ratio of uniform to well graded quartz sands. J. Geotech. Geoenviron. Eng. **139**, 59–72 (2013)

Yu, P., Richart, F.: Stress ratio effects on shear modulus of dry sands. J. Geotech. Eng. **110**, 331–345 (1984)

Author Index

A
Aggelis, Dimitrios G., 183
Ahmadi, Abdul Bashit, 183

B
Baeßler, Matthias, 355

C
Chrisopoulos, Stylianos, 281
Cudmani, Roberto, 323
Cuellar, Pablo, 355

D
Dahmen, Dieter, 221
Dausch, Gebhard, 347

F
Founta, Vasiliki, 267

G
Gerolymatou, Eleni, 297
Gortsas, Theodore V., 183
Goudarzy, Meisam, 383
Grandas Tavera, Carlos Eduardo, 13, 237

H
Herle, Ivo, 3
Hettler, Achim, 370

K
Karcher, Christian, 3
Knittel, Lukas, 13, 136
König, Diethard, 383

L
Livitsanos, Georgios, 183
Lüddecke, Falk, 355

M
Machaček, Jan, 237

N
Niemunis, Andrzej, 85

O
Osinov, Vladimir A., 199, 281

P
Petalas, Alexandros, 297
Polyzos, Demosthenes, 183

R
Rackwitz, Frank, 97
Rebstock, Daniel, 323

S
Schorr, Joshua, 323
Selvadurai, A. P. S., 44
Shi, X. S., 3

T
Tafili, Merita, 54
Triantafyllidis, Theodoros, 13, 54, 237, 370

V
v. Blumenthal, Achim, 383
Victor, Albrecht, 355

W
Whittle, Andrew J., 267
Wichtmann, Torsten, 113, 136

Y
Yang, J., 169

Z
Zachert, Hauke, 113

T. Triantafyllidis (Ed.): *Recent Developments of Soil Mechanics and Geotechnics in Theory and Practice*, LNACM 91, p. 409, 2020.
https://doi.org/10.1007/978-3-030-28516-6

The manufacturer's authorised representative in the EU is Springer Nature Customer Service Centre GmbH, Europaplatz 3, 69115 Heidelberg, Germany. If you have any concerns regarding our products, please contact ProductSafety@springernature.com

Printed and bound by CPI Group (UK) Ltd, Croydon, CR0 4YY
05/10/2026
02228130-0014